Immersive Sound

Immersive Sound: The Art and Science of Binaural and Multi-Channel Audio provides a comprehensive guide to multi-channel and binaural sound theory and applications. With contributions from leading recording engineers, researchers, and industry experts, *Immersive Sound* includes an in-depth description of the physics and psychoacoustics of spatial audio as well as practical applications. Chapters include the history of 3D sound, binaural reproduction through headphones and loudspeakers, stereo, surround sound, height channels, object-based audio, sound field (ambisonics), wave field synthesis, and multi-channel mixing techniques. Knowledge of the development, theory, and practice of spatial and multi-channel sound is essential to those advancing the research and applications in the rapidly evolving fields of 3D sound recording, augmented and virtual reality, gaming, film sound, music production, and post-production.

Agnieszka Roginska is Music Associate Professor and Associate Director of the Music Technology program at New York University. Her research focuses on the analysis, simulation, and applications of immersive and 3D audio.

Paul Geluso is Music Assistant Professor at New York University in the Music Technology program. His work focuses on the theoretical, practical, and artistic aspects of sound recording and reproduction.

Audio Engineering Society Presents

AUDIO ENGINEERING SOCIETY PRESENTS . . .

www.aes.org

Other titles in the Series:

Handbook for Sound Engineers, 5th Edition
Edited by Glen Ballou

Audio Production and Critical Listening, 2nd Edition
Authored by Jason Corey

Recording Orchestra and Other Classical Music Ensembles
Authored by Richard King

Recording Studio Design, 4th Edition
Authored by Philip Newell

Modern Recording Techniques, 9th Edition
Authored by David Miles Huber

Immersive Sound: The Art and Science of Binaural and Multi-Channel Audio
Edited by Agnieszka Roginska and Paul Geluso

Immersive Sound

The Art and Science of Binaural and Multi-Channel Audio

Edited by Agnieszka Roginska
and Paul Geluso

Routledge
Taylor & Francis Group
NEW YORK AND LONDON

First published 2018
by Routledge
711 Third Avenue, New York, NY 10017

and by Routledge
2 Park Square, Milton Park, Abingdon, Oxon OX14 4RN

Routledge is an imprint of the Taylor & Francis Group, an informa business

Library of Congress Cataloging-in-Publication Data
Names: Roginska, Agnieszka, editor. | Geluso, Paul, editor.
Title: Immersive sound : the art and science of binaural and multi-channel audio / edited by Agnieszka Roginska and Paul Geluso.
Description: New York ; London : Routledge, 2017.
Identifiers: LCCN 2017016918 | ISBN 9781138900011 (hardback) | ISBN 9781138900004 (pbk.)
Subjects: LCSH: Surround-sound systems.
Classification: LCC TK7881.83. I46 2017 | DDC 621.389/334—dc23
LC record available at https://lccn.loc.gov/2017016918

ISBN: 978-1-138-90001-1 (hbk)
ISBN: 978-1-138-90000-4 (pbk)
ISBN: 978-1-315-70752-5 (ebk)

Typeset in Sabon
by Apex CoVantage, LLC

Contents

Contributors

Durand R. Begault, NASA Ames Research Center, works in the area of research and development of 3D audio and multi-modal technologies for aeronautic and space applications, including psychoacoustic research, human factors evaluation, sound quality, acoustical modeling, and communications engineering. He has been associated with the Human Systems Integration Division of NASA Ames Research Center since 1988.

Braxton Boren, American University, is an acoustics and audio researcher specializing in applications of science to music and the humanities. He earned his PhD at the Music and Audio Lab at New York University, and is Assistant Professor in the Audio Technology Program at American University.

Karlheinz Brandenburg, Fraunhofer Institute for Digital Media Technology IDMT, is Full Professor at Technische Universität Ilmenau and Director of the Fraunhofer Institute for Digital Media Technologies in Ilmenau. He is best known for his contributions to audio coding (MP3, AAC). Among others, he is currently supervising research on binaural hearing, immersive audio, and music information retrieval.

Sandra Brix, Fraunhofer Institute for Digital Media Technology IDMT, joined the Fraunhofer Institute for Digital Media Technology IDMT in Germany as the head of the virtual acoustics area in 2000, where she has been leading the Acoustics Department since 2012. Dr. Brix is highly involved in the development of the Wave-Field-Synthesis technology. She has been teaching at the Technical University of Ilmenau since 2007.

Edgar Choueiri, Princeton University, is Professor of Applied Physics at Princeton University. He heads both the Electric Propulsion and Plasma Dynamics Laboratory, where he works on advanced spacecraft propulsion, and the 3D Audio and Applied Acoustics (3D3A) Laboratory, where he works on psychoacoustics and virtual reality 3D audio.

Paul Geluso, New York University, focuses his work on the theoretical, practical, and artistic aspects of sound recording and reproduction. He is an active recording engineer currently serving as an Assistant Music Professor in Music Technology at New York University where he teaches courses in music production, electronics, ear training, and immersive sound technologies.

Martine Godfroy-Cooper, NASA Ames Research Center, has been a research associate on a cooperative agreement between the Human Systems Integration Division at NASA Ames Research Center and the Psychology Department at San Jose State University since 2005. She is currently responsible for the development of 3D audio and multimodal interfaces for the US Army Aviation Development Directorate degraded visual environment mitigation program.

Sungyoung Kim, Rochester Institute of Technology, is a researcher, teacher, recording engineer (Tonmeister), and guitarist. He worked for Korea Broadcasting System (KBS) (1996–2001) and Yamaha Corporation (2007–2012) before joining Rochester Institute of Technology (RIT). His research topics are cross-cultural difference in spatial hearing and rehabilitation of spatial hearing through VR technologies.

Brett Leonard, BLPaudio, is an audio educator, researcher, consultant, and freelance audio engineer with a specialty in acoustic music production. His research includes collaborations with NHK, SwissAudec, and Skywalker Sound, amongst others. His research focuses on understanding the complex interactions of spatial audio, human perception, and small room acoustics.

Rozenn Nicol, Orange Labs, works in the area of research and development of acoustics, 3D audio and sound perception for telecommunication applications at Orange Labs since 2000. She takes part in classes on Spatial Audio for universities (Le Mans, Brest) and engineering schools (ENST, ENSATT) in France.

Agnieszka Roginska, New York University, has been Music Associate Professor and Associate Director of the Music Technology program at New York University since 2006. Her research focuses on the simulation and applications of immersive and 3D audio, including the capture, analysis and synthesis of auditory environments, auditory displays, and their applications in augmented acoustic sensing.

Francis Rumsey, Logophon Ltd., is a technical writer, organist and consultant, Chair of the AES Technical Council and Consultant Editor of the *Journal of the Audio Engineering Society*. Until 2009 he was a Professor in the Music and Sound Recording Department of the University of Surrey, leading a research group concerned with psychoacoustics and sound reproduction.

Christoph Sladeczek, Fraunhofer Institute for Digital Media Technology IDMT, works in the field of spatial audio and acoustics where he (co)-authored numerous papers. He is the head of the Virtual Acoustics research group at Fraunhofer IDMT. Christoph is responsible for the development of object-based audio technology.

Thomas Sporer, Fraunhofer Institute for Digital Media Technology IDMT, works in the area of research and development of 3D audio, audio quality assessment, and applications based on acoustics. He has been associated with Fraunhofer since 1988. He was involved in the development of mp3, aac and MPEG-H.

Nicolas Tsingos, Dolby, leads the virtual and augmented reality exploration group at Dolby Laboratories. Previously, he designed the authoring and rendering tools for the Dolby Atmos cinema sound system and format. Nicolas also holds a tenure research position at INRIA, the French National Institute for Computer Science.

Elizabeth M. Wenzel, NASA Ames Research Center, has been a Research Psychologist with the Human Systems Integration Division at NASA Ames Research Center since 1986, directing development of 3D audio and multimodal display technology and conducting basic and applied research in auditory perception, localization in virtual acoustic displays, and multimodal information presentation for aerospace applications.

Foreword

Immersive sound produced over loudspeakers or headphones has the capacity to deliver a seamless illusion of alternative reality and change the way we relate to and behave in sound. It can revolutionize how we interact with, listen to, and live with music. It can redefine how we entertain, how we communicate, and how we collaborate. It opens creative possibility never imagined before, and it has the potential to improve the quality of life. Immersive sound is strategically aligned with the future of communication and entertainment.

The book you are about the read offers the wealth of information any researcher or practitioner in the audio field would eagerly welcome into his or her personal library. Agnieszka Roginska and Paul Geluso have succeeded in bringing together a capable team of experts who are able to address with eloquence the important aspects of perception, technology, and applications involved in auditory communication through immersive sound.

The goal of this book is to expand the reader's understanding of principles and working methods in immersive sound presented over loudspeakers and headphones using multi-channel and binaural technologies. While spatial audio techniques have a rich historical context, only recently have researchers and developers been able to collectively offer consumers a new paradigm in auditory experience through signal processing and streaming of high-quality digital audio. We are soon to arrive at the forefront of big changes in how people will access and interact with audio information.

The book is aimed at students, practitioners, and researchers who seek a thorough grounding in theory and practice of immersive sound. The information is invaluable, each chapter adding a critical component to the whole. This timely publication will no doubt help readers to acquire an informed and compelling view of the field.

Wieslaw Woszczyk

Acknowledgements

We would like to thank the many people who inspired and contributed to this project. First, we express our gratitude to the amazing team at Routledge, including Lara Zoble and Kristina Ryan, for their constant support and guidance through this process. Our very special thanks to the reviewers and readers for their critiques and comments, especially Wieslaw Woszczyk for his invaluable mentorship, and for contributing his thoughtful review and the foreword of this book. We thank all the chapter authors for sharing their time, expertise, and insight, whose invaluable contributions collectively make up this book. Without their commitment, enthusiasm, and dedication, this book would not have been realized. Working with this incredibly talented and devoted group of experts has been a true honor and inspiration for us. Finally, we would like to give a personal thanks to our very supportive families, colleagues, and friends.

Introduction

Agnieszka Roginska and Paul Geluso

Immersive Sound

Through sound, vision, touch, smell, and taste, multi-sensory integration into a scene can create an immersive experience. Immersive sound can give the listener an experience of *being there* through sound. Compared to vision, sound provides a fully immersive experience and can be perceived from all directions simultaneously. In fact, sound has the ability to ground a listener in a fixed location while other sensory information changes simultaneously. Filmmakers are well aware of this effect, often using sound to establish a fixed location in a scene while having the visual perspective change frequently.

For a moment, let's consider an immersive experience generated using sound alone. The location relationships between the listener, sound source(s) and the boundaries of a room create auditory cues that convey a sense of space. A continuous yet seemingly directionless sea of sound can envelop a listener through the use of environmental elements such as reverberation. Using sound this way, the sense of being immersed can be accomplished through a constructed sound-scape of directional and non-directional sounds surrounding the listener.

For example, New York's Grand Central station creates a natural immersive audio environment through a combination of close directional sounds such as conversation and foot steps. These combine with farther sound sources such as announcements, and interact with the acoustics of the space to create the reverberant and enveloping environment. The combination of sound sources and the enveloping environment create the immersive auditory experience. A sound source can become an environmental and enveloping sound or remain a point source, depending on the focus and perception of the listener.

The Listening Experience

The natural listening environment can be defined as the acoustic space we occupy during our daily lives. A virtual auditory space is an acoustic environment created through the use of loudspeakers or headphones designed to replace or augment the natural listening environment. Using technology, sounds from our natural listening environment can be captured, processed, stored, and/or transmitted to be reproduced in a virtual auditory space. The natural listening environment and the virtual auditory space both contribute to the listener's experience.

Natural listening defines the way we hear normally, without the aid of loudspeakers. Very realistic virtual auditory spaces that approximate the natural listening environment can be created with the use of loudspeakers and headphones through various techniques described in this book. The goal of immersive sound may be to recreate a sound environment that is as close as possible to the real world, or it may be to create an experience that augments the real world and can only exist in the virtual space. To create a listener's virtual experience, natural sounds can be captured or synthesized, processed, and played back using immersive sound reproduction systems.

Listener's Perspective

In a virtual listening environment, we rely on both analytical and psychoacoustic abilities to make sense of the sounds that are being reproduced using loudspeakers or headphones. Consider a mono recording made with a single microphone played back through a single loudspeaker. In

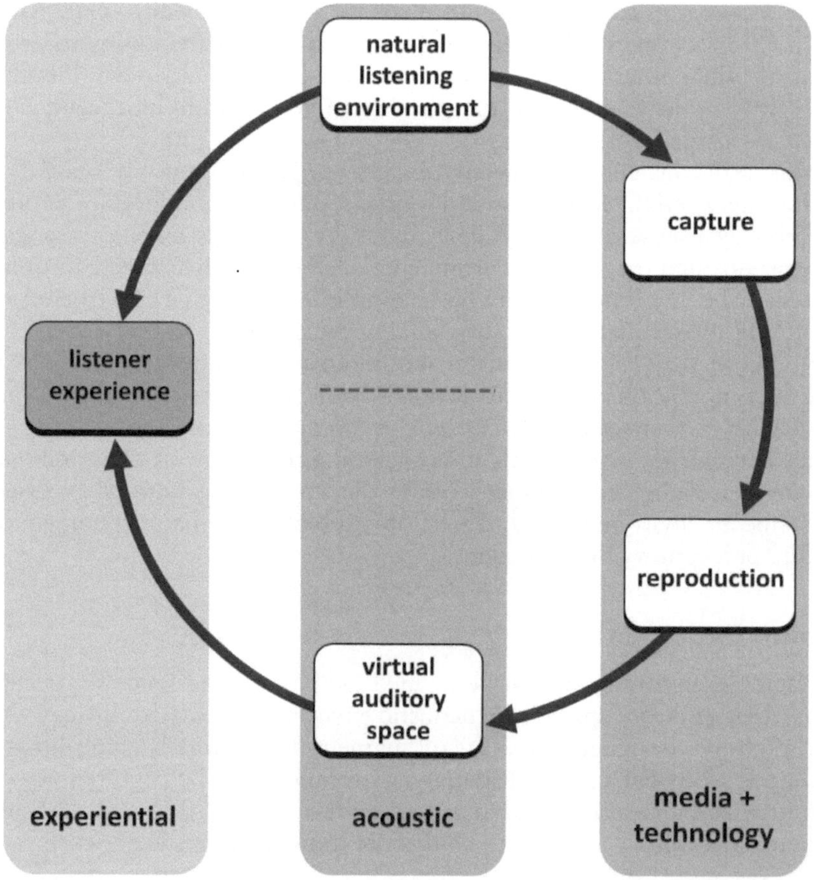

Figure 0.1 The listening experience.

this situation, we can imagine how distant the sound sources are from the microphone and get a general impression of the recording space by analyzing the relative volume, timbre, and room reflections for each sound source captured in the recording. If a second microphone and loudspeaker are added to the system, our auditory system now has two signals to analyze, thus more physical information to process in order to extract spatial information to inform the listener's perspective.

A fixed perspective can be achieved by creating a monophonic or stereophonic sound stage where the virtual sound stage is bounded by the speakers in the frontal plane. Surround sound expands to a panoramic virtual sound stage including the listening environment, or placing the listener on the sound stage surrounded by virtual sound sources. These systems are considered channel based and rely on a priori knowledge about the location of the loudspeakers and their relationship to the listener. Sound field and wave field systems move away from the channel-based model and aim to reproduce the physical wavefront as it would appear in a natural environment by utilizing multiple loudspeakers.

Binaural audio reproduction techniques take advantage of the human natural spatial auditory cues to recreate a virtual auditory environment through headphones or loudspeakers that emulate headphone reproduction. This results in a "you are there", first-person perspective, in contrast to the loudspeaker "they are here" systems described above.

As our recording and playback systems become more sophisticated, the immersive listening experience is guided by reproduction systems that can better approximate the natural listening environment. Thus, allowing the listener to rely less on their a priori knowledge to make sense of what they hear, and move toward a more natural listening experience. This creates a more compelling and *easier to listen to* virtual auditory environment.

About This Book

The intended audiences for this book are those enrolled in undergraduate, graduate study, and higher learning institutions in the areas of music technology, recording and production, sound design, sound art and film, as well as audiophiles, game designers, simulation and virtual reality professionals, post-production professionals, and entertainment professionals. We assume readers have a fundamental understanding of the physics of sound, digital audio, recording, and reproduction principles.

Chapters are written by experts in their corresponding field of research within immersive audio. Chapter authors are researchers in academic institutions, research laboratories in the industry, US government agencies. The body of the text is grounded in research and empirical work, with each chapter covering the evolution and historical perspective of the development of a reproduction technology.

The organization of the book proceeds with a chapter by Elizabeth Wenzel, Durand Begault, and Martine Godfroy-Cooper about the physiology, psychoacoustics, and acoustics of spatial hearing, including the perception of spatial sound, binaural cues, perceptual plasticity, distance perception, and environmental context for immersive sound. In Chapter Two, Braxton Boren takes the reader on a historical journey across immersive sound starting with the impact of immersive sound on our prehistoric ancestors, through the use of spatial separation of instruments and choirs as a compositional tool in the 15th century, to modern-age techniques and technologies of immersive sound. In Chapter Three, Paul Geluso introduces stereo loudspeaker systems and discusses techniques of sound capture, reproduction, and methods of stereo enhancement.

Chapter Four, written by Agnieszka Roginska, describes the capture, synthesis, and reproduction of binaural sound over headphones, including extended reproduction techniques and applications of binaural sound. In Chapter Five, Edgar Choueiri describes the principles of binaural audio reproduction over loudspeakers using crosstalk cancellation. Continuing in Chapter Six, Francis Rumsey discusses the evolution and principles of surround sound, with speakers located on the horizontal plane. These techniques are extended in Chapter Seven, where Sungyoung Kim describes the methods of surround sound reproduction with height speakers, including the psychoacoustics of height perception, configuration of loudspeakers, and recording techniques.

Nicolas Tsingos discusses the principles of object-based audio in Chapter Eight, where he describes audio objects, their representation, advanced metadata, audio object capture, and rendering. In Chapter Nine, Rozenn Nicol addresses the theory and practice of sound field capture and reproduction, starting from the initial development of the sound field approach to High Order Ambisonics. Wave Field Synthesis is described in Chapter Ten, where Thomas Sporer, Karlheinz Brandenburg, Sandra Brix, and Christoph Sladeczek describe the development of Wave Field Synthesis from one of the earliest examples of immersive sound through the acoustic curtain by Steinberg and Snow in 1934, through the theory and practice of WFS reproduction, and its limitations and applications using modern techniques and signal processing. The book concludes with Brett Leonard describing the applications of extended multi-channel techniques in Chapter Eleven, where he discusses broader concepts and introduces practical mixing techniques for engineers working with immersive sound.

Chapter 1

Perception of Spatial Sound

Elizabeth M. Wenzel, Durand R. Begault, and Martine Godfroy-Cooper

Immersion refers acoustically to sounds as coming from all directions around a listener, which normally is an inevitable consequence of natural human listening in an air medium. Audible sound sources are everywhere in real environments where sound waves propagate and reflect from surfaces around a listener. Even in the quietest of environments, such as an anechoic chamber, the sounds of one's own body will be audible. However, the common meaning of immersion in audio and acoustics refers to the psychological sensation of being surrounded by specific sound sources as well as ambient sound. Although acoustically a sound can reach a listener from multiple surrounding directions, its spatial characteristics may be judged as unrealistic, static or constrained. For example, good quality concert hall acoustics has traditionally been correlated with a listener's sensation of being immersed by the sound of the orchestra, as opposed to the sound seeming distant and removed. Spatial audio techniques, particularly 3D audio, can provide an immersive experience because virtual sound sources and sound reflections can be made to appear from anywhere in space around a listener. This chapter introduces a listener to the physiological, psychoacoustic and acoustic bases of these sensations.

Auditory Physiology

Auditory perception is a complex phenomenon determined by the physiology of the auditory system and affected by cognitive processes. The auditory system transforms the fundamental independent aspects of sound stimuli, such as their spectral content, temporal properties and location in space into distinct patterns of neural activity. These patterns will give rise to the qualitative experience of pitch, loudness, timbre and location. They will ultimately be integrated with information from the other sensory systems to form a unified perceptual representation and provide behavior guidance that includes orienting to acoustical stimuli and engaging in intra-species communication.

Auditory Function: Peripheral Processing

The functional auditory system extends from the ears to the brain's frontal lobes with successively more complex functions occurring as one ascends the hierarchy of the nervous system (Figure 1.1). The different functions performed by the auditory system are classically categorized as *peripheral auditory processing* and *central auditory processing*.

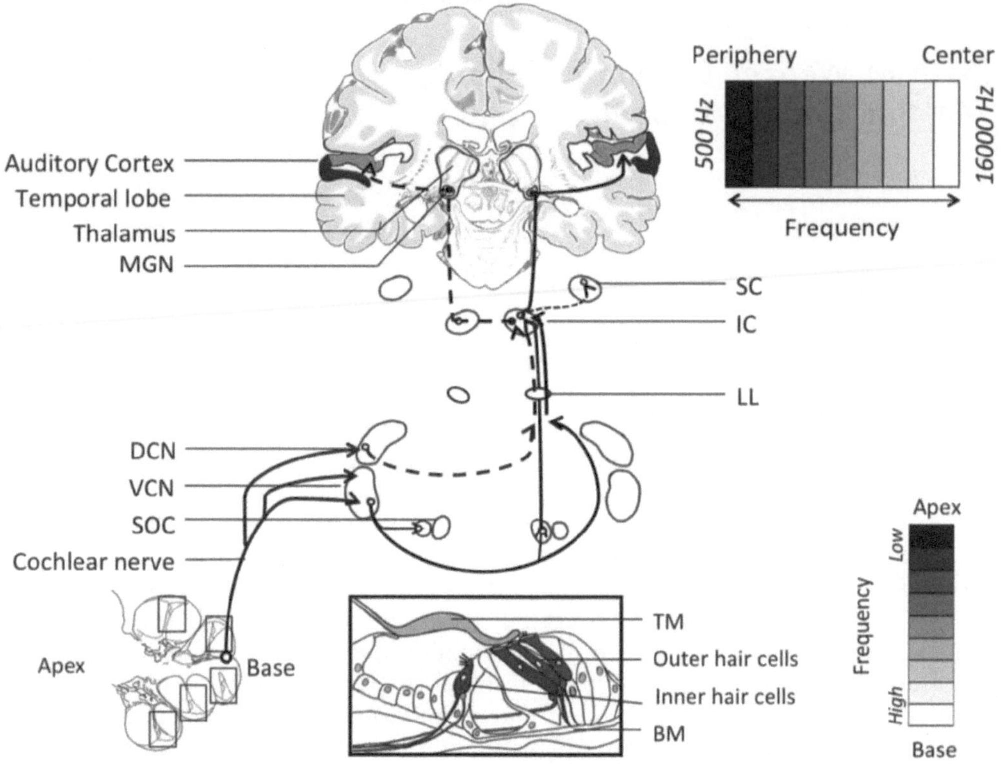

Figure 1.1 Diagram of the major auditory pathways. Note the auditory system entails several parallel pathways and that information from each ear reaches both sides of the system (dashed line: ipsilateral, continuous line: contralateral). Top Right: Tonotopic representation in the auditory cortex. High frequencies are represented at the center of the cortical map while low frequencies are represented at its periphery. Bottom Left: Cochlea. Bottom Center: Organ of Corti. Bottom Right: Cochleotopic coding along the basilar membrane (BM).

The peripheral auditory system includes processing stages from the outer ear to the cochlear nerve. A crucial transformation is performed within these early stages, which is often compared to a Fourier analysis of the incoming sound waves that defines how the sounds are processed at the later stages of the auditory hierarchy. Sound enters the ear as pressure waves. At the periphery of the system, the external and the middle ear respectively collect sound waves and selectively amplify their pressure, so that they can be successfully transmitted to the fluid-filled cochlea in the inner ear.

The *external ear*, which consists of pinna (plural, pinnae) and auditory meatus (or canal), gathers the pressure waves and focuses them on the eardrum (tympanic membrane) at the end of the canal. One consequence of the configuration of the human auditory canal is that it selectively boosts the sound pressure 30 to 100 fold for frequencies around 3 kHz via a passive resonance effect due to the length of the ear canal. This amplification makes humans especially sensitive to frequencies in the range of 2–5 kHz, which appears to be directly related to speech perception. A second important function of the pinnae is to selectively filter sound frequencies in order to provide cues about the elevation of a sound source: up/down and front/back angles (Shaw, 1974). The vertically asymmetrical convolutions of the pinna are shaped so that the external ear transmits more high frequency components from an elevated source than from the same source at ear level. Similarly, high frequencies tend to be more attenuated for sources in the rear than for sources in the front, as a consequence of the orientation and structure of the pinna (Blauert, 1997).

The *middle ear* is a small cavity, which separates the outer and the inner ear. The cavity contains the three smallest bones (hammer, anvil and stirrup) in the body called ossicles, connected more or less flexibly to each other. Its major function is to match the relatively low impedance (impedance in this context refers to a medium's resistance to movement) airborne sounds to the higher impedance fluid in the inner ear. Without this action, there would be a loss of transmission of 1000:1 corresponding to a loss of sensitivity of 30 dB. The middle ear has two small muscles known as the tensor tympani and the stapedius, which have a protective function. When sound pressure reaches a threshold loudness level (at approximately 85 dB HL[1] in humans with normal hearing), a sensory driven afferent signal is sent to the brainstem via the cochlear nerve, which initiates an efferent reflexive contraction of the stapedius muscle within the middle ears referred to as the stapedius reflex or acoustic middle ear reflex. The excitation of the muscle results in a stimulus level-dependent attenuation of low-frequency (< 1 kHz) ossicular chain vibration reaching the cochlea (Wilson & Margolis, 1999).

Transduction Mechanisms in the Cochlea

The cochlea in the *inner ear* is the most critical structure in the peripheral auditory pathway. The cochlea is a small-coiled snail-like structure that responds to the sound-induced vibrations and converts them into electrical impulses, a process known as mechanoelectrical transduction. Cochlear signal transduction involves amplification and decomposition of complex acoustical waveforms into their component frequencies.

Two membranes, the basilar membrane (BM) and the vestibular membrane (VM), divide the cochlea in three fluid-filled chambers. The organ of Corti sits on the BM and contains an array of sensory hair cells that contact with the tectorial membrane (TM), a structure that plays multiple, critical roles in hearing including coupling elements along the length of the cochlea, supporting a travelling wave and ensuring the gain and timing of cochlear feedback are optimal (Richardson, Lukashkin, & Russell, 2008). The sensory hair cells are responsible for the mechanoelectrical transduction, i.e., the transformation of mechanical stimulus into electrochemical activity. The acoustical stimulus initiates a traveling wave by displacing the hair, thus enabling the encoding of frequency, amplitude and phase of the original sound stimulus by the electrical activity of the auditory nerve fibers. As the BM displaces, it causes deflection in the hair bundles (stereocilliae, tiny processes that protrude from the apical ends of the hair cells) of the hair cells of the location-matched inner hair cells, which results in a current flow and ultimately an action potential. Because the stiffness of the BM changes throughout the cochlea, the displacement of the membrane induced by an incoming sound depends on the frequency of that sound. Specifically, the BM is stiffer near its "base" than near the middle of the spiral ("the apex"). Consequently, high frequency sounds (20 kHz) produce displacement near the base, while low frequency sounds (20 Hz) disturb the membrane near the apex. As a result, each locus on the BM is identified by its characteristic frequency (CF) and the whole BM can be described as a bank of overlapping filters (Patterson et al., 1987; Meddis & Lopez-Poveda, 2010). Because the BM displaces in a frequency-dependent manner, the corresponding hair cells are "tuned" to sound frequency. The resulting spatial proximity of contiguous preferred sound frequency ("place theory of hearing", von Békésy, 1960; "place code" model, Jeffress, 1948) is referred to as tonotopy or better, cochleotopy. This tonotopic organization is carried up through the auditory hierarchy to the cortex (Moerel et al., 2013, Saenz & Langers, 2014) and defines the functional topography in each of the intermediate relays.

Auditory Function: Central Processing

The central auditory system is composed of a number of nuclei and complex pathways that ascend within the brainstem. The earliest stage of central processing occurs at the cochlear nucleii (dorsal, DCN and ventral, VCN), where the tonotopic organization of the cochlea is maintained. Accordingly the output of the CN has several targets.

One is the superior olivary complex (SOC), the first point at which information from the two ears interacts. The best understood function of the SOC is sound localization. Humans use at least two different strategies, and two different pathways, to localize the horizontal position of sound sources, depending on the frequencies of the stimulus. For frequencies below 3 kHz, which the auditory nerve can follow in a phase-locked manner, interaural time differences (ITDs) are used to localize the source; above these frequencies, interaural intensity differences (IIDs) are used as cues (King & Middlebrooks, 2011; Yin, 2002). ITDs are processed in the medial superior olive (MSO), while IIDs are processed in the lateral superior olive (LSO). These two pathways eventually merge in the midbrain auditory centers. The elevation of sound sources is determined by spectral filtering mediated by the external pinnae. Experimental evidence suggests that the spectral notches created by the shape of the pinnae are detected by neurons in the DCN. See the section on human sound localization for additional discussion of these cues.

The binaural pathways for sound localization are only part of the output of the CN. A second major set of pathways from the CN bypasses the SOC and terminates in the nuclei of the lateral lemniscus (LL) on the contralateral side of the brainstem. These particular pathways respond to sound arriving at one ear only and are thus referred to as monaural. Some cells in the nuclei of the LL signal the onset of sound, regardless of its intensity or frequency. Other cells process other temporal aspects of sound such as duration.

As with the outputs of the SOC, the pathways from the LL project to the midbrain auditory center, also known as the inferior colliculus (IC). This structure is a major integrative center where the convergence of binaural inputs produces a computed topographical representation of the auditory space. At this level, neurons are typically sensitive to multiple localization cues (Chase & Young, 2006) and respond best to sounds originating in a specific region of space, with a preferred elevation and a preferred azimuthal location. As a consequence, it is the first point at which auditory information can interact with the motor system. Another important property of the IC is its ability to process sounds with complex temporal patterns. Many neurons in the

IC respond only to frequency-modulated sounds while others respond only to sounds of specific durations. Such sounds are typical components of biologically relevant sounds, such as those made by predators and in humans, speech.

The IC relays auditory information to the medial geniculate nucleus (MGN) of the thalamus, which is an obligatory relay for all ascending information destined for the cortex. It is the first station in the auditory pathway where pronounced selectivity for combinations of frequencies is found. Cells in the MGN are also selective for specific time intervals between frequencies. The detection of harmonic and temporal combination of sounds is an important feature of the processing of speech.

In addition to the cortical projection, a pathway to the superior colliculus (SC) gives rise to an organized representation of ITDs and IIDs in a point-to-point map of the auditory space (King & Palmer, 1983). Topographic representations of multiple sensory modalities (visual, auditory and somatosensory) are integrated to control the orientation of movements toward specific spatial locations (King, 2005).

Auditory Function: The Auditory Cortex

The auditory cortex (AC, Figure 1.2) is the major target of the ascending fibers from the MGC and plays an essential role in our conscious perception of sound, including speech comprehension, which is arguably the most significant social stimulus for humans.

Although the AC has a number of subdivisions, a broad distinction can be made between a primary area and a secondary area. The primary auditory cortex (BA41, core area) located on the superior temporal gyrus of the temporal lobe receives point-to-point input from the MGC and

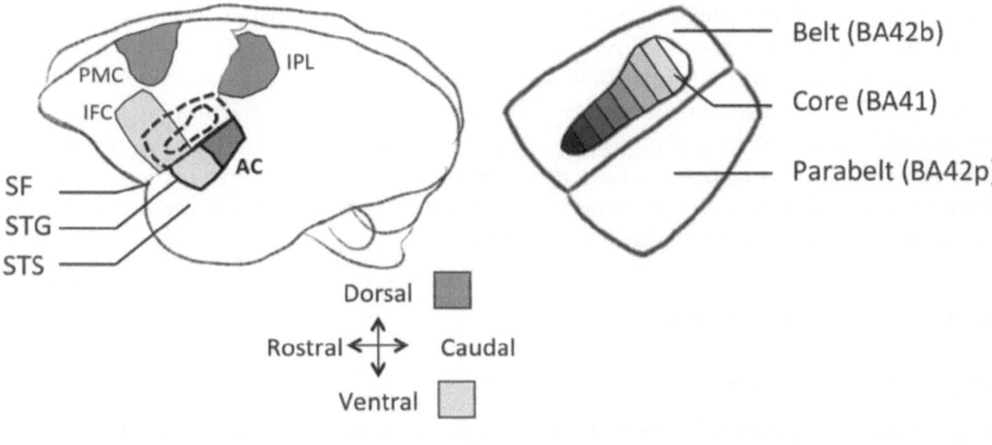

Figure 1.2 Left: Auditory cortex (AC) location in the human brain. The superior temporal gyrus (STG) is bordered superiorly by the Sylvian fissure (SF). The STG is bordered on the inferior side by the superior temporal sulcus (STS). Dorsal stream: intraparietal lobe (IPL), premotor cortex (PMC). Ventral stream: inferior frontal cortex (IFC). Right: The primary auditory cortex is denoted as the auditory Brodmann's area 41 (BA41) in humans. The auditory association cortex is subdivided into lateral belt and parabelt cortices (BA42). The view is from above the STG, after removing the overlying cortex, revealing the superior temporal plane.

comprises three distinct tonotopic fields (Saenz & Langers, 2014). Neurons in BA41 have narrow tuning functions and respond best to tone stimuli, supporting basic auditory functions such as frequency discrimination and sound localization. It also plays a role in processing of within-species communication sounds.

The belt areas of the auditory cortex receive more diffuse inputs from the MGC as well as inputs from BA41, and are less precise in their tonotopic organization. Neurons in the belt areas (BA42b) have broader frequency tuning functions and respond better to complex sounds, such as those that mediate communication. Lateral to the belt is a region of cortex denoted as the parabelt (BA42p) where neurons prefer complex stimuli including band-passed noise, moving stimuli and vocalizations (Rauschecker, Tian & Hauser, 1995).

The projections from the parabelt out of the auditory cortex to higher order cortical structures define the auditory dorsal processing stream and the ventral processing stream (see Figure 1.2). According to the auditory dual-stream model, spatial information ("where") is primarily processed within the dorsal stream and non-spatial information (object features, i.e., "what") within the ventral stream (Rauschecker & Tian, 2000; Romanski & Goldman-Rakic, 2002). The auditory ventral stream supports the perception and recognition of auditory objects and is involved in the processing of pitch changes, auditory working memory for words and tones, as well as semantic processing. There is less agreement regarding the functional role of the auditory dorsal stream. The earliest models argued for a role in spatial hearing, but recent research suggests that the auditory dorsal stream supports an interface with the motor system. In fact, this segregation between dorsal/ventral streams appears to be more relative than absolute. It seems that the functions of sound identification and spatial analysis are rather co-localized in the dorsal and ventral auditory streams (Gifford & Cohen, 2005; Lewald et al., 2008) and that human spatial processing is strongly linked with functions of pitch perception (Douglas & Bilkey, 2007).

Connections from the auditory cortex to the frontal lobes mediate a number of functions including language, object recognition and spatial localization. The frontal cortex is a heterogeneous region with multiple functional subdivisions, including the prefrontal cortex (PFC), and is part of the association cortices. The frontal association cortex is importantly involved in guiding complex behavior by planning responses, ongoing stimulation or remembered information. Collectively, the association cortices mediate the cognitive functions of the brain, including speech processing and executive functions that include attention, working memory, planning and decision-making (Fuster, 2009; Plakke et al., 2014).

Human Sound Localization

Primary Localization Cues

Auditory spatial perception refers to the ability to localize individual sound sources in 3D space even when multiple, simultaneous sources are present. Unlike the visual and somatosensory systems, spatial information is not directly represented at the sensory receptor in the auditory system. Instead, spatial locations are estimated by integrating neural binaural properties and frequency-dependent pinna filtering (binaural and monaural cues, Yost & Dye, 1997).

Sound source location is often specified in terms of azimuth, elevation and distance using a coordinate system in which a listener facing directly forward is defined as 0° azimuth and 0°

elevation. Azimuth is defined by the angle (θ) between the source location and the median plane at 0° azimuth (projected onto the horizontal plane) and elevation is the angle (δ) between the source location and the horizontal plane at 0° elevation (projected onto the median plane).

Azimuths to right of the listener are positive, to the left are negative, and the rear is defined as 180°.[2] Elevations are positive for upper directions and negative for lower directions relative to the listener. Distance is defined as the radius (r) projected along the vector formed by the azimuth and elevation of the source. Another important terminological distinction relevant to interaural cues is between the ipsilateral and contralateral ears. The ipsilateral ear is that closest to the sound source; sound thus arrives first and is greater in intensity at the ipsilateral ear. The contralateral ear is that farthest from the sound source, thus sound arrives later and with less intensity at the contralateral ear.

The localization of a sound source in the horizontal dimension (azimuth) results from the detection of left-right interaural differences in time of arrival and interaural differences in intensity at the two ears (Middlebrooks & Green, 1991). These cues also facilitate speech intelligibility in background noise in human listeners (Culling, Hawley & Litovsky, 2004). To localize a sound in the vertical dimension (elevation) and to resolve front-back confusions, the auditory system relies on the detailed geometry of the pinnae, causing acoustic waves to diffract and undergo direction-dependent reflections (Blauert, 1997; Hofman & Van Opstal, 2003). The two different modes of indirect coding of the position of a sound source in space (as compared to the direct spatial coding of visual stimuli) result in differences in spatial resolution in these two directions.

Sound Localization in the Horizontal Dimension

Much of the research on human sound localization in azimuth has derived from Lord Rayleigh's "duplex theory" (1907) which emphasizes the role of two primary cues (top two panels of Figure 1.3): interaural time differences (ITDs) and interaural intensity differences (IIDs, also referred to as Interaural Level Differences, ILDs, particularly when specified in dB SPL). Because the theory had been based primarily on experiments with single-frequency (sine wave) sounds, the original proposal was that IIDs resulting from head-shadowing determine localization at high frequencies (roughly 2000 Hz for larger human heads), while ITDs were thought to be important only for low frequencies because of the phase ambiguities occurring at frequencies greater than ~1000–1500 Hz. Recently, Brughera, Dunai and Hartmann (2013) have demonstrated that humans are sensitive to ITD fine structure in sound (temporal fine structure, TFS; the sound pressure waveform) up to a limit of 1400 Hz. For broadband sounds, the situation is more complex, and ITDs contribute to sound localization at higher frequencies, up to 4000 Hz (Bernstein, 2001).

The interaural differences present in natural spatial hearing can be understood by considering a simplified rigid sphere model of a listener with a perfectly round head and no outer ears (spherical head model, Woodworth, 1938) placed at a fixed distance in an anechoic chamber from a broadband sound source at eye level (see Figure 1.4).

Modeling this situation involves calculating two paths representing the sound source wavefront from its center of origin to two points representing the entrance to the ear canals. An additional simplification is the placement of these points exactly at the midline crossing the sphere, at the ends of the interaural axis. With the source at position A at 0° azimuth, the path lengths are equal, causing the wavefront to arrive at the eardrums at the same time and with equal intensity.

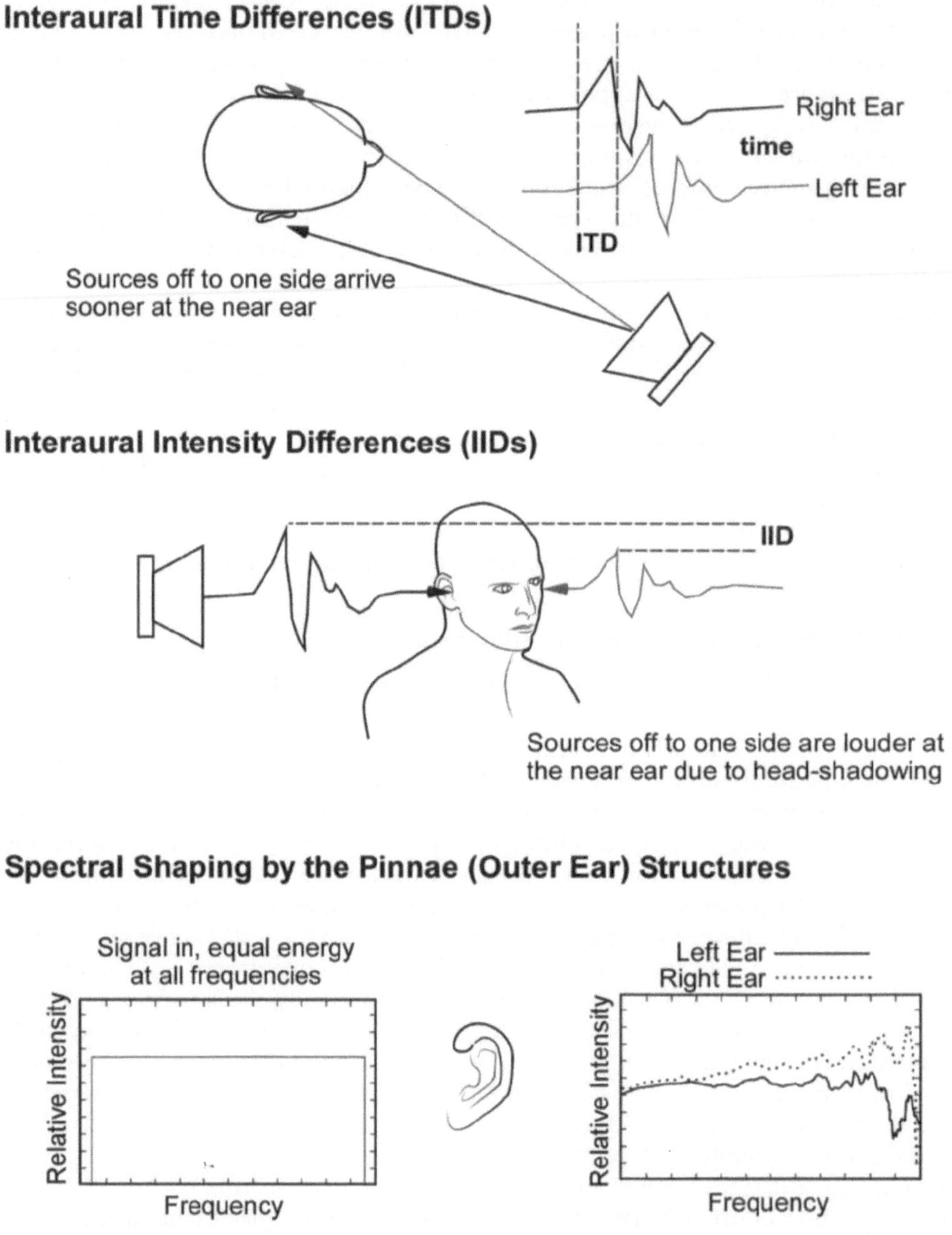

Figure 1.3 Illustration of the primary cues for human sound localization.

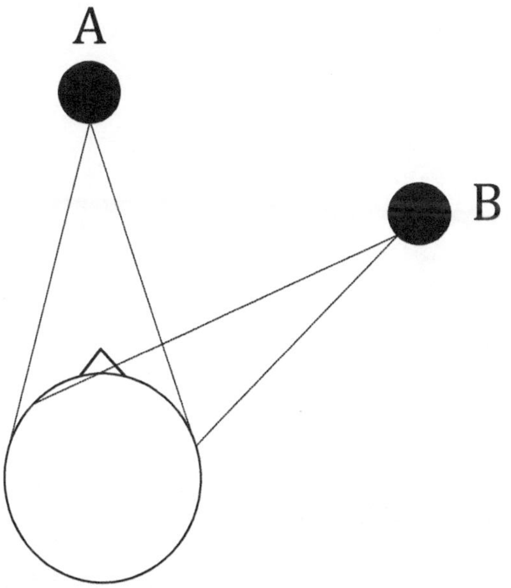

Figure 1.4 A listener in an anechoic chamber, with a sound source oriented directly ahead on the median plane (A, 0°) and displaced to +60° azimuth (B).

At position B, the sound source is at +60° azimuth to the right of the listener, and the paths are now unequal; this will cause the sound source wavefront to arrive later in time at the left ear relative to the right. This sound path length difference is the basis of the ITD cue, and it relates to the hearing system's ability to detect interaural phase differences (IPDs) below approximately 1,000 Hz. If the sounds are pure tones, a simple frequency factor relates the ITDs to the IPDs, for which there are known iso-IPD boundaries (90°, 180°. . .) defining regions of spatial perception. Although dependent upon the nature of the stimulus and the measurement technique, a value of around 650 µsec (750 µsec for low-frequency sounds, Kuhn, 1977) is a good approximation of the observed maximum value for an average human head. Figure 1.5 illustrates how the ITD changes as a function of the azimuthal angle of incidence.

The auditory system can phase-lock, or change the rate of neural firing corresponding to the peaks in a stimulus waveform, as long as the inter-peak interval is above about 1 msec. Such phase-locking can occur for the peaks in either sine waves or the envelopes of complex signals. One theory is that ITDs are estimated by the auditory system by comparing peaks in neural firing rates between the two ears in a manner known as coincidence detection (place code or topographic model, Jeffress, 1948), a neural mechanism that enables interaural cross-correlation. More recent data support the opponent channel coding of auditory space in humans (van Bergeijk, 1962; Magezi & Krumbholz, 2010; Salminen et al., 2010) where ITD is represented by a non-topographic population rate code, which involves only two opponent (left and right) channels,

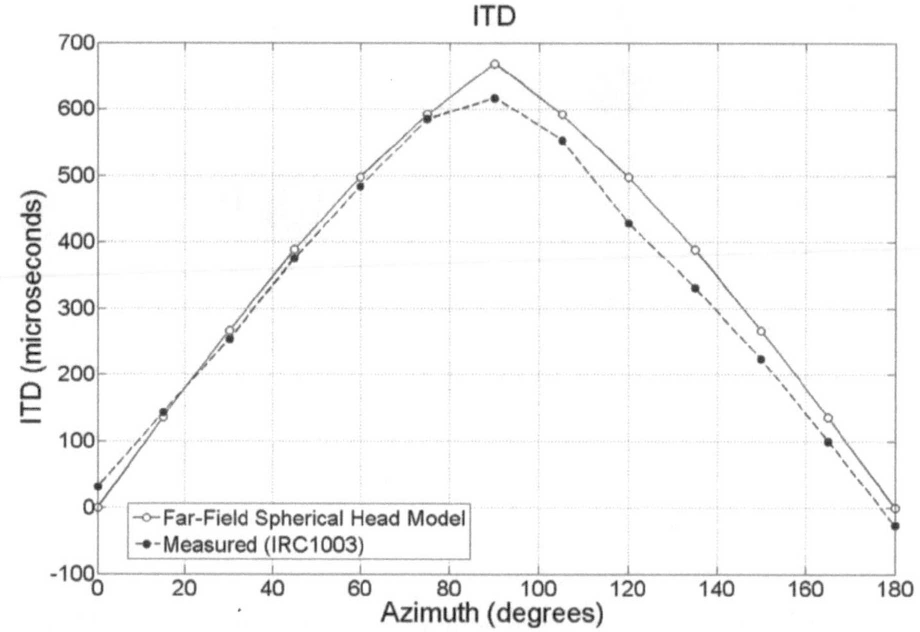

Figure 1.5 Interaural time differences (ITDs) plotted as a function of sound source azimuth (similar to data published by Feddersen et al., 1957). The open dots represent ITDs computed from the spherical head model and the closed dots are measured data at 0° elevation for subject IRC_1003 of the IRCAM Listen database (http://recherche.ircam.fr/equipes/salles/listen/). ITDs were computed using the method described in Miller, Godfroy-Cooper and Wenzel (2014).

broadly tuned to ITDs from the two auditory hemifields. The data suggest that the majority of ITD-sensitive neurons in each hemisphere are tuned to ITDs from the contralateral hemifield.

The sound source at position B in Figure 1.4 will also yield a significant interaural intensity difference cue, but only for those waveform components that are smaller than the diameter of the head, i.e., for frequencies greater than about 2,000 Hz. Higher frequencies will be attenuated at the left ear because the head acts as an obstacle, creating a "head shadow" effect at the opposite side. The relation of a wavefront to an obstacle of fixed size is such that the shadow effect increases with increasing frequency (i.e., decreasing size of the wavelength). However, below 2,000 Hz, the IID is no longer effective as a natural spatial hearing cue because longer wavelengths will diffract ("bend") around the obstructing surface of the head, thereby minimizing the intensity differences. Figure 1.6 illustrates this head-shadow effect by plotting the IID as a function of azimuth location and stimulus frequency. Measured data from the literature for IIDs shows that a 3,000 Hz sine wave at 90° azimuth will be attenuated by about 10 dB, a 6,000 Hz sine wave will be attenuated by about 20 dB, and a 10,000 Hz wave by about 35 dB (Feddersen et al., 1957; Middlebrooks & Green, 1991). Measured IIDs derived from an individual listener are also shown in Figure 1.6 (bottom). Note that the pattern of IIDs can be quite complex across both frequency and azimuth.

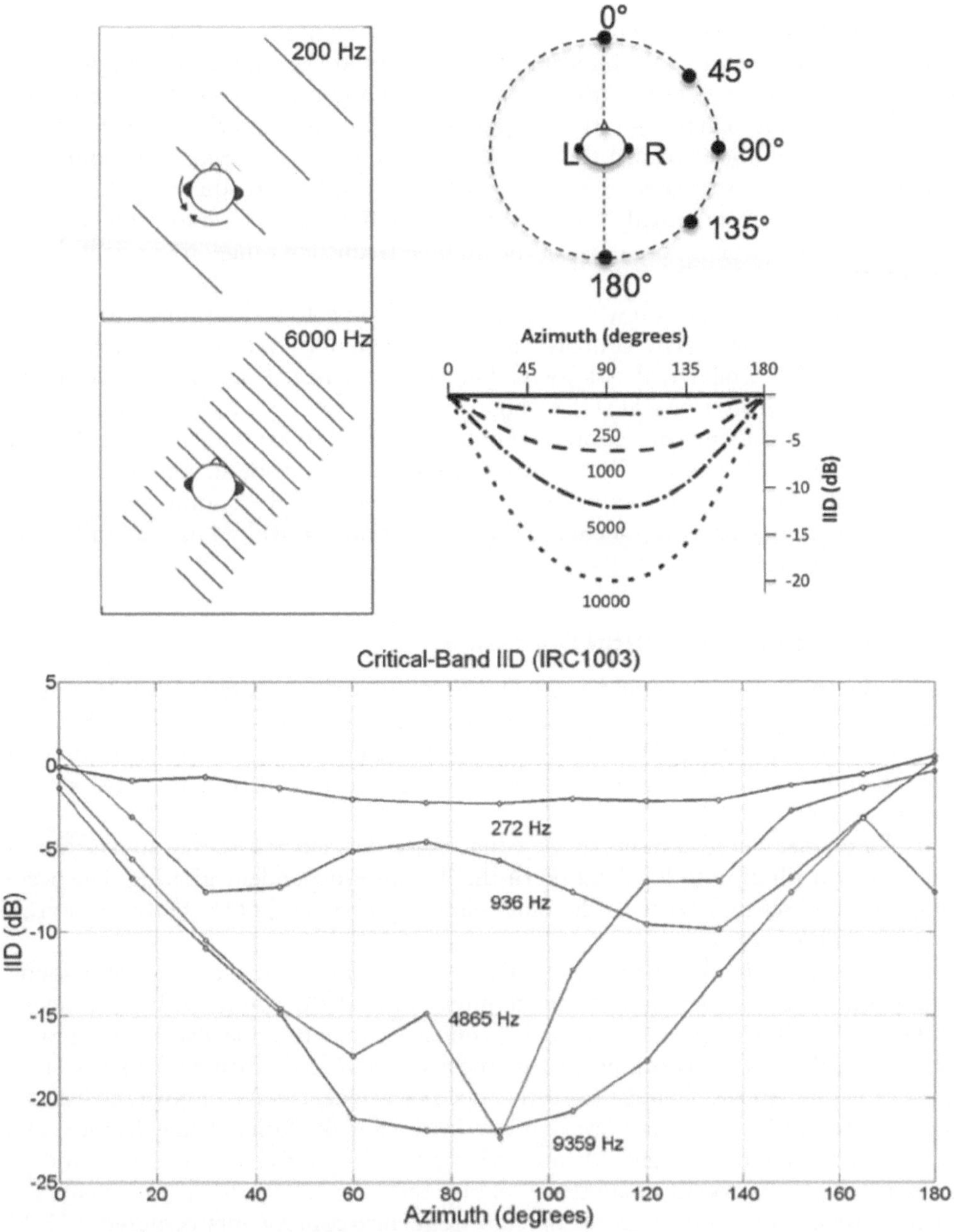

Figure 1.6 Top Panels: Illustration of the head-shadow effect and the resulting interaural intensity differences (IIDs) for 70 dB SPL tones of different frequencies plotted as a function of the azimuth of the source location (adapted from Gulick, Gescheider & Frisina, 1989, p. 324). Bottom Panel: IIDs as a function of critical-band center-frequency and azimuth are plotted for measured data at 0° elevation for subject IRC_1003 of the IRCAM Listen database. Critical bands were chosen to parallel the illustration in the top panel and were computed using the method described in Miller, Godfroy-Cooper and Wenzel (2014).

Independent of frequency content, variations in the overall difference between left and right intensity levels at the eardrum are interpreted as changes in the sound source position from the perspective of the listener. Consider the primary spatial auditory cueing device built into stereo recording consoles, the panpot (short for "panoramic potentiometer"). Over headphones, the panpot creates IIDs without regard to frequency, yet it works for separating sound sources in most applications. This is because the frequency content of typical sounds includes frequencies above and below the hypothetical "cut-off" points for IID and ITD, and listeners are sensitive to IID cues for localization across most of the audible frequency range, down to at least 200 Hz (Blauert, 1997).

Binaural research over the last few decades, however, points to serious limitations of the duplex theory. For example, for nearby sources, the IID is available even at low frequencies (Shinn-Cunningham, 2000). Similarly, it is known that ITD cues based on the relative timing of the amplitude envelopes ("envelope ITD") of high-frequency sounds can be used by a mechanism such as interaural coincidence detection (Henning, 1974, 1980; van de Par & Kohlrausch, 1997; Bernstein & Trahiotis, 2010). Finally, in theory, the azimuth of a sound source can also be determined monaurally because the high-frequency components of the sound are more attenuated compared to low-frequency components as the sound source moves contra-laterally (Shub, Durlach & Colburn, 2008).

Sound Localization in the Vertical Dimension

The duplex theory cannot account for the ability of subjects to localize sounds on the vertical median plane (directly in front of the listener), where interaural cues are minimal. Similarly, when subjects listen to stimuli over headphones, the sounds are perceived as being lateralized inside the head even though interaural temporal and intensity differences appropriate to an external source location are present.

The results of many studies now suggest that these deficiencies of the duplex theory reflect the important contribution to localization of the direction-dependent filtering that occurs when incoming sound waves interact with the outer ears or pinnae and other body structures such as the shoulders and torso.

The main cue the human auditory system uses to determine the elevation of a sound source is the monaural spectrum determined by the interaction of the sound with the pinnae (Wightman & Kistler, 1997). However, small head asymmetries may provide a weak binaural elevation cue. Specifically, there is a spectral notch that moves in frequency from approximately 5 kHz to 10 kHz as the source moves from 0° (directly ahead of listener) to 90° (above the listener's head) that is considered to be the main elevation cue (Musicant & Butler, 1985; Moore, Oldfield & Dooley, 1989). As sound propagates from a source to a listener's ears, reflection and refraction effects tend to alter the sound in subtle ways and the effect depends on frequency. For example, for a particular location, a group of high-frequency components centered at 8 kHz may be attenuated more than a different band of components centered at 6 kHz. Such frequency dependent effects or filtering also vary greatly with the direction of the sound source. Thus, for a different source location, the band at 6 kHz may be more attenuated than the higher frequency band at 8 kHz. It is clear that listeners use these kinds of frequency-dependent effects to discriminate one location from another. Experiments have shown that spectral shaping by the pinnae is highly direction-dependent, that the absence of pinna cues degrades localization accuracy, and that pinna cues are partially responsible for externalization or the "outside-the head" sensation (Gardner & Gardner, 1973; Oldfield & Parker, 1984a, b; Plenge, 1974; Shaw, 1974).

Other monaural cues are provided by the ratio of direct-to-reverberant energy that expresses the amount of sound energy that reaches our ears directly from the source versus the amount that is reflected off the walls in enclosed spaces (Larsen et al., 2008). In general, monaural cues are more ambiguous spatial cues than binaural cues because the auditory system must make *a priori* assumptions about the acoustic features of the original sound in order to estimate the filtering effects corresponding to the monaural spatial cues. Environmental cues will be discussed in more detail in a later section.

Factors Affecting Localization Performance

Localization performance generally refers to the degree of accuracy with which listeners can identify and/or discriminate the location of a sound source. It may be measured using a variety of experimental paradigms with different types of response measures. For example, in a direct localization task a sound source (real, recorded or virtual) is presented to a listener who is asked to report its location, perhaps in terms of estimates of azimuth, elevation and distance.

Several kinds of error are usually observed in perceptual studies of localization when listeners are asked to judge the position of a static sound source in the free field. One, which Blauert (1997) refers to as localization blur, is a relatively small error in resolution on the order of about 5° to 20°. A related measure of localization accuracy is the minimum audible angle (MAA), the minimum detectable angular difference between two successive sound sources. MAAs increase from about 1° for a sound directly ahead, to 20° or more for a sound directly to the right or left (Mills, 1958; Perrott & Saberi, 1990). The minimum audible movement angle (MAMA) is the minimum detectable angular difference of a continuously moving sound source (Perrott & Tucker, 1988). The MAMA depends on the speed of the moving sound source and ranges from about 8° for a velocity of 90°/s to about 21° for a velocity of 360°/s.

Another class of error observed in nearly all localization studies is the occurrence of front-back "reversals" (Figure 1.7, right). These are judgments indicating that a source in the front hemisphere was perceived by the listener as if it were in the rear hemisphere. Occasionally, back-to-front confusions are also found (e.g., Oldfield & Parker, 1984a, b). Confusions in elevation, with up locations heard as down, and vice versa, have also been observed (Wenzel, 1991).

Although the reason for such reversals is not completely understood, they are probably due in large part to the static nature of the stimulus and the ambiguities resulting from the so-called cone of confusion (Woodworth, 1938; Woodworth & Schlosberg, 1954; Mills, 1972). Assuming a stationary, spherical model of the head and symmetrically located ear canals (without pinnae), a given interaural time or intensity difference will correlate ambiguously with the direction of a sound source, with a conical shell describing the locus of all possible sources (Figure 1.7, left). Intersection of these conical surfaces with the surface of a sphere results in circular projections corresponding to contours of constant ITD or IID (i.e., considering sources at an arbitrary fixed distance). While the rigid sphere model is not the whole story, the observed pattern of such iso-IDT and iso-IID contours indicates that the interaural characteristics of the stimulus

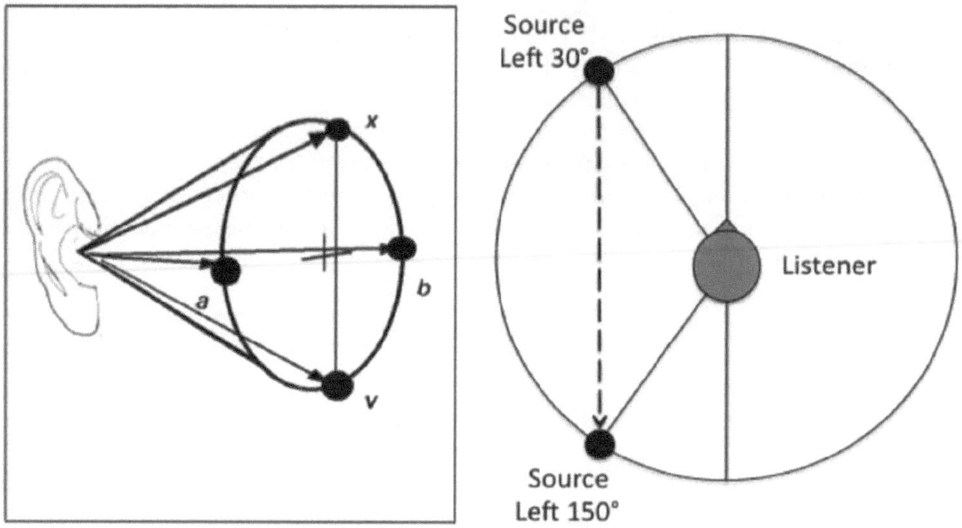

Figure 1.7 Cone of confusion effects (left panel) cause perceptual reversals in location for sound sources with identical or near-identical ITDs or IIDs (right panel).

are inherently ambiguous. In the absence of other cues, both front-back and up-down reversals would seem to be quite likely.

Several cues are thought to help in disambiguating the cones of confusion. One is the complex spectral shaping provided by the pinnae as a function of location that was described above. For example, because of the orientation and shell-like structure of the pinnae, high frequencies tend to be more attenuated for sources in the rear than for sources in the front [e.g., see Blauert's (1997) discussion of "boosted bands", pp. 111–116]. For the case of static sounds, such cues would essentially be the only clue to disambiguating source location. With dynamic stimuli, however, the situation improves greatly. A variety of studies have shown that allowing listeners to move their heads substantially improves localization ability and can almost completely eliminate reversals (e.g., Wallach, 1939, 1940; Thurlow & Runge, 1967; Fisher & Freedman, 1968; Wightman & Kistler, 1999; Begault, Wenzel & Anderson, 2001). With head-motion, the listener can apparently disambiguate front-back locations by tracking changes in the magnitude of the interaural cues over time; for a given lateral head movement, ITDs and IIDs for sources in the front will change in the opposite direction compared to sources in the rear (Wallach, 1939, 1940). Time-varying cues provided by moving sources may also aid in disambiguation, particularly if there is *a priori* knowledge about the direction of motion (Wightman & Kistler, 1999), although relatively little research has been done on the topic of source motion in this context.

In addition to the primary localization cues, the localizability of a sound also depends on other factors such as its spectral content: narrowband (pure) tones are generally difficult to localize while broadband, impulsive sounds are the easiest to locate. A closely related issue in the localizability of sound sources is their degree of familiarity. Logically, localization based on spatial cues other than the interaural cues, e.g., cues related to spectral shaping by the pinnae, is largely determined by a listener's *a priori* knowledge of the spectrum of the sound source. The listener must "know" what the spectrum of a sound is to begin with to determine that the same sound at different positions has been differentially shaped by the effects of his or her ear structures. Thus both the perception of elevation and relative distance, which depend heavily on the detection of spectral differences, tend to be superior for familiar signals like speech (e.g., Plenge & Brunschen, 1971, in Blauert, 1997, p. 104; Coleman, 1963). Similarly, spectral familiarity can be established through training (Batteau, 1967).

In an acoustical environment multiple acoustic objects can be present and room reverberation can also distort spatial cues. When the listener is in a room or other reverberant environment the direct sound received at the ears is combined with multiple copies of the sound reflected off the walls before arriving at the ears. Reverberation alters the monaural spectrum of the sound as well as the IIDs and IPDs of the signals reaching the listener (Shinn-Cunningham, Kopco & Martin, 2005). These effects depend on the source position relative to the listener as well as on the listener position in the room. On the other hand, the ratio of direct to reverberant energy itself can provide a spatial cue.

Such phenomena suggest that the auditory system exhibits neural plasticity, adapting over time in response to perceptual experience and changes in the sensory environment. Neural plasticity will be discussed in a later section.

Head-Related Transfer Functions (HRTFs) and Virtual Acoustics

The data on the primary localization cues suggest that perceptually veridical localization over headphones is possible if the spectral shaping by the pinnae and other body structures as well as the interaural difference cues can be adequately reproduced in a 3D sound system or virtual acoustic display. There may be many cumulative effects on the sound as it makes its way to the eardrum, but it turns out that all of these effects can be expressed as a single filtering operation much like the effects of a graphic equalizer in a stereo system. The exact nature of this filter can be measured by a simple experiment in which an impulse (a single, very short sound pulse or click) or other broadband probe stimulus is produced by a loudspeaker at a particular location. The acoustic shaping by the two ears is then measured by recording the outputs of small probe microphones placed inside an individual's ear canals (Figure 1.8). If the measurement of the two ears occurs simultaneously, the responses, when taken together as a pair of filters, include an estimate of the interaural differences as well. Thus, this technique allows one to measure all of the relevant spatial cues together for a given source location, a given listener and in a given room or environment.

The bottom panel of Figure 1.3 illustrates these effects for the transfer functions of the ears. The illustration on the left shows the frequency domain representation, derived from a mathematical operation known as the Fourier Transform, of an acoustic impulse in the time domain before interaction with the outer ear (and other body) structures. The illustration on the right shows what happens to the frequency response of an impulse delivered from a loudspeaker located directly to the right of a listener after interaction with the outer ear structures, as measured in the left (solid line) and right (dashed line) ear canals of a listener. The differences between the left and right intensity curves are the IIDs at each frequency. Spectral phase effects (frequency-dependent

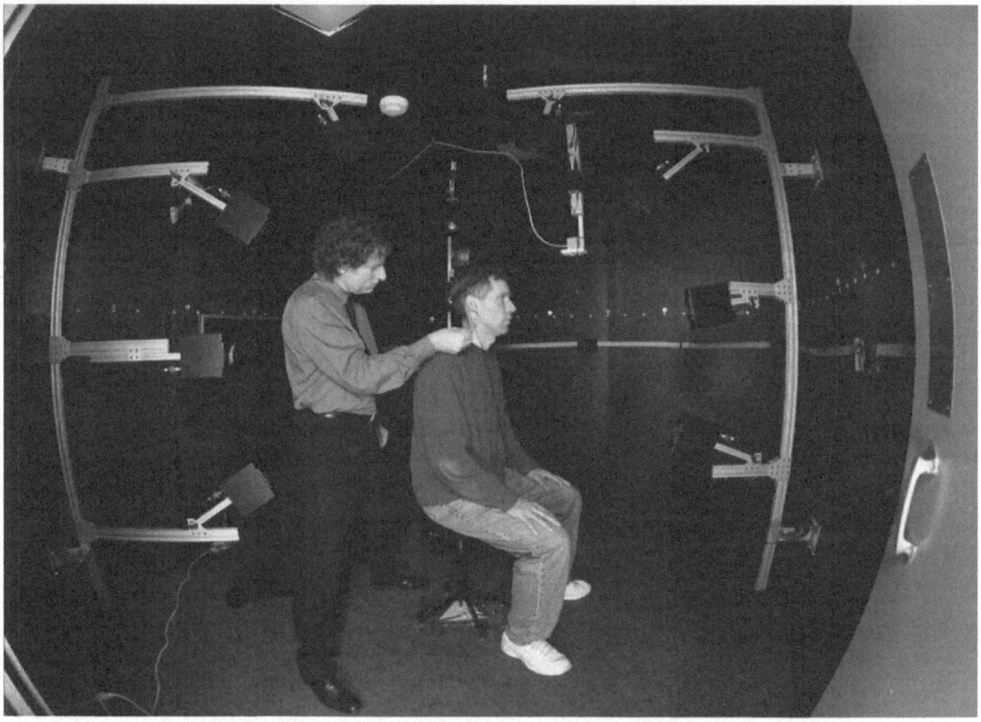

Figure 1.8 Facility at the NASA Ames Spatial Auditory Displays Lab for measuring HRIRs. The 12-speaker system in a double-walled soundproof booth can measure 432 locations at 10° intervals. The measurement signal is a Golay code (Foster, 1986) and responses are "quasi-anechoic", i.e., responses are windowed to remove possible reflections. The impulse response (down-sampled from 96 kHz to 44.1 kHz sample rate) is high-pass filtered at 700 Hz to avoid the influence of reflections. The amplitudes of frequencies below 700 Hz are not significantly affected by the head due to diffraction. The system calculates a frequency-independent ITD that is applied to the measured (minimum phase) HRIR that includes a flat extension of the frequency response below 700 Hz.

phase, or time delays) are also present in the measurements, but are not shown here for clarity. The filters constructed from these ear-dependent characteristics are examples of Finite Impulse Response (FIR) filters and are often referred to as Head-Related Impulse Responses (HRIRs) in the time domain, and Head-Related Transfer Functions (HRTFs) in the frequency domain. Filtering in the frequency domain is a point-by-point multiplication operation while filtering in the time domain occurs via a somewhat more complex operation known as convolution [see Brigham (1974) for a useful pictorial discussion of filtering and convolution]. By filtering an arbitrary sound with these HRTF-based filters, it is possible to impose spatial characteristics on the signal such that it apparently emanates from the originally measured location. If the filtering occurs in real-time, the effects of source motion and the listener's head motion can also be simulated.

Figure 1.9 provides examples of the magnitude responses for the left and right ears derived from measured HRTFs. The top panels show the frequency responses for azimuths of –90°, 0° and +90° at 0° elevation. Note that one can see how the magnitude responses are similar at 0° azimuth and are larger in the ipsilateral ear for the –90° (left ear) and +90° (right ear) source locations. The difference between the left and right ear responses represents the IIDs as a function of frequency. The overall IIDs averaged across frequency, as well as the ITDs, tend to be similar between individual subjects. Consequently, accuracy in azimuth perception is generally observed to be reasonably comparable when listening to stimuli generated from either individualized (one's own) or non-individualized HRTFs (Wightman & Kistler, 1989; Wenzel et al., 1993). The bottom panels show the frequency responses for elevations of –45°, 0° and +45° at +45° azimuth. Note that one can see how the center frequency of the notches in the magnitude spectra shift toward different frequencies as the elevation moves from –45 to +45° in the ipsilateral (right ear). Such frequency notches are thought to be a primary cue for elevation (Hebrank & Wright, 1974;

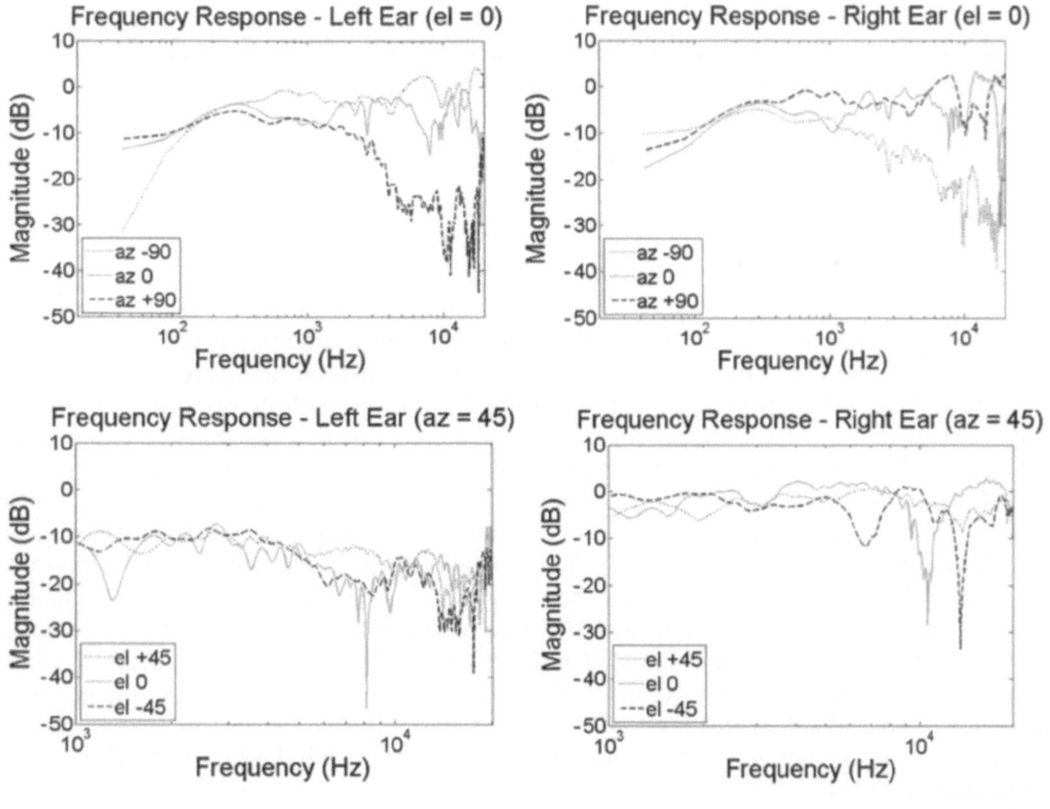

Figure 1.9 Examples of the magnitude responses for the left and right ears derived from measured HRTFs (subject IRC_1003 of the IRCAM Listen database). The top panels show the frequency responses for azimuths of –90°, 0° and +90° at 0° elevation. The bottom panels show the frequency responses for elevations of –45°, 0° and +45° at +45° azimuth.

Middlebrooks, 1992). Since these notch locations are highly dependent on specific pinna structures their particular frequency location may vary greatly between individuals. Thus, elevation perception is generally observed to be more accurate when listening to stimuli generated from individualized HRTFs.

It should be noted that the spatial cues provided by HRTFs, especially those derived from simple anechoic (free-field or echoless) environments, are not the only cues likely to be necessary to achieve veridical localization in a virtual display. Anechoic simulation is merely a first step, allowing a systematic study of the technological requirements and perceptual consequences of synthesizing spatial cues by using a less complex, and therefore more tractable, stimulus. The section on Distance and Environmental Context Perception will discuss the impact of environmental cues on localization and the perception of distance and immersion in more detail.

Neural Plasticity in Sound Localization

In natural environments, neural systems must be continuously updated to reflect changes in sensory inputs and behavioral goals. Recent studies of sound localization have shown that adaptation and learning involve multiple mechanisms that operate at different timescales and stages of processing, with other sensory and motor-related inputs playing a key role. In recent years, studies on sound localization have provided evidence to support the view that neural processing is rapidly updated to reflect changes in sensory conditions.

Spatial Cue Remapping

A popular approach to studying the plasticity of auditory spatial processing has been to reversibly alter the relation between stimulus location and the binaural cues available. This can be easily achieved by occluding one ear so that the acoustical input is attenuated and delayed, thereby changing the IID and ITD values corresponding to each direction in space. In the barn owl, this procedure leads to adaptation to the abnormal binaural cues, with neurons shifting their sensitivity to these cues in a way that compensates for the effects of monaural occlusion (Gold & E. I. Knudsen, 2000; E. I. Knudsen, P. F. Knudsen & Esterly, 1984). Recent work has shown that mammals also possess the ability to developmentally remap spatial position onto abnormal IIDs, which can be observed both behaviorally and in the responses of neurons in the primary auditory cortex (Keating, Dahmen & King, 2015). This capacity to accommodate altered spectral cues is not restricted to development and adult humans can learn to localize accurately using altered binaural (Bauer et al., 1966; Mendonça et al., 2013) or spectral localization cues (Hofman, Van Riswick & Van Opstal, 1998; Hofman & Van Opstal, 1998; Carlile & Blackman, 2014; Majdak, Walder & Laback, 2013). Interestingly, no aftereffect is seen in adult humans following adaptation to altered spatial cues, implying that different sets of spatial cues can be mapped onto the same location (Hofman et al., 1998; Hofman & Van Opstal, 1998; Carlile & Blackman, 2014; Majdak et al., 2013).

Spatial Cue Reweighting

In situations where some, but not all, of the spatial cues are altered, an alternative form of plasticity to cue remapping is to down-weight the spatial information provided by the altered cues and instead, to rely more on the cues that remain intact. In the specific case of monaural hearing loss, a number of studies have shown that sound localization behavior in mammals adapts both during development and adulthood by giving greater weight to the unchanged monaural spatial cues provided by the normal hearing ear (Agterberg et al., 2014; Kumpik, Kacelnik & King, 2010; Keating et al., 2013).

Importance of Behavioral Context

In addition to adapting to changes in the localization cues available, auditory spatial processing can be refined in situations where its behavioral importance is increased even if the acoustical input remains the same. Studies in humans have shown that training-induced improvement in spatial processing are specific to individual binaural cues, though cue specificity may be asymmetric, with one study showing that IID training generalizes to an ITD task, but not vice versa (Sand & Nilsson, 2014). Training in adulthood can even reverse the negative impact of abnormal developmental experience on sound localization accuracy and responses in the primary auditory cortex responses (Guo et al., 2012; Pan et al., 2011) and can improve sound localization performance in hearing-impaired populations (Firszt et al., 2015). Although training-dependent plasticity often takes place slowly, recent work indicates that feature-specific learning can occur rapidly in a task that involves spatial processing, which may reflect top-down biasing (Du et al., 2015).

Visual Influences on Auditory Spatial Plasticity

Sound sources are often visible as well as audible and the availability of visual information can improve the accuracy of sound localization estimates (Tabry, Zatorre & Voss, 2013) and even help to suppress echoes that are the consequence of listening in reverberant environments (Bishop, London & Miller, 2012). Binaural cue discrimination is also enhanced if subjects look toward the sound while keeping their head still (Maddox et al., 2014), adding evidence that eye position signals can modulate activity in the auditory system (Bulkin & Groh, 2012). Not surprisingly, auditory localization abilities can be altered if vision is impaired. The most commonly reported finding is that some blind individuals show superior auditory spatial perception relative to sighted control subjects (Hoover, Harris & Steeves, 2012; Lewald, 2013; Jiang, Stecker & Fine, 2014). Interestingly, as with adaptation to a unilateral hearing loss, more accurate sound localization in blind humans is associated with greater dependence on spectral cues (Voss et al., 2011). However, this superior use of spectral cues for localization in the horizontal plane appears to come at the cost of reduced ability to use these cues for localization in the vertical plane (Voss, Tabry & Zatorre, 2015).

Distance and Environmental Context Perception

Fundamental Cues to Sound Source Distance

The perceived distance of a sound source is mainly cued by the acoustic attributes of sound level and reverberation. Sound sources ordinarily reach a listener by not only an unobstructed wavefront (referred to as **direct sound**) but also by reflection off of and diffraction around nearby objects, such as the ground, walls or other surfaces (referred to as indirect sound, or

reverberation). In fact, distance judgments involve a process of integrating multiple perceptual cues, including loudness (perceived level), timbre (primarily driven by spectral content), amplitude envelope (attack and decay), reverberation and cognitive familiarity.

The perceived distance of a sound source is tied to the perceived environment in which the sound source is heard; hence, the overall immersive experience of a sound event's location relative to a listener is a multi-dimensional percept. We can model or identify the specific cause of reverberation in terms of the **environmental context** of a sound source: the surrounding physical surfaces that result in reverberation (either a room or outdoor environment). The environmental context typically consists of multiple sound sources, including background sounds ("noise") that combine with a specific sound source. It provides an important cognitive cue if the environmental context is familiar to the user.

An important distinction is made between **absolute** or **relative** perception of distance of a sound source. Absolute distance perception refers to a listener's accuracy in estimating the distance of a sound source from the listener themselves, upon an initial exposure to a new or familiar sound. Relative distance perception refers to judgments of the distance of one virtual source to another, and includes more strongly the benefits gained from listening to the source at different distances over time, perhaps within a particular environmental context. In most cases of spatial audio reproduction, one is typically more interested in relative distance judgments. Within a reasonable range, changing the overall volume level of a playback system still preserves the relative distance relationships between different virtual sound sources, an important advantage for audio production.

In the absence of other acoustic cues, the sound level of a sound source (and its interpretation as loudness) is the primary distance cue used by a listener. Coleman (1963, p. 302) stated in a review of cues for distance perception that, "It seems a truism to state that amplitude, or pressure, of the sound wave is a cue in auditory depth perception by virtue of attenuation of sound with distance." From one perspective, auditory distance is learned from a lifetime of visual-aural observations, correlating the physical displacement of sound sources with corresponding increases or reductions in sound pressure level. This is likely the primary means we use for many everyday survival tasks, for instance, knowing when to step out of the way of an automobile coming from behind us. In isolation from other cues, which happens rarely in the real world, sound level probably plays a more important role as a cue to distance with unfamiliar sounds than with familiar sounds. Exposure to a particular sound source at different distances allows an integration of multiple cues over time for distance perception; but without this exposure, cues other than level (loudness) fall out of the equation.

The relationship between sound source distance and level changes at a listener can be predicted under anechoic conditions via **inverse square law** for sound intensity reduction with increasing distance. In the absence of significant reflections, an omnidirectional point sound source's level will fall almost exactly 6 dB for each subsequent doubling of distance from a source. If the sound source was not omnidirectional but instead a **line source**, such as a freeway, then the level reduction is closer to a 3 dB per doubling of distance. This illustrates the importance of characterizing the sound power profile of a modeled sound source dimensionally.

A theoretical problem with the inverse square law as an effective cue to distance lies in the fact that perceptual scales of loudness are unaccounted for. Given that a perceptual scale is desired, when loudness is the only available cue for distance involved, a mapping where the relative estimation of doubled distance follows "half-loudness" rather than "half-level" may be more effective. Studies have shown that estimates of half-loudness to be equivalent to half of the auditory distance (Stevens & Guirao, 1962; Begault, 1991). Based on the sone scale of loudness, doubling of distance would require a 10 dB rather than a 6 dB reduction in level.

Level adjustments to create relative relationships between the loudness of different sounds are of course ubiquitous in the world of audio. A recording engineer, given the assignment to distribute a number of sound sources on different tracks of a multitrack recording to different apparent distances from a listener, would most probably accomplish the task intuitively by adjusting the volume of each track. Terminology taken from the visual world, such as "foreground" and "background", are usually used in these contexts; most audio professionals don't get more specific than this verbally, although in practice the distance-intensity relationships of a multitrack recording can be quite intricate.

In contrast to the study of physical acoustics or acoustical engineering, determining realistic relationships between the levels of different sound sources is frequently not desired in the art of sound design for film or music. A good sound designer must take into account both the emotional impact of a narrative and the constraints of a limited dynamic range, as opposed to representing realistically the sound levels of the real world. More often than not, the sound levels and corresponding auditory distance cues used in film sound express an artistic rather than realistic adjustment of speech, footsteps, gunshots, ambient sounds and the like. In music, recording engineers often use "close microphone" techniques in orchestral recordings to alter the balance of different instruments and vocalists, and in most popular music create completely synthetic distance relationships between sound sources that would never exist in the real world. Nevertheless, in order to be sensible to a listener there is a necessary reference to real-world conditions on which to base the boundaries within a specific artistic sound creation.

Familiarity and Cognitive Cues to Distance

Distance cues can be modified as a function of expectation or familiarity with the sound source, especially with speech (Coleman, 1962; Gardner, 1969). Even with non-spatial cues, one can tell something about the distance of different sounds and their context; for instance, you can experience distance listening with one ear, and you can sometimes tell in a telephone conversation where a person is calling from, based on the type of background noise. But the inclusion of 3-D sound techniques and spatial reverberation cues can greatly increase the immersiveness of a simulation, as well as the overall quality and the available nuance in representing different environmental contexts.

A good example of the role of familiarity is a comparison of the experience of listening to sounds just before going to sleep, in an unfamiliar versus a familiar environmental context. In the familiar environment, say an apartment in a city, you know the distance of the sound of the local bus passing by and of the ticking clock in the kitchen. Although the bus is louder than the clock, familiarity allows distance estimations that would be reversed if sound level were the only cue. But when for instance camping outdoors in an unfamiliar environment, the distance percepts of different unfamiliar animal noises are likely cued more by level.

Implementation of distance cues into a 3-D sound system requires an assessment of the cognitive associations for a given sound source. If the sound source is completely synthetic or unfamiliar, then a listener may need more time to familiarize themselves with the parametric changes in loudness and other cues that occur for different simulated distances. If the sound source is associated with a particular location from repeated listening experiences, the simulation of that distance will be easier than simulation of a distance that is unexpected or unfamiliar. Evidence exists that listeners are very good at distance estimates of talkers speaking at a particular level (Zahorik, 2002; Zahorik, Brungart & Bronkhorst, 2005). However, the manner of speaking can act as a cognitive cue to bias distance estimates. The average level of male speech at a distance of 1 meter ranges from 52 dB(A) for "casual" speech up to 89 dB(A) for a "shouted" voice (Pearsons, Bennett & Fidell, 1977), and there are spectral and phonetic differences resulting from different speaking levels that can act as cues. Whispering, caused by air pressure through the vocal tract without vibration of the vocal cords, can be as loud as speech, but is immediately associated with more intimate conversation at closer distances than normal speech.

Gardner (1969) conducted several studies for speech stimuli that illustrate the role of familiarity and expectation on estimated distance. In one experiment, categorical estimations were given by subjects of sound source positions at 0° azimuth, by choosing from numbered locations at 3, 10, 20 and 30 feet in an anechoic chamber. When loudspeaker playback of recorded normal speech was presented, the perceived distance was always a function of the sound pressure level at the listener instead of the actual location of the loudspeaker. But with a live person speaking inside the chamber, subjects based their estimates of distance on the manner of speaking rather than on the actual distance. Figure 1.10 shows an illustration of these results. Listeners overestimated the

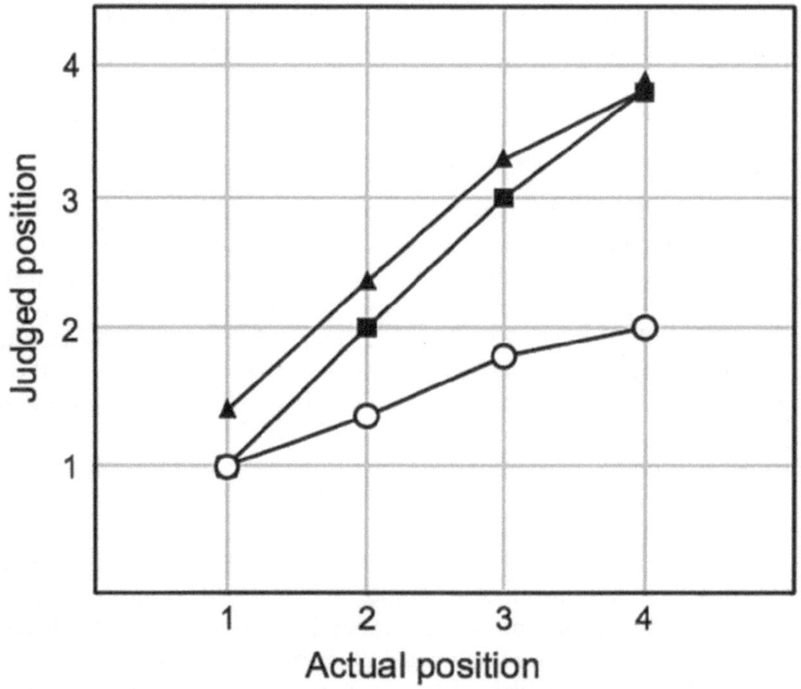

Figure 1.10 Results obtained by Gardner (1969) for a live speaker at 0° azimuth in an anechoic chamber. Open circles = whispering; solid squares = low-level and conversational-level speech; triangles = shouting.

distance of shouting in reference to normal speech and underestimated the distance of whispering, although the opposite should have been true if intensity were the relevant cue.

Reverberation Cues

In reverberant environmental contexts, the ratio of direct to reverberant sound changes as a function of distance between a sound source and listener. Typically the inverse square law reduction in sound pressure only operates in the acoustic "near field" of a sound source where the level of a direct sound significantly exceeds that of the indirect sound. With increasing distance, the overall sound pressure of a sound source at a receiving point becomes increasingly made up of indirect as well as direct sounds. At a certain point, the sound source reaches a **critical distance** (also termed "reverberation distance" or "reverberation radius"), where the level of direct and reflected sound are the same. At locations at or beyond the critical distance, the overall level tends to be the same because the sound at the receiving point is made up principally of indirect sound. See Figure 1.11.

To effect a relatively crude (but effective) audio simulation of distance, it is possible to change the ratio of reverberant to direct sound directly by adjusting level controls on an audio mixer. This ratio is a measurement of the proportion of reflected-to-direct sound energy at a particular receiver location. As one moves away from a sound source in an enclosure, the level of the direct

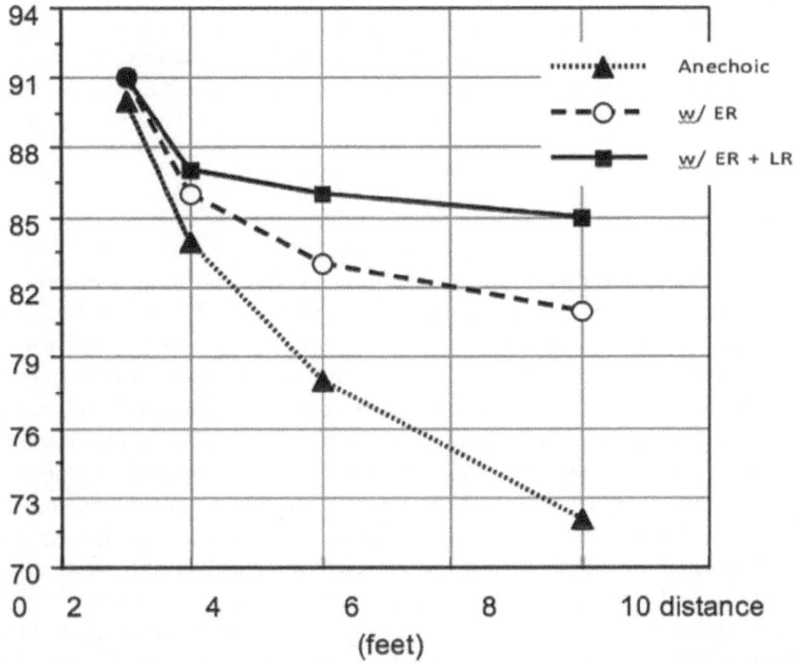

Figure 1.11 Level reduction under anechoic and reverberant conditions, at successive doublings of distances of 1, 2, 4 and 8 feet. ER = early reflections; LR = late reverberation. Note that with reverberation, the overall sound level as a function of the inverse square law is no longer useful at distances greater than 2 feet.

sound will decrease, while the reverberation level will remain constant. The interaction between reverberation and the direct sound level as a sound source varies in distance is too complex to allow a precise prediction of a distance percept as a function of the R/D ratio.

The reverberant-to-direct sound (R/D) ratio has been cited in many studies as a cue to distance, with varying degrees of significance attributed to it (Coleman, 1963; von Békésy, 1960; Sheeline, 1983; Mershon & King, 1975; Mershon & Bowers, 1979). Sheeline (1983, p. 71) found that reverberation was an important adjunct to intensity in the formation of distance percepts, concluding that "reverberation provides the 'spatiality' that allows listeners to move from the domain of loudness inferences to the domain of distance inferences."

Von Békésy (1960, p. 303) observed that when he changed the R/D ratio, the loudness of the sound remained constant, but a sensation of changing distance occurred. He noted that, "though this alteration in the ratio between direct and reverberant sound can indeed be used to produce the perception of a moving sound image, this ratio is not the basis of auditory distance. This is true because in [an anechoic room] the sensation of distance is present, and, in fact, is even more distinct and of much greater extensiveness than elsewhere." He also observed that the sound image's width increased with increasing reverberation: "Along with this increase in distance there was an apparent increase in the size or vibrating surface of the sound source . . . for the direct sound field the source seemed to be extremely small and have great density, whereas the reverberant field had a more diffuse character." This demonstrates the multidimensional nature of sound source "locatedness" as a function of both level and context. The **apparent source width** of a sound source is defined as its perceived extent or size of the sound source's image, and is related to the concept of **auditory spaciousness**. Blauert (1997, p. 348) describes auditory spaciousness to mean that ". . . auditory events, in a characteristic way, are themselves perceived as being spread out in an extended region of space," and cites low levels of varying interaural coherence over time ("temporal incoherence") as a primary cause.

IHL

An often-reported sensation for headphone listening is that the sound image appears to exist entirely within or at the edge of the head, instead of externalized outside the listener. **Inside-the-head locatedness (IHL)** can be considered a type of "externalization failure" for correct distance simulation, particularly for binaural and 3D sound simulations over headphones (the effect happens rarely with loudspeaker reproduction). For example, when binaural cues are nearly identical in a 3D audio simulation, as with a source directly in front of the listener, the case is similar to **diotic** sound presentation, where the sound presented to both ears is the same or nearly so. This is an artificial sound condition compared to the experience of sound sources outside a listener, leading to a cognitive conclusion that the sound originates within or near the body, like self-generated speech.

However, it is obviously possible even listening with one earphone, such as to a baseball game with an old-fashioned AM radio, that the sound events are in fact externalized. The illusion of an "internalized" versus "externalized" sound image can in many cases be willfully switched in a manner analogous to the Necker cube or other similar illusions (von Békésy, 1960). One study of 3D audio simulation of speech stimuli showed that inclusion of reverberation or head-tracking into a simulation helped mitigate internalized sound (Begault, Wenzel & Anderson, 2001).

Figure 1.12 shows how HRTF-processed speech presented at a level corresponding to a distance of ~15 in. is underestimated and varies as a function of direction. This underestimation has also been observed with actual as opposed to virtual sound sources (Holt & Thurlow, 1969; Mershon & Bowers, 1979; Butler, Levy & Neff, 1980). One reason for this underestimation may be the absence of reverberation in the stimulus. Underestimation in general may also be related to the bounds of perceptual space, i.e., the **auditory horizon**. Note that the standard deviation bars of Figure 1.12 indicate a high amount of variability between subjects, for what was essentially one target distance; one might have expected a more stable distance estimate among individuals, based on the common familiarity of speech.

The goal of eliminating IHL effects arose in the 1970s with the desire to make improved binaural (dummy head) recordings. Many who heard these recordings were disturbed by the fact that the sound remained inside the head, as with lateralization. Ensuring that the sound was filtered by accurate replicas of the human pinnae and head was found to be an important consideration. Plenge (1974) had subjects compare recordings made with a single microphone to those made with a dummy head with artificial pinnae; the IHL that occurred with the single microphone disappeared with the dummy head micing arrangement. Laws (1973) and others determined that part of the reason for this had to do with non-linear distortions caused by various parts of the communication chain, and the use of **free-field** instead of **diffuse-field** equalized headphones. Durlach and Colburn (1978, p. 374) have mentioned that the externalization of a sound source is difficult to predict (or even describe) with precision, but "clearly, however, it increases as the stimulation approximates more closely stimulation that is natural." The likely sources of these natural interaural attributes include the binaural HRTF, head movement and reverberation.

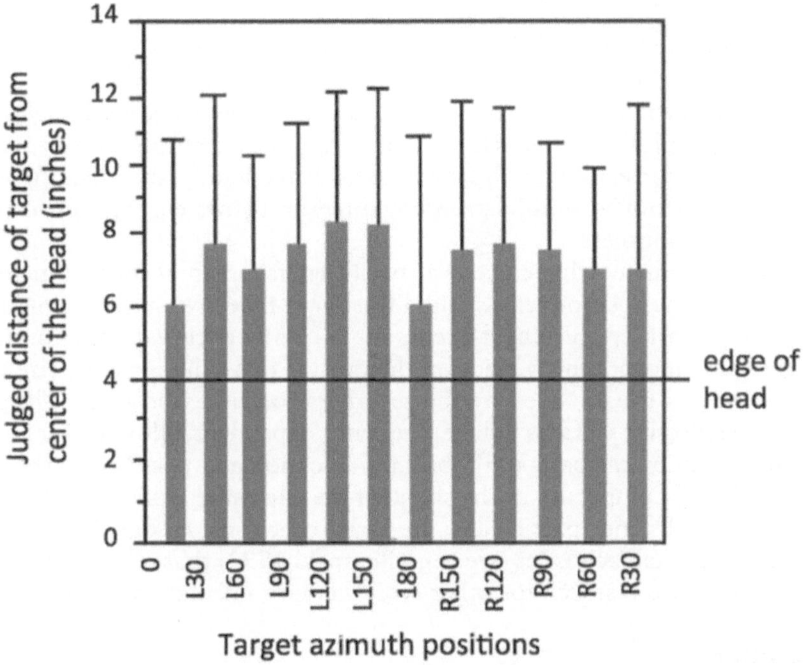

Figure 1.12 Means and standard deviations for distance judgments of non-reverberant speech stimuli (Begault & Wenzel, 1993).

Many researchers have discounted theories that IHL is a natural consequence of headphone listening (due to bone conduction or pressure on the head), simply because externalized sounds *are* heard through headphones in many instances.

Environmental Context Cues

The defining characteristics of an environmental context allow us to discriminate between the same source heard in different size enclosures or out-of-doors. Although the reverberation time and R/D ratio serve as cues, oftentimes the spatial distribution of early reflections over time from sounds heard within an environmental context can be just as important (Bronkhorst & Houtgast, 1999; Kendall & Martens, 1984). Such attributes can serve as cues for characterizing and identifying both the context of a sound source and its potential range of distances. The environment can also provide a characteristic background noise that can help define the range and context of sounds. Studies have also noted that listeners eventually adapt to reverberation in a room; compared to an initial exposure, both distance and azimuth estimates improve over time as the environmental context is "learned" (Shinn-Cunningham, 2000).

Different spatial-temporal patterns particularly from indoors can affect distance perception and image broadening, as well as the sensation of being "immersed" or surrounded by sound, a percept sometimes referred to "auditory spaciousness" (Beranek, 1992; Kendall, Martens & Wilde, 1990). By measuring the similarity of the reverberation over a specific time window at the two ears, a value for **interaural cross-correlation** can be obtained that is frequently cited in measures of immersion (Blauert & Cobben, 1978; Ando, 1985). Cross-correlation analysis indicates the degree of similarity of two time-domain waveforms over a given period of time. For example, the cross-correlation of a signal with a delayed version of itself will show a peak at the time lead or lag of the delay within an analysis window (Figure 1.13). In the case of interaural cross-correlation, the analysis window corresponds to lead or lag times corresponding to the maximum interaural time delay (typically 0.7 milliseconds); a "running" interaural cross-correlation refers to a succession of cross-correlation analysis windows. Perceptually, the magnitude of the overall differences between lead and lag within a running interaural cross-correlation corresponds to the percept of auditory spaciousness.

Relevant physical parameters that can cue a specific environmental context include the volume or size of an enclosure, the absorptiveness and diffusion of reflective surfaces, and the complexity of the shape of the enclosure. This effect occurs in "partially enclosed" environmental contexts as well, such as under an overhung surface on the outside of a building. The size (volume) of an environmental context is usually cued by the reverberation time and level. The absorptiveness and diffusion of the reflective surfaces will be frequency dependent, allowing for cognitive categorization and comparison on the basis of timbral modification and possibly on the basis of speech intelligibility. Finally, the complexity of the shape of the enclosure will shape the spatial distribution of reflections to the listener, particularly the early reflections. A considerable literature exists in the domain of concert hall acoustics (see, e.g., Beranek, 1992) that relates physical measures to percepts, while there is less research for other typical environmental contexts.

Late Reverberation

Late reverberation is informative perceptually as to the volume of a particular space occupied by a sound source. This is most noticeable when a sound source is "turned off", since one can hear

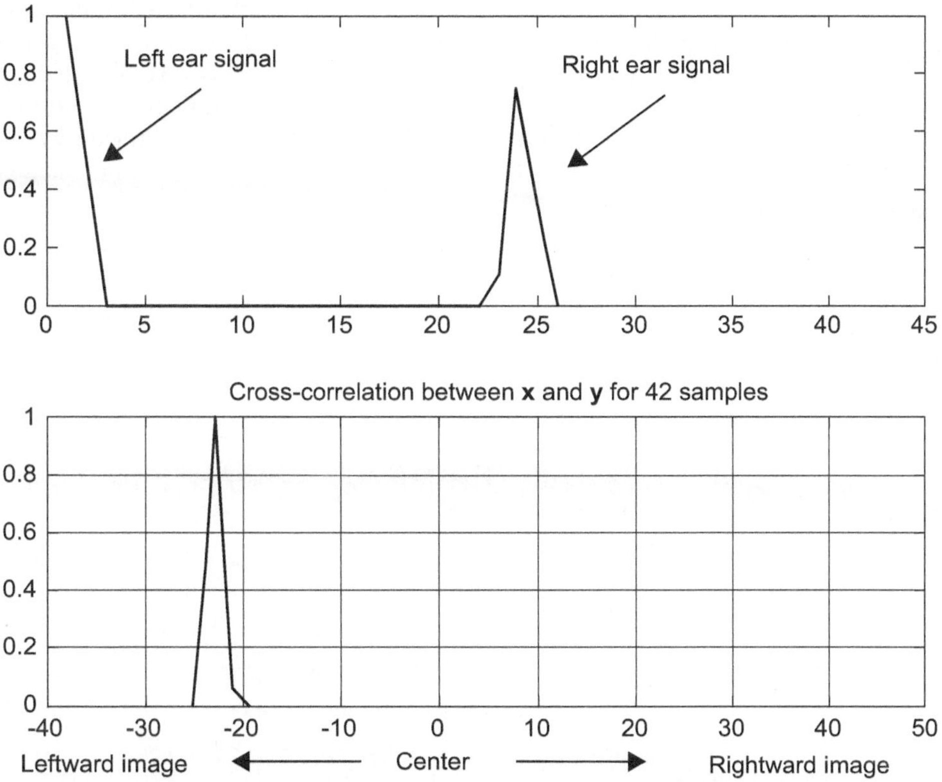

Figure 1.13 Top: Time of arrival of a signal **x** at the left ear followed by an attenuated copy of the signal **y** delayed by 24 samples the right ear (0.54 msec at an audio sampling rate of 44,100 samples/second, corresponding to a sound source image to the left of a listener). Bottom: the cross-correlation of the two signals shows a peak at −22 samples, corresponding to a leftward image. If signals **x** and **y** were reversed the bottom plot would show a peak at +22 samples.

the time it takes for the echoes to decay into relative silence, and the shape of the amplitude decay envelope over time. During continuous speech or music, only the first 10–20 dB of decay will be heard, since energy is constantly being "injected" back into the system. The long-term "room frequency response"—i.e., the reverberation time as a function of frequency—can also be significant in the formulation of a percept of the environmental context. The relative decay times for different frequency regions is affected by both the volume of the enclosure and the relative acoustic absorption of materials within the environment. For example, although a tiled bathroom has a smaller volume than a typical living room, the minimal acoustic absorption of the tiles causes the reverberation to be relatively brighter and the reverberation time to be longer compared to a living room with a carpet, couches and other absorptive surfaces.

Concert hall studies since the beginning of the 20th century have emphasized reverberation time as a strong physical factor affecting subjective preference, with particular frequency ranges being important for "warmth" or "clarity". Many enclosures in fact do not decay exponentially, although the use of reverberation time is often used as a standard approximation. Many small enclosures including automobiles do not have a proper "reverberation time"; their "sound" is a function of a complex early reflection field. Another characteristic of reverberation in complex spaces is the irregular and random nature of the fine structure of its decay, resulting in a "ragged" response that can make determination of a single reverberation time difficult. A smoother plot of the response can be obtained through the technique of **reverse integration** of the decay (Schroeder, 1965). Figure 1.14, top, shows the result of the "impulse response" of a room, obtained from a

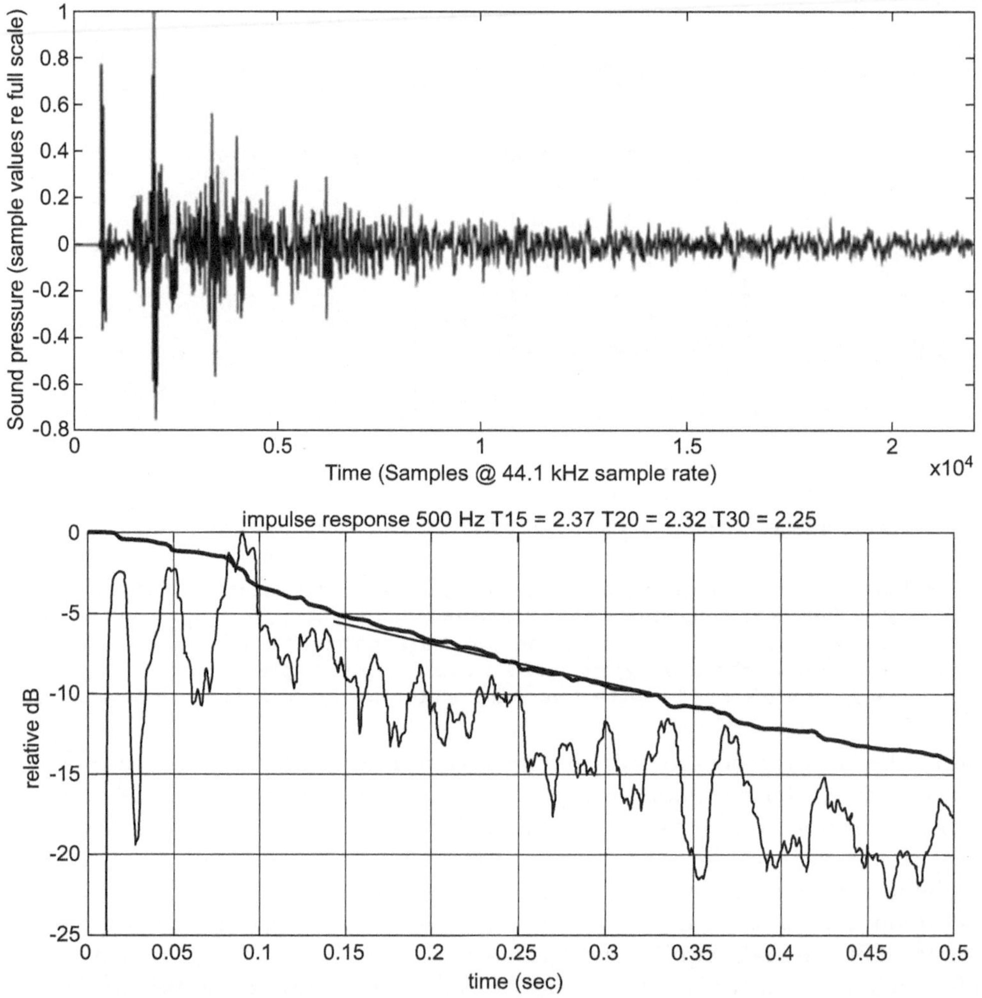

Figure 1.14 Top: Impulse response of a room for measurement of reverberant decay. Bottom: Thin line shows result from passing the impulse through a 500 Hz octave band filter and representing values on a decibel scale. Note the raggedness of the response. The thick line indicates the result of *reverse integration* of the same impulse response. The smoother line can be more easily fit to an estimate of the reverberation time.

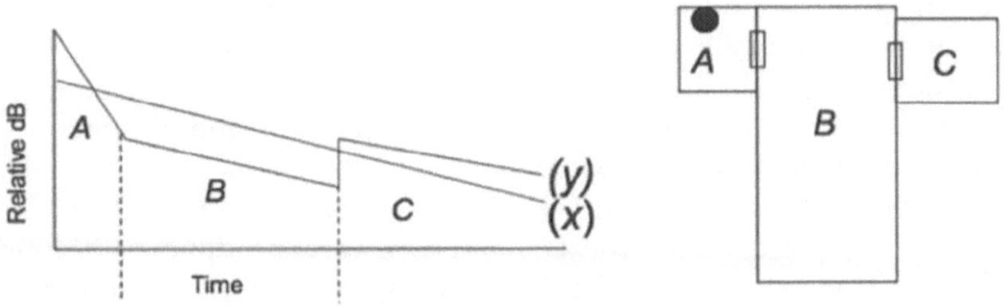

Figure 1.15 Left: Reverberation time slopes A, B, C from a hypothetical impulse response, and (right) corresponding room volumes, coupled via openings. The reverberation time might be averaged from the delay slope *x* but in fact consist of separate decays for A, B, C (slopes indicated as *y*).

balloon pop (a starter pistol or analytic signal such as a swept sine wave can also be used). Figure 1.14, bottom, shows the result of plotting the impulse response on a decibel scale ($10\log_{10}$ of the squared impulse response values) with the thin line; note the raggedness of the response. To determine a reverberation time, a straight line must be fit to 15, 20 or 30 decibels of decay (T15, T20 or T30). By reverse integration of the impulse response, a smoother decay is obtained, shown by the thick line, making the straight line fit more apparent. The result of reverse integration is equivalent to the average of multiple impulse responses.

An interesting phenomenon is the reverberation that occurs from the interaction of multiple enclosures that are connected to one another via hallways, atriums or other openings. This is referred to in acoustics as a "coupled space" phenomenon.

While reverberation time can be an adequate descriptor for a single enclosed space, the coupling of two volumes with different reverberation times can result in multiple decay slopes (Figure 1.15). This is most audible in large complex spaces that have adjoining volumes of different sizes. In some such spaces, such as a cathedral, the late reverberation is perceptibly modulated both in amplitude and spatial location (Woszczyk, Begault & Higbie, 2014).

For relatively distant sources, particularly out-of-doors, the influence of atmospheric conditions, molecular absorption of the air, and wavefront curvature can change the spectral content of a virtual sound. From a psychoacoustic standpoint, these cues are relatively weak, compared to loudness, familiarity and reverberation cues. Since the sound sources used in a spatial audio simulation are more than likely dynamically changing in location, their spectra are constantly changing as well, making it difficult to establish any type of "spectral reference" for perceived distance. "As one would expect on the basis of purely physical considerations, if one eliminates the cues arising from changes in loudness and changes in reverberant structure, the binaural system is exceedingly poor at determining the distance of a sound source" (Durlach & Colburn, 1978, p. 375).

Conclusion

A listener's perception of immersion is influenced by a complex set of interactions between humans and acoustic waves, beginning with the peripheral auditory system and ending with the information processing aspects of cognition. Each of these interactions contributes to overall judgments of the spatial attributes of these sensations, including the sense of acoustic immersion and the location of specific sound sources. Virtual simulations of spatial audio are best realized when an understanding of the relevant psychoacoustic and acoustical cues are implemented into signal processing design. This chapter has given an overview of the role of auditory processes and psychoacoustic data relevant to the execution of successful spatial audio techniques for providing listeners with many types of immersive experiences.

A recording engineer aims for conveyance of spatial imagery, and will be challenged by an ever-increasing number of distribution formats and hardware: headphones or loudspeakers, traditional two-channel systems, 22.2–9.1–7.1 and other systems, or wave field synthesis, streaming audio or high-quality archival compression and so on. If the creative imagination of the engineer is the source of spatial imagery, and the listeners are the receivers of this imagery, then we can judge the quality and the challenges presented by a particular system by how successfully the intended creative imagery is conveyed. All of the perceptual factors discussed in this chapter will contribute to the resulting listener experience in a virtual auditory environment using any of the methods of sound reproduction that will be discussed in the rest of the book. Further, they will pose different challenges to different fields of expertise and practice, including the audio engineer, sound designer and digital signal processing effects designer.

Notes

1 For a specified signal, the Hearing Level (HL) refers to the amount in decibels by which the hearing threshold for a listener, for either one or two ears, exceeds a specified reference equivalent threshold level. The reference is based on a large number of otologically normal individuals of both genders ranging in age from 18 to 25 years. (ANSI/ASA S1.1–1994, S3.6–2010)
2 Alternatively, sometimes azimuth is represented in a 360° coordinate system with 0° in front, 90° to the right, 180° in back and 270° to the left.

References

ANSI/ASA S3.6–2010 "Specification for Audiometers".

ANSI/ASA S1.1–1994 "Acoustical Terminology".

Agterberg, M. J., Hol, M. K., Van Wanrooij, M. M., Van Opstal, A. J., & Snik, A. F. (2014). Single-sided deafness and directional hearing: Contribution of spectral cues and high frequency hearing loss in the hearing ear. *Frontiers in Neuroscience*, *8*, 188.

Ando, Y. (1985). *Concert Hall Acoustics*. Berlin: Springer-Verlag.

Batteau, D. W. (1967). The role of the pinna in human localization. *Proceedings of the Royal Society of London B: Biological Sciences*, *168*(1011), 158–180.

Bauer, R. W., Matuzsa, J. L., Blackmer, R. F., & Glucksberg, S. (1966). Noise localization after unilateral attenuation. *Journal of the Acoustical Society of America*, *40*(2), 441–444.

Begault, D. R. (1991). Preferred sound intensity increase for sensation of half distance. *Perceptual and Motor Skills*, *72*, 1019–1029.

Begault, D. R., & Wenzel, E. M. (1993). Headphone localization of speech. *Human Factors*, *35*, 361–376.

Begault, D. R., Wenzel, E. M., & Anderson, M. R. (2001). Direct comparison of the impact of head tracking, reverberation and individualized head-related transfer functions on the spatial perception of a virtual speech source. *Journal of the Audio Engineering Society*, *49*, 904–916.

Békésy, G. von. (1960). *Experiments in Hearing*. New York: McGraw-Hill.

Beranek, L. L. (1992). Concert hall acoustics. *Journal of the Acoustical Society of America*, *92*, 1–39.

Bergeijk, W. A. van. (1962). Variation on a theme of Békésy: A model of binaural interaction. *Journal of the Acoustical Society of America*, *34*(9B), 1431–1437.

Bernstein, L. R. (2001). Auditory processing of interaural timing information: New insights. *Journal of Neuroscience Research*, *66*(6), 1035–1046.

Bernstein, L. R., & Trahiotis, C. (2010). Accounting quantitatively for sensitivity to envelope based interaural temporal disparities at high frequencies. *Journal of the Acoustical Society of America*, *128*, 1224–1234.

Bishop, C. W., London, S., & Miller, L. M. (2012). Neural time course of visually enhanced echo suppression. *Journal of Neurophysiology*, *108*(7), 1869–1883.

Blauert, J. (1997). *Spatial Hearing: The Psychophysics of Human Sound Localization,* Rev. ed. (J. Allen, Trans.). Cambridge, MA: MIT Press.

Blauert, J., & Cobben, W. (1978). Some consideration of binaural cross correlation analysis. *Acustica*, *39*, 96–104.

Brigham, E. (1974). *The Fast Fourier Transform*. New Jersey: Englewood Cliffs.

Bronkhorst, A., & Houtgast, T. (1999). Auditory distance perception in rooms. *Nature*, *397*, 517–520.

Brughera, A., Dunai, L., & Hartmann, W. M. (2013). Human interaural time difference thresholds for sine tones: The high-frequency limit. *Journal of the Acoustical Society of America*, *133*, 2839–2855.

Bulkin, D. A., & Groh, J. M. (2012). Distribution of eye position information in the monkey inferior colliculus. *Journal of Neurophysiology*, *107*(3), 785–795.

Butler, R. A., Levy, E. T., & Neff, W. D. (1980). Apparent distance of sounds recorded in echoic and anechoic chambers. *Journal of Experimental Psychology: Human Perception and Performance*, *6*(4), 745.

Carlile, S., & Blackman, T. (2014). Relearning auditory spectral cues for locations inside and outside the visual field. *Journal of the Association for Research in Otolaryngology*, *15*(2), 249–263.

Chase, S. M., & Young, E. D. (2006). Spike-timing codes enhance the representation of multiple simultaneous sound-localization cues in the inferior colliculus. *Journal of Neuroscience*, *26*(15), 3889–3898.

Coleman, P. D. (1962). Failure to localize the source distance of an unfamiliar sound. *Journal of the Acoustical Society of America*, *34*(3), 345–346.

Coleman, P. D. (1963). An analysis of cues to auditory depth perception in free space. *Psychological Bulletin*, *60*(3), 302–315.

Culling, J. F., Hawley, M. L., & Litovsky, R. Y. (2004). The role of head-induced interaural time and level differences in the speech reception threshold for multiple interfering sound sources. *Journal of the Acoustical Society of America*, *116*(2), 1057–1065.

Douglas, K. M., & Bilkey, D. K. (2007). Amusia is associated with deficits in spatial processing. *Nature Neuroscience*, *10*(7), 915–921.

Du, Y., He, Y., Arnott, S. R., Ross, B., Wu, X., Li, L., & Alain, C. (2015). Rapid tuning of auditory "what" and "where" pathways by training. *Cerebral Cortex*, *25*(2), 496–506.

Durlach, N. I., & Colburn, H. S. (1978). Binaural phenomena. In E. C. Carterette & M. P. Friedman (Eds.), *Handbook of Perception* (pp. 365–466). New York: Academic Press.

Feddersen, W. E., Sandel, T. T., Teas, D. C., & Jeffress, L. A. (1957). Localization of high frequency tones. *Journal of the Acoustical Society of America*, *29*(9), 988–991.

Firszt, J. B., Reeder, R. M., Dwyer, N. Y., Burton, H., & Holden, L. K. (2015). Localization training results in individuals with unilateral severe to profound hearing loss. *Hearing Research*, *319*, 48–55.

Fisher, H. G., & Freedman, S. J. (1968). The role of the pinna in auditory localization. *Journal of Auditory Research*, *8*(1), 15–26.

Foster, S. (1986). Impulse response measurement using Golay codes. *IEEE 1986 Conference on Acoustics, Speech and Signal Processing*, 2, 929–932. New York: IEEE.

Fuster, J. M. (2009). Cortex and memory: Emergence of a new paradigm. *Journal of Cognitive Neuroscience*, *21*(11), 2047–2072.

Gardner, M. B. (1969). Distance estimation of 0 degree or apparent 0 degree-oriented speech signals in anechoic space. *Journal of the Acoustical Society of America*, *45*, 47–53.
Gardner, M. B., & Gardner, R. S. (1973). Problem of localization in the median plane: Effect of pinnae cavity occlusion. *Journal of the Acoustical Society of America*, *53*, 400–408.
Gifford III, G. W., & Cohen, Y. E. (2005). Spatial and non-spatial auditory processing in the lateral intraparietal area. *Experimental Brain Research*, *162*(4), 509–512.
Gold, J. I., & Knudsen, E. I. (2000). Abnormal auditory experience induces frequency-specific adjustments in unit tuning for binaural localization cues in the optic tectum of juvenile owls. *Journal of Neuroscience*, *20*(2), 862–877.
Gulick, W. L., Gescheider, G. A., & Frisina, R. D. (1989). *Hearing: Physiological Acoustics, Neural Coding, and Psychoacoustics*. New York: Oxford University Press, Inc.
Guo, F., Zhang, J., Zhu, X., Cai, R., Zhou, X., & Sun, X. (2012). Auditory discrimination training rescues developmentally degraded directional selectivity and restores mature expression of GABA A and AMPA receptor subunits in rat auditory cortex. *Behavioural Brain Research*, *229*(2), 301–307.
Hebrank, J., & Wright, D. (1974). Spectral cues used in the localization of sound sources on the median plane. *Journal of the Acoustical Society of America*, *56*, 1829–1834.
Henning, G. B. (1974). Detectability of interaural delay in high-frequency complex waveforms. *Journal of the Acoustical Society of America*, *55*(1), 84–90.
Henning, G. B. (1980). Some observations on the lateralization of complex waveforms. *Journal of the Acoustical Society of America*, *68*, 446–454.
Hofman, P. M., & Van Opstal, A. J. (1998). Spectro-temporal factors in two-dimensional human sound localization. *Journal of the Acoustical Society of America*, *103*(5), 2634–2648.
Hofman, P., & Van Opstal, A. (2003). Binaural weighting of pinna cues in human sound localization. *Experimental Brain Research*, *148*(4), 458–470.
Hofman, P. M., Van Riswick, J. G. A., & Van Opstal, A. J. (1998). Relearning sound localization with new ears. *Nature Neuroscience*, *1*(5), 417–421.
Holt, R. E., & Thurlow, W. R. (1969). Subject orientation and judgement of distance of a sound source. *Journal of the Acoustical Society of America*, *46*, 1584–1585.
Hoover, A. E., Harris, L. R., & Steeves, J. K. (2012). Sensory compensation in sound localization in people with one eye. *Experimental Brain Research*, *216*(4), 565–574.
Jeffress, L. A. (1948). A place theory of sound localization. *Journal of Comparative and Physiological Psychology*, *41*, 35–39.
Jiang, F., Stecker, G. C., & Fine, I. (2014). Auditory motion processing after early blindness. *Journal of Vision*, *14*(13), 4.
Keating, P., Dahmen, J. C., & King, A. J. (2015). Complementary adaptive processes contribute to the developmental plasticity of spatial hearing. *Nature Neuroscience*, *18*(2), 185–187.
Keating, P., & King, A. J. (2013). Developmental plasticity of spatial hearing following asymmetric hearing loss: Context-dependent cue integration and its clinical implications. *Frontiers in Systems Neuroscience*, *7*, doi: 10.3389/fnsys.2013.00123.
Kendall, G. S., & Martens, W. L. (1984). Simulating the cues of spatial hearing in natural environments. *Proceedings of the 1984 International Computer Music Conference*. San Francisco: International Computer Music Association.
Kendall, G., Martens, W. L., & Wilde, M. D. (1990). A spatial sound processor for loudspeaker and headphone reproduction. *Proceedings of the AES 8th International Conference*. New York: Audio Engineering Society.
King, A. J. (2005). Multisensory integration: Strategies for synchronization. *Current Biology*, *15*(9), 339–341.
King, A. J., & Middlebrooks, J. C. (2011). Cortical representation of auditory space. In J. A. Winer & C. E. Schreiner (Eds.), *The Auditory Cortex* (pp. 329–341). New York: Springer.

King, A. J., & Palmer, A. R. (1983). Cells responsive to free-field auditory stimuli in guinea-pig superior colliculus: Distribution and response properties. *Journal of Physiology*, *342*(1), 361–381.

Knudsen, E. I., Knudsen, P. F., & Esterly, S. D. (1984). A critical period for the recovery of sound localization accuracy following monaural occlusion in the barn owl. *Journal of Neuroscience*, *4*(4), 1012–1020.

Kuhn, G. F. (1977). Model for the interaural time differences in the azimuthal plane. *Journal of the Acoustical Society of America*, *62*(1), 157–167.

Kumpik, D. P., Kacelnik, O., & King, A. J. (2010). Adaptive reweighting of auditory localization cues in response to chronic unilateral earplugging in humans. *Journal of Neuroscience*, *30*(14), 4883–4894.

Larsen, E., Iyer, N., Lansing, C. R., & Feng, A. S. (2008). On the minimum audible difference in direct-to-reverberant energy ratio. *Journal of the Acoustical Society of America*, *124*(1), 450–461.

Laws, P. (1973). Auditory distance perception and the problem of "in-head localization" of sound images [Translation of "Entfernungshören und das Problem der Im-Kopf-Lokalisiertheit von Hörereignissen." *Acustica*, *29*, 243–259]. NASA Technical Translation TT—20833.

Lewald, J. (2013). Exceptional ability of blind humans to hear sound motion: Implications for the emergence of auditory space. *Neuropsychologia*, *51*(1), 181–186.

Lewald, J., Riederer, K. A., Lentz, T., & Meister, I. G. (2008). Processing of sound location in human cortex. *European Journal of Neuroscience*, *27*(5), 1261–1270.

Lopez-Poveda, E., Fay, R. R., & Popper, A. N. (2010). *Computational Models of the Auditory System*. New York: Springer Verlag.

Lord Rayleigh (Strutt, J. W.) (1907). XII: On our perception of sound direction. *The London, Edinburgh, and Dublin Philosophical Magazine and Journal of Science*, *13*(74), 214–232.

Maddox, R. K., Pospisil, D. A., Stecker, G. C., & Lee, A. K. (2014). Directing eye gaze enhances auditory spatial cue discrimination. *Current Biology*, *24*(7), 748–752.

Magezi, D. A., & Krumbholz, K. (2010). Evidence for opponent-channel coding of interaural time differences in human auditory cortex. *Journal of Neurophysiology*, *104*(4), 1997–2007.

Majdak, P., Walder, T., & Laback, B. (2013). Effect of long-term training on sound localization performance with spectrally warped and band-limited head-related transfer functions. *Journal of the Acoustical Society of America*, *134*(3), 2148–2159.

Meddis, R., & Lopez-Poveda, E. A. (2010). Auditory periphery: From pinna to auditory nerve. In Meddis et al. (Eds.), *Computational Models of the Auditory System* (pp. 7–38). New York: Springer.

Mendonça, C., Campos, G., Dias, P., & Santos, J. A. (2013). Learning auditory space: Generalization and long-term effects. *PloS one*, *8*(10), e77900.

Mershon, D. H., & Bowers, J. N. (1979). Absolute and relative cues for the auditory perception of egocentric distance. *Perception*, *8*(3), 311–322.

Mershon, D. H., & King, L. E. (1975). Intensity and reverberation as factors in the auditory perception of egocentric distance. *Perception & Psychophysics*, *18*(6), 409–415.

Middlebrooks, J. C. (1992). Narrow-band sound localization related to external ear acoustics. *Journal of the Acoustical Society of America*, *92*, 2607–2624.

Middlebrooks, J. C., & Green, D. M. (1991). Sound localization by human listeners. *Annual Review of Psychology*, *42*(1), 135–159.

Miller, J. D., Godfroy-Cooper, M., & Wenzel, E. M. (2014). Using published HRTFS with Slab3D: Metric-based database selection and phenomena observed. *Proceedings of the International Conference on Auditory Display*, New York, June 2014.

Mills, A. W. (1958). On the minimum audible angle. *Journal of the Acoustical Society of America*, *30*, 237.

Mills, A. W. (1972). Auditory localization (Binaural acoustic field sampling, head movement and echo effect in auditory localization of sound sources position, distance and orientation). *Foundations of Modern Auditory Theory*, *2*, 303–348.

Moerel, M., De Martino, F., Santoro, R., Ugurbil, K., Goebel, R., Yacoub, E., & Formisano, E. (2013). Processing of natural sounds: Characterization of multipeak spectral tuning in human auditory cortex. *Journal of Neuroscience*, *33*(29), 11888–11898.

Moore, B. C. J., Oldfield, S. R., & Dooley, G. (1989). Detection and discrimination of spectral peaks and notches at 1 and 8 kHz. *Journal of the Acoustical Society of America*, *85*, 820–836.

Musicant, A. D., & Butler, R. A. (1985). Influence of monaural spectral cues on binaural localization. *Journal of the Acoustical Society of America*, *77*(1), 202–208.

Oldfield, S. R., & Parker, S. P. (1984a). Acuity of sound localization: A topography of auditory space: I: Normal hearing conditions. *Perception*, *13*, 581–600.

Oldfield, S. R., & Parker, S. P. (1984b). Acuity of sound localization: A topography of auditory space: II: Pinna cues absent. *Perception*, *13*, 601–617.

Pan, Y., Zhang, J., Cai, R., Zhou, X., & Sun, X. (2011). Developmentally degraded directional selectivity of the auditory cortex can be restored by auditory discrimination training in adults. *Behavioural Brain Research*, *225*(2), 596–602.

Par, S. van de, & Kohlrausch, A. (1997). A new approach to comparing binaural masking level differences at low and high frequencies. *Journal of the Acoustical Society of America*, *101*(3), 1671–1680.

Patterson, R. D., Nimmo-Smith, I., Holdsworth, J., & Rice, P. (1987). An efficient auditory filterbank based on the gammatone function. *A Meeting of the IOC Speech Group on Auditory Modelling at RSRE*, *2*(7). Retrieved from http://www.pdn.cam.ac.uk/other-pages/cnbh/files/publications/SVOSAnnexB1988.pdf.

Pearsons, K. S., Bennett, R. L., & Fidell, S. (1977). *Speech Levels in Various Noise Environments*. Office of Health and Ecological Effects, Office of Research and Development, US EPA.

Perrott, D. R., & Saberi, K. (1990). Minimum audible angle thresholds for sources varying in both elevation and azimuth. *The Journal of the Acoustical Society of America*, *87*(4), 1728–1731.

Perrott, D. R., & Tucker, J. (1988). Minimum audible movement angle as a function of signal frequency and the velocity of the source. *Journal of the Acoustical Society of America*, *83*, 1522.

Plakke, B., & Romanski, L. M. (2014). Auditory connections and functions of prefrontal cortex. *Frontiers in Neuroscience*, *8*, 199. Retrieved from http://doi.org/10.3389/fnins.2014.00199

Plenge, G. (1974). On the differences between localization and lateralization. *Journal of the Acoustical Society of America*, *56*, 944–951.

Rauschecker, J. P., & Tian, B. (2000). Mechanisms and streams for processing of "what" and "where" in auditory cortex. *Proceedings of the National Academy of Sciences*, *97*(22), 11800–11806.

Rauschecker, J. P., Tian, B., & Hauser, M. (1995). Processing of complex sounds in the macaque nonprimary auditory cortex. *Science*, *268*(5207), 111–114.

Richardson, G. P., Lukashkin, A. N., & Russell, I. J. (2008). The tectorial membrane: One slice of a complex cochlear sandwich. *Current Opinion in Otolaryngology & Head and Neck Surgery*, *16*(5), 458.

Romanski, L. M., & Goldman-Rakic, P. S. (2002). An auditory domain in primate prefrontal cortex. *Nature Neuroscience*, *5*(1), 15–16.

Saenz, M., & Langers, D. R. (2014). Tonotopic mapping of human auditory cortex. *Hearing Research*, *307*, 42–52.

Salminen, N. H., Tiitinen, H., Yrttiaho, S., & May, P. J. (2010). The neural code for interaural time difference in human auditory cortex. *Journal of the Acoustical Society of America*, *127*(2), 60–65.

Sand, A., & Nilsson, M. E. (2014). Asymmetric transfer of sound localization learning between indistinguishable interaural cues. *Experimental Brain Research*, *232*(6), 1707–1716.

Schroeder, M. R. (1965). New method of measuring reverberation time. *Journal of the Acoustical Society of America*, *37*, 409–412.

Shaw, E. A. G. (1974). The external ear. In W. D. Keidel & W. D. Neff (eds.), *Handbook of Sensory Physiology, Vol. 5/1, Auditory System* (pp. 455–490). New York: SpringerVerlag.

Sheeline, C. W. (1983). *An Investigation of the Effects of Direct and Reverberant Signal Interaction on Auditory Distance Perception*, Doctoral dissertation, Stanford University.

Shinn-Cunningham, B. (2000). Learning reverberation: Considerations for spatial auditory displays. *Proceedings of the International Conference on Auditory Display, Atlanta, Georgia USA, April 2000*, 126–134.

Shinn-Cunningham, B. G., Kopco, N., & Martin, T. J. (2005). Localizing nearby sound sources in a classroom: Binaural room impulse responses. *Journal of the Acoustical Society of America*, *117*(5), 3100–3115.

Shub, D. E., Durlach, N. I., & Colburn, H. S. (2008). Monaural level discrimination under dichotic conditions. *Journal of the Acoustical Society of America*, *123*(6), 4421–4433.

Stevens, S. S., & Guirao, M. (1962). Loudness, reciprocality, and partition scales. *Journal of the Acoustical Society of America*, *34*, 1466–1471.

Tabry, V., Zatorre, R. J., & Voss, P. (2013). The influence of vision on sound localization abilities in both the horizontal and vertical planes. *Frontiers in Psychology*, *4*, 932. doi: 10.3389/fpsyg.2013.00932

Thurlow, W. R., & Runge, P. S. (1967). Effect of induced head movements on localization of direction of sounds. *Journal of the Acoustical Society of America*, *42*, 480–488.

Voss, P., Lepore, F., Gougoux, F., & Zatorre, R. J. (2011). Relevance of spectral cues for auditory spatial processing in the occipital cortex of the blind. *Frontiers in Psychology*, *2*(48). doi: 10.3389/fpsyg.2011.00048

Voss, P., Tabry, V., & Zatorre, R. J. (2015). Trade-off in the sound localization abilities of early blind individuals between the horizontal and vertical planes. *Journal of Neuroscience*, *35*(15), 6051–6056.

Wallach, H. (1939). On sound localization. *Journal of the Acoustical Society of America*, *10*(4), 270–274.

Wallach, H. (1940). The role of head movements and vestibular and visual cues in sound localization. *Journal of Experimental Psychology*, *27*, 339–368.

Wenzel, E. M. (1991). *Three-dimensional Virtual Acoustic Displays*. NASA-Ames Research Center, NASA Technical Memorandum 103835.

Wenzel, E. M., Arruda, M., Kistler, D. J., & Wightman, F. L. (1993). Localization using nonindividualized head-related transfer functions. *Journal of the Acoustical Society of America*, *94*, 111.

Wenzel, E. M., Miller, J. D., & Abel, J. S. (2000). Sound lab: A real-time, software-based system for the study of spatial hearing. *Proceedings of the 108th Audio Engineering Society Convention*. New York: Audio Engineering Society.

Wightman, F. L., & Kistler, D. J. (1989). Headphone simulation of free-field listening: II: Psychophysical validation. *Journal of the Acoustical Society of America*, *85*, 868–878.

Wightman, F. L., & Kistler, D. J. (1997). Monaural sound localization revisited. *Journal of the Acoustical Society of America*, *101*(2), 1050–1063.

Wightman, F. L., & Kistler, D. J. (1999). Resolution of front-back ambiguity in spatial hearing by listener and source movement. *Journal of the Acoustical Society of America*, *105*, 2841–2853.

Wilson, R. H., & Margolis, R. H. (1999). Acoustic-reflex measurements. In F. E. Musiek & F. E. Rintelmann (Eds.), *Contemporary Perspectives in Hearing Assessment*, *1*, 131. Boston, MA: Allyn and Bacon.

Woodworth, R. S. (1938). *Experimental Psychology*. New York: Holt.

Woodworth, R. S., & Schlosberg, H. (1954). *Experimental Psychology*. New York: Holt.

Woszczyk, W., Begault, D. R., & Higbie, A. G. (2014). Comparison and contrast of reverberation measurements in Grace Cathedral San Francisco. *Audio Engineering Society 137th Convention*, ebrief 178.

Yin, T. C. (2002). Neural mechanisms of encoding binaural localization cues in the auditory brainstem. In D. Oertel, R. R. Fay & A. N. Popper (Eds.), *Integrative Functions in the Mammalian Auditory Pathway* (pp. 99–159). New York: Springer.

Yost, W. A., & Dye, R. H. (1997). Fundamentals of directional hearing. *Seminars in Hearing*, *18*(4), 321–344. New York: Thieme Medical Publishers.

Zahorik, P. (2002). Assessing auditory distance perception using virtual acoustics. *Journal of the Acoustical Society of America*, *111*, 1832–1846.

Zahorik, P., Brungart, D. S., & Bronkhorst, A. W. (2005). Auditory distance perception in humans: A summary of past and present research. *Acta Acustica United with Acustica*, *91*, 409-420.

Chapter 2

History of 3D Sound

Braxton Boren

Introduction

The history of 3D sound is complicated by the fact that, despite how much the concept may appear to be a late 20th-century technological buzzword, it is not at all new. Indeed, just as Jens Blauert famously reminded us that "there is no non-spatial hearing" (Blauert, 1997), so too due to the nature of our world all sound is inherently three-dimensional (Begault, 2000). For the majority of human history the listener—the hunter in the field, a singing congregant within a cavernous stone church, or the audience member at a live performance—perceived sound concomitantly with its spatial setting.

In this sense it is the late 20th-century view that is out of step with historical sound perception. The advent of audio recording in the 19th century led to the development of zero-dimensional (mono) sound, and later one-dimensional (stereo), and two-dimensional (quad and other surround formats) reproduction techniques. Due to the greater sensitivity of the human auditory system along the horizontal plane, early technology understandably focused on this domain. Our capability to mechanically synthesize full 3D auditory environments is relatively recent, compared to our long history of shaping sound content and performance spaces.

The effect of physical space is not limited to the perceived physical locations of sounds—different spaces can also affect music in the time domain (e.g., late reflection paths) or frequency domain (by filtering out high-frequency content). Often a listener's experience of a space's effect on sound—such as singing in the shower or listening to a choir in a reverberant stone church—is describing primarily non-localized qualities, which could be captured more or less in a monaural recording. Though space has always been an integral part of live performance, it has rarely served as more than an additional ornamentation on this spectral/temporal palette for most composers (with some notable exceptions), though technological advances are leading to progress on this front.

The other area focusing on 3D sound currently is the field of virtual auditory spaces (VAS), which may be used for either virtual reality simulations or augmented reality integrations of 3D sound into the listener's existing auditory environment. In contrast to the musical front, these applications' goals fundamentally require convincing spatial immersion, elevating the importance of space above that of frequency or time in many cases. In a sense, the rapid developments in these fields seek to re-establish that connection between sound and space, which was to some extent severed by early audio recording and reproduction. It follows then, that to look forward to the future of 3D sound, we should first look backward to the various uses and experiences of sound and space throughout history.

Prehistory

Because sound is inherently transient, most sounds of the past are lost to present-day observers. For the prehistoric period—that is, the time before written language or history—we lack even subjective descriptions of sounds and thus must rely on the tools of archaeology to reconstruct the acoustic world experienced by early humans. Hope Bagenal famously stated that the acoustics of all auditoria originated from either the open air or the cave (Bagenal, 1951). For the case of our earliest ancestors this was literally true as they spent most of their time hunting and gathering in outdoor environments that approximated free field listening conditions. Their 3D sound localization was honed in this largely non-reverberant context, allowing them to evade predators and find their own prey.

However, when these early hunter-gatherers stepped inside the diffuse sound field of a cave, they would have found an acoustic environment wholly different from that outside. The reflective walls and enclosed spaces would have generated audible echoes, modal resonances, and reverberation—causing listeners to be surrounded by thousands of copies of their own voices, an immersive, amplified, and somewhat mystical experience. Even today when we are more used to moderate amounts of reverb in popular music, entering a large stone church with a long reverberation time for the first time yields a sense of 3D envelopment unrivaled by the best surround sound system. For our prehistoric ancestors who had no such experience or knowledge, such a space would have sounded otherworldly—a complete removal from the world they knew outside.

How did this experience shape the early humans' use of caves? Analysis of Paleolithic caves in Britain, Ireland, and France suggests many connections and correlations between the locations of rock drawings and strong acoustic resonances, particularly those in the range of the male singing voice (Jahn et al., 1996; Reznikoff, 2008). The discovery of 20 shell trumpets at Chavín de Huántar in Peru suggests that this space was used for musical performance during ritual ceremonies. Because Chavín consists primarily of stone-cut galleries (constructed around 600 BC), these smaller interior spaces possess lower reverberation times but still provide acoustic immersion because of widely distributed reflection patterns and non-coherent energy density (Abel et al., 2008). This unique acoustic environment, which does not occur in either the free field or most natural caves, can be seen as a precursor to the modern concert hall, which minimizes the correlation of the pressure signals at both ears, resulting in an enveloping sound field, while still maintaining sufficient acoustic clarity to understand the content of the music being played (Kendall, 1995b).

Ancient History

With the transition from nomadic hunter-gatherers to settled agricultural societies, architectural spaces also transitioned from natural or roughly hewn stone caves to more advanced ceremonial spaces. The open-air condition evolved into the Greek amphitheaters, which had roots in the Minoan theaters from as early as 20,000 BC but found their apotheosis in the Classical amphitheaters around the 5th century BC (Chourmouziadou & Kang, 2008). These spaces set the standard for theater design in Western culture, maintaining excellent speech intelligibility through strong early reflections while avoiding confusion stemming from reverberation. Recent analysis has also shown that diffraction effects from the corrugated seating structure in Greek amphitheaters preferentially amplify the frequencies associated with speech intelligibility while attenuating low frequency noise (Declercq & Dekeyser, 2007). Vitruvius, a Roman architect whose writings are the best preservation of Roman and Greek architectural knowledge, mentions that in some theaters bronze vessels were placed beneath the seats whose resonances amplified the actors' speaking voices (Vitruvius, 1914). Though it is doubtful Vitruvius actually saw these vessels in action (Rindel, 2011b), the approach is a striking early example of distributing sound sources throughout a performance space.

As the Classical theaters gave way to Hellenistic and later Roman theaters, changes were made that added reverberation and decreased intelligibility (Chourmouziadou & Kang, 2008; Farnetani, Prodi & Pompoli, 2008). This in effect made these outdoor theaters somewhat of a bridge between Classical amphitheaters and *odea*, the interior halls built throughout Greece specifically for the performance of music (Navarro, Sendra & Muñoz, 2009). These featured higher reverberation times and lower musical clarity, but greater sound reinforcement for relatively weak instruments such as the Greek lyre (Rindel, 2011a). The Greek *odeon* was a singular example of architecture designed around music, as this pattern was not generally followed throughout the rest of the ancient world: elsewhere music had to adapt itself to performances in theatres and other public spaces that were constructed based on non-musical criteria, such as optimal speech intelligibility, and sometimes non-acoustic criteria, such as maximizing seating or using the cheapest materials possible.

Because architectural acoustics as a discipline was largely developed in the West, so too the majority of historical acoustical analysis has focused on Western civilization from Greek antiquity onward. However, enclosed immersive temples are found throughout the world, and recent analysis has begun to examine acoustic phenomena particular to non-Western cultural and religious traditions (Prasad & Rajavel, 2013; Soeta et al., 2013). Meanwhile, in the West the Christian church would serve as the primary site of musical performance for over a thousand years (Navarro, Sendra & Muñoz, 2009).

Space and Polyphony

To understand the effect of space on musical development in the first millennium AD, it is first helpful to understand the basic context surrounding the rise of the Christian church, which so profoundly shaped music composition during this period. Originally a splinter sect of Judaism, Christian worship had its roots in the Jewish synagogue, which focused on readings and exhortations, which were generally spoken or chanted as a monotone (Burkholder, Grout & Palisca, 2006). Early Christian worship retained this spoken liturgy, which was appropriate for their physical setting: since Christians refused to worship the Roman emperor, they were not allowed recognition by the empire and thus met in small groups within house churches whose dry acoustics matched their spoken liturgy. However, after the Emperor Constantine issued the Edict of Milan in 313, confiscated Christian property was returned, and Roman architects began building large stone basilicas for Christian worship with long reverberation times. Around this same point the Christian liturgy became more focused on a sung liturgy, which slowed the rate of spectral variation over time in the highly time-dispersive communication channel from the priest to the congregation (Lubman & Kiser, 2001). With this change we see the first large swing of the history of Western music on Bagenal's continuum: while the house churches had resembled the clarity of the "open air," the church now found itself in the "cave," which shifted worship away from the semantic content of the spoken word toward the aesthetic experience of being surrounded by a line of chanted music and its many reflections.

As the Christian church grew in size and influence, many varieties of sung chant took root, but the best-known repertory, codified in the early eighth century, was known as Gregorian chant. Though attributed to Pope Gregory I (r. 590–604) it is more likely that the standardization of chant took place under Pope Gregory II (r. 715–31) (Burkholder, Grout & Palisca, 2006). The proto-Romanesque churches being built during this period were made of reflective stone materials and possessed large volumes that ensured that any monophonic sung line would be heard along with the reflection of one or more previous tones. Navarro argues that over time this "ensured a certain education and familiarisation with polyphonic sound. Indeed, the persistence of the sound of the different notes of the melody [led] to melodic developments, structured in such ways that the simultaneity of notes [was] resolved over certain harmonies" (Navarro, Sendra & Muñoz, 2009, p. 782). Lubman and Kiser (Lubman & Kiser, 2001) go even further and argue that the piling up of multiple notes at once in reverberant churches was a catalyst for the development of polyphonic music in the West.

The simplest polyphony—that of a single 'drone' under the melody—had existed in both the East and the West since antiquity. However, the earliest music with multiple moving parts is known generally as *organum*. The simplest and earliest form of organum, in which a parallel tone was sung along with the main melody of the chant, was apparently already a standard practice by the ninth century (Burkholder, Grout & Palisca, 2006). Indeed, the term *organum* does not refer to polyphony *per se*, but rather to voices which co-existed in a way consistent with natural law (Fuller, 1981). Indeed, modern listeners might instead use the word 'organic' to describe this relationship, which also suggests an early connection between organum and reverberation, which blends direct and indirect sound into a single perceptual object. Without direct documentary evidence, the hypothesis of modern polyphony's origins in reflected sound cannot be proved explicitly, yet it is still a compelling candidate for the first experimental environment for the comparison of multiple tones moving at once. If true, this would indicate that the immersive component of sound has affected not only our spatial arrangements of music but also the very fabric of how music theory itself has developed.

Even after the development of polyphony, church design continued to strongly influence music composition: the Gothic churches, which followed the Romanesque period, often possessed a single resonance that was known as the 'note' of the church. Since institutional Christianity at this point generally opposed instrumental accompaniment during the Mass (Burkholder, Grout & Palisca, 2006), the church's natural 'note' served as their reference tone and thus the church itself was a sort of natural accompaniment to the *a cappella* vocal music by reinforcing the choir within a single key (Bagenal, 1951). A more detailed history of the influence of architectural space on Western music composition is given in Forsyth (1985).

Spatial Separation in the Renaissance

The word 'renaissance' denotes *rebirth*, and was meant to signify the rediscovery of Classical Greek and Roman culture after the supposed backwardness of the Middle Ages. At least in the case of music, this narrative is not accurate: as we have already seen, the Medieval organum style expanded upon the musical traditions of the Classical world, and during the Renaissance advances in polyphony would surpass both the Medieval and Classical styles in scope and complexity. Indeed the complex polyphony of the Venetian Renaissance seems at first glance to be rather ill-suited to the large reverberant churches Palladio and other architects designed in Venice during this period (Howard & Moretti, 2010). However, computational simulations show that on the festive occasions for which such music was composed, large crowds and wall tapestries could have reduced the reverberation by as much as a factor of one half, drastically increasing the clarity of the performance (Boren & Longair, 2011; Boren, Longair & Orlowski, 2013).

But despite the advances in polyphonic style during this period, perhaps more significant to our modern ears was the practice of *cori spezzati*, the composition of music for multiple choirs that were separated in space. This practice originated in the late 15th century in northern Italy, spread throughout the region in the early 1500s, and came to its apex in Venice under Adrian Willaert, Andrea Gabrieli, and Claudio Monteverdi (Arnold, 1959; Howard & Moretti, 2010). Willaert was the first major composer to adopt this style, and the considerations he made in his compositions for this ensemble resemble those made by an engineer mixing a stereo recording: he made sure to include a wide spectral range in both choirs in case a listener was too near a single choir, and he implemented one of the earliest documented uses of doubled bass lines to keep his ensembles together. Good examples of this polychoral style include Willaert's *Vespers* of 1550 and Gabrieli's three-choir mass for the visit of the ambassadors of Japan in 1585 (Zvonar, 2006; Arnold, 1959). Again we see that independent of localization effects, spatial factors significantly affected the tonal development of Western music even at this early juncture.

But besides these tonal effects, it seems clear that the spatialization of separated choirs was an integral part of the aesthetic of *coro spezzato* music. Simple call-and-response antiphony, both recited and sung, dates back to antiquity (Slotki, 1936), but the more complex composition of music for spatially separated ensembles was fully realized for the first time in the Venetian Renaissance. Though we have evidence that on certain occasions the choirs performed in a single location (Fenlon, 2006), Gioseffo's Zarlino's comments on the genre suggest that physical separation of the choirs "is a structural requisite for this particular genre of composition . . . and not merely one of various possibilities" (Moretti, 2004, p. 154). Because the ruler of Venice, Doge Andrea Gritti, became too fat to reach his old elevated throne in the Basilica San Marco, in 1530 he moved his seat into the chancel, where previously the high priest had resided (Howard & Moretti, 2010). After this move, Jacopo Sansovino, the chief architect of the church, constructed two *pergoli*, or raised singing galleries, on either side of the doge's new throne because former galleries lower down had become obstructed by wooden seating in the chancel. Moretti argues that these galleries were used to give a stereo effect for the performance of split-choir music to the doge's position (Moretti, 2004). On-site acoustic measurements showed that these positions produced a near-perfect blend of clarity and reverberation at the doge's position, while other congregants in the nave received a much muddier sound (Howard & Moretti, 2010). Further analysis showed that this effect resulted from Sansovino's galleries maintaining a direct line-of-sight to the doge's position, without which the galleries would instead produce the same unclear acoustics heard by the parishioners (Boren et al., 2013).

Outside of Italy, one notable polychoral work from this period is Thomas Tallis's *Spem in alium*, a 40-piece motet composed in England around 1570 for eight separate choirs of five voices each. Tallis may have been inspired to compose this piece by a similar 40-voice motet by Alessandro Striggio, an Italian composer who visited London in 1567. Many details of the original performance of Tallis's piece are unknown, but it is thought to have been performed for the first time in Arundel House, London, around 1571 (Stevens, 1982). Though we have no indications of the spatial arrangement of the choirs on this occasion, such a large array of choirs inevitably would have created some spatial separation. At least by 1958 *Spem in alium* had been staged with the eight choirs arranged in a circle, and the audience enclosed in the center (Brant, 1978). Tallis's motet, however, is most significant to the history of spatial sound because it served as the basis for a 2001 sound installation, *40 Part Motet*, by Janet Cardiff. This exhibit featured a 40-channel close-miked recording of each of the voices of Tallis's motet, played on 40 raised loudspeakers arranged in a circle (MacDonald, 2009). This installation, which debuted when most consumers had access to only 5-channel surround sound, was one of the earliest exposures of many outside the professional audio community to the possibilities of simulating the full sound field of a real performance distributed in space.

Spatial Innovations in Acoustic Music

Baroque Period

At the dawn of the Baroque period (1600–1750), modern-day Germany was in a state of transition due to the success of the Lutheran Reformation, begun by Martin Luther in 1517. This catalyzed a conflict between the Catholic Church and the Lutherans, who stressed, among other things, the importance of preaching and singing in the vernacular German language, since many congregants could not understand Latin. This led the Reformers to stress the importance of clarity and speech intelligibility in their churches: as formerly Catholic churches were taken over by Lutherans, the spaces were altered to improve the clarity of the spoken word. After the Lutherans took over the Thomaskirche in Leipzig in 1539, they made a variety of alterations to the space that greatly reduced the reverberation (Lubman & Kiser, 2001) and made it more like "a religious opera house" (Bagenal, 1930, p. 149). This church, of course, is most famous as the compositional home of Johann Sebastian Bach (1685–1750) from 1723 until the end of his life. This extreme swing away from the "cave" and back toward the "open air" thus led to "the acoustic conditions that made possible the seventeenth century development of cantata and Passion" (Bagenal, 1930, p. 149). While the more reverberant churches had been appropriate for chant music at a slow, steady tempo, the drier acoustics introduced by the Reformation allowed Bach to make use of dramatic shifts in tempo with works such as the famous St. Matthew Passion (Bagenal, 1951). Thus theological considerations shaped architectural development, which in turn shaped musical development through the singular musical career of Bach, one of the most influential composers in Western history.

Meanwhile, in the southern Catholic regions of Europe, the situation was quite different but no less spatially interesting: it is interesting to note that St. Peter's Basilica in Rome, whose immense expense necessitated the selling of indulgences that spurred Luther to write his 95 theses, later provided the space necessary for the immense scale of the "Colossal Baroque" style, which featured performances by 12 separate choirs, each of which had a separate organ accompanying it (Dixon, 1979). The best-known example of this opulent genre is the late 17th-century *Missa*

Salisburgensis by Heinrich Biber, which featured five separated choirs, each with different accompanying instruments, as well as two separated ensembles of brass and timpani positioned in the galleries of the Salzburg Cathedral (Hintermaier, 1975; Holman, 1994). Whereas in Leipzig the drier acoustics allowed Bach to explore music in tonal and temporal dimensions, in the less clear churches of southern Europe, physical separation helped listeners perceive the different ensembles despite the spaces' longer reverberation.

Classical Period

It is tempting to hypothesize the dramatic spatial effects that might have been used by the Salzburg Cathedral's most famous composer, Wolfgang Mozart (1756–1791), when he wrote for the space a century after Biber. However, perhaps in reaction to the dramatic spatial effects employed during the Baroque period, the Prince-Archbishop of Salzburg issued a decree in 1729 that forbade the wide separation of ensembles and confined all music to the main choir in front (Rosenthal & Mendel, 1941). Visitors to the Salzburg Cathedral often remark at the site of the church's five organs: four near the chancel in the front (though not the same as the organs from Mozart's time), and the larger main organ at the rear. Yet based on Mozart's father's account of performance practice, it does not seem likely that multiple organs were used simultaneously (Harmon, 1970). Though Mozart was unable to experiment within the interior setting, he did employ spatial separation in his secular music: in the score for his opera *Don Giovanni* (1787), Mozart wrote parts for three separate orchestras: one in the pit, one onstage, and one backstage. Each ensemble plays highly differentiated material in separate meters, requiring very precise temporal coordination (Brant, 1978). Mozart also employed a larger distance separation in his *Serenade for Four Orchestras, K. 286*, which likely was composed for an outdoor gathering in Salzburg (Huscher, 2006). This work employs an echo effect between the separate orchestras that can be somewhat confusing in a reverberant interior hall, but is well suited to a free-field outdoor performance, since each orchestra provides 'reflections' for the others' initial motives.

Echo effects were a popular spatial effect during the Classical period—Mozart's friend Joseph Haydn (1732–1809) also employed non-spatially separated echoes in the second movement of his Symphony No. 38, sometimes called the "Echo" symphony for this reason. In this case the echo originates from violins played at normal strength, followed by muted violins. Haydn elaborates this concept more fully in "Das Echo" (Hob. II:39), a string sextet for two trios spatially separated, traditionally in different rooms (Begault, 2000). In this piece Haydn changes the length of time before the echo is heard, beginning with a whole measure delay, then a half note delay, quarter note delay, and eventually the echo is shortened to only an eighth note. This has the aesthetic effect of changing not the spacing of the ensemble but the size of the virtual room Haydn is simulating—perhaps the earliest-known example of altering a performance space dynamically as a musical parameter. Echo effects can still be heard today, either real or simulated, at many funerals: the playing of Taps issues from a single bugle, and then is heard from another bugle, farther away. In areas where brass players are scarce, many a bugler has had to perform a simulated echo by either turning around and playing away from the funeral site or quickly running to the other side of a hill and playing again.

Romantic Period

While the Classical period used echo effects as a sedate gesture towards an abstract acoustic space, the Romantic trend toward programmatic music that told a distinct story also led to more radical uses of spatialization. Perhaps no composer would make better use of spatial storytelling than that master of Romantic program music, Hector Berlioz (1803–1869). Earlier, François Joseph Gossec had surprised and alarmed the audience for his *Grande messe des morts* (1760) by using a separate brass ensemble hidden in upper reaches of the church, which suddenly gave the call of the last judgment. This is thought to have inspired Berlioz to go further and use four antiphonal brass choirs, each placed in one of the cardinal directions, to sound the call to judgment in his *Requiem* in 1837 (Broderick, 2012). Instead of a surprising call from afar, the audience is surrounded by the brass ensembles' sharp attacks that give illusions of surrounding the audience through the time differences between ensembles as well as the inevitable reflections of each ensemble through the performance space. In fact, we know that Berlioz was aware of this spatial effect as well, for he wrote only two years earlier:

> Many fail to recognize that the very building in which music is made is itself a musical instrument, that it is to the performers what the sound board is to the strings of the violin, viola, cello, bass, harp, and piano stretched above.
>
> (Bloom, 1998, p. 84)

Thus it seems that Berlioz conceived the *Requiem* not only as a set of sound sources in space, but rather as a single 3D immersive environment whose spatial characteristics could be controlled to some extent through careful orchestration and ensemble placement.

Later in his *Symphonie Fantastique*, Berlioz uses offstage instruments and a very specific narrative program to prime his audience to hear music "here" and "there" (Begault, 2000). His offstage oboe in the third movement echoes back the tune of the English horn, representing two shepherds piping to each other across a valley. In addition to the spectral low-pass effect of being behind the stage curtain, the oboe also makes a slight change to the initial theme, a clear demarcation between the shepherd nearby and his friend far-off. At the end of the movement, the horn repeats its call but is not answered. This primes the audience to listen for something far away, and instead of the friendly oboe they are instead greeted with the ominous terror of the famous March to the Scaffold in movement 4, as the executioner's procession begins to approach the protagonist (Ritchey, 2010). Berlioz's offstage invocation of a far-off sound source would also be adopted by late Romantic composers such as Giuseppe Verdi (1813–1901), who used an offstage ensemble in his own *Requiem* (1874), and Gustav Mahler (1860–1911), who used offstage brass in 1895 at the premiere of his "Resurrection" Symphony No. 2 (Zvonar, 2006).

20th-Century Acoustic Music

Though many of the contemporary advances in spatialization occurred through technological advances, there were also some 20th-century composers who continued in the purely acoustic spatial tradition that dated back to Willaert and the Renaissance. Thus it will be more helpful to cover these composers first, and then discuss electroacoustic spatialization after the history of spatial technology from the 19th century onward. The spatial storytelling of the Romantic Era was continued within the American Experimental movement, as typified by Charles Ives (1874–1954).

Charles Ives grew up learning from his father George Ives, who was a bandmaster during the American Civil War and had much experience with another archetypal form of moving sound sources: the marching band. Indeed, George at one point conducted an experiment in which he led two separate bands, marching in opposite directions through the town square (Zvonar, 2006). His son Charles, known for his juxtapositions of contrasting musical material, would add spatial separation to his compositional toolkit in *The Unanswered Question* (1908). In this piece Ives places a trumpet and woodwinds onstage, which respectively pose "The Perennial Question of Existence" and various answers. Meanwhile, Ives places a separate string quartet offstage, which represents "The Silences of the Druids—Who Know, See and Hear Nothing," and more broadly can be seen to be "representatives of the unfathomable cosmos beyond" (McDonald, 2004, pp. 270–271). Thus Ives uses spatial separation as a stand-in for the cosmic separation between Us—the artists, the thinkers, the ones asking and answering questions—and It—the cosmos, which will go on in silence long after we have ceased our questioning. A later performance of the piece by Henry Brant also separated the Questioner and the Answerers, perhaps indicating some metaphysical distance between those personalities as well (Brant, 1978).

Brant (1913–2008) was strongly influenced by Ives's use of space, particularly in *The Unanswered Question*. Whereas the previous examples have largely been major composers for whom space was a minor effect, Brant is in many ways the opposite: while not a dominant figure in 20th-century music, Brant's body of work embraced and explored spatial music more than any acoustic composer before him. Beginning with *Antiphony I* (1953), Brant would compose spatially organized music for separated ensembles for the next 50 years (Harley, 1997). Though he admitted that electric reproduction could add flexibility to spatial performance, Brant disliked loudspeakers because their directivity was markedly different from the live instrumentalists or vocalists (Brant, 1978). Where Ives indicated a general separation between ensembles, Brant rigorously specified positions for different ensembles within a performance space (Harley, 1997).

Though his compositional contribution is significant, Brant was also active as a theorist of spatial music theory. He believed that physical space provided freedom to the composer by allowing an ensemble to be more compressed in tonal space: whereas in a single ensemble a unison between different instruments led to confusion, when those instruments are spread out a shared tone may be a benefit rather than a hindrance (Brant, 1978). Brant possessed no formal scientific education, but he was aware of many critical psychoacoustic concepts (see Chapter 1 for further discussion) through a lifetime of musical experiments. In particular, he, like Ives before him, alludes to the "Cocktail Party Effect," whereby a listener may more easily shift attention between highly contrasting material when the sound sources are spatially separated (Harley, 1997). In addition, Brant anticipates later research on localization blur, when he suggests that loudspeakers' directivity is less of a problem when placed on the ceiling, and also the elevation-dependent spectral variation of the head-related transfer function (HRTF), when he states that high and low pitches are "enhanced" by high and low elevations (see Chapter 7), respectively (Brant, 1978). Though his music is called 'spatial,' Brant's writings do not indicate that space is the primary organizational dimension for his work. Rather, he consistently saw space from a utilitarian perspective, whereby traditional constraints of tonal composition could be alleviated through a more expansive exploration of physical space.

3D Sound Technology

By the time we reach the 19th century, the history of 3D sound begins to be tied more and more closely to the development of modern science and technology. Though research into acoustics dates back to antiquity (Lindsay, 1966), a rigorous theory of sound localization in 2D would not be put forth until the late 19th century and gradually refined for a century afterward to account for 3D localization (Strutt, 1875; Blauert, 1997; Kendall, 1995a). However, as is often the case, the technology necessary to achieve spatial auditory effects often plunged forward far before the science behind the process was well understood.

Binaural or Stereo?

It may be argued that 3D audio technology begins and ends with binaural (see Chapter 4), though as we shall see, we must take some care with how we define the term. The word 'binaural' refers, at the most basic level, to hearing with two ears, but it later came to include all the spatial cues from the ears, head, and body of a listener. This odd trajectory stems from the fact that binaural audio is perhaps the easiest spatial effect to capture, but the hardest to realize in post-production. Only four years after Alexander Graham Bell's invention of the telephone, he did some early experiments using two telephones receivers and transmitters (Davis, 2003). The next year a French engineer named Clément Ader devised a system for the spatial transmission of the Paris Opera over a mile away to a listening booth at the International Exposition of Electricity (Hospitalier, 1881; MacGowan, 1957; Torick, 1998; Paul, 2009). This invention, dubbed the 'Theatrophone,' used an array of pairs of transmitters across the stage that were then routed to pairs of telephone receivers at the Exposition's listening booth. Attendees held both receivers to their ears and were able to perceive the spatial position of sound sources through the interaural differences transmitted over the lines. Though the system suffered from insufficient amplification and vibration damping, the service proved to be successful enough to merit a home-subscription service. Among the best-known Theatrophone subscribers were Marcel Proust and Great Britain's Queen Victoria. Though popular among the well-to-do in the early 20th century, the advent of cheaper monophonic wireless broadcasting opened up a narrower sound to a wider segment of the population, and Theatrophone broadcasts ceased in 1932. It would be another 30 years before stereo broadcasting would bring back this basic spatial feature that had been discovered so early on (Collins, 2008).

In discussing Ader's significance, it is useful to consider what his technology actually represented: because it transmitted many points along the wavefronts emitting from the Opera's stage, it could be thought of as an early form of wave field synthesis (see Chapter 10). However, because it conveyed interaural differences to be conveyed to both ears of the listener, some have classified this as the earliest example of 'binaural' sound (Sunier, 1986), or as it was called at the time, 'binauricular,' perhaps too much of a tongue-twister to attain popular acceptance (Collins, 2008). It is important to note that at the time, binaural was principally used to mean hearing with two ears, rather than recording with a real or synthetic human head. The modern distinction between binaural and stereo was not even suggested until the 1930s, and widespread usage of these separate definitions would not follow until the 1970s (Paul, 2009). Thus it is probably safest to say that Ader's achievement was the earliest reproduction of binaural sound under the 19th-century understanding of the term. Since the transmissions did not make use of a dummy head to obtain level differences, time differences, or spectral cues corresponding to the actual filtering effects of the head, a present-day understanding might instead classify the Theatrophone as a very effective form of 2-channel stereo, distributed over several listening points.

Despite this distinction, advances in stereophony and proper binaural sound were fast-coming. By World War I (1914–1918), two-ear listening devices were being used to track enemy planes (Sunier, 1986) and also to track submarines using inputs from dual hydrophones (Lamson, 1930).

Some have claimed that artificial heads were used as early as 1886 at Bell Laboratories, but this seems doubtful for several reasons and has not been confirmed (Paul, 2009). Indeed, the earliest definite cases of binaural transmissions using some form of primitive artificial head were both patented in 1927, one by Harvey Fletcher and Leon Sivian, and another system for recording and reproduction by W. Bartlett Jones (Fletcher & Sivian, 1927; Paul, 2009). These both used very basic spheroid objects as a dummy head, but Fletcher's research at Bell Labs would later develop a more sophisticated binaural recording device in 1931 using a tailor's manikin nicknamed 'Oscar.' The 1.4-inch microphones placed in Oscar were too large to fit into the ear canal, so the microphones were mounted instead on the manikin's cheekbones directly in front of the ears. Listeners to the transmissions from Oscar from the Philadelphia Academy of Music were astounded at the degree of localization that could be achieved—Fletcher stated that "the mechanism by which this location is accomplished is not altogether understood, but the interaction of the two ears seems to have much to do with it, for stopping up one ear destroys the ability almost completely" (Fletcher, 1933, pp. 286–287). Despite not fully understanding the mechanisms of spatial hearing, there was widespread agreement that binaural listening was more pleasant: over a third of listeners in Fletcher's experiments preferred binaural to monaural even when the binaural content was low-pass filtered with a cutoff frequency of 2.8 kHz (Fletcher, 1933). Oscar was later used at an exhibition at the Chicago World's Fair, where listeners were amazed at being able to hear moving sources when there were none around them (Paul, 2009). Despite Fletcher's bold statement that "there is no longer any limitation, except expense, to the acoustic fidelity which electrical transmission systems can achieve" (Fletcher, 1933, p. 289), there were in fact many front-back confusions and distance errors, as would be expected with a non-individualized static binaural transmission, especially given the limitation effects of the pinnae at Oscar's microphones. Later famous binaural dummy heads such as KEMAR and the Neumann KU-100 would be developed, along with many others, but despite many advances binaural technology remained a niche area of audio for most of the 20th century. A detailed history of binaural recording devices is given by Stephan Paul (2009).

Loudspeakers: From Stereo to Multichannel

While the recording methods of these early 'binaural' methods varied, they had in common an early headphone-based reproduction that allowed most of the coarse interaural differences to be conveyed directly to the ear canals of listeners. However, loudspeaker technology was also rapidly developing during this time, and many spatial experiments would be carried out during this era. As early as 1911 Edward Amet filed a patent on a device to pan a mono record, synchronized with a film projector, around a series of loudspeakers such that the sound of an actor's voice would follow his position on the screen (Amet, 1911). As Davis notes, "this was a remarkably far-sighted invention, as it would be another dozen years before even mono-synchronized sound was commercially employed in the cinema" (Davis, 2003, p. 556). Thomas Edison's phonograph was used for strictly monophonic reproduction, but an audience in 1916 perceived it as very realistic, perhaps due in part to the reverberant acoustics of Carnegie Hall where the demonstration took place (Davis, 2003).

Efforts to store and reproduce stereo sound abounded, partly due to the success of the various binaural experiments mentioned above. A radio engineer in Connecticut named Franklin Doolittle filed patents for 2-channel recording (1921) and broadcasting (1924), and the radio station he owned, WPAJ, began broadcasting sound captured with two microphones, emitting on two separate radio frequencies (Paul, 2009). In 1931, British engineer Alan Blumlein (1903–1942) filed a patent that is widely considered to mark the birth of stereo (see Chapter 3), as we understand it today (Blumlein, 1931). As we have already seen, Blumlein was not the first to record or broadcast two audio channels at once, though his patent covered both of these things. The impact of Blumlein's patent should instead be seen in the comprehensiveness with which he envisioned the transformations that stereo sound could bring to the world of audio. Especially innovative were his creation of the 90-degree XY stereo microphone technique (the so-called Blumlein pair), his system of amplitude panning between two output channels, and a special disk-cutting technique for recording two channels of audio into either side of a single groove of a record.

Blumlein's work was so far ahead of its time that when he died in an airplane crash at the age of 38, he and his work were still largely unknown. His stereo disk-cutting technique, relatively unknown and ahead of its time, would later be re-invented twice more, in separate patents by Bell Labs and Westrex Corporation, respectively, before becoming commercially viable (Davis, 2003).

As even a monophonic wireless radio was a big investment for the average household at this time, much of the commercial research into multichannel audio reproduction came from motion picture companies who could afford to invest in cutting-edge technology to draw listeners to a media experience that surpassed anything they could hear at home. However, many early attempts made use of expensive prototypes and were often abandoned afterward. In New York City, a rehearsal screening of *Fox Movietone Follies of 1929* made use of the same concept that Amet patented 18 years before: a monitoring device was used to pan the monophonic movie sound track back and forth between the left and right loudspeakers, but after this, Fox gave up on the idea (MacGowan, 1957). Conductor Leopold Stokowski, who had been involved with Harvey Fletcher's earlier experiments using the binaural dummy Oscar, also worked with Fletcher in 1933 to produce a 3-channel transmission of the Philadelphia Orchestra, reproduced for an audience in Washington, D.C. (Torick, 1998). This three-microphone-to-three-loudspeaker recreation was a more perfect antecedent to wave field synthesis even than was Clement's Theatrophone. But Stokowski's experiment would also pave the way for the first use of surround sound in film through the 1940 Walt Disney film *Fantasia*, for which Stokowski was the conductor (MacGowan, 1957). For this occasion a new audio system was designed called Fantasound, which used three tracks of audio similar to that transmitted during the 1933 experiment. However, the system also used a separate optical control track to pan the three audio tracks to any of 10 groups of loudspeakers: nine surrounding the audience horizontally and one on the ceiling

(Torick, 1998). The rear and elevated speakers were not used excessively, but during the film's finale, Franz Schubert's *Ave Maria*, they were used to give the sensation of a chorus progressing from the rear to the front of the audience, yielding perhaps the most complete surround immersion yet achieved in a commercial audio system (Malham & Anthony, 1995). Yet again, the system was largely abandoned after this technical achievement, and the playback equipment was unfortunately lost at sea thereafter (Davis, 2003). Stokowski remained an avid believer in audio exploration for the rest of his life, proclaiming to the Audio Engineering Society in 1964 that audio technology represented "the greatest mass presentation of musical experience so far known to man" (Torick, 1998, p. 27). Though he had no technical training, Stokowski's reputation and influence greatly aided Fletcher and other audio scientists and engineers during the early development of 3D sound technology.

As stereo began to reach commercial viability in the late 1950s, work was already progressing on more ambitious multichannel audio formats. By the early 1960s commercial efforts had been made to market a system that extracted out-of-phase signal components from a standard stereo recording to a separate pair of rear loudspeakers. By 1968 the earliest 'quad' system was proposed by Peter Scheiber, who developed a system to compress four analog channels into just two for storage purposes, while reconstructing the original four channels under certain constraints of channel separation and phase artifacts (Torick, 1998; Davis, 2003). A variety of quad matrixing formats followed, possibly motivated by "commercial one-upmanship, [which] arguably result[ed] in products and systems being rushed into the marketplace prematurely" (Davis, 2003, p. 561). At any rate, despite the aggressive marketing of quad formats throughout the 1970s, the format failed to attain commercial success, resulting in a widespread loss of confidence about the future of 3D sound technology.

However, out of the ashes of the commercial failure of these early quadrophonic systems rose several technologies that would contribute substantially to the rise of spatial audio as we know it today. In 1976 Dolby Laboratories took the idea of 4–2–4 channel matrixing and instead applied it to motion picture sound (Davis, 2003). Instead of following a symmetrical speaker arrangement, Dolby used three channels in front and a single surround channel (Torick, 1998). The investment in this technology, particularly from Dolby, would "prove to be the gateway through which consumer audio evolved from stereo to surround" (Davis, 2003, p. 563). As storage capacity increased, it became feasible to use discrete rather than matrixed channels, and the number of output channels increased: the 1978 release of *Superman* marked the first use of a 5.1 channel soundtrack with a motion picture (Allen, 1991). Dolby continued to lead in the expansion of multichannel audio, providing surround encoding formats for both theatrical and home settings. More recently, even larger numbers of discrete channels have been proposed and implemented to varying degrees, including 7.1, 10.2, and 22.2 channels (Davis, 2003; Hamasaki et al., 2005). Mark Davis, an engineer at Dolby, gives a thorough history of the development of spatial coding formats through the 20th century (2003).

Outside of the commercial realm (that is, mostly within academia) other more generalizable multichannel formats have been put forward that have found success within certain niches. Chiefly among these are Ville Pulkki's Vector Base Amplitude Panning or VBAP (1997), and Distance-Based Amplitude Panning (DBAP), proposed independently by Lossius and Pascal Baltazar (2009) as well as Kostadinov and Reiss (Kostadinov, Reiss & Mladenov, 2010). VBAP applies Blumlein's amplitude panning in full 3D, forming a vector basis composed of each loudspeaker in an enclosing array, allowing highly accurate spatial gestures. DBAP, though less rigorous, is more flexible and does not constrain the arrangement of speakers or listeners, making it popular for use in sound installations.

Wave Field Methods

While VBAP and DBAP can achieve convincing spatial effects under certain conditions, both of these methods fundamentally encode audio output in relation to a specific arrangement of loudspeakers, and are thus classified as *multichannel* methods. We will reserve the term *wave field* methods for those spatial formats, which seek to encode an entire sound field, independent of the arrangement of output transducers. This is done via Huygen's Principle, which states that each point on a progressing wavefront may instead be considered as a separate source, and its frequency-domain formulation, the Kirchoff-Helmholtz integral (Berkhout, de Vries & Vogel, 1993). Wavefront methods tend to be broadly separable into 1) Ambisonics (see Chapter 9), which is concerned with reproducing the incoming sound field around the listener, and 2) wave field synthesis, which is concerned with reproducing the outgoing sound field emitted by one or more acoustic sources.

Ambisonics

Ambisonics came into being as another gem from the ruins of quadrophonic sound: in 1973 a mathematician and audio enthusiast named Michael Gerzon (1945–1996) put forward an encoding scheme that stood out from the crowd of competing matrixing schemes (Davis, 2003). Gerzon's scheme, later named Ambisonics, called for the use of spherical harmonic basis functions that could encode the portions of a sound field originating from many different directions around a listener's position (Gerzon, 1973). Though Gerzon worked out this system for an arbitrary number of bases (*n*th-order Ambisonics), the famous Sound Field Microphone Gerzon would later help develop was limited to first-order Ambisonics, including an omnidirectional signal and three orthogonal dipole terms (Gerzon, 1975). Lower orders of Ambisonics constitute a more severe truncation of the spherical harmonic decomposition and yield less spatial precision in reproduction. Since Gerzon's initial work, the increasing miniaturization of audio technology has allowed the development of 32-channel (Manola, Genovese & Farina, 2012) and 64-channel (O'Donovan & Duraiswami, 2010) Ambisonic microphones, as well as software tools for spatialization using Higher-Order Ambisonics (Malham, 1999; Kronlachner, 2014). Since Ambisonic encoding is independent of any specific playback system, today's renewed interest in 3D sound has led to calls for using Ambisonics as a flexible production and distribution format for 3D audio content (Frank, Zotter & Sontacchi, 2015).

Wave Field Synthesis

Wave field synthesis uses the same principle as Ambisonics to achieve the opposite aim: given an infinite amount of microphones around an acoustic source and an infinite number of loudspeakers in the same arrangement, each driven by the signal from its respective microphone, the wave fields in both cases should be identical. As mentioned earlier, Ader's Theatrophone was arguably a very simple implementation of this idea, and Fletcher and Stokowski's early 3-channel orchestral transmissions were a better example of one-dimensional, highly truncated wave field synthesis. The basic theory behind wave field synthesis was outlined by William Snow (1955), and it received its modern formulation by Berkhout et al. (1993). In practice when a finite number of transducers is used, this leads to spatial aliasing above the spatial Nyquist frequency, which is around 1.7 kHz for practical transducer arrays (Berkhout et al., 1993).

Many methods have since been employed to reduce these shortcomings and allow practical implementations of wave field synthesis in two or three dimensions (Spors, Rabenstein & Ahrens, 2008).

Back to Binaural

Binaural methods are the simplest 3D audio technology conceptually, yet they contain subtle difficulties and problems that still await a solution. It is because of this tension that binaural audio was the earliest 3D sound technique to be explored yet is only beginning to reach maturity. Though the late 20th century showed improvements in manikins and recording equipment (Paul, 2009), the dummy heads being used today are not categorically different from the binaural transmissions made by Fletcher in the 1930s. While they more accurately model a single averaged HRTF through the dummy's pinnae, this HRTF will inevitably deviate from that of the end listener, leading to degradations of front-back and up-down perception (Wenzel et al., 1993) as well as sound source externalization over headphones (Hartmann & Wittenberg, 1996). Because measuring an individualized HRTF has traditionally been a time-consuming and expensive process, much work has been done to quickly obtain individualized HRTFs for end-users of 3D audio simulations. Current approaches to this include wave equation simulations (Katz, 2001; Meshram, Mehra & Manocha, 2014), database matching techniques (Andreopoulou, 2013), and reciprocal HRTF measurements, in which an emitter rather than a receiver is placed in the subject's ear canal (Zotkin et al., 2006).

Transaural

Another route by which binaural content may be delivered bypasses headphones completely: the *transaural* technique (see Chapter 5) uses crosstalk cancellation to deliver binaural content directly to a single listener over stereo loudspeakers. This requires using each channel of the stereo system to send delayed, phase-inverted versions of the sound coming from the other speaker, cancelling the acoustic 'crosstalk' that occurs when the left ear hears the signal from the right speaker and vice versa (Schroeder, Gottlob & Siebrasse, 1974). The first crosstalk cancellation system was invented by Atal & Schroeder (1962) for the purposes of making instant A/B comparisons of different concert halls' acoustics to study listener preferences for music performance. The system was somewhat unstable as slight head movements caused it to induce severe spectral coloration upon the signals being reproduced. In 1985 Kendall and others used transaural properties to encode the earliest 3D sound broadcasts for CBS's *The Twilight Zone* (Gendel, 1985; Wolf, 1986; Kendall, 2015). Later improvements by Bauck and Cooper (1996) and Choueiri (2008) addressed these issues, and today it is possible to achieve spectrally uncolored transaural reproduction while maintaining sufficient interaural level difference for convincing 3D sound effects without the externalization problems that often accompany headphone listening. It must be noticed that the various systems mentioned above are not mutually exclusive—a single 3D audio system might use Ambisonic content as the basis for binaural synthesis using a listener's HRTF, but play back the content over a transaural system, creating a synergy out of the strengths of these different technologies to achieve the most convincing 3D sound possible.

Technology and Spatial Music

Art Music

The advent of electroacoustic technology—microphones, recordings, amplifiers, and loudspeakers—had a huge and far-reaching impact, not only on the frequency and temporal content of music composition, but also on the ways in which music could be spatialized. John Cage (1912–1992), another inheritor of Ives's American Experimental tradition, quickly saw the potential of phonograph recordings and wireless radios, which he employed with spatial separation respectively in his Imaginary Landscape Nos. 1 and 4 from 1939 to 1951, or the later installation Writings Through the Essay: On the Duty of Civil Disobedience, in 1985. Both Cage and another aleatoric composer, Morton Feldman, would later make use of multiple tape recorders, routing each audio tape to a separate spatialized loudspeaker (Zvonar, 2006).

In contrast to Cage's and Feldman's use of space to embrace chaos, the European electroacoustic tradition began to see space as another parameter that could be serialized within highly deterministic musical structures. Pierre Schaeffer and Pierre Henry constructed a tetrahedral four-loudspeaker playback system for their *musique concrete*, which employed a potentiometer allowing them to route different channels of audio to specific speakers (Zvonar, 2006). Meanwhile, in Germany Karlheinz Stockhausen arguably began the modern electroacoustic tradition with his landmark *Gesang der Jünglinge*, which employed five loudspeakers around the audience, and included spatial direction among the many parameters that Stockhausen rigidly serialized in the piece (Stone, 1963). Indeed, Stockhausen, in opposition to Brant's spatial philosophy, believed that distance effects should not be employed because they affected sound timbre. For this reason, Stockhausen believed that sound direction was "the only spatial feature of sound worthy of compositional attention because it could be serialized" (Harley, 1997, p. 74). Despite Stockhausen's large compositional influence, his ideas about space were not widely adopted, partially because the high tide of serialism began to ebb, and also because it was later understood that sound direction is also indelibly tied to sound timbre through the spectral filtering of the listener's HRTF.

After these initial explorations of multichannel spatial exploration, more ambitious playback environments were built. Probably the most famous example is the Phillips Pavilion at the 1958 World's Fair, designed by Iannis Xenakis (1922–2001), which hosted the tape piece *Poeme Electronique* by Edgard Varèse (1883–1965). Varèse recorded the piece on four separate tape recorders, which gradually desynchronized over time due to differences in playing speeds (Kendall, 2006). The final piece was presented over 425 loudspeakers in the finished pavilion, featuring nine different predetermined 'routes' for the sound to travel over (Zvonar, 2006). After this landmark installation, more ambitious and more flexible facilities were constructed, including those at IRCAM, the University of California at San Diego (UCSD), and Stanford University (Forsyth, 1985; Zvonar, 2006). The composer Roger Reynolds (1934–) explored spatial arrangements at UCSD, beginning with quadrophonic playback and pushing forward to 6- and 8-channel works (Zvonar, 2006). At Stanford, John Chowning created software to control the motion of a sound recording over a multichannel loudspeaker array, including a simulated Doppler shift using frequency modulation (Chowning, 1977). As personal computers and multichannel sound cards proliferated, many of the different spatialization techniques described earlier were adopted by composers and sound installation artists. The exploration and understanding of spatial music is a continuing area of music research: composer and theorist Denis Smalley has put forth the idea that since acousmatic musical motion "always implies a direction" (Smalley, 1986, p. 73), that, "rather than being the final frontier of investigation, space should now move to centre stage to become the focal point of analysis" (Smalley, 2007, p. 54).

Popular Music

It is well known that the Beatles were influenced by Stockhausen, as seen by his face's inclusion on the cover of *Sergeant Pepper's Lonely Hearts Club Band* (Richardson, 2015). Yet this did not extend to sound spatialization—in 1967 the Beatles were not even present for the stereo mixes of *Sergeant Pepper*, which were conducted far more quickly than the mono mixes. George Harrison reported that using two speakers seemed unnecessary and even made the music sound "naked" (Komara, 2010). Economic growth and the proliferation of stereo broadcasting and reproduction equipment would eventually make stereo playback the norm in pop music. Composer/producer Brian Eno sought to embrace and enhance the natural spatial characteristics of playback environments and landscapes with his ambient albums such as "Music for Airports" (1978) and "On Land" (1982), though these albums used only 2-channel stereo for reproduction. Experimental/psychedelic band The Flaming Lips would go farther, performing early experiments in a parking lot full of their fans playing different tapes of prerecorded content through car stereos. This would later lead to their adventurous, though commercially unsuccessful, album *Zaireeka* (1997), which was released as four separate compact disks that had to be synchronously played on separate CD players. This distribution format made *Zaireeka* communal, almost like a concert recital, as at least four people had to be present to listen to the album. Slight asynchronies in disk reading speeds, as with Varese's tape recorders used for *Poeme Electronique*, made every experience of *Zaireeka* different, giving the album a small but dedicated fan base.

The lack of a compact format for multichannel audio led to the dual releases in the late 1990s of the Super Audio CD format from Phillips/Sony, and the DVD Audio format from JVC, which featured 6- and 8-channel encoding, respectively (Verbakel et al., 1998; Funasaka & Suzuki, 1997). Indeed, the Flaming Lips' next album, *Yoshimi Battles the Pink Robots* (2003), included a DVD version in Dolby Digital 5.1, taking advantage of the growing market in home theater systems to allow a single listener to hear immersive content from her living room (Rickert & Salvo, 2006). However, both formats failed to catch on outside the audiophile market, and by 2007 *The Guardian* called DVD Audio 'extinct' and Super Audio CDs 'dying' (Schofield, 2007). Today, larger multichannel formats for home and theatrical reproduction are multiplying, including object-based systems (see Chapter 8) such as Dolby Atmos and DTS:X, as well as more traditional channel-based formats like Auro 3D (Dolby, 2014; Claypool et al., n.d.; Fonseca, 2015).

Conclusions and Thoughts for the Future

The experience and use of 3D sound in human culture has always been tied to the technological capabilities of the generation hearing it: for most of human history, this technology was limited to architectural spaces and music composition. Since the 19th century, more advanced technologies have allowed both the more accurate representation of real-world soundscapes as well as sonic spaces that have no correlate in physical reality. The former trend is now being used in the rapid development of audio for virtual reality applications, which use head-mounted displays and headphones and are thus reliant on binaural reproduction methods (Begault, 2000; Xie, 2013). The latter trend of exploring new non-physical spatial audio has been manifested in augmented reality systems, which use spatial auditory content to represent information beyond that encountered in the real world. Though these systems may use multichannel or wave field loudspeaker methods within certain controlled environments (Boren et al., 2014), they also require binaural reproduction for deployment in day-to-day life, especially given the ubiquity of in-ear headphones or 'earbuds' in modern society (Sundareswaran et al., 2003). Multichannel methods persist for film audio and home theater systems, but in the future these markets may also face increased competition from binaural or transaural content as HRTF individualization methods improve.

What can the history of 3D sound tell us about the future of this field? Specific predictions are difficult because history shows us many contrasting trends that may lead to either rapid progress and standardization or else atomization and stagnation. Speculative investment without patient development may lead to another failure like quadrophonic sound, which lost the attention of the wider public and arguably delayed the implementation of more ambitious spatial audio formats. Perhaps some poor soul will outline the next century's worth of progress but languish in obscurity, as did Blumlein. On the other hand, co-operation between scientists and content creators, as with the example of Fletcher and Stokowski, may lead to faster development than otherwise would have been expected. A hopeful sign is the recent collaboration between many different audio research institutions to create the Spatially Oriented Format for Acoustics (SOFA), a standardized file format that allows for the consolidation of many years' and ears' worth of HRTF research from around the world (Majdak et al., 2013). If this file format is adopted by musicians, game designers, sound engineers, and other audio content creators, it might usher in a new renaissance in the use and experience of 3D sound. History contains many examples that elicit pessimistic predictions for the future. Thus it is best to continue to hope for the optimistic vision, yet those of us in the field, whether scientists or artists, must also continue working to make that vision a reality.

Acknowledgments

Many thanks to Durand Begault, Gary Kendall, and Agnieszka Roginska, who provided many starting points for the exploration of this large subject. Thanks also to John Krane for introducing me to spatial audio many years ago through a *Zaireeka* listening party with four exceptionally synchronized play-button-pushers.

References

Abel, J., Rick, J., Huang, P., Kolar, M., Smith, J., & Chowning, J. (2008). On the acoustics of the underground galleries of ancient Chavin de Huantar, Peru. *Acoustics '08*, Paris.

Allen, I. (1991). Matching the sound to the picture. *Proceedings of the 9th Audio Engineering Society International Conference* (pp. 177–186). Detroit, Michigan.

Amet, E. H. (1911). *Method of and Means for Localizing Sound Reproduction.* US Patent 1,124,580.

Andreopoulou, A. (2013). *Head-Related Transfer Function Database Matching Based on Sparse Impulse Response Measurements*, Doctoral Dissertation, New York University.
Arnold, D. (1959). The significance of "cori spezzati." *Music & Letters*, *40*(1), 4–14.
Atal, B. S., & Schroeder, M. R. (1962). *Apparent Sound Source Translator*. US Patent 3,236,949.
Bagenal, H. (1930). Bach's music and church acoustics. *Music & Letters*, *11*(2), 146–155.
Bagenal, H. (1951). Musical taste and concert hall design. *Proceedings of the Royal Musical Association*, *78*(1), 11–29.
Bauck, J., & Cooper, D. H. (1996). Generalized transaural stereo and applications. *Journal of the Audio Engineering Society*, *44*(9), 683–705.
Begault, D. R. (2000). *3-D Sound for Virtual Reality and Multimedia*. Moffett Field, CA: National Aeronautics and Space Administration.
Berkhout, A. J., de Vries, D., & Vogel, P. (1993). Acoustic control by wave field synthesis. *The Journal of the Acoustical Society of America*, *93*(5), 2764–2778.
Blauert, J. (1997). *Spatial Hearing: The Psychophysics of Human Sound Localization* (3rd ed.). Cambridge, MA: The MIT Press.
Bloom, P. (1998). *The Life of Berlioz*. Cambridge, UK: Cambridge University Press.
Blumlein, A. D. (1931). *Improvements in and Relating to Sound-transmission, Sound-recording, and Sound Reproducing Systems*. Great Britain Patent 394,325.
Boren, B. B., & Longair, M. (2011). A method for acoustic modeling of past soundscapes. *Proceedings of the Acoustics of Ancient Theatres Conference*. Patras, Greece.
Boren, B. B., Longair, M., & Orlowski, R. (2013). Acoustic simulation of renaissance Venetian Churches. *Acoustics in Practice*, *1*(2), 17–28.
Boren, B., Musick, M., Grossman, J., & Roginska, A. (2014). I HEAR NY4D: Hybrid acoustic and augmented auditory display for urban soundscapes. *Proceedings of the 20th International Conference on Auditory Display (ICAD)*. New York, NY.
Brant, H. (1978). Space as an essential aspect of musical composition. In E. Schwartz & B. Childs (Eds.), *Contemporary Composers on Contemporary Music* (pp. 223–242). New York: Da Capo Press.
Broderick, A. E. (2012). *Grand Messe Des Morts: Hector Berlioz's Romantic Interpretation of the Roman Catholic Requiem Tradition*, Master's Thesis, Bowling Green State University.
Burkholder, J. P., Grout, D. J., & Palisca, C. V. (2006). *A History of Western Music* (7th ed.). New York, NY: W. W. Norton and Company.
Choueiri, E. (2008). *Optimal Crosstalk Cancellation for Binaural Audio with Two Loudspeakers*. Princeton University. Retrieved from www.princeton.edu/3D3A/Publications/BACCHPaperV4d.pdf, Accessed 6/23/2015, at 5:21 pm.
Chourmouziadou, K., & Kang, J. (2008). Acoustic evolution of ancient Greek and Roman theatres. *Applied Acoustics*, *69*(6), 514–529.
Chowning, J. M. (1977). Simulation of moving sound sources. *Computer Music Journal*, *1*(3), 48–52.
Claypool, B., Van Baelen, W., & Van Daele, B. (n.d.). *Auro 11.1 versus Object-based Sound in 3D*. Retrieved from www.barco.com.cn/~/media/Downloads/White papers/2012/WhitePaperAuro 111 versus object-based sound in 3Dpdf.pdf, Accessed 1/12/16, at 3:03 pm.
Collins, P. (2008). Theatrophone: The 19th-century iPod. *New Scientist*, January 12, 44–45.
Davis, M. F. (2003). History of spatial coding. *Journal of the Audio Engineering Society*, *51*(6), 554–569.
Declercq, N. F., & Dekeyser, C. S. A. (2007). Acoustic diffraction effects at the Hellenistic amphitheater of Epidaurus: Seat rows responsible for the marvelous acoustics. *The Journal of the Acoustical Society of America*, *121*(4), 2011–2022.
Dixon, G. (1979). The origins of the Roman "colossal baroque." *Proceedings of the Royal Musical Association*, *106*(1), 115–128.

Dolby Laboratories. (2014). *Authoring for Dolby Atmos Cinema Sound Manual*. San Francisco, CA. Retrieved from www.dolby.com/us/en/technologies/dolby-atmos/authoring-for-dolby-atmos-cinema-sound-manual.pdf, Accessed 1/12/16, at 2:38 pm.

Eno, B. (1978). *Liner Notes, "Ambient 1: Music for Airports."* Retrieved from www.iub.edu/~audioweb/T369/enoambient.pdf, Accessed 6/25/2015, at 3:25 pm.

Eno, B. (1982). *Liner Notes, "Ambient 4: On Land."* Retrieved from www.iub.edu/~audioweb/T369/eno-ambient.pdf, Accessed 6/25/2015, at 3:25 pm.

Farnetani, A., Prodi, N., & Pompoli, R. (2008). On the acoustics of ancient Greek and Roman theaters. *The Journal of the Acoustical Society of America*, *124*(3), 1557–1567.

Fenlon, I. (2006). The performance of cori spezzati in San Marco. In D. Howard & L. Moretti (Eds.), *Architettura e Musica Nella Venezia Del Rinascimento*, 79–98. Bruno Mondadori, Milan.

Fletcher, H. (1933). An acoustic illusion telephonically achieved. *Bell Laboratories Record*, *11*(10), 286–289.

Fletcher, H., & Sivian, L. J. (1927). *Binaural Telephone System*. US Patent 1,624,486.

Fonseca, N. (2015). Hybrid channel-object approach for cinema post-production using particle systems. *Proceedings of the 139th Audio Engineering Convention*. New York, NY.

Forsyth, M. (1985). *Buildings for Music: The Architect, the Musician, and the Listener from the Seventeenth Century to the Present Day*. Cambridge, MA: The MIT Press.

Frank, M., Zotter, F., & Sontacchi, A. (2015). Producing 3D audio in Ambisonics. *Proceedings of the 57th AES International Conference*. Hollywood, CA.

Fuller, S. (1981). Theoretical foundations of early Organum Theory. *Acta Musicologica*, *53*(1), 52–84.

Funasaka, E., & Suzuki, H. (1997). DVD-Audio format. *Proceedings of the 103rd Audio Engineering Society Convention*. New York, NY.

Gendel, M. (1985, September 24). Hearing is believing on new "twilight zone." *Los Angeles Times*. Retrieved from http://articles.latimes.com/1985-09-24/entertainment/ca-18781_1_twilight-zone, Accessed 6/23/2015, at 5:40 pm.

Gerzon, M. A. (1973). Periphony: With-height sound reproduction. *Journal of the Audio Engineering Society*, *21*(1), 2–10.

Gerzon, M. A. (1975). The design of precisely coincident microphone arrays for stereo and surround sound. *Proceedings of the 50th Audio Engineering Society Convention*. London, UK.

Hamasaki, K., Hiyama, K., & Okumura, R. (2005). The 22.2 multichannel sound system and its application. *Proceedings of the 118th Audio Engineering Society Convention*. Barcelona, Spain.

Harley, M. A. (1997). An American in space: Henry Brant's "spatial music." *American Music*, *15*(1), 70–92.

Harmon, T. (1970). The performance of Mozart's church sonatas. *Music & Letters*, *51*(1), 51–60.

Hartmann, W. M., & Wittenberg, A. (1996). On the externalization of sound images. *The Journal of the Acoustical Society of America*, *99*(6), 3678–3688.

Hintermaier, E. (1975). The Missa Salisburgensis. *The Musical Times*, *116*(1593), 965–966.

Holman, P. (1994). Mystery man: Peter Holman celebrates the 350th anniversary of the birth of Heinrich Biber. *The Musical Times*, *135*(1817), 437–441.

Hospitalier, E. (1881). The telephone at the Paris opera. *Scientific American*, *45*, 422–423. Retrieved from http://earlyradiohistory.us/1881opr.htm, Accessed 6/17/2015, at 3:54 pm.

Howard, D., & Moretti, L. (2010). *Sound and Space in Renaissance Venice*. New Haven and London: Yale University Press.

Huscher, P. (2006). *Program Notes: Wolfgang Mozart, Notturno in D, K.286, Chicago Symphony Orchestra*. Retrieved from https://cso.org/uploadedFiles/1_Tickets_and_Events/Program_Notes/ProgramNotes_Mozart_Notturno.pdf, Accessed 6/15/2015, at 6:00 pm.

Jahn, R. G. (1996). Acoustical resonances of assorted ancient structures. *The Journal of the Acoustical Society of America*, *99*(2), 649–658.

Katz, B. F. G. (2001). Boundary element method calculation of individual head-related transfer function: I: Rigid model calculation. *The Journal of the Acoustical Society of America*, *110*(5), 2440.

Kendall, G. (1995a). A 3-D sound primer: Directional hearing and stereo reproduction. *Computer Music Journal*, *19*(4), 23–46.

Kendall, G. (1995b). The decorrelation of audio signals and its impact on spatial imagery. *Computer Music Journal*, *19*(4), 71–87.

Kendall, G. S. (2006). Juxtaposition and non-motion: Varèse bridges early modernism to electroacoustic music. *Organised Sound*, *11*(2), 159–171.

Kendall, G. (2015). *Personal communication*.

Komara, E. (2010). The Beatles in mono: The complete mono recordings. *ASRC Journal*, *41*(2), 318–323.

Kostadinov, D., Reiss, J. D., & Mladenov, V. (2010). Evaluation of distance based amplitude panning for spatial audio. *Proceedings of the IEEE International Conference on Acoustics, Speech, and Signal (ICASSP)*. Dallas, TX.

Kronlachner, M. (2014). *Spatial Transformations for the Alteration of Ambisonic Recordings*, Master's Thesis, University of Music and Performing Arts, Graz, Austria.

Lamson, H. W. (1930). The Use of Sound in Navigation. *The Journal of the Acoustical Society of America*, *1*(3), 403–40.

Lindsay, B. (1966). The story of acoustics. *Journal of the Acoustical Society of America*, *39*(4), 629–644.

Lossius, T., & Pascal Baltazar, T. (2009). DBAP-Distance-Based Amplitude Panning. *Proceedings of International Computer Music Conference (ICMC)*. Montreal, Quebec.

Lubman, D., & Kiser, B. H. (2001). The history of western civilization told through the acoustics of its worship spaces. *Proceedings of the 17th International Congress on Acoustics*. Rome, Italy.

MacDonald, C. (2009). Scoring the work: Documenting practice and performance in variable media art. *Leonardo*, *42*(1), 59–63.

MacGowan, K. (1957). Screen wonders of the past: And to come? *The Quarterly of Film Radio and Television*, *11*(4), 381–393.

Majdak, P., Iwaya, Y., Carpentier, T., Nicol, R., Parmentier, M., Roginska, A., . . . Noisternig, M. (2013). Spatially oriented format for acoustics. *Proceedings of the 134th Audio Engineering Society Convention*. Rome, Italy.

Malham, D. G. (1999). Higher order Ambisonic systems for the spatialisation of sound. *Proceedings of the International Computer Music Conference* (pp. 484–487). Beijing, China.

Malham, D. G., & Anthony, M. (1995). 3-D sound spatialization using Ambisonic techniques. *Computer Music Journal*, *19*(4), 58–70.

Manola, F., Genovese, A., & Farina, A. (2012). A comparison of different surround sound recording and reproduction techniques based on the use of a 32 capsules microphone array, including the influence of panoramic video. *Audio Engineering Society 25th UK Conference: Spatial Audio in Today's 3D World*. York, UK.

Meshram, A., Mehra, R., & Manocha, D. (2014). Efficient HRTF computation using adaptive rectangular decomposition. *AES 55th International Conference*. Helsinki, Finland.

McDonald, M. (2004). Silent narration? Elements of narrative in Ives's the unanswered question. *19th-Century Music*, *27*(3), 263–286.

Moretti, L. (2004). Architectural spaces for music: Jacopo Sansovino and Adrian Willaert at St Mark's. *Early Music History*, *23*(2004), 153–184.

Navarro, J., Sendra, J. J., & Muñoz, S. (2009). The Western Latin church as a place for music and preaching: An acoustic assessment. *Applied Acoustics*, *70*(6), 781–789.

O'Donovan, A., & Duraiswami, R. (2010). Audio-visual panoramas and spherical audio analysis using the audio camera. *Proceedings of the 16th International Conference on Auditory Display (ICAD2010)* (pp. 167–168). Washington, DC.

Paul, S. (2009). Binaural recording technology: A historical review and possible future developments. *Acta Acustica United with Acustica*, *95*, 767–788.

Prasad, M. G., & Rajavel, B. (2013). Acoustics of chants, conch-shells, bells and gongs in Hindu worship spaces. *Acoustics 2013* (pp. 137–152). New Delhi, India.

Pulkki, V. (1997). Virtual sound source positioning using vector base amplitude panning. *Journal of the Audio Engineering Society*, *45*(6), 456–466.

Reznikoff, I. (2008). Sound resonance in prehistoric times: A study of Paleolithic painted caves and rocks. *Acoustics '08* (pp. 4137–4141), Paris.

Richardson, C. E. (2015). *Stockhausen's Influence on Popular Music: An Overview and a Case Study on Björk's Medúlla*, Master's Thesis, Texas State University.

Rickert, T., & Salvo, M. (2006). The distributed Gesamptkunstwerk: Sound, worlding, and new media culture. *Computers and Composition*, *23*(3), 296–316.

Rindel, J. H. (2011a). The ERATO project and its contribution to our understanding of the acoustics of ancient theatres. *The Acoustics of Ancient Theatres Conference*. Patras, Greece.

Rindel, J. H. (2011b). Echo problems in ancient theatres and a comment to the "sounding vessels" described by Vitruvius. *The Acoustics of Ancient Theatres Conference*. Patras, Greece.

Ritchey, M. (2010). Echoes of the Guillotine: Berlioz and the French fantastic. *19th-Century Music*, *34*(2), 168–185.

Rosenthal, K. A., & Mendel, A. (1941). Mozart's Sacramental litanies and their forerunners. *The Musical Quarterly*, *27*(4), 433–455.

Schofield, J. (2007). No taste for high-quality audio. *The Guardian*. Retrieved from www.theguardian.com/technology/2007/aug/02/guardianweeklytechnologysection.digitalmusic, Accessed 1/12/16, 2:02 pm.

Schroeder, M. R., Gottlob, D., & Siebrasse, K. F. (1974). Comparative study of European concert halls: Correlation of subjective preference with geometric and acoustic parameters. *Journal of the Acoustical Society of America*, *56*(4), 1195–1201.

Slotki, I. W. (1936). Antiphony in ancient Hebrew poetry. *The Jewish Quarterly Review*, *26*(3), 199–219.

Smalley, D. (1986). Spectro-morphology and structuring processes. In S. Emmerson (Ed.), *The Language of Electroacoustic Music*, pp. 61–93. New York: Harwood Academic.

Smalley, D. (2007). Space-form and the acousmatic image. *Organised Sound*, *12*(1), 35–58.

Snow, W. (1955). Basic principles of stereophonic sound. *IRE Transactions on Audio*, *3*(2), 42–53.

Soeta, Y., Shimokura, R., Kim, Y. H., Ohsawa, T., & Ito, K. (2013). Measurement of acoustic characteristics of Japanese Buddhist temples in relation to sound source location and direction. *The Journal of the Acoustical Society of America*, *133*(5), 2699–2710.

Spors, S., Rabenstein, R., & Ahrens, J. (2008). The theory of wave field synthesis revisited. *Proceedings of the 124th Audio Engineering Society Convention*. Amsterdam, The Netherlands.

Stevens, D. (1982). A songe of fortie parts, made by MR. Tallys. *Early Music*, *10*(2), 171–182.

Stone, K. (1963). Karlheinz Stockhausen: Gesang der Junglinge (1955/56). *The Musical Quarterly*, *49*(4), 551–554.

Strutt, J. W. (1875). On our perception of the direction of a source of sound. *Proceedings of the Musical Association*, *2*, 75–84.

Sundareswaran, V., Wang, K., Chen, S., Behringer, R., McGee, J., Tam, C., & Zahorik, P. (2003, October). 3D audio augmented reality: Implementation and experiments. *Proceedings of the 2nd IEEE/ACM International Symposium on Mixed and Augmented Reality* (p. 296). IEEE Computer Society.

Sunier, J. (1986). A history of binaural sound. *Audio*, March, 36–44.

T, H. W. (1930). The use of sound in navigation. *The Journal of the Acoustical Society of America*, *1*(3), 403–409.

Torick, E. (1998). Highlights in the history of multichannel sound. *Journal of the Audio Engineering Society*, *372*, 368–372.

Verbakel, J., van de Kerkhof, L., Maeda, M., & Inazawa, Y. (1998). Super audio CD format. *Proceedings of the 104th Audio Engineering Society Convention*. Amsterdam, The Netherlands.

Vitruvius, M. (1914). *Vitruvius: The Ten Books on Architecture* (M. Morgan, Ed.). Cambridge, MA: Harvard University Press.

Wenzel, E., Arruda, M., Kistler, D. J., & Wightman, F. L. (1993). Localization using nonindividualized head-related transfer functions. *Journal of the Acoustical Society of America*, *94*(1), 111–123.

Wolf, R. (1986, April 18). At Northwestern, they're reshaping world of sound. *Chicago Tribune*. Retrieved from http://articles.chicagotribune.com/1986-04-18/entertainment/8601280476_1_spatial-sound-outer-ear, Accessed 6/26/2015, at 4:28 pm.

Xie, B. (2013). *Head-related Transfer Function and Virtual Auditory Display* (2nd ed.). Boca Raton, FL: J Ross.

Zotkin, D. N., Duraiswami, R., Grassi, E., & Gumerov, N. A. (2006). Fast head-related transfer function measurement via reciprocity. *The Journal of the Acoustical Society of America*, *120*(4), 2202–2215.

Zvonar, R. (2006). A history of spatial music. *eContact*, *7*(4). Retrieved from http://cec.sonus.ca/econtact/7_4/zvonar_spatialmusic.html, Accessed 6/20/2015, at 1:28 pm.

Chapter 3

Stereo

Paul Geluso

Stereo Systems

Two-channel stereo has been the mainstay for hi-fidelity recording and playback systems since the first wave of stereo media was brought to the marketplace in the 1950s. Stereo systems are designed to create the illusion of a spatial sound scene with directional sound sources localized between two or more loudspeakers placed in front of the listener. Snow observed that binaural systems transport the listener to the scene of the recording whereas *stereo* systems transport the sound sources to the listener's room (Snow, 1953). This chapter will explore methods used to capture, create, playback, and enhance stereo programs.

Blumlein's Patent

Alan Blumlein laid the groundwork for modern 2-channel stereo recording and reproduction systems in his landmark 1933 British patent. He referred to his stereo invention as a binaural transmission system and explained that a realistic, multi-directional sound impression can be created by using two acoustic pathways. Further, he stated that acoustic phase and amplitude information captured by two directional microphones can be reconstructed using just two loudspeakers (Blumlein, 1933). Based on these principles, Blumlein explained that a 2-channel system was capable of producing a near-replica of the original directional sound image. More elaborate directional sound system concepts (see Chapter 2) were developed in his time that used a great number of speakers, but eventually, 2-channel stereo became standard.

It should be noted that Blumlein's ideas extended beyond 2-channel stereo reproduction. He suggested methods to capture sound vertically as well as horizontally simultaneously, thus laying the foundation for the immersive sound systems to come (Eargle, 1986).

Monophonic Systems and Distance Location

When a single loudspeaker is used to reproduce sound, the system is considered monophonic. Snow described the effect of monophonic systems as if "sound (is) coming through a hole in the wall" (Snow, 1953, p. 44). Compared to stereo and binaural systems, extreme spatial errors can exist using monophonic systems. When recording multiple directional sound sources across the sound stage with a single microphone, sounds arriving from many directions are captured and fused together into a single signal. When the mono signal is later reproduced through a single loudspeaker, the directional information from the recording environment may appear to be lost. But remarkably, a monophonic system can give the illusion of depth and space if the sound recording has captured enough spatial information such as wall reflections, reverberation, and location-dependent spectral information. This is certainly the case with many early wax cylinder and disk recordings where a single transducer was used to record several sound sources. The high-frequency content and the relative balance between the direct and diffused sound for each sound source can give the listener a good idea of how far away the sound sources were from the microphone. In general, sources recorded close are characterized by a high level of timbre detail, whereas sources recorded from a distance will have a considerable alteration in timbre and a general lack of definition (Moylan, 2007). Captured early reflections and reverberation can provide spatial cues with information about the size and treatment of the space in which the recording was made. Even so, during playback from a monophonic system, the perceived sound image will stay with the loudspeaker. In other words, monophonic systems are very speaker-centric. Ironically, unlike stereo and surround sound reproduction, monophonic systems have a wide listening area where the program balance is perceived correctly.

Stereo Monitoring

A stereo system can create a realistic illusion of directional sounds arriving from across the horizontal plane in front of the listener, bounded by two or more loudspeakers—and even beyond (see Chapter 5). The ideal stereo listening position for 2-channel stereo is known as the *sweet spot*. A listener placed in the sweet spot is centered in between the loudspeakers and is directly facing the loudspeaker's baseline (Dickreiter, 1989). The ITU[1]stereo sound evaluation specification recommends locating the loudspeakers 30 degrees off the center line, at equal distance, and in front of the listener for 2-channel stereo monitoring (see Figure 3.1). The direction of the sound image and the sense of spaciousness experienced by a listener outside of the ideal listening area can be distorted and unstable. Even a subtle turn of the head can alter the acoustic pathways to the ears enough to affect the spectral, directional, and spatial attributes of a stereo program. The stereo listening area and stereo imaging can be adapted and optimized by adjusting the loudspeaker spacing, direction (also known as *toe-in* angle), directional radiation characteristics, and the frequency response (Eargle, 1986). The size, design, and acoustic treatment of the listening room greatly affect the perception of stereo images as well. Griesinger (1985) found that good low-frequency response is very important for creating the sense of spaciousness and that a poor choice of speaker location in a room can cause a stereo signal to sound more monaural than desired. If a listening space suffers from severe spectral filtering due to modal frequencies, early reflections, or reverberation in the room, it is difficult for the listener to tell if the spectral coloring and/or room sounds they hear are in the signal or are being generated and superimposed onto the signal by the listening room. In this situation, headphone monitoring can be used to get a better sense of what is actually captured in the sound recording. On the other hand, a listening space can constructively add a sense of spaciousness and immersion to enhance the listener's experience. Surely, if a stereo program is monitored loud enough in a small room or car, the sound image can envelop the listener even though sound is being emitted by only two loudspeakers.

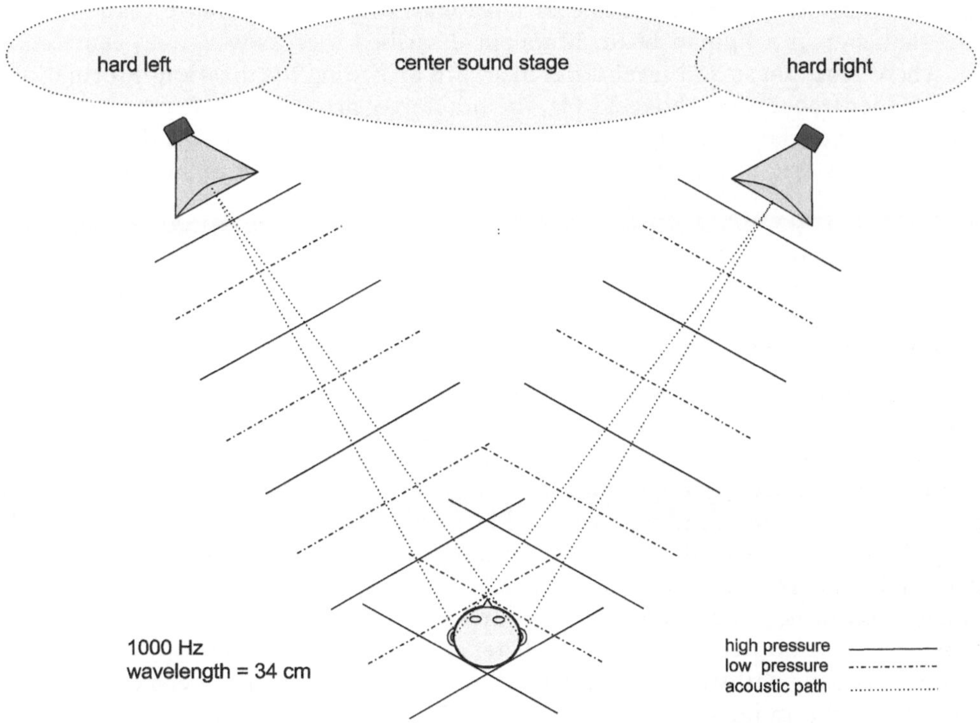

Figure 3.1 Stereo monitoring with the loudspeakers positioned + and −30 degrees off center. Phantom images make up the center sound stage in between the loudspeakers. Sounds hard-panned in the left or right channel only are perceived at the respective loudspeaker. The spacing of 1 kHz high- and low-pressure wave fronts is indicated for reference (solid and dot-dashed lines) (wavelength = 34.32 cm in air). The *direct* acoustic pathways to the ears are indicated as well (dashed).

Directional information can be encoded as inter-channel time difference (ICTDs) and/or inter-channel level differences (ICLDs) into a stereo program. ICLDs are less effective at low frequencies due to their long wavelengths in comparison to the average size of the human head (see Chapter 1). For 2-channel stereo systems with speakers located 30 degrees off axis, Cooper suggests that the amplitudes of frequencies below 327 Hz are not affected by head shadowing based on wave front head shadow models (Cooper, 1987), as such waves will diffract easily around the head. However, inter-channel level difference in stereo loudspeaker projection will result in inter-aural time difference, due to speakers being positioned to the sides of the listener, and this will produce a directional auditory sensation. This conversion from inter-channel amplitude difference to inter-aural time difference cannot occur in headphone listening. Of course, directional sensation will be stronger for non-stationary, transient sounds compared to stationary, continuous sounds (see Chapter 1).

ICTDs alone start to provide effective localization for frequencies below 1 kHz for stereo systems. This includes the frequencies that have wavelengths in air more than 34 cm, about twice the diameter of a human head. Blumlein described a crossover area, centered around 700 Hz, where both phase and level differences are providing localization information to the brain. For higher frequencies above 2 kHz, the brain may produce a false reading of phase difference as the size of the wavelengths falls below the size of the human head, but at this point, the head shadow starts to take significant effect and will attenuate higher frequencies at the opposite ear (Blumlein, 1933). Albeit, research suggests that panning systems that use ICTDs in conjunction with ICLDs provide the most natural sense of source localization (Theile, 1991; Lee & Rumsey, 2013).

Phantom Sound Images

Signals of mono sound sources are often mixed into 2-channel stereo programs creating multi-mono signals delivered through two loudspeakers. By duplicating a mono signal and routing it to both loudspeakers of a 2-channel stereo system, a phantom centered sound image appears between the loudspeakers. A phantom image may be perceived as a virtual point source, or be spread to exhibit some degree of width. Electronic methods to create stereo width of an object are discussed later in this chapter. Although a phantom image of a source produced with two loudspeakers is less clear and spatially less precise than the source image produced by a single loudspeaker, phantom sound sources may appear very realistic to the listener located precisely in the sweet spot. As the listener moves laterally out of the sweet spot, the left and right acoustic pathways from the two loudspeakers to the listener's ears are no longer equal and the phantom image appears to move and follow the listener to the closest speaker (see Figure 3.2). At some point, when the sound level and arrival times from each loudspeaker are no longer perceived as equal at the listener's ears, the phantom center program image will become unstable, less focused, and appear to follow the listener to the closest speaker. At the extreme, as the listener moves farther to one side of the listening area, all phantom centered images will now be perceived as coming from only one loudspeaker. Unlike phantom sound images, produced by stereo loudspeaker pairs, sounds panned hard left or hard right will behave in a monophonic way, staying with their assigned single loudspeaker despite the listener's location in the room.

LCR Stereo

The stability of the center portion of a 2-channel stereo image can be greatly improved by the addition of a dedicated center channel. Left-center-right (LCR) systems are commonly used in larger venues to accommodate a wide seating area. In film houses, the center channel keeps dialogue and other on-screen sounds firmly centered for all viewers. For this reason, LCR speaker configurations are used for the front channels of many surround sound systems (see Chapter 6). As illustrated in case D in Figure 3.2, the center channel is locked to the center of the sound stage (and to the center of the picture stage, defined by the screen) therefore any signal sent to the center channel will appear in that position regardless of where the listener/viewer is seated. Dialogue

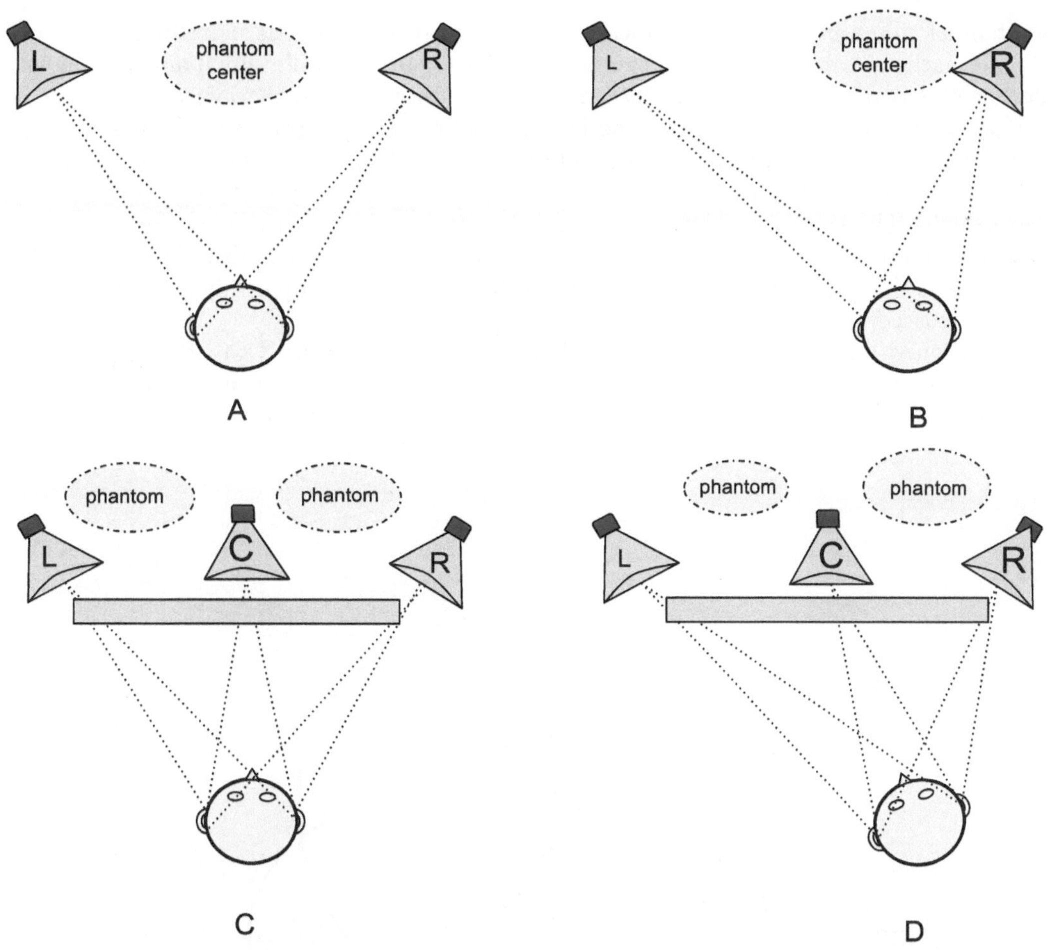

Figure 3.2 An illustration representing how phantom images are affected by the listener's location in the sweet spot (A and C) and outside of the sweet spot (B and D) for 2-channel stereo systems (A and B), and for 3-channel stereo systems (C and D). The dotted lines represent the acoustic pathways from the speakers to each ear.

and music can be integrated into an LCR stereo image by simultaneously using the center channel for mono dialogue signals while routing 2-channel stereo signals like music and some sound effects to the left and right speakers. In practice, some of the 2-channel music and sound effects can be mixed into the center as well for stability. In either case, with the center channel employed, excellent spatial focus, stability, and clarity are achieved for the center material, such as dialogue, while a wider stereo image is created for music and effects.

Middle-Side Stereo

Two-channel stereo programs are typically stored and broadcasted as paired left and right signals. Alternatively, a stereo program can be stored as a paired middle signal and side signal. The middle signal is derived from a unidirectional microphone oriented toward the center of the sound stage or electronically by summing left and right stereo program signals, whereas the side signal is derived from a bidirectional microphone oriented laterally with its null facing the center of the sound stage (see Figure 3.3), or electronically by taking the difference of the left and right stereo signals. The side signal effectively cancels out sounds arriving from the front and rear center. The phase information stored in the side signal can be used to decode directional information across the stereo field when paired with the middle signal by using a sum and difference matrix (discussed below).

As mentioned above, to convert any 2-channel Left-Right stereo (XY stereo) program to a middle-side stereo program (MS stereo), the middle signal is obtained by summing the left and right signals whereas the side signal is obtained by polarity-inverting the right signal and summing

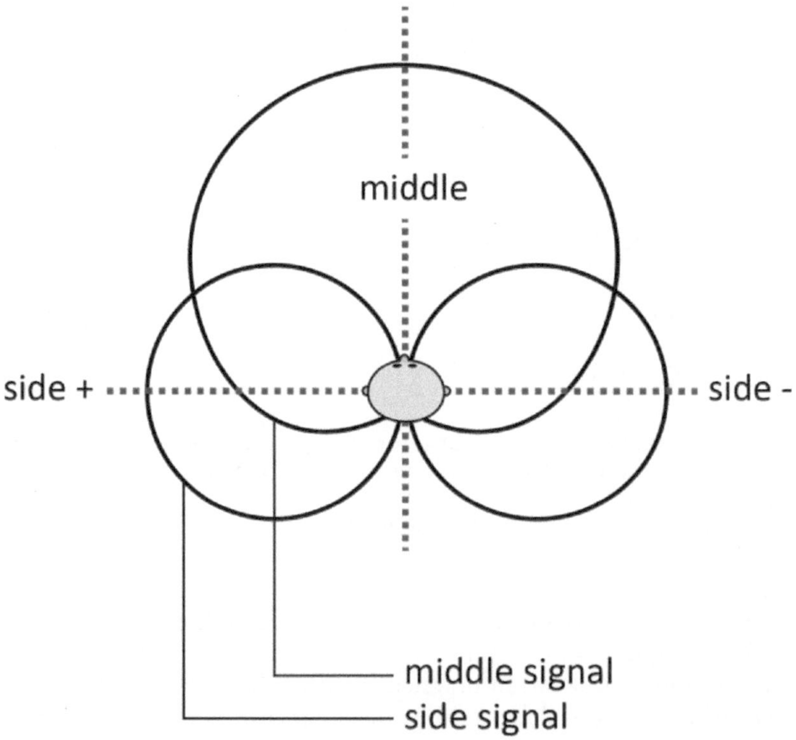

Figure 3.3 Polar patterns for a cardioid-based middle-side system. The side signal is bi-polar, a requirement for middle-side systems.

it with the left signal, effectively creating the difference between left and right signals. The relative intensities of the middle and side signals will vary depending on the stereo width of the program material.

Middle = Left + Right

Side = Left – Right

In theory, this process can be reversed without signal loss. The left signal can be restored by summing the middle and the side signals. Similarly, the right signal can be restored by finding the difference between middle and side signals.

Left = Middle + Side

Right = Middle – Side

The middle signal emphasizes the center of the stereo image. Any multi-mono signals representing phantom centered images in the stereo program will be summed perfectly in the middle signal. At the same time, these perfectly centered signals will be cancelled out of the side signal thus leaving only lateral, differential, or de-correlated stereo program material like stereo reverberation, stereo delays, and panned sources in the side signal. Middle-side recording and processing techniques will be discussed in detail later in this chapter.

Phase Correlation Metering

The term *phase* is often used in audio to describe the time-based relationship of two or more elementary audio signals, such as sine waves. For complex signals containing multiple frequencies, *group delay* defines relative signal delay, and *signal's polarity* determines its waveform orientation. The term *phase flip* or *phase inversion* implies that the polarity of an audio signal has been reversed, creating a mirror image when viewing the signal's waveform. A phase shift implies that a displacement in time has occurred. Phase correlation relates to the amount of or lack of phase inversion detected between two signals. For example, an in-phase, dual-mono signal, meaning the same signal in each channel, will have a phase correlation of 1 (see Figure 3.4).

If the left signal has little in common with right signal, the phase correlation value is zero. If the left signal contains a phase-inverted copy of the right signal, the phase correlation will go to negative 1 (see Figure 3.5).

In other words, phase correlation describes just how in-phase, or out-of-phase, two signals are; or similarly, just how correlated or de-correlated they are; or how similar or dissimilar two signals are. The amount of correlation measured between the left and right signals can be monitored with a phase correlation meter (see Figures 3.4, 3.5, and 3.6). The meter range is from –1 to 0 to +1. For example, a correlation reading of +1 indicates the stereo program is dominated by dual-mono signals. A reading of +.5 indicates a mixture of left, right, and phantom centered images. A reading of 0 indicates a lack of correlation or random correlation thus indicating a wide stereo image—for example, when two very different signals are panned to the opposite channels. This

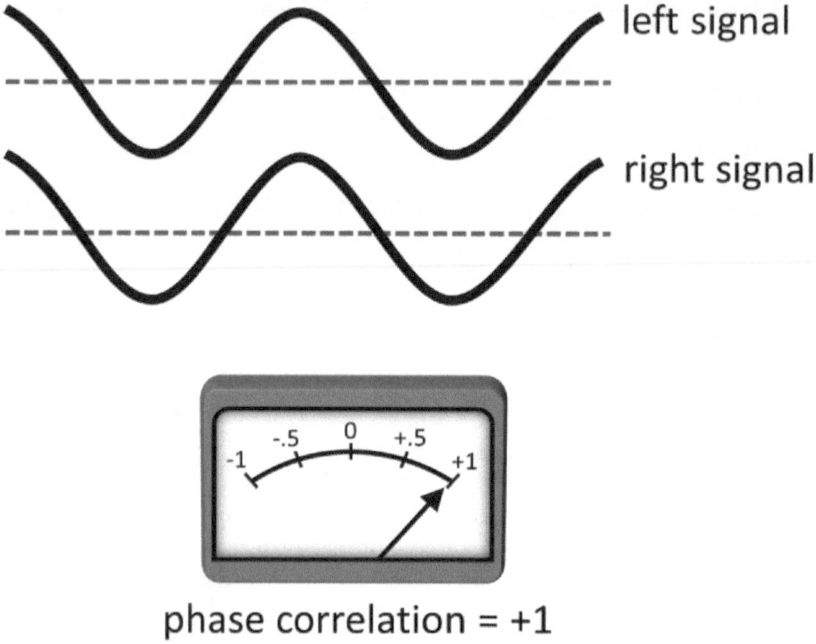

Figure 3.4 Two signals perfectly in phase with a phase correlation of +1 (Huber, 1992).

can also occur when there is high degree of similar musical content in both channels, but not identical sound recordings—as is the case when a musical part is doubled and each performance is panned hard to opposite sides. If the program content in each channel is too different, a hole in the middle of the stereo image may occur due to the absence of a phantom center image caused by the lack of common elements in the left-right signals. If the phase correlation falls below zero, it is an indication that a portion of the stereo program is phase-inverted in one of the channels. When inter-channel phase-inverted portions of the stereo program are present, the mono downmix will be compromised and can suffer from spectral coloration and/or severe gain loss. If a good mixture of phantom center and split stereo program material exists, the phase correlation meter will hover around +.5 (see Figure 3.6). In general, this is a good sign indicating that most likely a wide stereo image without a hole in the center exists.

Since stereo programs can contain a complex mixture of dual-mono, split, and de-correlated stereo signals, the reading of the correlation meter does not always correspond precisely to the perception of stereo width. Therefore, it is safe to say that using your ears is still the best way to adjust the stereo program width during production. Even so, for mixing, mastering, and broadcast engineers, the correlation meter is a very useful metering device to warn of potential holes in the middle of the stereo image, an overly mono mix, or mono-compatibility issues.

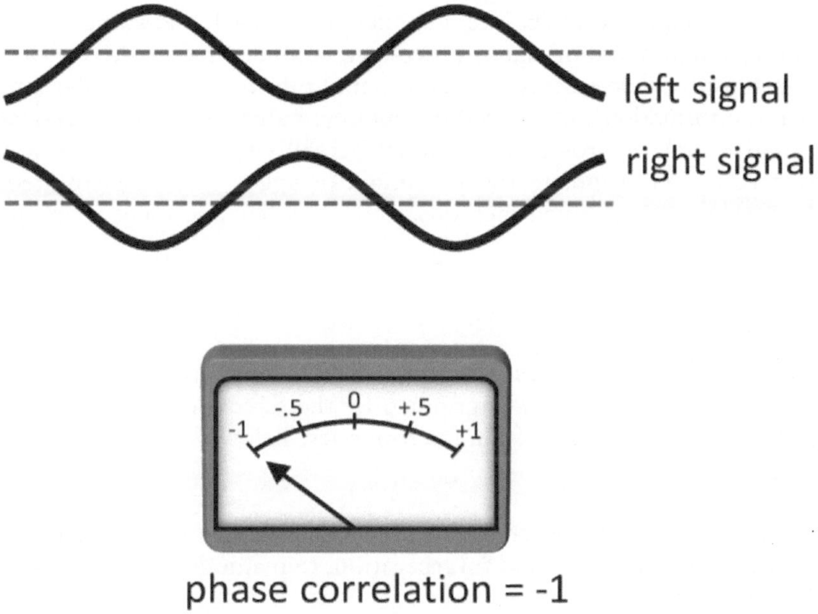

Figure 3.5 Two signals perfectly out of phase with a phase correlation of −1 (Huber, 1992).

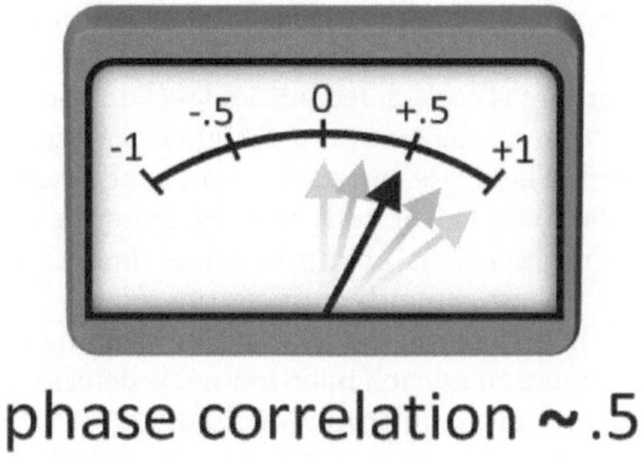

Figure 3.6 Phase correlation of +.5 (Huber, 1992).

Creating a Stereo Image

Stereo programs can contain complex inter-channel phase, level, and spectral relationships. Stereo microphone techniques and/or signal processing equipment can use one or a combination of these relationships to encode directional and spatial information into a 2-channel stereo program. This spatial information can be used to imitate a natural listening experience or to create perceived alternative sonic environments for a listener. Surely, the possibilities are endless and a great degree of artistic license and creative potential exists that should be explored. Fundamental stereo recording techniques, panning methods, and stereo enhancement techniques will be discussed next.

Stereo Microphone Techniques

By placing two or more microphones in a sound field, a directional sound image can be captured. The effective recording area is determined by the directional characteristic, the distance between, and the angle orientation of the microphones. Below are some general guidelines for stereo recording.

- Both microphones should be placed to capture an abundance of directional sounds, including direct sounds, room reflections, and reverberation, to maintain a certain degree of decorrelation between the microphones.
- The microphones should not be placed in total isolation from one another; all recorded sounds should be effectively *heard* by both microphones.
- The pick-up angle between the microphones should be no greater than 180 degrees.

XY Recording

The XY recording technique uses a matched pair of directional microphones positioned so that level information is captured with minimal acoustic time of arrival differences between the two capsules. Therefore, XY recording is considered a pure level difference recording system. In practice, an angle of 90 degrees between the axes of the highest sensitivity of the microphones (their 0°—on axis) is normally used. The angle can be varied at the discretion of the engineer to narrow or widen the recorded sound image. Sounds arriving from the center will be captured evenly by both microphones and will appear as phantom center images when reproduced through a 2-channel stereo loudspeaker system. Sounds arriving off-center will be attenuated at the opposite microphone thus providing a stereo impression when listening to the loudspeaker playback. The amount of off-center attenuation at each microphone is determined by the angle of incidence and by directional characteristics of the microphone used.

When the left signals and the right signals from an XY recording are summed, the resulting mono signal takes on a similar directional characteristic as the microphones being used. For example, the summation of an XY pair of cardioid signals yields a single mono cardioid signal (see Figure 3.7). Similarly, the summation of an XY pair of figure-of-eights (a Blumlein Pair) yields a single figure-of-eight signal (Dickreiter, 1989) (see Figure 3.8). When using XY recording techniques, the mono down-mix signal is without unwanted comb-filter effects (but with boosted center information), and will be perceived a bit closer to the center sound stage than its stereo counterpart.

MS Recording

Middle-side recording is another coincident stereo recording technique. The system consists of a directional *middle* microphone oriented toward the center of the sound stage paired with a figure-of-eight *side* microphone oriented laterally 90 degrees with its positive lobe facing the left side (see Figure 3.9). The middle microphone is typically a cardioid, although virtually any

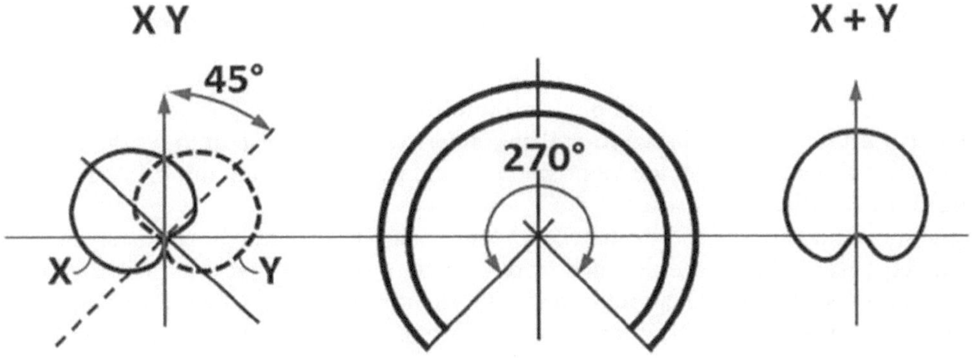

Figure 3.7 Cardioid pair XY configuration and its equivalent mono signal (Dickreiter, 1989).

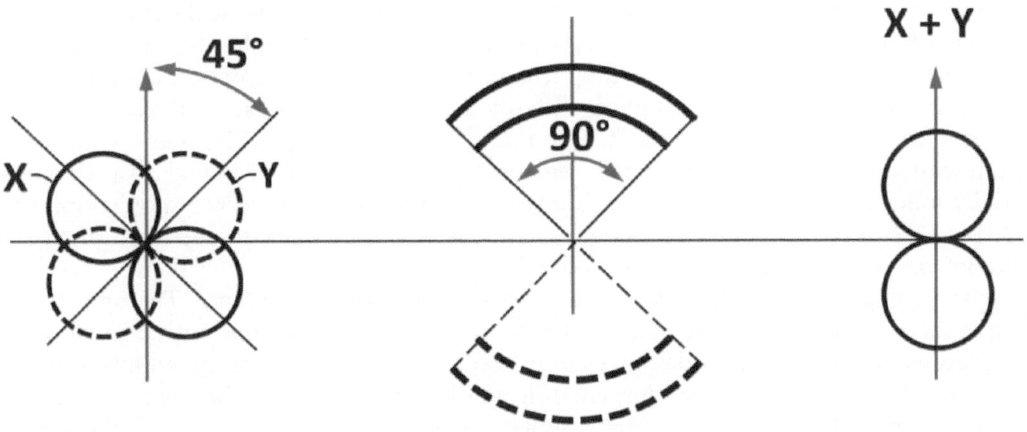

Figure 3.8 Blumlein Pair configuration and its equivalent mono signal (Dickreiter, 1989).

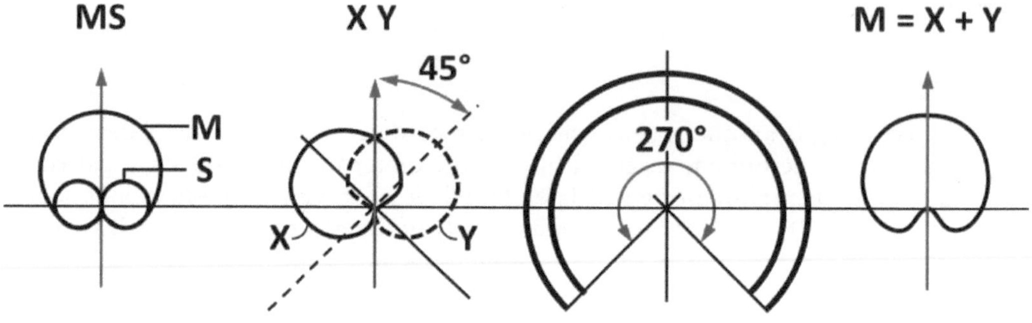

Figure 3.9 MS pair configuration and its equivalent XY and mono signal (Dickreiter, 1989).

type of microphone can be used, including an omni-directional microphone. In any case, the side microphone must have a true bi-directional characteristic for the system to work. The MS signal can be converted to a 2-channel left-right stereo signal by using a sum and difference MS matrix (Hibbing, 1989) as discussed earlier.

Using a sum and difference MS matrix, the side signal level can be adjusted to obtain the desired stereo-base width. If the decoded stereo signal is summed back to mono, all of the side information is cancelled out thus totally restoring the middle signal. Therefore, the mono signal will have significantly less lateral information and appear much closer to the center sound stage than its stereo counterpart (see Figure 3.9, right side).

Spaced Pair Recording

Using a spaced pair of microphones, time-of-arrival differences are captured. A small AB configuration, with a spacing of 16.5 cm to 30 cm, keeps the ICTDs within the scale of natural hearing. Due to the subtle differences in level between the closely spaced microphones, small AB is considered a pure stereo time-of-arrival recording technique. When working with microphones spaced farther apart, greater amplitude and phase differences are captured. Spacing the AB system 1 to 2 meters or more apart will enhance the ICTDs and introduce greater ICLDs thus creating a stereo program with an enhanced width. The placement of microphones used to make an AB recording essentially imitates the placement of the loudspeakers for 2-channel stereo monitoring (Zielinsky, 2016). The great advantage of using this system is in the ability to use high-quality omni-directional microphones. Omni-directional microphones are known for their natural sound quality, lack of proximity effect, and extended bass-frequency response. The spacing of the microphones provides an exciting spacious stereo image but must be done with care. Poor microphone placement can cause sound sources to inadvertently jet across the stereo image or lead to a severe loss of channel correlation thus creating a hole in the middle of the stereo image. Despite these challenges, AB recording is a very popular technique among professional classical music producers (see Figure 3.10).

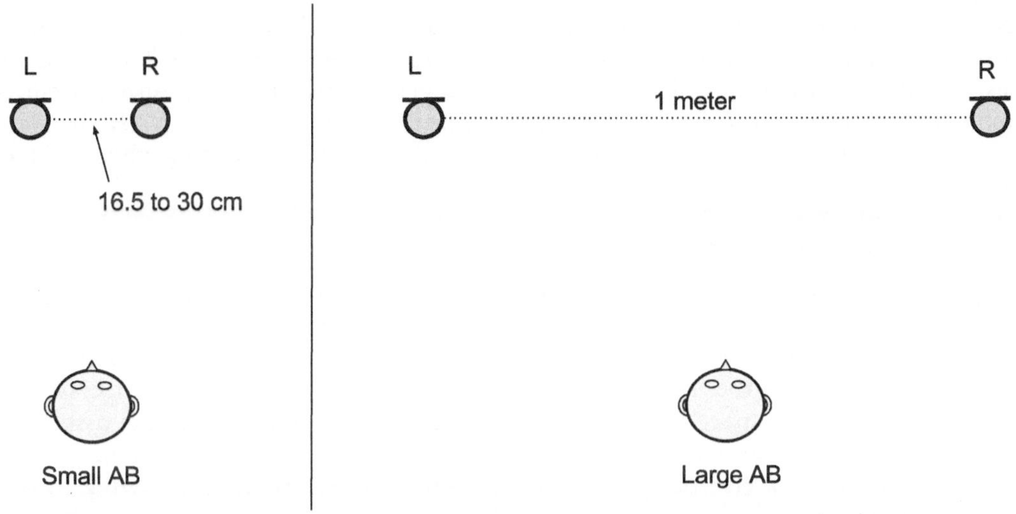

Figure 3.10 Spaced pair systems. Small AB (left) relies on head-related time of arrival cues with little level difference being captured between the microphones. Large AB is spaced farther (1–2 meters) thus enhancing both ICTDs and ICLDS for a more spacious effect.

Near-Coincident Recording

Amplitude and time-of-arrival recording techniques can be combined using a pair of near-coincident directional microphones. Taking a mixed ICLD and ICTD approach, a stereo image with stable localization and a good sense of depth and spaciousness can be achieved while maintaining excellent mono compatibility.

O.R.T.F

The O.R.T.F. stereo microphone technique was developed by French broadcasters in the 1960s. This near-coincident stereo recording technique uses a pair of cardioid microphones spaced 17 cm and oriented 110 degrees apart (see Figure 3.11 left side). Engineers have relied on this system for decades to deliver a spacious yet fully mono compatible stereo image. Since 17 cm is close to the average spacing between human ears, the system delivers a very familiar and natural sense of spaciousness in headphone and loudspeaker listening. Similarly, the 110-degree orientation causes sufficient attenuation from off-axis sounds, not unlike the effect of head shadowing, delivering natural localization cues for directional sounds. When summing the left and right channels to create a mono signal, the overall level and the direct-to-diffuse sound mix heard in the stereo program is preserved with minimal spectral coloration making excellent mono compatibility a unique feature of the O.R.T.F. system.

N.O.S

The N.O.S. system (see Figure 3.11 right side) is a similar near-coincident microphone technique developed by the Dutch Broadcasting Foundation. This system consists of 2 cardioid microphones spaced 30 cm apart, with an angle of 90 degrees between them. Since 30 cm is the approximate path from ear to ear going around the head, like O.R.T.F., the system is related to the way we naturally hear and therefore delivers a natural sounding recording while maintaining excellent mono compatibility, like the O.R.T.F. system.

Acoustic Barrier Recording: Sphere and OSS

The stereo image produced by using near-coincident microphone techniques can be enhanced with the addition of an acoustic barrier placed between the microphones. Theile (1991) found that working with a solid sphere placed between omni-directional microphones delivered a natural (related to the human hearing system) inter-aural correlation necessary for a good sense of depth and space in a sound recording. An acoustic barrier, like a sphere or a disk placed in between a near-coincident pair of microphones, will cause a human head–like acoustic shadow. The barrier physically blocks high frequencies while allowing lower frequencies to diffract around it, thus creating a human head shadow-like effect. The amount of low-pass filtering depends on the angle of incidence and the size of the barrier. Recording with a binaural microphone system with the addition of pinnae can encode all directional information including height, but can introduce steep notch filtering into both left and right signals. Although less precise, recording with an acoustic barrier alone (no pinnae) can offer a natural sounding direction-dependent low-pass filtering effect without severe coloration (see Figure 3.12).

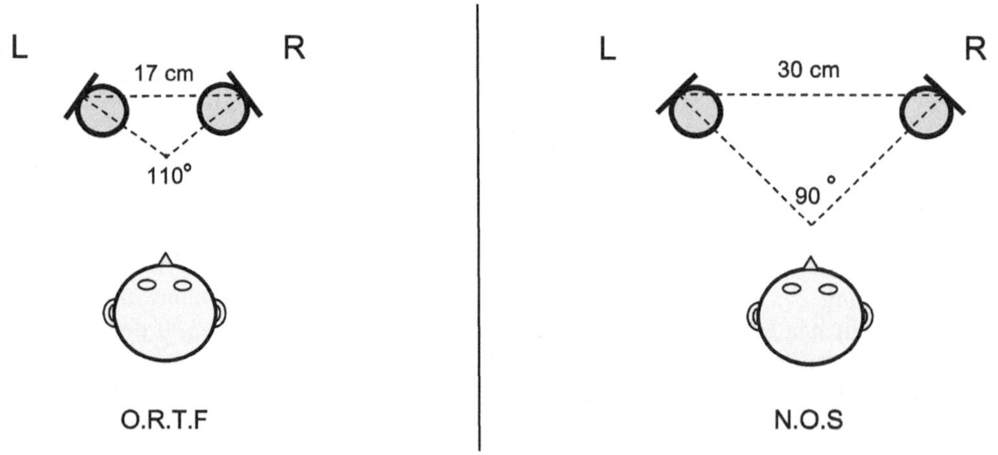

Figure 3.11 O.R.T.F. (left) and N.O.S. (right) stereo recording microphone configuration are shown in relation to the size of the average human head.

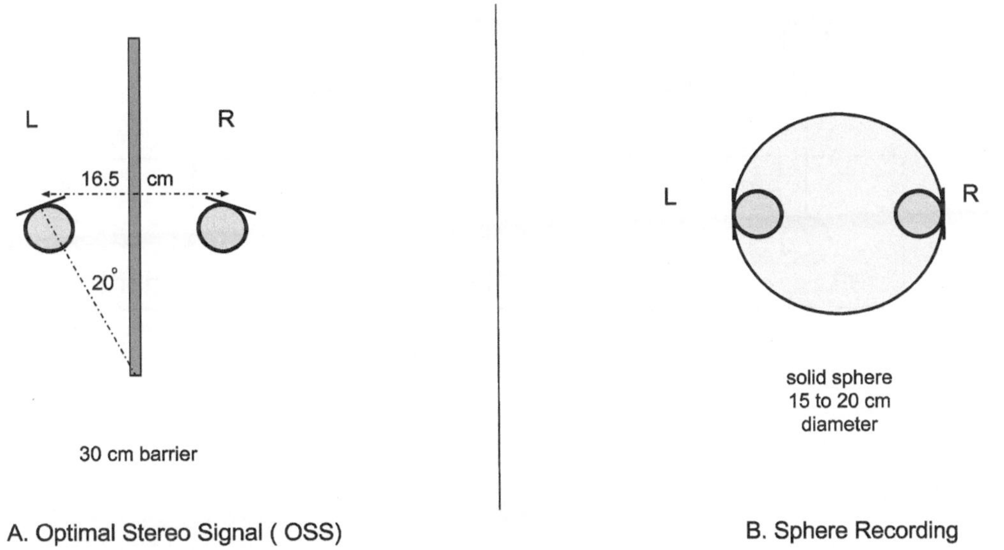

Figure 3.12 Examples of acoustic barrier-based stereo recording systems using an Optimal Stereo System (OSS) disk (left), and a solid sphere (right).

LCR Recording

With the addition of a dedicated center microphone, the phantom center image produced by a spaced microphone system can be enhanced. In addition, the spacing of the left and right microphones can be greater than a conventional 2-channel recording system thus capturing greater acoustic separation. If the intended playback system has a dedicated center channel, the center signal can be routed directly into the center loudspeaker, creating a unified wave front working together with the left and right channels (see Figure 3.13).

Decca Tree and OCT

The Decca Tree system consists of a central microphone flanked by two opposed left and right microphones. Rather than using cardioid microphones like the OCT system (a similar 3-channel recording system discussed in detail in Chapter 7), Neumann M50 or M150 microphones are used. Both microphones have an omni-pressure capsule mounted flush to a small sphere designed to enhance the directional characteristics of the microphone only in the mid and higher frequencies while maintaining an omni-directional response in the lower frequencies. For orchestral recording, the system is typically placed directly above the conductor (see Figure 3.14).

LCR Recording With a Coincident Stereo Center Channel

The center mono microphone in an LCR recording system can be substituted with a coincident stereo pair of microphones like XY, Blumlein Pair, or MS. The center stereo system can provide

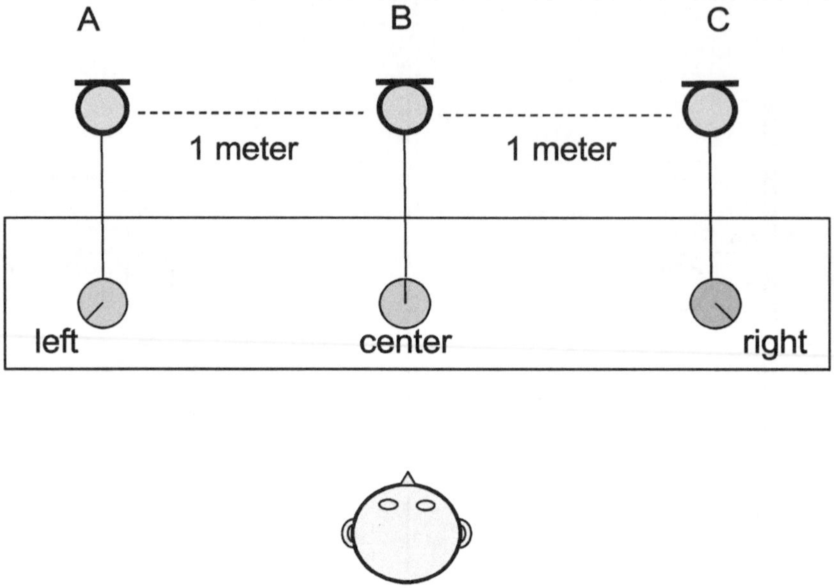

Figure 3.13 ABC recording technique. A = left channel, B = center channel, and C = right channel. The center B signal can be distributed evenly into the left and right channels for 2-channel stereo reproduction.

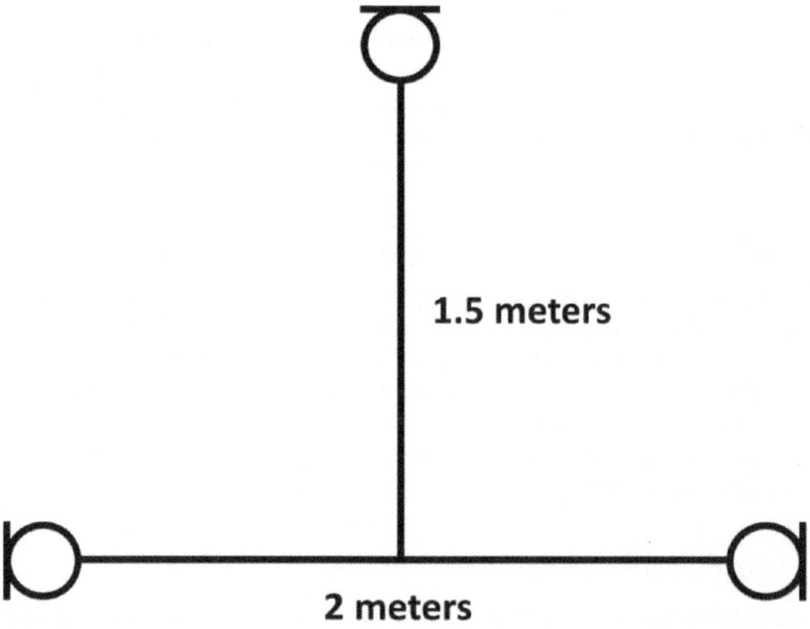

Figure 3.14 Microphone configuration for the Decca Tree recording method (Huber, 1992).

excellent control over the width of the center image in post-production. Based on the author's experience, a coincident stereo system is desirable, as opposed to a spaced pair, because it can deliver a highly focused uncolored mono signal if desired later during post-production.

Flanking Microphones (Outriggers)

An auxiliary pair of widely spaced microphones can be used to augment any mono or stereo recording system. Normally, a stereo-paired set of microphones should contain a certain balance of correlated and de-correlated sounds, but when using a pair of flanking microphones, the goal is to capture highly de-correlated sound to widen the recording area and/or to enhance the sense of spaciousness. The flanking system should always be used with a center counterpart to prevent a hole in the middle of the stereo image. Flanking microphones are able to deliver highly

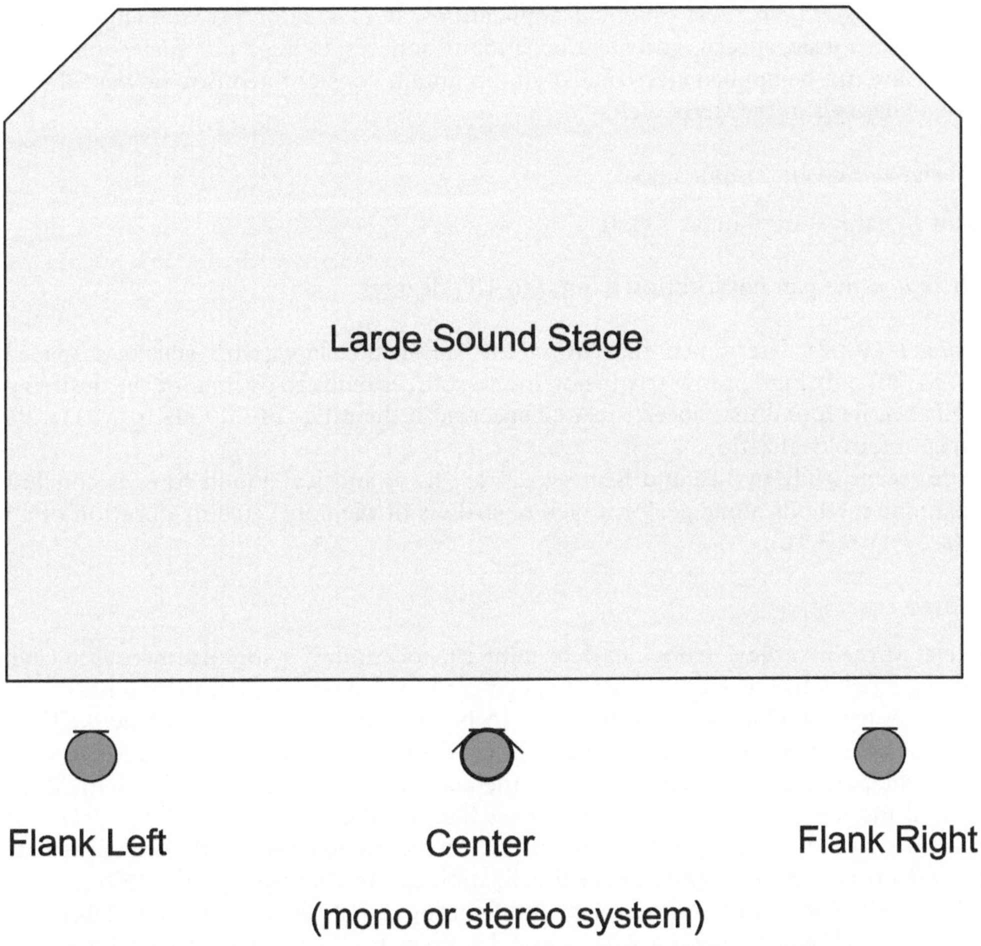

Figure 3.15 Flanking microphones used with a main center mono or stereo system to capture highly de-correlated sounds.

de-correlated low frequencies information that is difficult to obtain using closely spaced microphones. Typically, the signals from the flanking microphones are panned hard left and hard right, respectively, and attenuated about 6 dB lower than the main center system (see Figure 3.15).

Stereo Panning

Level Based Panning

A good panning system should produce a sharp phantom image, and provide a smooth and continuous illusion of fixed or moving sound sources between the loudspeakers without holes or images bouncing abruptly (Gerzon, 1992). The stereo panning effect can be achieved using channel level differences, delays, or equalization.

If the level of one side of mono 2-channel signal is attenuated, the image will start to migrate to the opposite side. For most practical applications, ICLDs alone provide enough directional information for music, speech, and most broadband sources. Using a panning potentiometer, the sine/cosine law can be applied electronically to maintain constant acoustic power when panning across two channels in the stereo field:

Left signal = $\cos(w)$*Input signal

Right signal = $\sin(w)$*Input signal

Where w = the pan pot location from 0 to 180 degrees.

Griesinger (2002) determined that using the sine/cosine law (with speakers spaced +/–45 degrees) to pan only high or low frequency sound sources tends to overshoot the desired panning angle. Further, he found that speech-related spectra, in the range of 700 Hz to 1 kHz, dominate our perception of localization.

A more recent study by Lee and Rumsey (2013) using musical sound sources concluded that ICLD panning methods alone perform well regardless of the note pitch or duration of a musical source (see Figure 3.16).

Delay Based Panning

The arrival of the first wave front can determine the location of a sound source (see Chapter 2). By applying a delay to one side of phantom centered dual-mono signal, the phantom image will migrate away from the side that is delayed. To be effective, the delay must be well below the threshold of echo detection, otherwise the delayed occurrence will be perceived as a discrete sound and the panning effect will be lost. In the author's experience, working with short delays from .2 to 2 ms is effective. By using a low-pass filter on the delayed signal, the delay range can be extended without an audible doubling effect. As mentioned before, the transient nature and the spectral properties of sounds affect localization accuracy, thus have an effect on panning methods as well. For musical sources, Lee and Rumsey (2013) determined that delay based panning performs well when working with musical sources for all but higher sustained pitches (see Figure 3.17). The drawback of using a delay to pan a signal within a stereo program is that the mono version may have unwanted audible comb filtering effects.

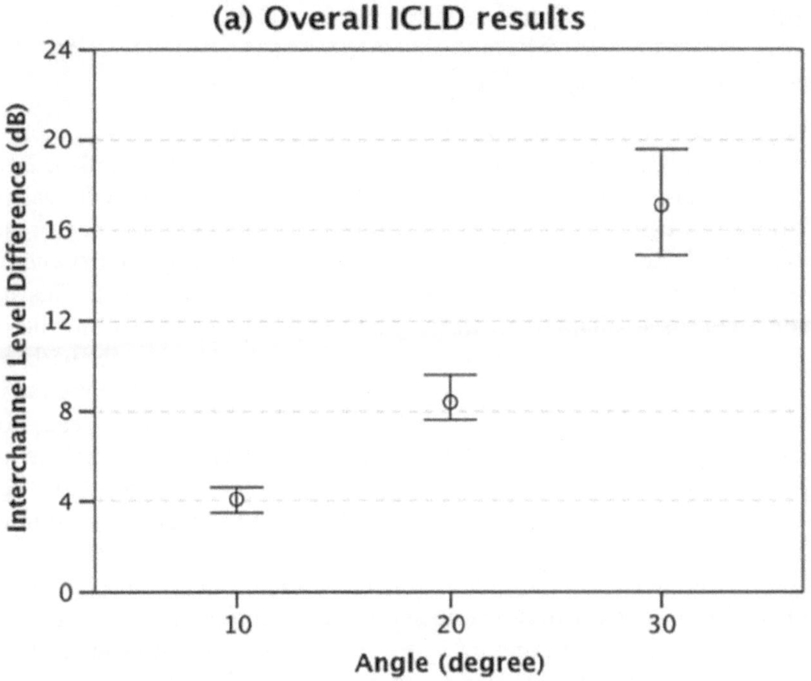

Figure 3.16 Lee and Rumsey (2013) ICLD study for speakers +/−30 degrees using musical sources.

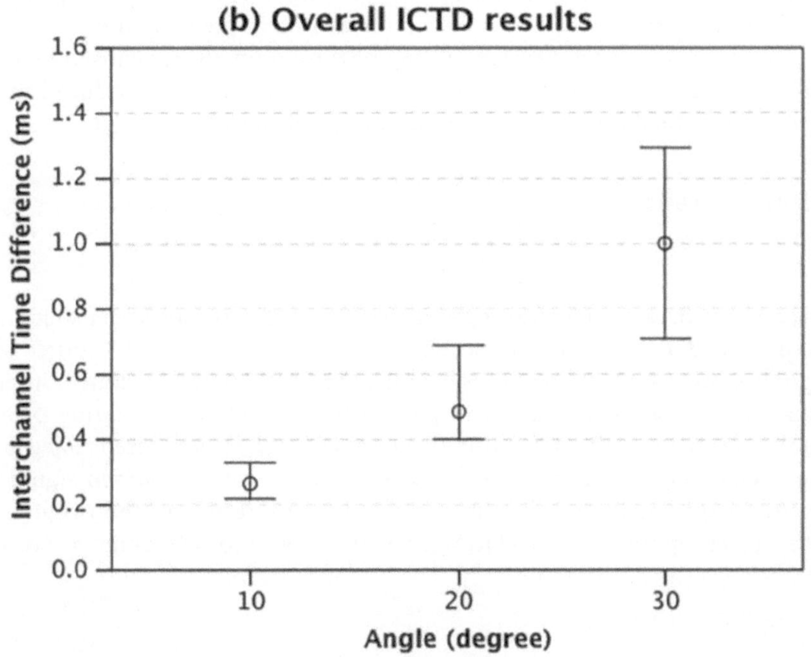

Figure 3.17 Lee and Rumsey (2013) ICTD study for speakers +/−30 degrees using musical sources.

Combined Level and Delay Panning Methods

As mentioned earlier, using a combination of ICLDs and ICTDs can create a more natural localization effect than using ICLDs or ICTDs alone (Theile, 1991; Lee & Rumsey, 2013). In natural listening environments, our ears rarely receive identical acoustic signals whereas when monitoring stereo programs through headphones or loudspeakers, it is a common occurrence. As a sound source moves closer to one side of our head in the natural listening environment, the opposite ear receives a delayed, attenuated, and colored copy of the acoustic signal. This signal is known as the cross-talk signal (see Chapter 5). In stereo headphone reproduction, the acoustic cross-talk signal can be simulated by pairing a signal in the opposite channel with an appropriate signal delay and a low-pass filter applied to a copy. Delays up to .5 milliseconds best simulate natural acoustic cross-talk (Nacach, 2014). Unlike monitoring through headphones—through loudspeakers, a listener will experience a double cross-talk signal when introducing head-related panning delays (electronically generated and real) since the acoustic cross-talk signal still exists from the opposite loudspeaker at each ear. Even so, through loudspeakers, combining level and delay panning methods creates a spacious and natural sounding panning effect.

A combined level and delay panning method can use longer delay times as well. Haas (1941, 1951) determined that a 5- to 35-millisecond signal delay applied to one of the separated loudspeakers changes the perceived panning location of a sound source to the direction of the first-arriving (precedent) sound. As much as 10 dB of gain is required to balance the perceived loudness of the later arriving sound, through a stereo pair of loudspeakers. When a longer delay is applied in combination with some or even no make-up gain in the delayed channel, the results include increased spaciousness, broadening of source image, and eventually splitting of the image and the emergence of echo perception. A low-pass filter can be implemented on the delayed channel to reduce the perception of an echo or a doubling of the sound source caused by panning delays. In addition, phase inversion, reverberation, and pitch-shift can be introduced into the delayed channel for effect to further enhance the spatial quality of the stereo image.

Stereo Enhancement

Pseudo Stereo

Mono signals can be processed to create a pseudo stereo spatial impression using middle-side processing techniques (Faller, 2005). In this scenario, the unprocessed mono source signal becomes the middle signal of the pseudo stereo middle-side pair. Processing the mono source signal generates an artificial side signal. The side signal path serves as the de-correlating processor as indicated in Figure 3.18 below. The final pseudo stereo signal is generated using a conventional middle-side to 2-channel (MS to XY) stereo decoder. If the pseudo stereo signal is summed to mono, the pseudo side signal cancels out completely thus restoring the original mono signal. Three methods of stereophonic decorrelation by using a pseudo side signal are discussed below.

Split Equalization Effect

A filter or equalizer can be used effectively for decorrelation to create a pseudo stereo image (Janovsky, 1948). For example, using a low-pass filter will create a stereo frequency split with

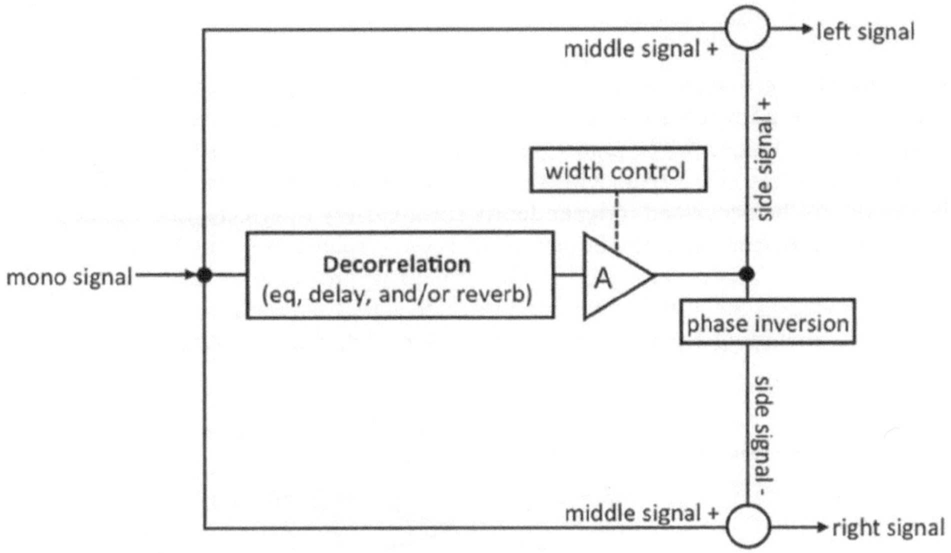

Figure 3.18 Block diagram of a mono to pseudo stereo processor.

a low-passed version of the mono input signal in the left channel and a high-passed version of the mono input signal in the right channel. A frequency splitting effect can be accomplished using virtually any type of spectral processor including a multi-band, parametric, or graphic equalizer.

Split Comb Filter Effect

Using a similar approach, a short delay can be inserted into the pseudo side signal path to create a stereo split comb filter effect. The first split occurs at Fc = 1 / 2*delay time (sec) and continues upward at *n**Fc intervals where *n* is a whole number; 1, 2, 3 . . . etc. This technique applies a comb filter to the left signal with a perfectly inverted version of the comb filter applied to the right signal—thus splitting the spectral content of the input signal at regular intervals between the left and right stereo channels. If the resulting left and right channels are summed to create a mono version of the pseudo stereo program, the comb filter effect will be completely cancelled out thereby restoring the original mono signal.

Delay and Reverberation

Similarly, a pseudo side signal can be created using longer delays, for example in the 20–50 msec range, and/or reverberation as well. Using these methods, a spacious stereo effect is created. As with all MS-based pseudo stereo processing effects, the mono version of the program (when left and right channels are combined) will cancel out the spacious effect and appear much dryer than the stereo version.

Stereo Width Enhancement

MS Processing

Once a 2-channel stereo signal is converted to a middle-side pair (XY to MS conversion), the stereo width can be adjusted to a certain degree by altering the balance between the middle and side signals (see Figure 3.19). Boosting the side signal will cause the stereo image to widen whereas attenuating the side signal will cause the stereo image to narrow. Similarly, the middle and side signals can be processed independently for a variety of effects, and then converted back into 2-channel stereo. For example, boosting the high frequencies of the middle signal alone will emphasize the centered program material when the program is converted back to 2-channel stereo. Similarly, boosting the low frequency in the side signal will enhance the sense of space. Virtually any effect, including equalization, compression, delay, and reverberation, can be applied

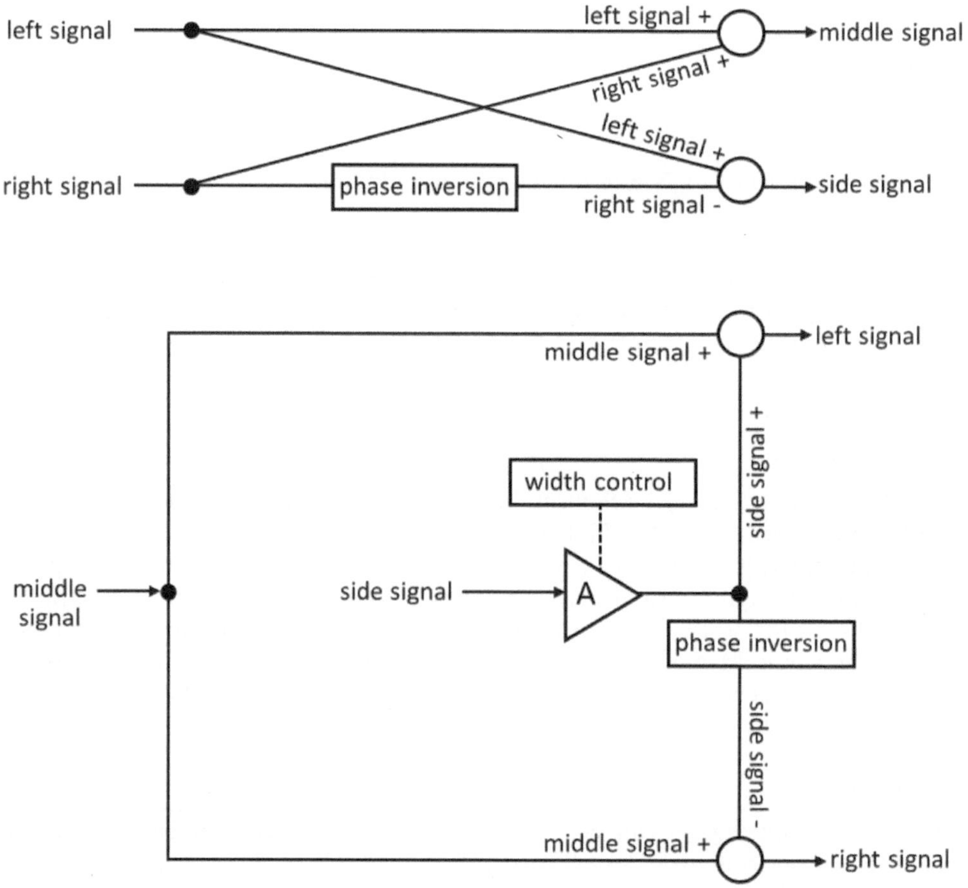

Figure 3.19 A 2-channel stereo to middle-side encoder (XY to MS converter) (top). A middle-side to 2-channel stereo decoder with stereo width control (MS to XY converter) (bottom).

discretely to the middle and/or side signals to create a dynamic or static stereo effect. The effectiveness of middle-side processing is highly program dependent. For example, if a stereo program is dominated by dual-mono signals, it will have an anemic side signal.

Delay Based Effects and Reverberation

To increase the sense of envelopment, reverberation and delays have been used to enhance mono and stereo recordings since the 1950s when engineers began to experiment regularly with reverberation chambers and tape delays. Often, these time-based effects introduced subtle pitch variations because unstable tape speeds and reverberation chambers can effect spectral balance and the perception of pitch. For example, using a dedicated reverberation and/or delay channel, dry signals can be panned independently and opposite of the effect to create a natural stereo width sensation (see Figure. 3.20). Phase inversion can be applied to one side of the stereo processed effects as well to enhance the size perceived space. Griesinger (1985) observed that some phase differences in 2-channel stereo signals are necessary for a good sense of spaciousness.

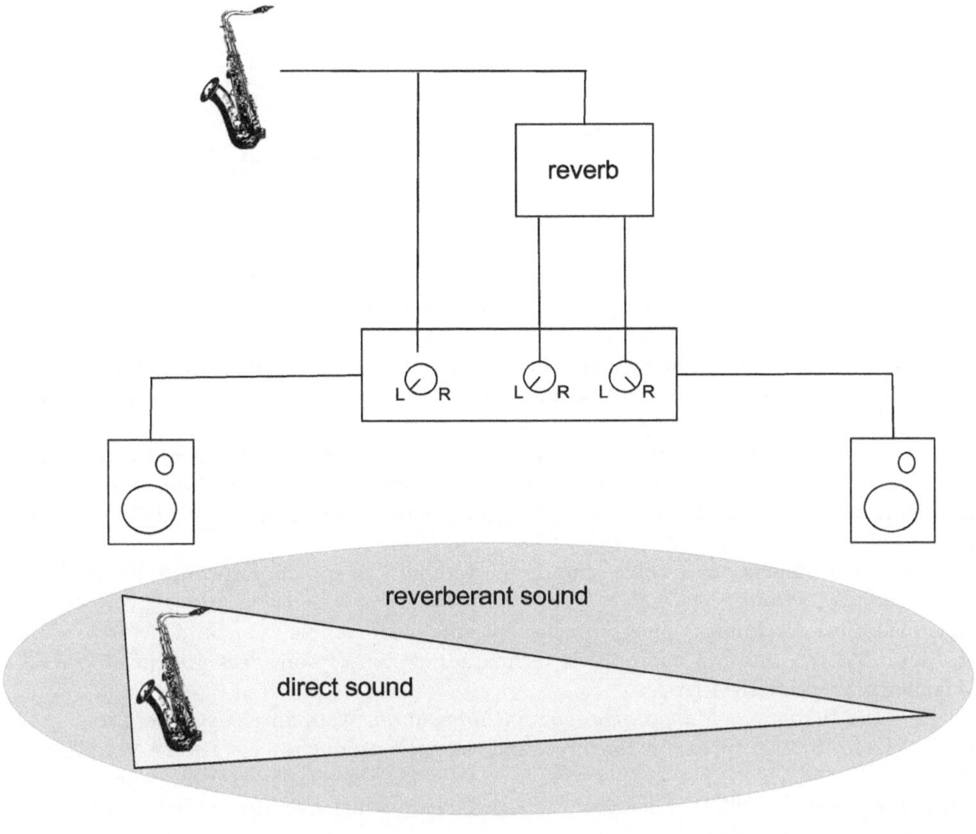

Figure 3.20 Reverberation applied to a hard-panned mono source to create a stereo effect.

Summary

As discussed in this chapter, immersive effects can be created and perceived in stereo. They can be captured acoustically by recording a combination of directional, distance, spectral, phase, and environmental information—or be created electronically using stereo or middle-side processing techniques in combination with spectral, delay, phase, and reverberation processing. In any case, spatial and directional stereo effects are based on the listener's binaural perception of sound, and the illusion of virtual sources and space created with loudspeakers placed in front of the listener, not by physically surrounding the listener with real acoustic sources. Alternatively, when multiple loudspeakers surround the listener (see Chapters 6, 7, 8, and 9), a physical immersion in acoustic signals can exist. Even then, due to technical limitations of the number of loudspeakers available, immersive effects must still rely on psychoacoustic principles to perceive them. Therefore, stereo principles can be applied to more complex multichannel systems if we consider that each pair of loudspeakers can function as a stereo sub-system within the larger configuration.

Note

1 The International Telecommunication Union is a body coordinating standardization for telecommunications.

References

Blauert, J. (1997). Spatial Hearing: The Psychophysics of Human Sound Localization, Rev. ed. (J. Allen, Trans.). Cambridge, MA: MIT Press.ntations," German Federal Republic.

Blumlein, A. (1933). British Patent Specification 394,325. Reprinted, *Journal of Audio Engineering Society*, *6*(2), 91.

Cooper, Duane H. (1987). Problems with Shadowless Stereo theory: Asymptotic spectral status. *Journal of Audio Engineering Society*, *35*(9), 629–642.

Dickreiter, Michael. (1989). *Tonmeister Technology*. New York: Temmer Enterprises.

Eargle, J. (1986). An analysis of some off-axis stereo localization problems. *Presented at the 79th AES Convention*, New York.

Faller, C. (2005). Pseudostereophony revisited. *Presented at 118nd AES Convention*, Barcelona.

Gerzon, M. (1992). Panpot laws for multispeaker stereo. *Presented at 92nd AES Convention*, Vienna. Preprint 3309, Audio Engineering Society.

Griesinger, D. (1985). Spaciousness and localization in listening rooms: How to make a coincident recording sound as spacious as a spaced microphone arrays. *Presented at 79th AES Convention*, New York.

Griesinger, D. (2002). Stereo and surround panning in practice. *Presented at the 112th AES Convention*, Munich.

Hass, H. (1949). The influence of a single echo on the audibility of speech. Reprint, *Journal of Audio Engineering Society*, *20*, 145–159.

Haas, H. (1951). Uber den Einfluss eines Einfachechos auf die Horsamkeit von Sprache. *Acustica*, *1*, 49–58.

Hibbing, M. (1989). XY and MS microphone techniques in comparison. *Presented at 86th AES Convention*, Hamburg. Preprint 2811 (A-5).

Huber, D. (1992). *Microphone Manual: Design and Application*. Waltham, MA: Focal Press.

Janovsky, W. H. (1948) "An apparatus for three dimensional reproduction ..." Patent No. 973570 (cited in Blauert 1997).

Lee, H.-K., & Rumsey, F. (2004). Elicitation and grading of subjective attributes of 2-channel phantom images. Presented at 116th AES Convention.

Lee, H.-K., & Rumsey, F. (2013). Level and time panning of phantom images for musical sources. *Journal of Audio Engineering Society*, *61*(12).

Moylan, W. (2007). *Understanding and Crafting the Mix: The Art of Recording*. Burlington: Focal Press.

Nacach, S. (2014). The Duplex Panner: Comparative testing and applications of an enhanced stereo panning technique for headphone reproduced commercial music. *Presented at the 137th AES Convention*, Los Angeles.

Snow, W. (1952). Basic principles of stereophonic sound. *Journal of Society of Motion Pictures and Television Engineers*, *61*, 567–589.

Snow, W. (1953). Basic principles of stereo sound. *Society of Motion Pictures and Television Engineers*. Reprint, *Journal of SMPTE*, *61*, 567–587.

Theile, G. (1991). On the naturalness of two-channel stereo sound. *Journal of Audio Engineering Society*, *39*(10), 761–767.

Zielinsky, G. (2016). *Personal communication with the author*.

Chapter 4

Binaural Audio Through Headphones

Agnieszka Roginska

Headphones provide us the most direct way of delivering audio content. In properly positioned traditional headphones, this means the left channel to the left ear, and the right channel to the right ear. That's it. And because we only have two ears, this method of delivering the audio should (in theory) be able to reproduce an aural experience that is virtually identical to the sound we experience in the natural environment. Yet our experiences of listening to sound through headphones range from disappointing to ones depicting absolute realism. This chapter focuses on the many acoustic and psychoacoustic factors, theories, and processing methods used to create immersive audio over headphones, as well as their impact and contribution to the resulting listening experience.

Binaural sound refers to the two-channel sound that enters a listener's left and right ears. Although many could argue that all stereo sound is binaural, the term *binaural* is reserved for sound where the two-channel sound entering a listener's ears has been filtered by a combination of time, intensity, and spectral cues intended to mimic human localization cues. These cues could be superimposed naturally (e.g. binaural recordings), or through signal processing.

As was discussed in Chapter 1, Head-Related Transfer Functions (HRTFs) are superimposing binaural cues on the sound before it reaches the eardrum. Before reaching the listener's ears, the acoustic waves emitted by a source are filtered by the interaction with the listener's head, torso, and the pinnae. This results in a directionally dependent spectral coloration of the sound. This systematic "distortion" of a sound's spectral composition acts as a unique fingerprint defining the location of a source. The auditory system uses this mapping between spectral coloration and physical location to disambiguate the points found on a cone of confusion, thus enabling accurate localization of a sound source.

The perception of binaural sound relies on Interaural Time Difference (ITD) and Interaural Intensity Difference[1] (IID) cues. Together, the ITD and IID are the foundation of Lord Rayleigh's Duplex theory (1907). In addition to the interaural cues, spectral information and variations as a function of location provide invaluable information to the listener about the position of a sound source. The composite of the ITD, IID and the spectral coloration characteristics are captured in Head-Related Transfer Functions (HRTF). The HRTF, Figure 4.2, is the frequency-domain representation of the Head-Related Impulse Response (HRIR), Figure 4.1. Even though HRTFs are very rich in acoustic information, perceptual research shows that the auditory system is selective in the acoustic information that it uses in making judgments of the originating direction of a sound source (Wenzel, 1992).

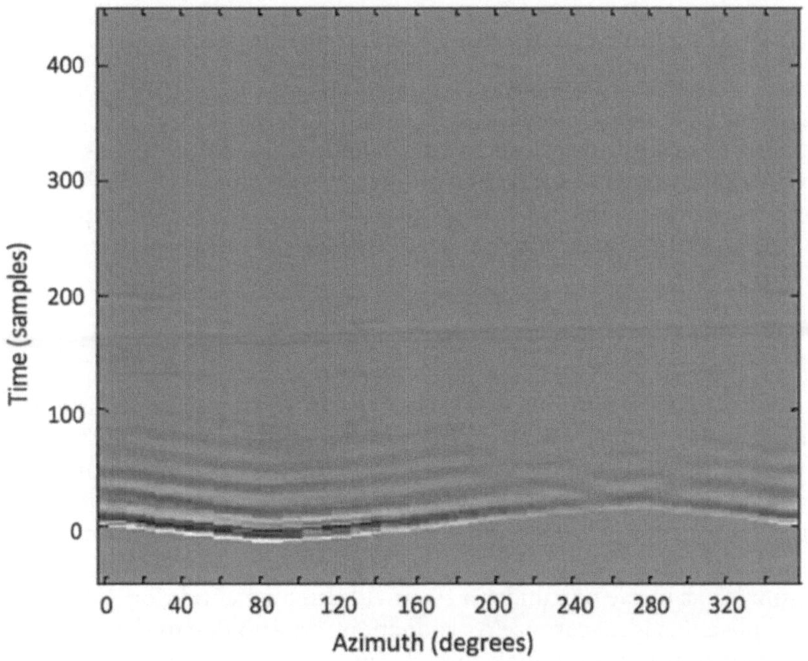

Figure 4.1 Head-Related Impulse Responses of KEMAR, right ear, 0° elevation ring, sampled at 44.1 kHz.

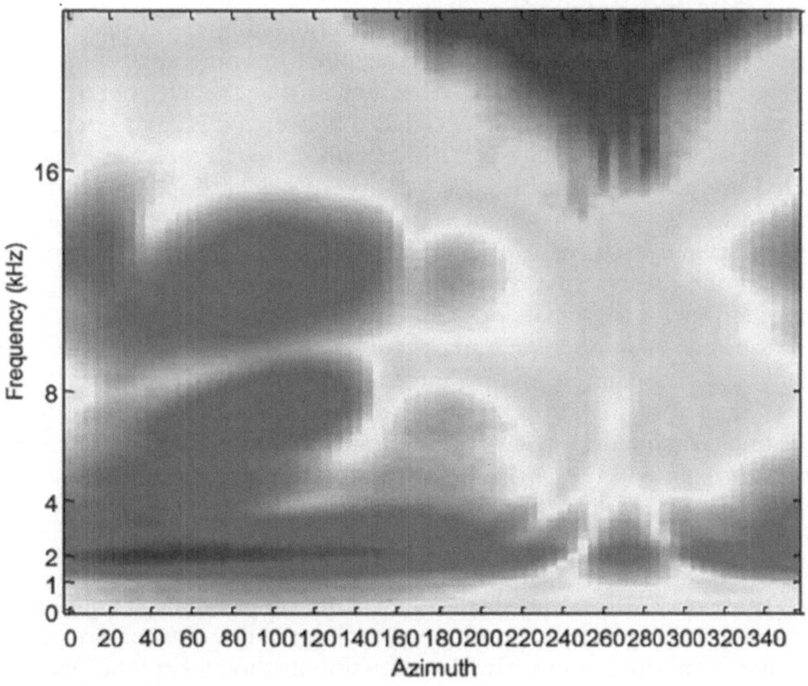

Figure 4.2 Head-Related Transfers Function of the Knowles Electronic Manikin for Acoustic Research (KEMAR), right ear, 0° elevation ring.

Due to physical differences between individuals, HRTFs vary greatly in both general shapes and detail (Middlebrooks, 1999; Moller, Sorensen & Hammershoi, 1995; Shaw, 1982). As a result, serious perceptual distortions can occur while listening using HRTFs that were either synthesized or measured on another individual (Fisher & Freedman, 1968; Wenzel, 1992). Nevertheless, research shows that some individuals' experience equaled, even improved (Wightman & Kistler, 1989), localization accuracy with non-individualized HRTFs—particularly when HRTFs of a "good localizer" are used.

Headphone Reproduction

The ultimate goal of reproducing binaural signals is to re-create an acoustic signal at the eardrums of the listener that would be equivalent to the signal that a listener would hear under natural listening conditions. One way to accomplish this is to record a sound and reproduce it at the same location. For example, a signal recorded at the entrance to the ear canal in the left and right ears that is later reproduced at the entrance to the ear canal would create an equivalent signal reaching the eardrum as was created by the real source and thus result in a listening experience that would be similar as to the one under a free-field listening condition. Although it may not be feasible to accomplish this ideal capture/reproduction scenario due to the transducers' limitation, the fact remains that transmitting signals directly into a listener's ears can be the most direct and effective way of delivering high-quality spatial audio signals. Headphones are the most convenient and preferred reproduction method to accomplish this goal and, with proper calibration and equalization, can result in a very convincing spatial auditory image.

There are many advantages to headphone reproduction of binaural signals. The most significant advantage is that headphones provide a controlled listening environment. The controlled environment results from two main factors. The first is that the left channel signal is delivered directly to the left ear, and the right channel signal is delivered directly to the right ear. Thus, the signals that are *intended* to reach each ear are in fact *presented* to each ear. Headphone reproduction is unaffected by such factors as listener location or orientation. Unless the head is tracked and additional processing is used to compensate for listener placement, the signal presented will be independent of listener location or orientation, and will result in the listener always being in the *sweet spot*. Another benefit that comes with the direct delivery of the signals to the ears is the prevention of any crosstalk signals reaching the unintended ear. Crosstalk is a common phenomenon in loudspeaker reproduction, and can lead to a complete collapse of a spatial auditory image. (This will be discussed in detail in Chapter 5.)

The second significant factor affecting signal control in headphone reproduction is acoustic isolation from ambient sounds. Unless highly controlled, any listening environment contains background noise, which will be superimposed on to the reproduced binaural signals, and can cause audible interference resulting in a distorted auditory image. The leakage of environmental sounds is diminished when using headphones that are intended to isolate the listener from the acoustic environment, such as closed headphones (discussed below).

Headphones, however, do not offer the ultimate solution to binaural audio reproduction and there are several disadvantages that can negatively affect the listening experience. As pointed out above, headphones are a listener-centering reproduction method where the listener can always be in the sweet spot. This is primarily caused by the fact that signals to the ears do not change as a listener moves their head. Although this can be considered an advantage, it is also a disadvantage because it results in some loss of interactivity with the environment. When using headphones, the entire acoustic environment moves with the listener as they move their head, thus the signal is no longer affected by the position of the listener and can result in an unnatural listening experience.

A common experience of headphone listening is that all sounds tend to be perceived as coming from inside the head. Termed as "inside-the-head locatedness" (or IHL), it is a phenomenon usually attributed to headphone listening, although it has been documented with loudspeaker reproduction as well. IHL (as discussed in detail below) is sometimes an undesirable result that can lead to fatigue especially during long exposure to headphones listening.

Lastly, just as acoustic isolation can have a positive impact by providing listeners with a controlled listening environment, extreme acoustic isolation can also have a negative effect. Some types of headphones (e.g. open, semi-open) alleviate this problem by being acoustically transparent and passing environmental sounds directly to the listener.

Mono

In the free-field listening environment, a monophonic signal refers to a single point of transmission of a sound. When the listener is facing a single loudspeaker, the signals propagating towards the left and right ears are identical, and are perfectly correlated. The perception of a monophonic auditory image under headphone listening conditions occurs in much the same way—when the signal sent to the left ear is identical to the one at the right ear. The Inter-Aural Cross-Correlation (IACC) is the measure of the similarity of the signals at the two ears. When the IACC is 1, the two signals are perfectly correlated and are identical. The result is an image that appears to be coming from the center of the head. When the cross-correlation is –1, the signals are identical but opposite in phase. As the cross-correlation index approaches 0, the two signals become more dissimilar—this is often associated as a measure of the quantification of signals that appear to have a higher degree of spaciousness and envelopment. Thus, signals reaching the left and right ears that are highly correlated will result in an image that is perceived to be less spacious and, when played over headphones, more internalized.

Stereo

One of the most commonly used tools by the audio engineer to position sounds in the stereophonic sound field is the pan pot. Traditional stereo mixing methods use the –3 dB sine-cosine equal-power amplitude panning law to pan sources. By adjusting the left and right output gains, a phantom image can be created between two speakers in loudspeaker reproduction, or along the axis between the ears in headphone reproduction (so-called lateralization).

Although most mixing engineers generally use a loudspeaker environment for mixing, the end product is often listened to over headphones. The two modes of reproduction, however, are categorically different, and the principal difference between the two is crosstalk. When a listener hears a stereo signal over headphones, there is no crosstalk and the left and right channels are completely isolated from one another—referred to as a "biphonic" signal. Thus, the listening experience intended by an audio engineer mixing under loudspeaker conditions will be significantly different due to the lack of crosstalk. As a result, sound sources are lateralized, and there's typically an energy depression in the center of the sound stage, when compared to loudspeaker reproduction of the same recording.

Several studies were performed comparing the sound quality, spatial attributes, and user preferences of the original biphonic signal to headphone listening enhancement algorithms, and how re-introducing crosstalk affects the listening experience. In the two experiments presented by Manor et al. (2015) listeners compared original biphonic signals of popular music to those processed using *near-field crosstalk simulation* where the simulated crosstalk is that of loudspeakers located near the listener. The near-field crosstalk introduces time and level differences at low frequencies that are consistent with cues a listener would experience when sources are very close to their head. Subjective test results based on overall preference and Ensemble Stage Width (ESW) suggest that some listeners may prefer the stereo image with near-field crosstalk simulation added, in comparison to the original biphonic signal.

Other algorithms of far-field crosstalk simulations have been investigated by Lorho (2005), where stereo enhancement algorithms and systems were subjectively rated. Results showed that, almost unanimously, subjects preferred the original (unprocessed) version when compared to the processed signal. In other words, the stereo enhancement did not improve the listening experience over headphones.

Binaural

Binaural signals consist of two channels. However, in contrast to *stereo* reproduction over headphones, binaural signals contain embedded spatial cues in the form of time, intensity and spectral coloration—cues that mimic and enhance natural human localization. The target reproduction system for binaural signals is headphones, or loudspeakers equipped with a crosstalk cancellation system (as described in Chapter 5). When correctly captured, synthesized, and reproduced, binaural signals create a powerful impression of spatial sounds as they appear in the natural listening environment.

Virtual Surround Sound/Virtual Multichannel

In addition to headphone reproduction of binaural signals containing spatial cues, headphones with extended capabilities can create a spatial listening environment. The extended capabilities can be divided into two categories—acoustic or hardware enhancement, and extension through signal processing.

Virtual loudspeakers over headphones aim at emulating the experience of listening to one or more real loudspeakers in a real listening environment. Imagine that you are sitting in a room with a stereo loudspeaker system located in that room. The room acoustics, the placement of the loudspeakers, and where you are located in relationship to them will affect the stereo image that you perceive. The sound arriving at the listener's eardrums is a combination of the sound emanating from the speakers, propagating through and interacting with the room, and reaching the listener's ears. In effect, each speaker can be thought of as a source. Capturing the combined characteristic of the loudspeaker, room acoustics and listener's HRTFs will result in the Binaural Room Impulse Response (BRIR). Superimposing BRIRs onto the signal to be played from the loudspeaker through the process of convolution will result in a *virtual* loudspeaker in a *virtual* room reproduced over headphones. This is depicted in Figure 4.3, where the BRIRs representing the location of the loudspeaker (b_L and b_R) are used to process a dry sound source $x(t)$. The resulting signals at the listener's ears and the perceived image of the *virtual* loudspeaker will be equivalent to the *real* loudspeaker in the room.

Reproducing more than one virtual loudspeaker via headphones will generate a virtual phantom image representative of the phantom image that would be perceived from the real loudspeakers in the room. Figure 4.4a depicts two loudspeakers with a phantom image perceived in the center. Two virtual loudspeakers presented over headphones will result in the same phantom image being perceived (Figure 4.4b).

Implementation

Virtual surround sound over headphones is based on the theory described in the section above. Figure 4.6 shows an example configuration of a 5-channel surround sound system (Figure 4.6a), and its virtual surround system representation (Figure 4.6b). The signals to be output by the loudspeakers are processed where each speaker is treated as an individual sound source. The fundamental way to create the headphone representation of surround sound system is to process each channel's audio signal by the impulse response corresponding to the location of the loudspeaker through which that channel would be played. An example of a 5-channel virtual surround system diagram is represented in Figure 4.5. In this figure the five channels (Center, Left, Right, Left

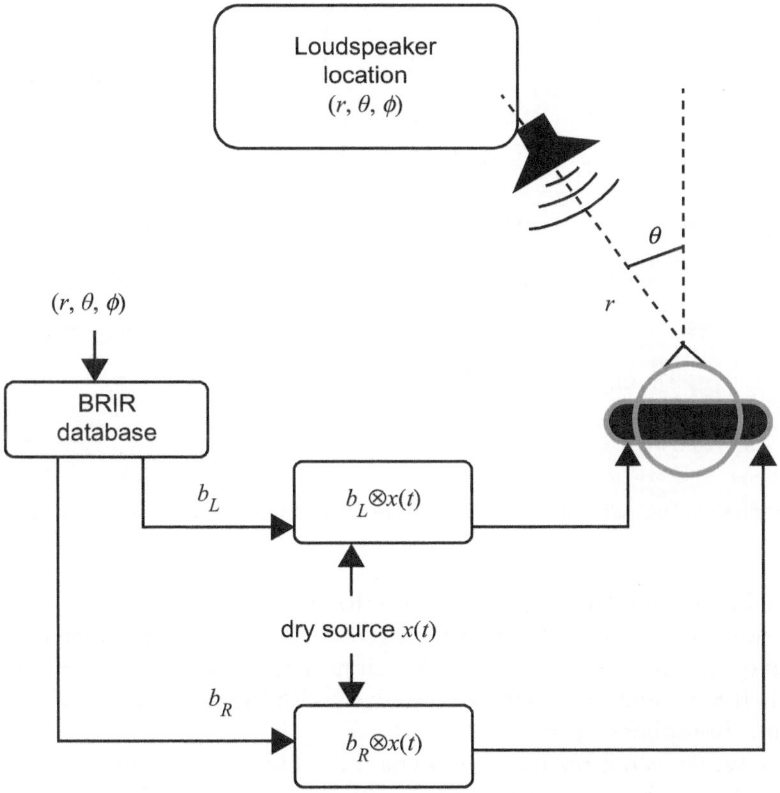

Figure 4.3 Rendering of virtual loudspeaker source. BRIR filters b_L and b_R are convolved with dry source *x(t)*.

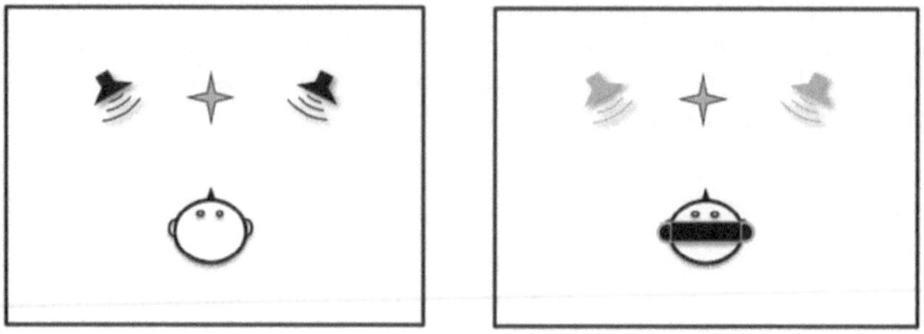

(a) (b)

Figure 4.4 Listener perceiving a phantom image between two loudspeakers (a), and the listener perceiving the same phantom image from two virtual loudspeakers reproduced over headphones (b).

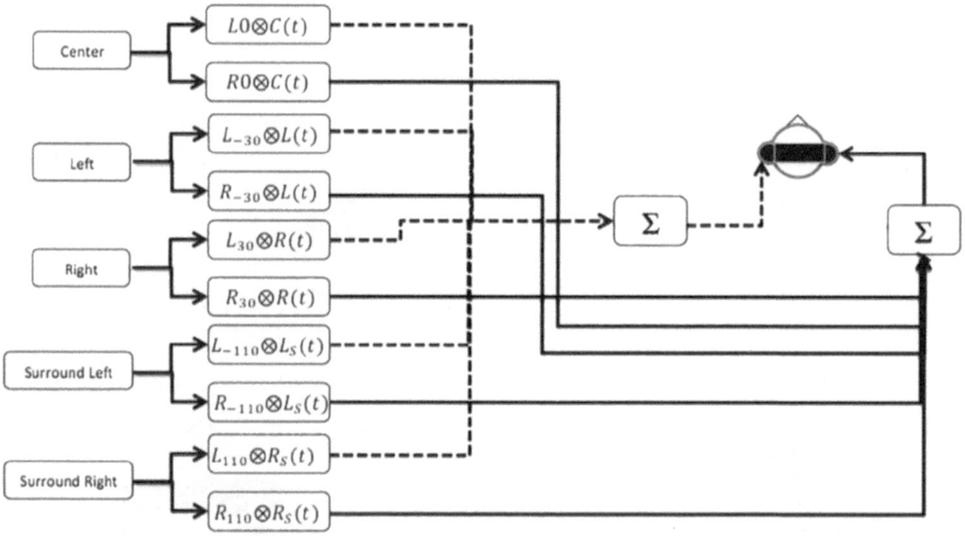

Figure 4.5 Five-channel virtual surround sound implementation diagram.

Surround, and Right Surround) are convolved with the left and right HRIRs corresponding to the loudspeaker location. For example, to create the virtual Left Surround channel, the Left Surround track will be convolved with the HRIR at –110° azimuth and sent to the left channel headphone output, and the Left Surround track will be convolved with the *right* –110° azimuth IR and sent to the right channel headphone output.

This concept, however, is not restricted to 5 channels and can be applied to creating a virtual rendering of 7.1, 10.2, 22.2, or any other configuration of loudspeakers.

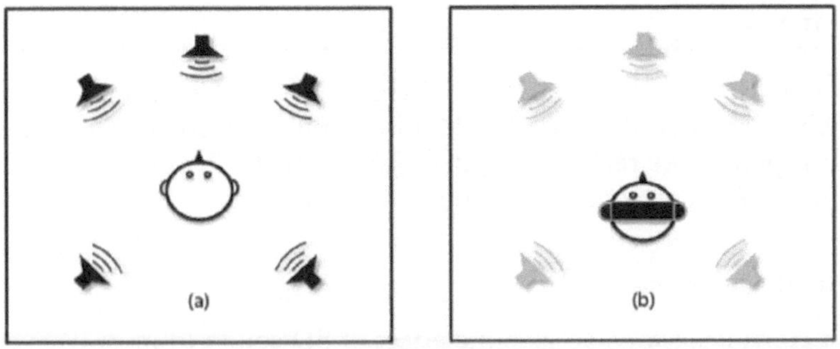

Figure 4.6 Representation of a real surround sound configuration with 5 physical speakers (a); and a virtual surround configuration with 5 virtual speakers presented over headphones (b).

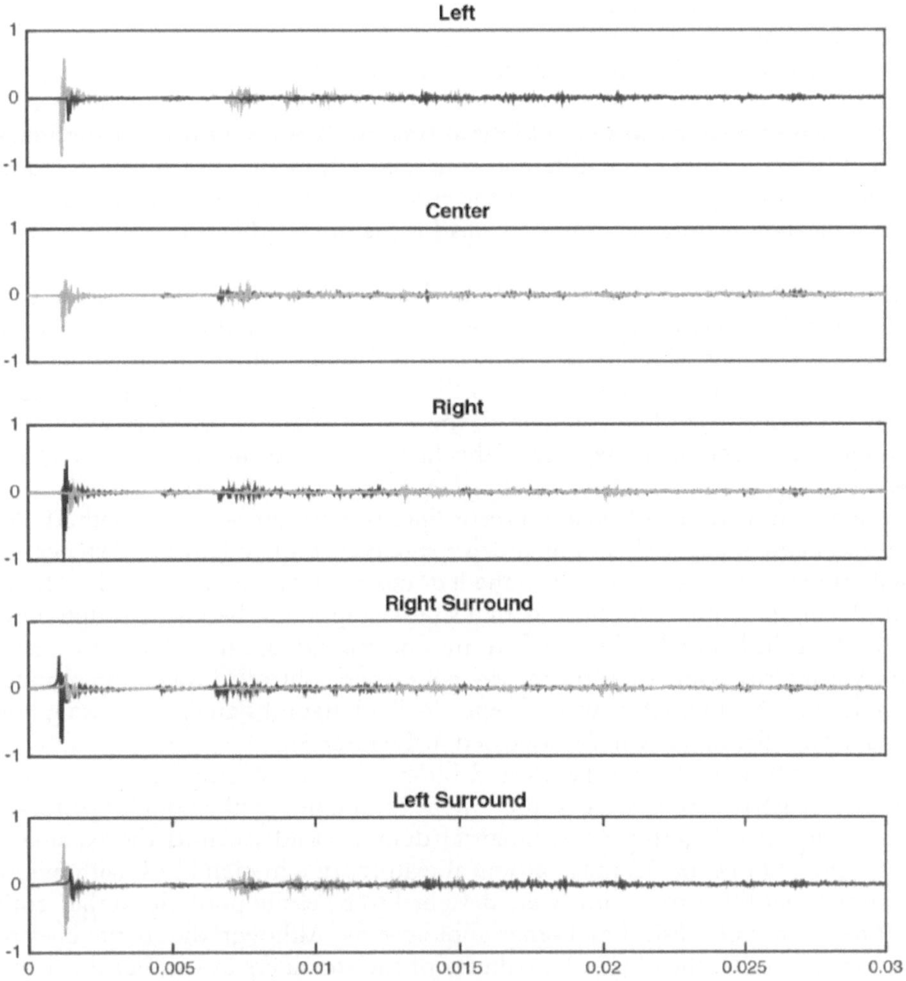

Figure 4.7 Impulse responses of the first 3 msec of the left, center, right, right surround, and left surround channels of a virtual surround sound technology.

Examples of impulse responses used in a 5-channel virtual surround system are presented in Figure 4.7, where the first 3 msec of a 16 msec impulse response are shown. The impulse responses contain spatial cues representing the location of the five surround speakers, as well as the binaural room impulse response representing a small room.

Binaural Sound Capture

One of the most convincing ways to give the effect of 3D sound to most listeners is through the use of dummy heads and binaural recordings. A dummy head is a physical representation of the human head, with anatomical features similar to an adult head, including head size and shape, location of ears, pinnae, and, in some cases, shoulders and/or torso. The role of the dummy head is to capture sound as it appears at the two ears of a listener. From that perspective, dummy heads function as binaural microphones.

The first documented demonstration of a binaural recording using a head-based system was at the 1933 Chicago World's Fair, where the AT&T dummy head "Oscar" was presented (see Figure 4.8a). The mechanical dummy contained microphones at the location of the two ears. The sounds at the two ears were transmitted in real time to listeners wearing headphones. Dummy heads available today are not much different from this early model, and are equipped with microphones at the location of the ears, either at the entrance or somewhere along the ear canal. One of the most important features of a dummy head is the pinnae that are located on either side of the head.

These are representative of the pinnae that may be found on human heads and, in some dummy head models, can be interchanged. Localization cues are superimposed on the recorded sound signal by having the pinnae color the sound before being captured by the microphones. The result is a 2-channel recording that represents the signals that would be captured or heard by a listener. This 2-channel recording contains all the spatial auditory cues that would be present if a human listener had been there. Because of this fact, these recordings are referred to as *binaural recordings*.

An important consideration of binaural recordings is their reproduction method. Because binaural recordings contain spatial cues intended for playback directly into a listener's ear canals, these signals are meant to be presented to the left and right ears independently. The most direct method of playing binaural recordings is by using headphones, because headphones allow the direct playback of the left and right signals to the corresponding ear with the greatest amount of isolation between the ears and thus, the most accurate and controlled way of presenting the audio signals to a listener. An alternative way of reproducing binaural signals is by using loudspeakers with crosstalk cancellation (as will be discussed in Chapter 5).

There are a number of dummy heads and binaural sound capture devices available today. These may be categorized into 3 types: dummy head, dummy with shoulder or torso, and binaural microphones (see Figure 4.8). A binaural dummy head (such as the Neumann KU-100 depicted in Figure 4.8b) contains the anatomical features of a human head, with microphones at the ears, but without the torso. These were designed to be easily portable so they could be used in binaural recordings of musical and other audio events. Although the spatial cues of the head and pinnae are captured, the physical attributes of the shoulders are absent due to the missing torso. This shortcoming may reduce the strength of elevation cues and thus limit the extent of spatial capture of sound.

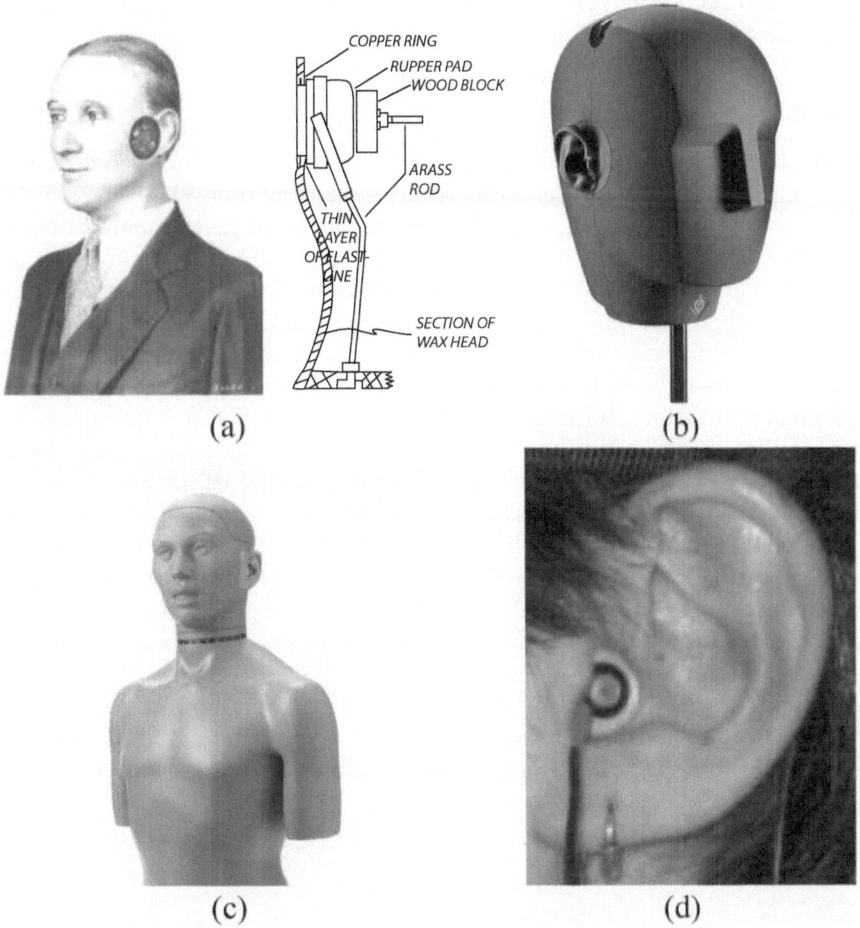

Figure 4.8 Methods used for binaural recordings, the (a) AT&T dummy head "Oscar" (Hammer & Snow, 1932), (b) Neumann KU-100, (c) KEMAR, and (d) a binaural microphone in a listener's ears.

Complete binaural cues that contain the head and torso have been used mainly for acoustic research and measurements, and include such mannequins as the Knowles Electronic Manikin for Acoustic Research (KEMAR), Figure 4.8c. The KEMAR has been extensively used for measurements of HRTFs, many of which are available for public use and research. These include measurements performed at the Massachusetts Institute of Technology (Gardner & Martin, 1994), near-field HRTFs from the South China University of Technology (Xie et al., 2013), and reverberant Binaural-Room Impulse Responses (BRIRs) from TU Berlin (Wierstorf et al., 2011). Lastly, binaural microphones inserted into a listener's ears provide the highest level of customized spatial cues. Binaural microphones consisting of small capsules can be placed at the entrance to the ear canal and kept in place using, for example, a foam ring (Figure 4.8d). This is referred to as the *blocked meatus* method. The signal entering the ear canal is captured by the microphones and stored as a 2-channel recording. The blocked meatus method is also often used during measurements of personalized HRTFs.

Interactive Binaural Capture

Although many consider binaural recordings to be the most convincing methods of capturing 3D sound and spatial cues, one of their constraints is that the image captured represents a fixed perspective. In other words, once the recording is captured, the listener cannot rotate their head and interact with the environment. Motion-Tracked Binaural (MTB) is a method introduced by Algazi, Duda and Thompson (2004) to capture and reproduce binaural sound while allowing the listener to move their head in order to get a listener-centric sound image. Based on the concept of dummy head sound capture, MTB uses microphones distributed over a spherical-shaped surface which approximates a listener's head. An array of microphones is distributed along the horizontal plane of this surface, where the signals are captured. Capturing sounds at points along the entirety of the horizontal plane gets us away from capturing a fixed-point listener perspective. Rather, it captures the perspective of the listener irrespective of the orientation of their head. Thus, during sound reproduction, the listener's perspective in the sound scene can be adjusted by using head-tracking information of the listener's head. Once the orientation of the listener is obtained from the head tracker, the signal from the nearest microphone pair representative of the listener's ear locations along the sphere is selected and played back. If the orientation of the listener and their ears falls between two microphones, the signals can be interpolated between the two adjacent microphones.

A method based on a similar concept to MTB, where an array of microphones is positioned on the horizontal plane, is shown in Figure 4.9. The Omni Binaural Microphone[2] is a headless system that uses four pairs of pinnae arranged along a square configuration. Each side is

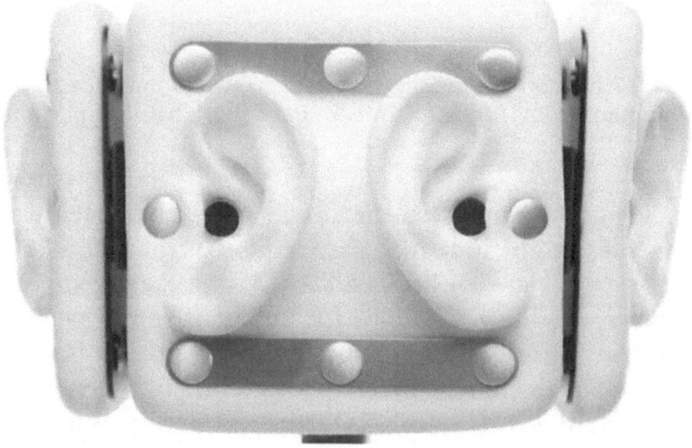

Figure 4.9 An omnidirectional binaural microphone array capture method utilizing multiple pinnae.

equipped with a left and right pinna allowing the capture of binaural recordings from four different perspectives—north, east, south, and west (or 0°, 90°, 180°, 270°). Thus, this method aims at capturing a binaural recording where the head is pointing four directions concurrently and, similarly to MTB, where the perspective can be adjusted during sound reproduction based on the orientation of the listener.

HRTF Measurement

The most reliable and accurate method to acquire HRTFs is through acoustic measurement. Acoustic measurements involve the use of binaural microphones inserted into a subject's ears. A test signal is played from a loudspeaker positioned at an azimuth and elevation location relative to the subject for which the HRTF (or the HRIR time domain equivalent) will be measured. The signal is recorded at the left and right ears and, from this, the HRIR is extracted.

In general, the relationship between the sound source and the signal reaching a listener's ears can be represented by:

$$Y_L(\theta,\phi,d,\omega) = H_L(\theta,\phi,d,\omega)X(\omega)$$

and

$$Y_R(\theta,\phi,d,\omega) = H_R(\theta,\phi,d,\omega)X(\omega)$$

where,
θ = *azimuth*
Φ = *elevation*
d = *distance*
ω = *angular frequency*
Y_L, Y_R = *spectra of acoustic signals at listener's ears*
H_L, H_R = *HRTF*
X = *spectrum of sound source*

The extraction of the HRTF is performed by cross-correlating the input signal with the resulting output. The system response is defined by:

$$H_L(\theta,\phi,d,\omega) = Y_L / X(\omega)$$

and

$$H_R(\theta,\phi,d,\omega) = Y_R / X(\omega),$$

Where the process of localizing a sound source can thus be described as the extraction of (θ, Φ, d) based on the information contained in Y_L (θ, Φ, d, ω) and Y_R (θ, Φ, d, ω).

A number of HRTF measurement systems have been implemented. Some have used a fixed subject, rotating speaker setup (e.g. CIPIC, ISVR), while others use a rotating subject, fixed speaker configuration (e.g. IRCAM, MARL). Since it is the directionally dependent filter that is being measured between the *relative* location of the source (loudspeaker) and listener, both types of systems can achieve the same result with careful calibration and considerations about the measurement equipment and space. Although some HRTF measurements are taken at multiple distances from the subject, HRTFs are usually measured at a distance of at least 1 m, where the sound source is said to be in the far field.[3] The behavior of a sound source, and HRTFs, in the near field is significantly different (for a discussion about the near field, see the Appendix).

When measuring HRTFs on a human subject, binaural microphones are inserted into the left and right ears. There are currently two principal microphone methods used—blocked meatus, and probe tube. The blocked meatus method uses miniature microphones placed at the entrance to the ear canal. A foam ring is used to block the ear canal, create a seal, and maintain the microphone in place (e.g. Figure 4.8d). Because of the fixed location of the microphone, the blocked meatus method is more repeatable and can result in a better signal to noise ratio (SNR).[4] Alternatively, a probe tube can be inserted into the ear canal, connected to a miniature microphone on the outside of the ear. This open ear canal configuration measures the HRTF inside the ear canal, thus it includes the ear canal characteristic, which some argue contain individual features.

HRTFs aim to represent the relevant directionally dependent spatial cues without any room acoustics or information about the space. Therefore, HRTFs are typically measured in an anechoic environment. However, they can be measured in a non-anechoic environment provided that the room is large enough so that reflections do not interfere with the direct response of the HRTF and that reflections are removed from the measurement (e.g. Algazi et al., 2001a).

The density of measurements varies greatly, with intervals anywhere between 2°–30° in azimuth, and 4°–30° in elevation. However, typical measurements have a density of 5°–15° in azimuth and elevation. Careful consideration must be given to the placement of the subject and loudspeaker(s), particularly for the denser measurements where a small positioning error can result in severe spatial distortion. To ensure correct positioning of the subject's head, a tracking device can be used to adjust the subject's position or to correct for errors.

All reproduction and signal capture equipment have a spectral characteristic that can impact the HRTF measurement. Therefore, microphones, speakers, pre-amplifiers, A/D, D/As, and any other equipment used must be carefully calibrated.

The signal measured is thus represented by:

$$Y(\omega) = X(\omega)S(\omega)M(\omega)H(\omega)$$

Where $Y(\omega)$ is the recorded signal, $X(\omega)$ is the test signal, $S(\omega)$ is the transfer function of the loudspeaker and amplifier, $M(\omega)$ is the transfer function of the microphone and pre-amplifier, and $H(\omega)$ is the HRTF. To capture the pure HRTF, we must eliminate all system characteristics, except for the HRTF. In other words, the desired HRTF $H(\omega)$ is represented by:

$$H(\omega) = \frac{Y(\omega)}{X(\omega)S(\omega)M(\omega)}$$

HRTF Format

Until recently, there has been no standardized format to store measured HRTFs. These have traditionally been stored in the. wav or. mat formats, or other proprietary formats. Information about the measurement setup, locations measured, sampling rate, and other pertinent information, was common to all the datasets in a database and could be hard coded. As the distribution of HRTFs is becoming more universal and their use more frequent in diverse applications, a common format is necessary to facilitate the exchange of HRTF datasets between institutions. In the recent past, several formats have been proposed (e.g. Andreopoulou & Roginska, 2011; Schwarz & Wright, 2000). In 2015, the Audio Engineering Society standardized the Spatially Oriented Format for Acoustics (SOFA) as AES69–2015. SOFA is a format designed to represent spatial data including HRTFs, BRIRs, HpIRs, and more complex datasets such as microphone array responses, directional room impulse responses, and others. The format supports flexibility to describe the measured data regardless of the conditions under which it was measured, allows an arbitrary geometry of measured locations, and contains specifications of object parameters that are pertinent to the measurement (receiver, listener, emitter, source, and room).

Binaural Synthesis

Binaural synthesis can be used to simulate a binaural signal that represents a sound source, as it would appear at a location around a listener. As discussed above, a left and right ear pair of HRTFs contain the spatial cues necessary for a listener to perceive a sound source at that location. When we superimpose the cues contained in these HRTFs onto a sound signal, we create a representation of the left and right ear signals that would appear at the entrance to the listener's ears, thus creating the illusion that the sound appears to originate at the location represented by the HRTFs.

Binaural—Static Source

Binaural synthesis is accomplished by taking a monophonic signal of a sound source, and convolving it with the filters that contain the left and right ear HRIRs representative of the location at which the binaural sound source is to be perceived, as is illustrated in Figure 4.10. For example, to synthesize a virtual sound source at 30° azimuth, 0° elevation, a monophonic sound *S* must be convolved with the left ear HRIR representative of 30° azimuth, 0° elevation to create the left channel signal; and the monophonic sound *S* must be convolved with the right ear HRIR representative of 30° azimuth, 0° elevation to create the right channel signal.

The HRIRs used for binaural synthesis can be generic, individually measured, selected from a database, and/or customized. Since many HRIRs are measured in an anechoic environment, artificial reverberation may be added to the synthesized virtual sound source in order to create a room effect, and minimize IHL.

Binaural With Motion

The natural world is a constantly changing environment with respect to both visual and auditory events. The relative position of sound sources to the listener can be modified in one of two

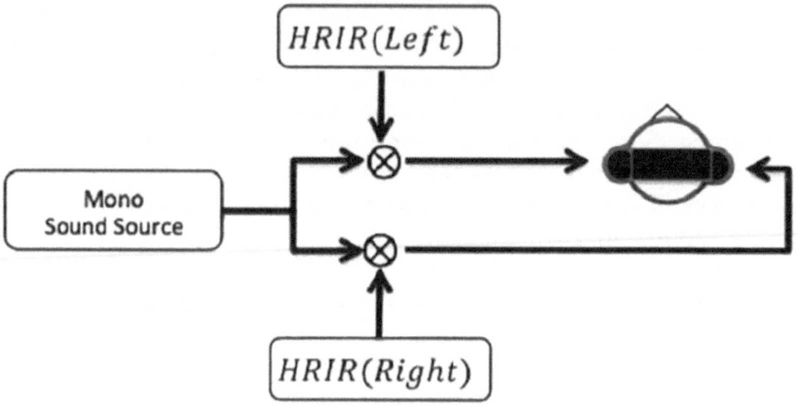

Figure 4.10 Binaural synthesis: convolution of a monophonic sound source with the left and right HRIRs.

ways: 1) due to the change in location of the source, 2) or due to a shift in the listener's location or orientation. Human listeners take advantage of the interactivity with sound sources to more accurately localize them.

A listener can acquire much information about the location of a source by moving their head. Such head motions are used for two reasons. First, head rotation may be used as a pointing device. A human listener could adjust the head orientation until the source appears to be located in front, where the localization accuracy is greatest. Second, head movement plays a dominant role in resolving confusions for sources located on the cone of confusion. To demonstrate this phenomenon, let us consider the case of a stationary source located at 180° in azimuth, as shown in Figure 4.11. This situation may create ambiguity as to whether the sound source is located in the back at 180° or in front at 0° due to the ITD and ILD being at 0 for both locations (Figure 4.11a). If the sound source is located in back, a head rotation to the left will cause the sound to arrive earlier and with a greater intensity at the left ear (Figure 4.11b). If, however, the sound is located in front, the same head rotation to the left will cause the sound to arrive earlier and with a greater intensity at the right ear (Figure 4.11c).

Motion in a Virtual Auditory Environment

In a virtual auditory environment, motion can be a result of a combination of one of two elements in binaural synthesis: listener motion and sound source motion. By rotating their head, a listener can change its orientation in yaw, pitch, and roll (Figure 4.12). The change in head orientation will affect the relative position of the sound source in relation to the listener's head. For example, a change in yaw will result in a change in the relative azimuth of the perceived sound source. A listener can change their physical location within a space by displacing their body and moving to another x, y, z location. A change in listener location not only affects the relative position of the sound source to the listener, but will also impact the relationship between the listener and the

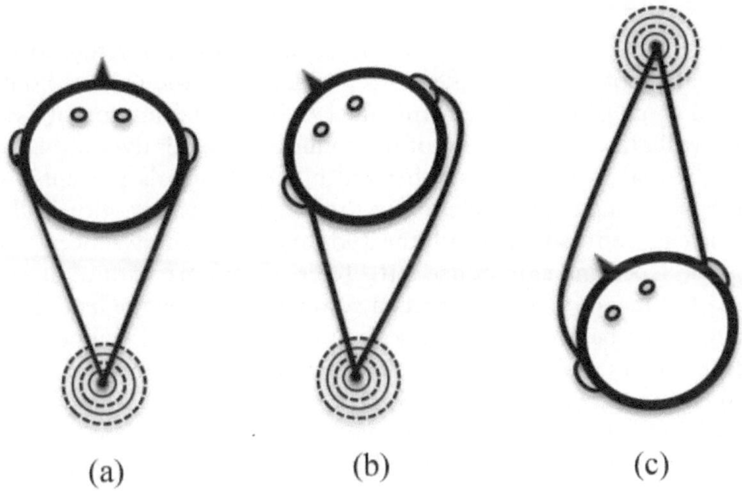

Figure 4.11 Head movements used to disambiguate the location of sources on the cone of confusion. Refer to text for explanation.

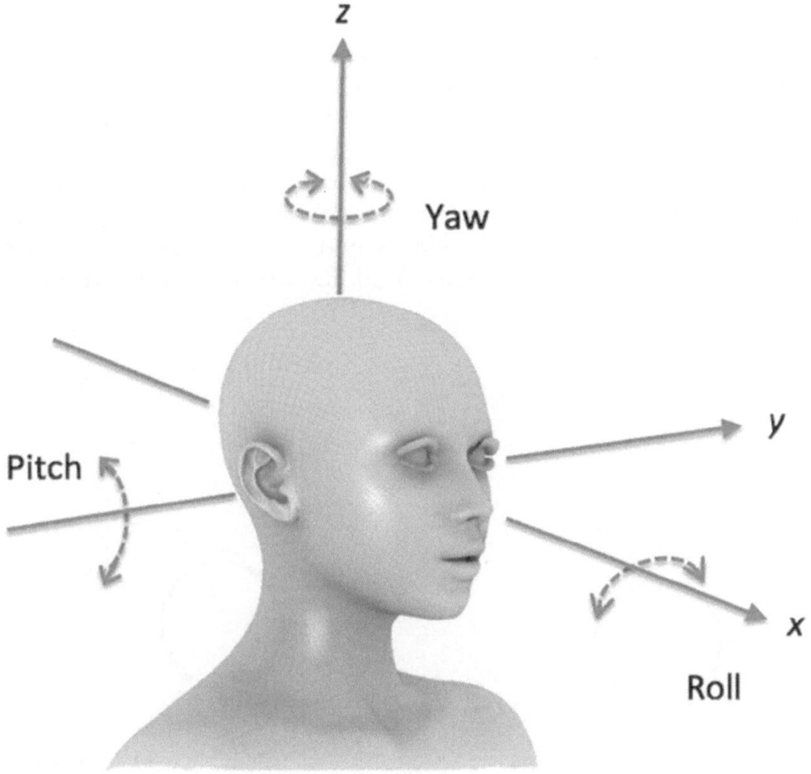

Figure 4.12 Three-dimensional head rotation: yaw, pitch, and roll.

room. This results in a reconfiguration of the direct sound and early reflections with respect to the listener. The location of the early reflections in relation to the listener contributes to source localization within a room (Rakerd & Hartmann, 1985). A combination of listener location and orientation change will affect the location of the sound source relative to the listener.

Without any additional processing, synthesized binaural signals presented over headphones will result in a static listening experience, where the listener's head motion will not be taken into account and as a listener moves their head, the auditory scene follows along. This leads to a less natural listening experience and can induce IHL. To create a more natural listening environment, and a better sense of immersion and presence, a compensation for the listener's head orientation and location change must be accounted for so that a virtual sound source will remain firmly in place when a listener moves their head—as it does in the natural listening environment. To create a dynamic virtual environment, the relative location of the virtual sound source to the listener must change according to their position (Figure 4.13).

In a virtual auditory environment, knowledge about the listener's location and orientation can be used to compensate for any location change. To do this, the position of the head must be tracked, and the location of the virtual sound source must be updated according to the position of the listener, and the desired location of the source.

Dynamic Listener

There are many techniques and technologies available to track the position of a listener. Head-trackers measure the position of a listener relative to a reference point. Three degrees-of-freedom (3DOF) trackers measure the orientation of the head (yaw, pitch, and roll) only. While six degrees-of-freedom (6DOF) trackers measure the orientation as well as the x, y, z location of the listener. Current trackers can be divided into four main tracking technologies. Inertial sensors use accelerometers and gyroscopes to extract positional information. Accelerometers are used to measure linear acceleration by doing a double integration—integrating the accelerometer information to find the velocity and integrating again to find the position relative to an initial point.

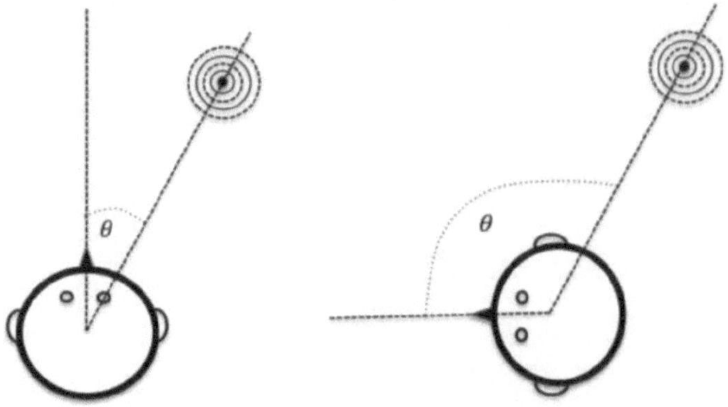

Figure 4.13 Change of relative source location as a listener changes their head orientation.

Inertial sensors are accessible and very common in mobile devices and other personal devices because of their low cost, high update rate, and low latency, but they tend to experience significant drift, which must be accounted for and compensated to maintain position accuracy.

Magnetic trackers rely on using a fixed emitter to create a magnetic field. A magnetic sensor (or receiver) is mounted on the top of a listener's head and senses the changes in the magnetic field that are caused by the location and movement of the listener. It determines the intensity and angle of the magnetic field, and thus the location and orientation of the sensor in relation to the emitter. Magnetic trackers have a very high accuracy but are sensitive to external electromagnetic disturbances, and typically have a short range of operation.

Acoustic trackers function by analyzing the time it takes for an acoustic signal to reach a receiver. Typically, multiple transmitters emit ultrasound in an environment and multiple receivers are placed on the head to track the location and orientation of the listener.

Optical trackers rely on a variety of image-based analysis methods to extract positional information from a camera. Optical tracking methods use a combination of techniques including using objects fitted with passive or active markers arranged in a known pattern, or marker-less tracking of objects that have a known geometry (e.g. a head or face) to extract the features of the head and use these for continuous tracking.

The information obtained from the tracker about the position of the listener is used to adapt the HRTFs applied to synthesize a binaural environment in real-time. In comparison to a static listening mode, an interactive system involving head tracking has a significant benefit not only in creating a more realistic and immersive listening environment, but can also improve localization accuracy and externalization, and decrease front/back and up/down confusions (Brungart et al., 2004; Begault et al., 2001).

Dynamic Source

When the listener is fixed, and a virtual sound source is moving, it is a similar situation to the listener changing position in relation to a fixed source. In both cases, the HRTFs must be continuously adapted to create the impression of motion when the source's position changes dynamically in relation to the listener. The impression of motion is a result of a smooth and frequent update in source location between the starting and ending points of a trajectory. To successfully do this, a virtual sound source must be processed at frequently updated positions with changing filters that represent the shift in spatial position, without abrupt spectral or temporal changes. To create the uniform transition across space, HRTFs must be interpolated continuously between two or more discretely measured HRTF locations.

HRTF interpolation can be used to create a trajectory, or to display a virtual source at a location for which an HRTF is not available. A binaural filter can be *estimated* from measured HRTFs that are in proximity of the target location. Several methods and techniques have been proposed and are used for interpolation of time and frequency domains (e.g. Christensen et al., 1999; Hartung, Braasch & Sterbing, 1999; Nishino et al., 1999), Principal Component Analysis (PCA), pole-zero model, and other data. Interpolation methods include interpolation between two (linear), three (triangular), or four (bi-linear) measured HRTFs, reconstruction through basis functions (such as the Spherical Thin Plate Splines, STPS, Wahba, 1981), functional representation of discrete basis functions, or multi-pole expansion (Duraiswami et al., 2004).

Inside-the-Head Locatedness

In a natural listening environment, listeners perceive most sound sources to be externalized, or appearing to be originating from *outside* their head. Sounds reproduced over headphones, however, often create the perception to be originating from *inside* the head of a listener. Inside-the-head locatedness (IHL) refers to the false impression that sound sources originate from inside the listener's head and move left and right along the interaural axis. This process is called *lateralization*. IHL is not an uncommon phenomenon. Although IHL has been reported since the late 19th century (Thompson, 1877, 1878, 1881), recent research has given it particular attention due to the advancement of binaural sound simulation and reproduction technology. An example of IHL is the sound of one's voice when the ears are blocked. This particular case leads to a very strong intracranial image perception. Even without blocked ears, the sound of one's voice leads to an inside-the-head image. IHL can also occur during sound reproduction over loudspeakers: Hanson and Kock (1957) have demonstrated IHL using two loudspeakers in an anechoic chamber driven by identical, but phase-inverted signals. Perhaps the most common example of IHL with external sources is the reproduction of sounds over headphones. Sounds reproduced over headphones are often perceived to be located inside the head.

This phenomenon is not only specific to stereo sounds reproduced over headphones; it also occurs in reproduction of 3D sound over headphones. Due to the perceived inside-the-head image, headphone reproduction is considered less natural. Externalized sound images are perceived to be more natural. In the 1970s, researchers started looking for ways to eliminate inside-the-head images by employing the dummy-head recording technique, instead of the conventional mono/stereo method. In a study performed by Plenge (1974), subjects were asked to compare recordings made with binaural technology to those made monophonically. The study showed that inside-the-head images present in the single microphone arrangement disappeared in the dummy-head recording. Furthermore, the study also showed that externalization and lateralization are not exclusive states of the auditory system. Both can occur at the same time.

There are many theories speculating the cause of IHL. These theories include: 1) an unnatural bone conduction and pressure on the head caused by the headphones (Sone, Ebata & Tadamoto, 1968), 2) the invariability of the signal under non-static head conditions, and 3) natural resonances of microphones and headphones (Blauert, 1983) as well as various theories of reproduction equipment, e.g. coupling between the ear and the headphone. The understanding of the cause of IHL was advanced with studies performed by Reichart and Haustein in 1968, as reported by Blauert (1983). The two scientists concluded that IHL occurs under two conditions: 1) when both ear signals are similar enough that the sound is fused into one auditory event, and 2) when each of the two sources must be perceived to be originating close to the ear. They also suggested that the alteration or the elimination of the acoustical effect of the pinnae contributes to this effect. In other words, a highly distorted pinna cue may lead to IHL. A binaural reproduction system may distort a spatially processed signal for two reasons. First, a headphone reproduction system of even the highest quality has its own characteristic. Second, placing headphones over a listener's ears produces an acoustic cavity that has its own transfer function. Thus, compensating for the effect of the headphone and ensuring a true free-field representation of a signal can minimize (or eliminate) IHL. There is not one theory currently that can fully explain IHL, or predict its occurrence. However, presenting a more natural sound to the listener can minimize IHL. This includes realistic reverberation, individualized HRTF cues, and an interactive environment.

Figure 4.14 depicts three modes of headphone listening—stereo, binaural, and binaural with tracking. In stereo headphone reproduction, sounds are perceived lateralized, with the auditory images appearing along a line between the two ears. When binaural cues are added, the typical perception is that of a source that moves around the listener, but frontal perception is challenging and most listeners perceive frontal sources as either coming from the back, or passing above their head. When motion tracking is combined with binaural cues, sounds tend to be perceived as originating from outside the head, around the listener.

Eliminating IHL

There are three factors contributing to externalization over headphones. The most important factor is the presence of spatial sound cues and the triggering of the perception of spaciousness, including adding reverberation (artificial or natural) (Begault, 1992; Sakamoto, Gotoh & Kimura, 1976; Toole, 1970). Secondly, the simulation of 3D space using individualized HRTFs improves the quality and accuracy of the perception of the simulated environment. Finally, lateralization occurs most often when head movement is not accounted for. In a non-interactive listening environment, the listener perceives the audio world from a frozen perspective. In this non-interactive situation, having the audio world move in synchrony with the listener's movements results in an unnatural listening situation. In the natural environment there's only one sound source that does not change any of its characteristics as a function of head rotation—one's own voice. The voice is also the only sound in the natural environment that originates from inside the head. Theories have suggested that the reason a non-interactive environment may be perceived as originating from inside the head is because it is similar in behavior to the only source that originates from inside our head under natural listening conditions. However, this is not the only explanation as many listeners experience an externalized image in static headphone listening conditions. Nevertheless, by accounting for head movements, the perception of inside-the-head images can be minimized (Griesinger, 1998).

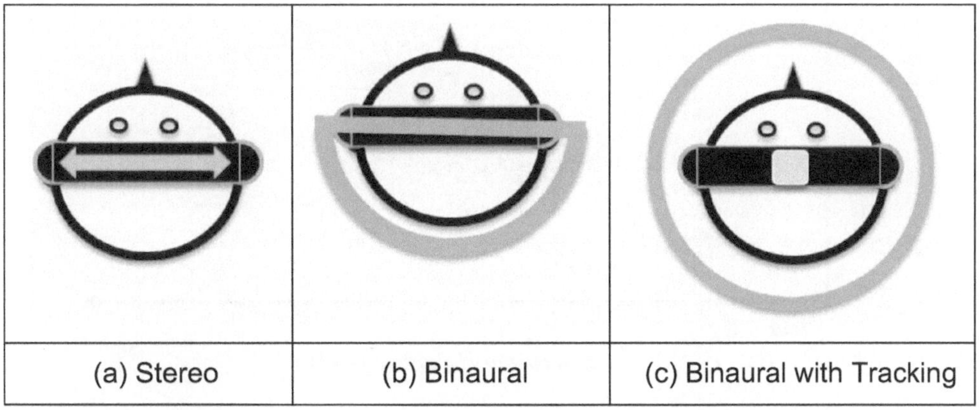

Figure 4.14 Three modes of headphone listening—(a) stereo, (b) binaural, and (c) binaural with head tracking. Image locations are represented by thick grey-color line.

Advanced HRTF Techniques

As has been discussed in Chapter 1, there exists a high degree of variability in the shape and size of people's heads, and in particular in the shape of the pinnae. Because of this, HRTFs are unique and vary significantly between listeners. Using non-individualized HRTFs in binaural reproduction can lead to an unconvincing experience due to a distorted spatial impression suffering from poor localization, an increase in front/back and up/down confusions, and decreased externalization.

Many studies have confirmed that individualized HRTFs lead to a better spatial impression and a more realistic listening environment. The improvements include better localization accuracy (Wightman & Kistler, 1989), including fewer errors along the cone of confusion, and an improved externalized image. Significant efforts are being made to improve the quality and accuracy of the spatial image by providing listeners with personalized, or close to personalized, HRTFs in order to provide a more natural listening environment over headphones.

There are a number of methods available to approximating a listener's individualized HRTFs. These methods include individualized HRTF measurement, numerical modeling using high- and low-resolution photographic-based methods, HRTF customization, perceptual selection, and using Binaural Room Impulse Responses (BRIRs).

Currently, acoustic measurement is the most accurate and reliable method of capturing the time, intensity, and spectral characteristics encompassed in HRTFs. The process of measuring individualized HRTFs can be time-consuming, requiring the listener to remain seated for an hour (or more), which may introduce errors due to noise, fatigue, listener motion, or other errors associated with human participation.[5] These errors can lead to distorted auditory images and unusable HRTFs. The HRTF process can also be costly, requiring specialized equipment and facilities. Making high-quality binaural 3D audio more available to everyone will require moving away

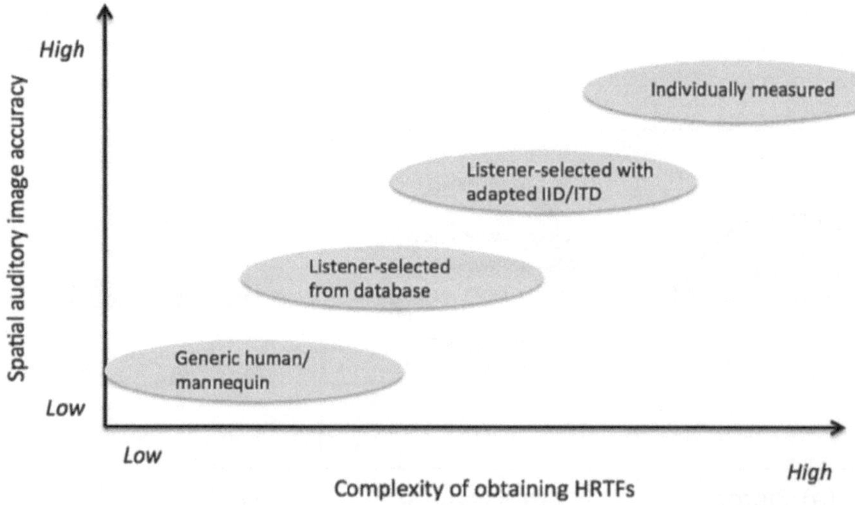

Figure 4.15 HRTF spatial auditory image quality represented along a continuum, from generic HRTFs yielding a lower degree of quality, to individually measured HRTFs resulting in higher quality of spatial impression.

from acoustically measured individualized HRTFs for every person, while somehow maintaining the high performance of individually measured HRTFs (see Figure 4.15).

There exists a tradeoff between the quality of the spatial auditory image and the complexity of obtaining HRTFs. Figure 4.15 represents this tradeoff and continuum of the spatial quality of the auditory image as a function of the method of obtaining HRTFs. Typically, HRTFs that are easily obtained and require no listener participation, like a generic HRTF measured on a mannequin, will generally result in a lower virtual source spatial image quality. A generic HRTF dataset will usually lead to a virtual sound source with poorer localization accuracy, has increased front/back confusions, and may lead to artifacts such as spectral distortions or an unnatural sound, or an image that is perceived inside the head, or any combination of these. On the other end of the spatial auditory image quality spectrum are the individually measured HRTFs. These require highly specialized equipment, facilities, and significant time, but result in an accurate spatial impression of the virtual auditory space. Between these two extremes can be found levels of customized or adapted filter sets. Adding a level of customization through either input from the listener or individual measurements will result in an improved quality image.

Evaluating HRTFs

There are two main trends in the evaluation of HRTFs. Quantitative methods aim at evaluating the match between physical parameters of the HRTF dataset and that of a target listener—such as the interaural cues and spectral fit, or look at the localization accuracy criterion, where a subject is asked to identify the apparent location of a target source in a virtual sound source condition (played over headphones) in comparison to localization under the free-field condition (Bremen, van Wanrooij & Van Opstal, 2010; Jeppesen & Moeller, 2005; Hofman & Van Opstal, 2003). Qualitative assessment based on user preference has been used as an alternate method of evaluating HRTFs as well, and can include subjective criteria such as the perception of externalization, spatial realism, front/back and up/down discrimination (e.g. Roginska, Santoro & Wakefield, 2010; Seeber & Fastl, 2003), and projected motion and trajectory (e.g. Katz & Parseihian, 2012). Clearly, both types of assessment, qualitative and quantitative, may be appropriate and useful in some cases.

In order to successfully pick an HRTF dataset when non-individualized HRTFs are used, there must be synergy between how the HRTF dataset is chosen and the end-goal of the listener. Applications where the focus of using binaural sound is to accurately localize sound sources in the virtual auditory environment (such as mission-critical applications) will require an HRTF selection process that will include localization accuracy as part of the evaluation criteria. Where the spatial image quality and overall auditory experience take precedence over other qualities like localization, the HRTF dataset selection process can focus on evaluating criteria that are directly relevant to the end goal. In the past 10 years, a number of approaches and methods have been investigated to select non-individually measured HRTFs including database matching (Zotkin et al., 2003), HRTF customization (Katz, 2001; Xu, Li & Salvendy, 2008; Faller, Barreto & Adjouadi, 2010), and HRTF modeling (Durant, Member & Wakefield, 2002; Kulkarni & Colburn, 2004; Hu, Chen & Wu, 2008).

Customized HRTFs

The complexity of measuring personalized HRTFs has made it a challenge to use high-quality binaural technologies in consumer applications such as virtual reality, gaming, and entertainment. To overcome this challenge, there have been many approaches explored to find a way to improve the spatial image quality and get as close as possible to the perception quality we associate with personalized HRTFs with reduced or eliminated personalized acoustic measurements. In this section we will explore some of the methods aimed at HRTF individualization.

Individualized HRTFs based on geometrical data aims at reconstructing, or modeling, HRTFs based on data that captures the detailed geometry of a listener's head and torso. If we consider HRTFs as a scattering problem in the free-field, there should exist a way to model HRTFs using sound simulation methods based on a 3D mesh of the head and torso. One of the methods used to accomplish this is the Boundary Element Method (BEM), e.g. Katz (2001). The main principle of BEM is to use a discrete mesh to represent the surface of a geometrical object. The mesh is acquired from a high-precision 3D laser scanner, which captures the detailed geometry of a listener's head and torso shape. The individual transfer functions are then modeled based on incoming and outgoing sound pressure waves on the surface geometry—any propagation through the head is ignored. Results of this method have been promising, especially when compared to acoustically measured HRTFs. Some argue, however, that the complexity and time needed to acquire the data from the scan is comparable to the acoustic measurements, thus may not offer a significant advantage over the traditional method.

Another approach to individualization is through HRTF decomposition. This can be done by decomposing the HRTFs into filters that represent different characteristics of the physiology of the upper body (e.g. Algazi et al., 2001b). The ILDs and ITDs are mainly affected by the physical characteristics of the head and can be modeled based on the size and shape of the head; whereas the shape and size of the pinnae and its cavities mainly affect the spectral variations. Thus, the HRTF can be reconstructed by using two independent, but complementary, filters representing the head and pinnae, respectively.

Perceptually based methods through user-selected HRTFs have been proposed by Seeber & Fastl (2003) and Roginska et al. (2010). These aim at having listeners "find the best fit" through a listening evaluation. An analogy can be made to going to a local pharmacy to try on reading glasses until you find the one that fits best versus getting a lens prescription from an optometrist. Although the prescription lenses will no doubt give a better result, the self-selected glasses will be quite good, and better than generic reading glasses chosen without experimentation. In Roginska et al. (2010) and subsequent studies, listeners were given a selected pool of HRTFs from which they were asked to select the one(s) they preferred and which resulted in a good spatial auditory image. Listeners focused on a specific criterion in a three-stage evaluation procedure, where the three criteria were *externalization*, *front/back*, and *elevation* discrimination. Because only a small collection of pre-selected HRTF datasets were presented, the task was relatively fast and practical for listeners to go through. An example of results is presented in Figure 4.16 (from Andreopoulou & Roginska, 2014). In this study, a total of 16 HRTFs were presented to each subject: an individually measured HRTF for the subject, 12 similar HRTFs[6] from publicly available datasets (CIPIC, LISTEN, FIU), one *least* similar HRTF, the MIT-KEMAR dataset, and a catch-trial. Although personally measured HRTFs give the overall best results across all criteria, there are HRTF datasets that approach (or are equal to) the performance of a personally measured dataset.

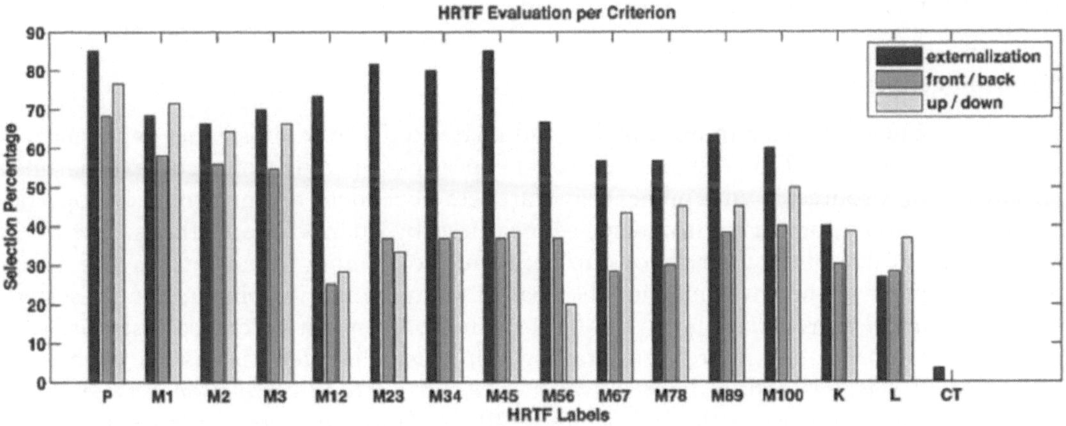

Figure 4.16 Results of user-selected responses per evaluation criterion. *P* corresponds to the personally measured HRTF, *Mi* to the *i*th HRTF in the returned ranked list, *K* to the KEMAR set, *L* to the least similar set, and *CT* to the catch-trial (from Andreopoulou & Roginska, 2014).

Quality Assessment

The number and variety of methods and technologies simulating binaural sound over headphones is continually increasing, and the quality of the virtual auditory experience is getting to the point that it closely resembles the natural listening environment. With so many technologies reaching the consumer market, there's a necessity to assess the quality of the immersive audio experience in order to quantify and optimize the technologies and improve on the realism of the listener's experience.

Although quantitative methods are useful in measuring a signal's physical attributes reaching a listener's ears (e.g. spectral content, time attributes, etc.), it is a listener's perceived judgment of the quality of a spatial auditory image that is more relevant to the actual listening experience. Quality assessment aims at measuring a listener's perception of the virtual auditory environment. Letowski (1989) describes that the complete listening experience is affected by two main sound attributes—the *sound quality*, which evokes a sentiment or an emotional and preferential response; and the *sound character*, which is linked to the judgment of a sound and is a result of a comparison between two stimuli.

When it comes to binaural reproduction over headphones, traditionally two dimensions have been used to assess the quality: spectral coloration of the sound, and accuracy of source location (Le Bagousse et al., 2011). However, as we further our understanding of the factors affecting binaural simulation, a more comprehensive method of assessment is needed to fully quantify the quality measurement of a binaural reproduction technology or system.

In the paper by Nicol et al. (2014), the authors describe a blueprint for the evaluation of binaural sound over headphones. Much like Letowski (1989), Nicol et al. (2014) distribute the assessment into two categories of parameters. The *affective* parameters are principally emotion-driven and relate directly to the listening experience. The *physical* parameters are based on acoustic or physical characteristics of the sound sources and environment.

Affective Parameters

Affective attributes are much more complex and relate to the overall listening experience. Such judgments may be related to a listener's personal experience and emotional context that may be associated with a source or environment, and are less dependent on the acoustic properties of the sound, the room, or the environment, as measured by quantitative methods. The affective attributes aim at describing the relationship between one or more sound sources and a listener within the context of the environment. The goal of many immersive binaural audio systems is to create a *natural* representation of the auditory environment. In perceptual listening tests of auditory environments, and in preference testing, *naturalness* is often used as one of the parameters evaluated. Past evaluations have shown there exists a strong correlation between naturalness and a listener's preference (Berg & Rumsey, 2000). Thus, *naturalness* is a desirable quality that creates the impression of a listener being in an environment that is indistinguishable from a simulated one, regardless of whether the environment has been captured using binaural recording techniques or simulated.

Perhaps no other affective attribute is more strongly related to a listener's experience than the *emotional* response. Many studies looking at music and emotion measure a listener's emotional involvement or music's emotional impact by the dimensions of *arousal* and *valence*—where valence relates to a person's well-being (happy to sad), and arousal represents a person's state of activation level (from calm to exciting). These can be measured physiologically—using skin conductance, cardiovascular, facial muscle, or brain activity—or through a declaration of subjective feeling, by using a rating interface.

The third affective attribute relates to how easily a listener can attend to and discriminate between different sources in an environment, and is closely related to auditory scene analysis (Guastavino & Katz, 2004). In a natural listening environment, listeners can selectively direct their attention to any of the sources around them, only affected by the auditory system's limitations of discerning sound sources based on their spectral content, spatial location, similarity, and other gestalt principles as described in Bregman (1990). In a binaural reproduction, however, a listener's ability to attend to sound sources within this environment will be significantly impacted by the techniques used to capture (or simulate) the auditory environment, and the technologies used to reproduce it.

Physical Attributes

Physical attributes relate to the measurable parameters of a sound or environment. When evaluating the quality of a binaural immersive environment, we must consider the more prominent physical characteristics including the sound source location accuracy, timbral quality and coloration, apparent source width (ASW) and source spread, and room acoustics and environment-related attributes.

Sound source location is the ability for a listener to localize a sound source by determining its azimuth, elevation, and distance. As mentioned earlier, localization is greatly impacted by the HRTFs used in the spatial audio processing (including sound reproduction and capture/

simulation). However, other parameters that affect localization include listener interaction with the environment, visual cues, sound source familiarity, and signal effort. Although the goal is always to have localization accuracy in a simulated environment be equivalent (or superior) to localization accuracy in the natural listening environment, the acceptable localization accuracy requirement will be different depending on the application and purpose of the binaural sound. For example, it may be critical to have the highest level of sound source localization accuracy for mission-critical applications, where the difference between a source perceived 1°–2° apart may mean the difference between life and death. However, for entertainment applications (e.g. virtual surround sound simulation), the listening experience will not be impacted, as long as the sound source or object is perceived within an acceptable range that may be as large as 10°–15°.

Spectral coloration has a direct impact on the spatial perception of an auditory environment, can influence a sound source's location, and affect the perception of the room or environment a sound is in. In binaural sound reproduction, it is most desirable to maintain a minimal level of unwanted timbral coloration and keep a most *natural* sounding environment. There are several factors that may affect the spectral color of the auditory image. These include the sound capture method (in the case of binaural recordings, the pinnae used), the HRTFs used in a simulation, sound reproduction method, and the spectral coloration of headphones.

The apparent source width (ASW) is a measure of the perceptual attribute that describes the apparent auditory width of a sound source or a sound field. It describes the listener's perception of how wide the source image is, regardless of any environment or room characteristics. A listener expresses the ASW as an angular measure. This subjective attribute has been related to an objective measure—most notably to the Inter-Aural Cross-Correlation (IACC)[7] function (e.g. Okano, Beranek & Hidaka, 1998; Vries, Hulsebos & Baan, 2001; Rumsey, 2002).

Finally, the room acoustics and the space in which a binaural environment is recorded or simulated will have a direct impact on the perceived audio quality and add to the spatial perception of sound sources within a room. The early reflections contribute to a listener's understanding of where a sound source is within the room, whereas the late reverberation gives information about the room itself. Many of the measured perceptual attributes related to room acoustics are based on seven attributes, as described by Beranek (1992).[8]

Binaural Reproduction Methods

There are several form factors of reproduction methods that supply binaural signals directly to the ears of a listener. Due to the physical configuration and some technical limitations, each one of these will have an impact on the audio quality and fidelity of the reproduced binaural signal and perceived spatial image. Binaural reproduction methods include fitted insert earphones that go inside a person's ear, over-the-ear headphones, headphones with multiple drivers, bone conduction headphones, and proxim-aural speakers.

Headphones

Over-the-ear headphones can be divided into two main categories based on how they couple to the ear. Circum-aural ("around the ear") headphones use large cups that completely surround and enclose the ear, and typically have pads around the earcups. Because these headphones completely surround the ear, they fully seal the ear from the environment and, thus, can provide complete acoustic isolation (in the case of *closed* headphones, as discussed below).

Supra-aural headphones (literally meaning "on top of the ear") are also knows as "earpad" headphones. They fit directly on top of the ear and are typically equipped with pads that hug the ear. The earcups are small and do not surround the ear.

Open back headphones have the back of the earcups open or vented. This leaks more sound out of the headphone and also lets more ambient sounds into the headphone, but gives a more natural or speaker-like sound and more spacious "soundstage"—with the perception of distance from the source.

Sealed headphones are somewhat more commonly called "closed" headphones, and are designed to block out environmental noise using a passive acoustic seal. Full-sized closed headphones provide about 10 dB of isolation, mostly in the higher frequencies. Earpad closed headphones provide somewhat less isolation, strongly dependent on their design and the shape of individual ear. In-ear headphones are a closed design in which the earpieces seal in the ear canal providing about 23 dB of outside noise attenuation and provide the highest isolation of any passive headphone type. Noise-cancelling headphones are typically closed designs that add electroacoustic techniques to attenuate noise more than a passive seal would provide.

In-Ear Monitors

In-ear monitors (IEM), also referred to as ear canal headphones, in-ear headphones, or canalphones, have been available since the 1980s. Their principal purpose is to provide an airtight seal between the ear and the outside acoustic environment in order to provide the maximum possible amount of isolation from external noise. This is an important consideration especially when listeners are in noisy environments. In such circumstances, IEMs provide a controlled listening environment and the acoustically isolating seal reduces the noise floor, thus improving the perceived Signal-to-Noise Ratio (SNR). The result is that the overall volume can be maintained at a lower level without sacrificing the intelligibility of the signal. The acoustic isolation of IEMs is superior to any headphone, including noise-cancelling headphones, with an acoustic isolation ranging from 15 dB to 23 dB, depending on the type of seal. Shallow sealing IEMs, where the seal is near the entrance to the ear canal, provide a lower level of acoustic isolation, while deep-sealing IEMs have the tip of the earphone sitting about halfway in the ear canal, and provide a much greater level of acoustic isolation.

Earbud Headphones

Earbud headphones (also commonly referred to as *earphones*, or *earbuds*) are quickly becoming one of the most popular methods of audio reproduction in portable music, in part due to their affordability. Earbud headphones are small in size and are positioned at the entrance to the ear canal in the largest cavity in the pinna—the concha. In contrast to in-ear monitors (described above), earphones do not seal the ear canal even though they are similar in size. Most of them use the concha for support. However, some are equipped with small ear clips to secure and stabilize them in the ear, and prevent them from moving. Although their sound quality is relatively good, earbuds suffer from response variances due to variability of fit and misfit, aggravated by close proximity to the eardrum.

Multi-Driver Headphones

In contrast to virtual surround sound presented over headphones, as discussed above, multi-driver headphones utilize multiple drivers inside the earcup to deliver surround sound to the listener. The drivers are located in each earcup, often in a configuration that mimics a loudspeaker surround sound configuration in a room. The number of drivers depends on the desired configuration and can be anywhere from 6 drivers for a 5.1 system, and above. For example, to represent a 7.1 surround system, a typical multi-driver headphone system will be equipped with 10 drivers, 5 in each cup. Each cup will be equipped with a driver to represent the center, front (left or right), surround, surround back, and the LFE (see Figure 4.17).

Frontal projection headphones (Sunder, Tan & Gan, 2013) contain a driver that is located at the front of the headphone cup. In contrast to a traditional headphone, where the driver is located in the middle of a headphone cup, the driver of the frontal projection headphone is physically located at the front of the headphone and projects the sound directly onto the pinna from the front direction. This mirrors the playback from a loudspeaker. Studies comparing the side and frontal projection headphones have shown that frontal projection headphones result in fewer front-back reversals, better frontal localization performance, and reduced timbral coloration.

Ear Speakers

Although headphone reproduction offers many advantages to the listener and environment, the disadvantages listed above may lead the reader to wonder whether delivering a binaural signal using loudspeakers may be a simpler and better solution. Delivering a binaural signal over

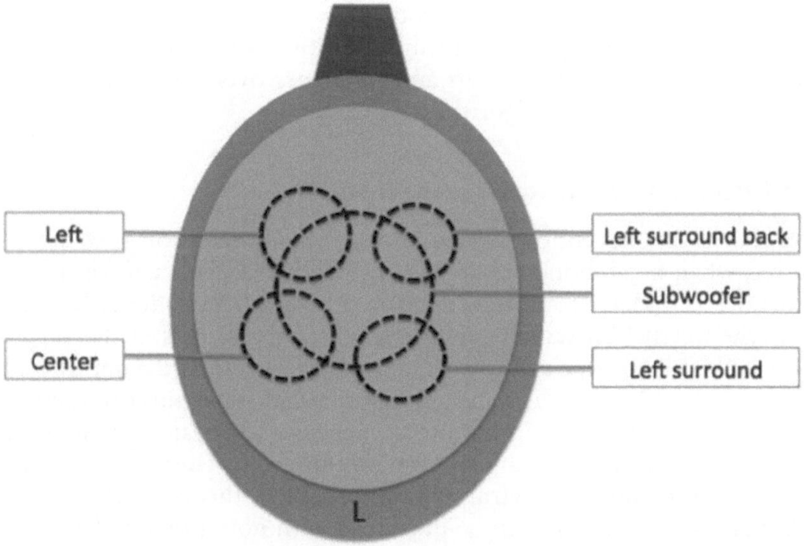

Figure 4.17 Example of the position of drivers on the left channel of a 7.1 multi-driver headphone.

loudspeakers poses one very significant challenge. In loudspeaker reproduction, there is no longer complete control over the signal reaching the left and right ears, as is case with headphones. The actual signal at the ears will be a result of the position and orientation of the listener, and their relationship to the location of the loudspeakers. Furthermore, the signal intended for the left ear will not only be received by the left ear of the listener but will also cross over to the right ear, and vice versa. This is referred to as *crosstalk* and will be discussed in detail in Chapter 5. For the purposes of this discussion, it is important to note that crosstalk is detrimental to creating a localized binaural auditory image. When binaural signals are played over conventional stereo loudspeakers (with the inherent crosstalk), the 3D image collapses. Fortunately, there are several ways of minimizing, eliminating, or cancelling crosstalk—either through physical or signal processing methods.

In order to successfully present a binaural signal to a listener that preserves spatial cues, we must isolate the left and right signals. In the case of loudspeaker presentation, this can be accomplished physically by creating an acoustic baffle that would prevent the left channel signal from reaching the right ear, and vice versa. This is the goal of ear speakers. Ear speakers, also referred to as *nearphones* or *proximaural speakers* (Tappan, 1964), are speakers that are located close to a listener's head, but they do not touch the listener. They are stereo loudspeakers that are typically embedded into a headrest of a movie theater or car seat. The close proximity of the speakers to the head takes advantage of having the head provide an acoustic barrier between the left and the right ears in order to minimize crosstalk. Because of the physical configuration, ear speakers have several advantages. Ear speakers do not touch the ears of a listener and therefore provide an increased wearing comfort and eliminate any potential fatigue from earphones or headphones. Their open nature allows ambient sounds to reach the listener and not isolate the listener from the surrounding environment. This allows creating hybrid reproduction methods combing, for example, a larger theatrical loudspeaker reproduction system with a more individualized nearphone presentation. The most significant advantage, however, is that a binaural signal containing spatial cues can be delivered to the left and right ears over loudspeakers without requiring sophisticated crosstalk cancellation algorithms.

Headphone Equalization and Calibration

Binaural listening and the perception of spatial cues heavily relies on specific spectral cues that may get distorted when headphone coloration is introduced. This can result in a poor spatial image. A number of studies have confirmed that the ideal signal reproduced by headphones should be equivalent to the free-field listening condition, where the signals at the eardrum reproduced by headphones would be the same as those generated by sources in the natural environment (e.g. Olive, Welti & Mcmullin, 2014). The reality is that the acoustic emitter of the headphones introduces its own frequency response. In addition, because of the physical configuration, resonances are created between the headphones and the cavities of the listener's pinna. These, in turn, can lead to significant and unnatural spectral coloration. When this coloration is minimized, studies have shown that externalization can be improved with non-individualized HRTFs (Boren & Roginska, 2011). Furthermore, incorrect compensation for headphone coloration can result in poorer localization (Schonstein, Ferr & Katz, 2008).

The combination of the spectral reproduction characteristics of the headphones, a listener's individual morphology (structural characteristic), the headphone amplifier, and all other components in the headphone reproduction chain can be characterized by the Headphone Transfer Function (HpTF). The HpTF describes both the headphone response and the coupling to a listener's ear. For binaural reproduction, HpTFs can be measured using a system with a dummy head or a person's individual ears. Research by Pralong and Carlile (1996) suggests that individual equalization for every listener is necessary. In testing with 10 subjects, they found significant inter-individual differences in the 4 kHz–10 kHz range.

In order to compensate for the headphone coloration, the transfer function can be inverted and result in a flat reproduction method eliminating the effect of the headphones. There are a number of frequency discrimination methods and algorithms commonly used today to compensate for the spectral coloration of headphones, including regularization methods (e.g. Lindau & Brinkmann, 2012), frequency-domain peak compression (e.g. Hiekkanen, Makivirta & Karjalainen, 2009), and statistical methods.

In addition to the physical characteristics of the headphones, the HpTF is subject to the position of the headphones on a listener's ears, and variations have been observed as a listener re-seated the headphones. The spectral re-positioning effect is low to moderate at low frequencies, but can result in significant coloration differences at high frequencies. The re-seating effect is more noticeable in supra-aural than circum-aural headphones. In research looking at the differences in characteristics of the spectral filtering of headphones that have been re-fitted on the same subject, the most significant change is observed around 4 kHz (Lindau & Brinkmann, 2012). Thus, although there may be significant inter-individual differences in HpTFs, some of these differences may be attributed to headphone re-fitting, and it may be feasible to assume that a *good*, although not a *perfect*, headphone compensation filter may be created to reduce coloration, though not eliminate it entirely. After a period of adaptation, some coloration may be effectively eliminated through perceptual adaptation that takes place with every re-fitting.

An archive of a publicly available library of individually measured HpTFs from multiple databases is available at Boren et al. (2014).

Conclusions

Binaural audio reproduction over headphones is an effective way of delivering spatial audio to a listener. To create a convincing auditory experience, and one that closely approximates a natural listening environment, careful considerations must be made with respect to binaural synthesis, capture, reproduction methods, and tuning. Applications of binaural reproduction over headphones are many, and examples of binaural audio technology and applications date back to the early 1800s (Chapter 2). In the recent past, however, there has been a fast surge in the quantity and breadth of these applications, which range from Virtual Reality (VR) and Augmented Reality (AR) and telepresence, music reproduction, virtual acoustics, and simulation, to mission-critical applications. Furthermore, personal auditory space, and the sound quality and effective realism of virtual auditory environments, are becoming increasingly important to listeners. As far as we have come with binaural reproduction over headphones, there are still many challenges that must be addressed, both technically and creatively.

Notes

1 Interaural Intensity Difference (IID) is also known as Interaural Level Difference (ILD).
2 The Omni Binaural Microphone is developed by 3Dio.
3 In physical acoustics, the far field is frequency-dependent and is the distance that is at least one wavelength. However, it is agreed in the spatial audio community that sound sources farther than 1 m are considered to be in the far field.
4 The subject's ears are sealed in the blocked meatus method and thus protected from loud sounds. Therefore the level of the stimulus can be increased without causing any discomfort or damage to the subject's hearing, resulting in a better SNR.
5 Digital multiplexing methods can cut down the measurement time substantially.
6 Similarity was measured based on spectral proximity using Linear Discriminant Analysis (LDA).
7 The IACC is a measure of the similarity of the signals at the two ears, and ranges between −1 and +1. An IACC of +1 indicates a perfect correlation between two signals, where the two signals are identical (e.g. a mono signal). An IACC of −1 signifies that the two signals are identical but out of phase, or in opposite polarity. When signals have a lower level of similarity, or when there is a random correlation, the IACC approaches 0.
8 The seven attributes described in Beranek (1992) include an acceptable reverberation time, adequate loudness, a short pre-delay, a first reflection immediately followed by early lateral reflections, a diffuse sound field, the ratio of energy in the first 80 msec equivalent to that in the next 2 sec, and the warmth the sound created by shaping the reverberation-time curve at low frequencies.

References

Algazi, V. R., Duda, R. O., & Thompson, D. M. (2004). Motion-tracked binaural sound. *Journal of Audio Engineering Society*, *52*(11), 1142–1156.

Algazi, V. R., Duda, R. O., Thompson, D. M., & Avendano, C. (2001a). The CIPIC HRTF database. *Proceedings of the 2001 IEEE Workshop on Applications of Signal Processing to Audio and Electroacoustics*. New Paltz, NY.

Algazi, V., Duda, R., Thompson, D., & Morrison, R. (2001b). Structural composition and decomposition of HRTFs. *Proceedings of the IEEE Workshop on Applications of Signal Processing to Audio and Acoustics*. New Paltz, NY.

Andreopoulou, A., & Roginska, A. (2011). Towards the creation of a standardized HRTF repository. *Proceedings of the 131st Audio Engineering Society Convention*. New York, NY.

Andreopoulou, A., & Roginska, A. (2014). Evaluating HRTF Similarity through subjective assessments: Factors that can affect judgment. *Proceedings of the 40th ICMC—11th SMC Conference*. Athens, Greece.

Begault, D. R. (1992). Perceptual effects of synthetic reverberation on three-dimensional audio systems. *Journal of Audio Engineering Society*, *40*(11), 895–904.

Begault, D. R., Wenzel, E. M., & Anderson, M. R., (2001). Direct comparison of the impact of head tracking, reverberation, and individualized head-related transfer functions on the spatial perception of a virtual speech source. *Journal of Audio Engineering Society*, *49*, 904–916.

Beranek, L. L. (1992). Concert hall acoustics. *Journal of Acoustical Society of America*, *92*, 1–39.

Berg, J., & Rumsey, F. (2000). Correlation between emotive, descriptive and naturalness attributes in subjective data relating to spatial sound reproduction. *Proceedings of the 109th Audio Engineering Society Convention*. Los Angeles, CA, USA.

Blauert, J. (1983). *Spatial Hearing*. Cambridge, MA: MIT Press.

Boren, B., Geronazzo, M., Majdak, P., & Choueiri, E. (2014). PHOnA: A public dataset of measured headphone transfer functions. *Proceedings of the 137th Audio Engineering Society Convention*. Los Angeles, CA.

Boren, B., & Roginska, A. (2011). The effects of headphones on listener HRTF preference. *Proceedings of the 131st Audio Engineering Society Convention*. New York, NY.

Bregman, A. S. (1990). *Auditory Scene Analysis: The Perceptual Organization of Sound*. Cambridge, MA: MIT Press.

Bremen, P., van Wanrooij, M. M., & Van Opstal, J. (2010). Pinna cues determine orienting response modes to synchronous sounds in elevation. *The Journal of Neuroscience: The Official Journal of the Society for Neuroscience*, *30*(1), 194–204.

Brungart, D., & Rabinowitz, W. (1996). Auditory localization in the near-field. *Proceedings of the 4th International Conference on Auditory Displays (ICAD)*. Palo Alto, CA.

Brungart, D., Simpson, B., McKinley, R., Kordik, A., Dallman, R., & Overshire, D. (2004). The interaction between head-tracker latency, source duration, and response time in the localization of virtual sound sources. *Proceedings of the 10th International Conference on Auditory Display (ICAD)*. Sydney, Australia.

Christensen, F., Moller, H., Minnaar, P., Plogsties, J., & Olsen, S. K. (1999). Interpolating between head-related transfer functions measured with low directional resolution. *Proceedings of the 107th Audio Engineering Society Convention*. New York, NY.

Duda, R.O. & Martens, W.L. (1997). Range-dependence of the HRTF of a spherical head. Appl. Signal Process. Audio Acoust. 104. 5 pp. 10.1109/ASPAA.1997.625597.

Duraiswami, R., Zotkin, D.N., and Gumerov, N.A. (2004). "Interpolation and range extrapolation of HRTFs". In Proceedings of 2004 IEEE International Conference on Acoustics, Speech and Signal Processing, Montreal, Quebec, Canada. Vol.(4), 45–48.

Durant, E. A., Member, S., & Wakefield, G. H. (2002). Efficient Model fitting using a Genetic Algorithm: Pole-Zero approximations of HRTFs. *IEEE Trans on Speech and Audio Processing*, *10*(1), 18–27.

Faller, K. J., Barreto, A., & Adjouadi, M. (2010). Decomposition of head-related transfer functions into multiple damped and delayed sinusoidals. In K. Elleithy (Ed.), *Advanced Techniques in Computing Sciences and Software Engineering* (pp. 273–278). Netherlands: Springer.

Fisher, H., & Freedman, S. J. (1968). The role of the pinnae in auditory localization. *Journal of Auditory Research*, *8*, 15–26.

Gardner, B., & Martin, K. (1994). *HRTF Measurements of a KEMAR Dummy-head Microphone*. Cambridge: Massachusetts Institute of Technology.

Griesinger, D. (1998). General overview of spatial impression, envelopment, localization, and externalization. *Proceedings of the 15th Audio Engineering Society Conference*. Copenhagen, Denmark.

Guastavino, C., & Katz, B. F. G. (2004). Perceptual evaluation of multi-dimensional spatial audio reproduction. *Journal of Acoustical Society of America*, *116*(2), 1105–1115.

Hammer, K., & Snow, W. (1932). *Binaural Transmission System at Academy of Music in Philadelphia*. Memorandum MM3950. Bell Laboratories.

Hanson, R. L., & Kock, W. E. (1957). Interesting effect produced by two loudspeakers under free space conditions. *Journal of Acoustical Society of America*, *29*, 145.

Hartung, K., Braasch, J., & Sterbing, S. J. (1999). Comparison of different interpolation methods for the interpolation of head-related transfer functions. *Proceedings of the 16th Audio Engineering Society Conference on Spatial Sound Reproduction*. Rovaniemi, Finland.

Hershkowitz, R.M., Durlach, N.I. (1969) "Interaural Time and Amplitude JND's for a 500Hz tone". J. Acoust. Soc. Am. 46, 1464–1467.

Hiekkanen, T., Makivirta, A., & Karjalainen, M. (2009). Virtualized listening tests for loudspeakers. *Journal of Audio Engineering Society*, *57*(4), 237–251.

Hofman, P. M., & Van Opstal, J. (2003). Binaural weighting of pinna cues in human sound localization. *Exp Brain Res*, *148*(4), 458–470.

Hu, H., Chen, L., & Wu, Z. Y. (2008). The estimation of personalized HRTFs in individual VAS. *Proceedings of the 4th International Conference on Natural Computation*, pp. 203–207. Washington DC, USA.

Jeppesen, J., & Moeller, H. (2005). Cues for localization in the horizontal plane. *Proceedings of the 118th Audio Engineering Society Convention*. Barcelona, Spain.

Katz, B. F. G. (2001). Boundary element method calculation of individual head-related transfer function. *Journal of Acoustical Society of America, 110*(5), 2440–2455.

Katz, B. F. G., & Parseihian, G. (2012). Perceptually based head-related transfer function database optimization. *Journal of Acoustical Society of America, 131*(2), EL99-EL105.

Kulkarni, A., & Colburn, H. S. (2004). Infinite-impulse-response models of the head-related transfer function. *Journal of Acoustical Society of America, 115*(4), 1714–1728.

Le Bagousse, S., Paquier, M., Colomes, C., & Moulin, S. (2011). Sound quality evaluation based on attributes—application to binaural contents. *Proceedings of the 131st Audio Engineering Society Convention*. New York, NY, USA.

Letowski, T. (1989). Sound quality assessment: Cardinal concepts. *Proceedings of the 87th Audio Engineering Society Convention*. Hamburg, Germany.

Lindau, A., & Brinkmann, F. (2012). Perceptual evaluation of headphone compensation in binaural synthesis based on non-individual recordings. *Journal of Audio Engineering Society, 60*(1), 54–62.

Lorho, G. (2005). Evaluation of spatial enhancement systems for stereo headphone reproduction by preference and attribute rating. *Proceedings of the 118th Audio Engineering Society Convention*. Barcelona, Spain.

Manor, E., Martens, W., Marui, A., & Cabrera, D. (2015). Nearfield crosstalk increases listener preference for headphone-reproduced stereophonic imagery. *Journal of Audio Engineering Society, 63*(5), 324–335.

Middlebrooks, J. C. (1999). Individual differences in external ear transfer functions reduced by scaling in frequency. *Journal of Acoustical Society of America, 106*, 1480–1492.

Moller, H., Sorensen, M., & Hammershoi, D. (1995). Head-related transfer function of human subjects. *Journal of Audio Engineering Society, 43*(5), 300–321.

Nicol, R., Gros, L., Colomes, C., Noisternig, M., Warusfel, O., Bahu, H., Katz, B., & Simon, L. (2014). A roadmap for assessing the quality of experience of 3d audio binaural rendering. *Proceedings of the EAA Joint Symposium on Auralization and Ambisonics*. Berlin, Germany.

Nishino, T., Kajita, S., Takeda, K., & Itakura, F. (1999). Interpolating head related transfer functions in the median plane. *Proceedings of the 1999 IEEE Workshop on Applications of Signal Processing to Audio and Acoustics*. New Paltz, New York.

Okano, T., Beranek, L. L., & Hidaka, T. (1998). Relations among interaural cross-correlation coefficient (IACCE), lateral fraction (LFE), and apparent source width (ASW) in concert halls. *Journal of Acoustical Society of America, 104*(1), 255–265.

Olive, S. E., Welti, T., & Mcmullin, E. (2014). The influence of listeners experience, age, and culture on headphone sound quality preferences. *Proceedings of the 137th Audio Engineering Society Convention*. Los Angeles, CA.

Plenge, G. (1974). On the difference between localization and lateralization. *Journal of Acoustical Society of America, 56*, 944–951.

Pralong, D., & Carlile, S. (1996). The role of individualized headphone calibration for the generation of high fidelity virtual auditory space. *Journal of Acoustical Society of America, 100*(6), 3785–3793.

Rakerd, B., & Hartmann, W. M. (1985). Localization of sound in rooms, II: The effects of a single reflecting surface. *Journal of Acoustical Society of America, 78*(2), 524–533.

Rayleigh, Lord [Strutt, J.W.]. (1907). On our perception of sound direction. *Philosophical Magazine, 13*, 214–232.

Roginska, A., Santoro, T., & Wakefield, G. H. (2010). Stimulus-dependent HRTF preference. *Proceedings of the 129th Audio Engineering Society Convention*. San Francisco, CA, USA.

Rumsey, F. (2002). Spatial quality evaluation for reproduced sound: Terminology, meaning, and a scene-based paradigm. *Journal of Audio Engineering Society*, *50*(9), 651–666.

Sakamoto, N., Gotoh, T., & Kimura, Y. (1976). On out-of-head localization in headphone listening. *Journal of Audio Engineering Society*, *24*, 710–716.

Schonstein, D., Ferr, L., & Katz, B. F. G. (2008). Comparison of headphones and equalization for virtual auditory source localization. *Acoustics '08*, Paris.

Schwarz, D., & Wright, M. (2000). Extensions and applications of the SDIF sound description interchange format. *Proceedings of the International Computer Music Conference.*

Seeber, B., & Fastl, H. (2003). Subjective selection of nonindividual head-related transfer functions. *Proceedings of the 2003 International Conference on Auditory Display*. Boston, MA, USA.

Shaw, E. A. G. (1982). External ear response and sound localization. In R. W. Gatehouse (Ed.), *Localization of Sound: Theory and Applications* (pp. 30–41). Groten, CT: Amphora.

Sone, T., Ebata, M., & Tadamoto, N. (1968). On the difference between localization and lateralization. *Proceedings of the 6th International Congress on Acoustics*. Tokyo, Japan.

Sunder, K., Tan, E. L., & Gan, W. S. (2013). Individualization of Binaural Synthesis Using Frontal Projection Headphones. *Journal of Audio Engineering Society*, *61*(12), 989–1000.

Tappan, P. W. (1964). Proximaural loudspeakers ("Nearphones"). *Proceedings of the 16th Audio Engineering Society Convention*. New York, NY, USA.

Thompson, S. P. (1877). On binaural audition, Part I. *Philosophical Magazine*, *4*, 274–277.

Thompson, S. P. (1878). On binaural audition, Part II. *Philosophical Magazine*, *6*, 383–391.

Thompson, S. P. (1881). On binaural audition, Part III. *Philosophical Magazine*, *12*, 351–355.

Toole, F. E. (1970). In-head localization of acoustic images. *Journal of Acoustical Society of America*, *48*, 943–949.

Vries, D. de, Hulsebos, E. M., & Baan, J. (2001). Spatial fluctuations in measures for spaciousness. *Journal of Acoustical Society of America*, *110*(2), 947–954.

Wahba, G. (1981). Spline interpolation and smoothing on the sphere. *SIAM Journal of Scientific and Statistical Computing*, *2*, 5–16.

Wenzel, E. M. (1992). Localization in virtual acoustic displays. *Presence*, *1*, 80–107.

Wierstorf, H., Geier, M., Raake, A., & Spors, S. (2011). A free database of head-related impulse response measurements in the horizontal plane with multiple distances. *Proceedings of the 130th AES Convention*, eBrief. London, UK.

Wightman, F. L., & Kistler, D. J. (1989). Headphone simulation of free-field listening II: Psychophysical validation. *Journal of Acoustical Society of America*, *85*(2), 868–878.

Xie, B., Zhong, X., Yu, G., Guan, S., Rao, D., Liang, Z., & Zhang, C. (2013). Report on Research Projects on Head-related transfer functions and virtual auditory displays in China. *Journal of Audio Engineering Society*, *61*(5), 314–326.

Xu, S., Li, Z., & Salvendy, G. (2008). Improved method to individualize head-related transfer function using anthropometric measurements. *Acoustical Science and Technology*, *29*(6), 388–390.

Zotkin, D., Hwang, J., Duraiswaini, R., & Davis, L. (2003). HRTF personalization using anthropometric measurements. *IEEE Workshop on Applications of Signal Processing to Audio and Acoustics, Institute for Advanced Computer Studies*, pp. 157–160. University of Maryland, College Park: IEEE.

Appendix

Near Field

The near field, in physical acoustics, is defined as "the region of space within a fraction of a wavelength away from a sound source" (Brungart & Rabinowitz, 1996). Considering the audible frequency range is between 20 Hz and 20 kHz, this causes the physical near field to vary immensely with frequency, ranging from 1.7 cm to 17 m. Concerning human localization, it has been accepted that the near field refers to the region of space closer than 1 m from the center of a listener's head, and the far field is the space that is more than 1 m away from the listener. From a localization point of view, the near field is important, as it is the only space where localization cues change as a function of distance. Researchers have found a fundamental difference in localization of sources near the head of a listener versus localization of sources farther away. Studies have shown that when a source is within 1 m of the listener, the HRTF changes dramatically with distance (e.g. Brungart & Rabinowitz, 1996; Duda & Martens, 1997). Other acoustic cues such as environmental sound are involved in distance perception beyond 1 m.

The greatest change in the near field occurs in the ILD. There are two factors that contribute to this change: increased head shadow and attenuation of sound over distance. As a sound source approaches the head, the ratio of distances from the source to the contralateral ear and from the source to the ipsilateral ear increases dramatically. This causes an increased head shadow effect and, thus, an increase in ILD. The second factor contributing to an increased ILD is the attenuation of sound over distance. As the distance from the sound source is decreased, the amplitude at the ipsilateral ear increases much faster than at the contralateral ear—leading to a "proximity effect". Studies by Brungart and Rabinowitz (1996) show an increase in ILD to 15 dB at 500 Hz for a source at 90° azimuth, 25 cm away from the head. This is a substantial increase in ILD from the average 5–6 dB.

Contrarily, the ITD is roughly independent of source distance, though studies by Hershkowitz and Durlach (1969) have shown a small ITD increase at very close distances. At close distances, there also exists a difference between the relative angle of the source to the head and between the source to the pinna. This results in an auditory parallax, as demonstrated in Figure 4.18. As can be seen, the relative angle of the source to the ipsilateral ear is much different from the relative angle of the source to the contralateral ear.

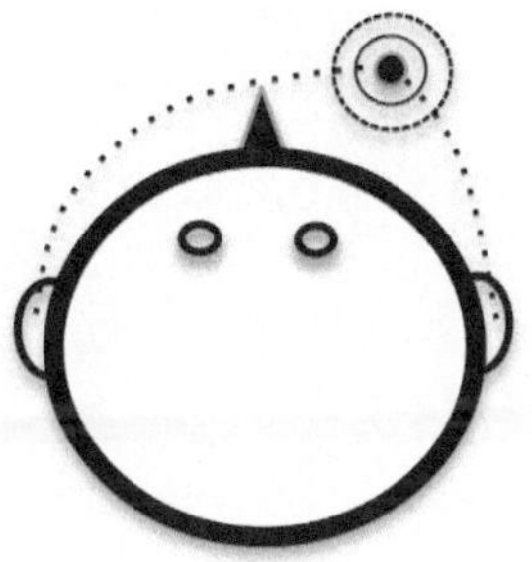

Figure 4.18 Auditory parallax effect for a near-field source.

Thus, the near field is a unique place in auditory localization. The above-mentioned factors combined make the near field the only space where a listener is able to estimate the distance of a sound source without any prior information about the intensity or spectrum of a source.

Chapter 5

Binaural Audio Through Loudspeakers

Edgar Choueiri

Introduction

Background and Motivation

The ultimate goal of binaural audio with loudspeakers (BAL), also known as transauralization (Cooper & Bauck, 1989), is to reproduce, at each of the listener's eardrums, the sound pressure signals recorded on only the ipsilateral channel of a stereo signal. If the stereo signal[1] was encoded with the head-related transfer function (HRTF) of the listener, and includes the proper ITD (interaural time difference) and ILD (interaural level difference) cues, then delivering the signal on each channel of the stereo recording to the ipsilateral ear, and only to that ear, would ideally guarantee that the listener's ear-brain system receives the cues it needs to perceive an accurate three-dimensional reproduction of the recorded sound field. Since, with playback from two loudspeakers, each of the cues is also heard by the contralateral ear (crosstalk), accurate 3D audio reproduction through BAL requires an effective cancellation of this unintended crosstalk. Without such crosstalk cancellation (XTC), the ITD and ILD cues will inevitably be corrupted.

In addition to XTC, effective BAL requires an abatement of sound reflections in the listening room, since such reflections directly degrade the integrity of the binaural cues at the listener's ears (Damaske, 1971; Sæbø, 2001). While this problem can be somewhat alleviated through prescriptions that increase the ratio of direct to reflected sound, accurate sound localization through BAL has been shown to require XTC levels[2] above 20 dB (Parodi & Rubak, 2011b), which are difficult to achieve practically even under anechoic conditions (Akeroyd et al., 2007).

Therefore, it would seem that the goal stated in the first paragraph could be more naturally reached with binaural audio through headphones (or earphones), as both crosstalk and room reflections would be non-existent. However, with headphones or earphones, the location of the playback transducers in or very near the ears means that non-idealities (e.g., mismatches between the HRTF of the listener and that used to encode the recording, movement of the perceived sound image with movement of the listener's head, lack of bone-conducted sound, transducer induced resonances in the ear canal, discomfort, etc.), when above a certain threshold, can lead to difficulties in perceiving a realistic three-dimensional image and to the perception that the sound (or some of its spectral components) is inside, or too close to, the listener's head (see Chapter 4 and Nicol, 2010, for a more thorough discussion of this issue).

Binaural playback through loudspeakers is largely immune to this head internalization of sound because, even when non-idealities in binaural reproduction are present, the sound originates far enough from the listener to be perceived to come from outside the head. Furthermore, cues such as bone-conducted sound and the involvement of the listener's own head, torso, and pinnae in sound diffraction and reflection during playback (even if it departs from, or interferes with, the diffraction-induced coloration represented in the HRTF used to encode the binaural recording) could be expected to enhance the perceived realism of sound reproduction relative to that achieved with earphones. These potential advantages have, implicitly or explicitly, motivated the development of XTC-enabled BAL since the earliest work on the subject (Atal, Hill & Schroeder, 1966; Bauer, 1961; Damaske, 1971 and Chapter 2).

Some applications of BAL, such as immersive virtual reality environments or scientific studies of spatial hearing, require binaural cues to be transmitted to a listener with a high degree of fidelity and reliability. Such transparency and robustness often require anechoic (or semi-anechoic) environments (or equivalently, high-directivity loudspeakers that abate the prominence of reflected sound), individualization of the XTC system for the listener and the playback set-up, precise matching of the listener's HRTF with that used in the recording, and either constraining the position of the listener's head in the area of equalization (the "sweet spot") (Akeroyd et al., 2007; Majdak, Masiero & Fels, 2013; Moore, Tew & Nicol, 2010; Parodi & Rubak, 2011b) or adding the complexity of a head-tracking system. However, in many less stringent applications, modest levels of XTC, even of a few dB over a limited range of frequencies, have the potential to significantly enhance the three-dimensional realism of the reproduction of recordings containing binaural cues. This is because, by definition, localization cues in a binaural recording represent differential interaural information that is intended to be transmitted to the ears with no crosstalk. In other words, crosstalk cancellation, at any level, is a reduction of unintended corruption in the loudspeaker playback of recordings containing significant binaural cues.

This reduction of unintended corruption through XTC should also apply to the loudspeaker playback of most stereo recordings,[3] especially those made in real acoustic spaces, and even to recordings made using standard stereo microphone techniques without a dummy head, because these techniques all rely on preserving in the recording a good measure of the natural ITD and ILD cues needed for enhancing the accuracy of spatial localization and the realism of hall reverberation during playback (Hugonnet & Walder, 1997). We should therefore expect that effecting even a relatively low level of XTC in the playback of such standard stereo recordings, even those lacking HRTF encoding, should enhance image localization compared to playback with full crosstalk, as well as the perception of width and depth of the sound field, since these binaural features are always, to some degree, corrupted by crosstalk.[4]

Before addressing the most fundamental challenge in XTC filter design, we list some of the practical challenges encountered when implementing an effective XTC-enabled BAL playback system and refer to the literature that discusses effective solutions to these practical problems. As alluded to above, such XTC systems are typically sensitive to room reflections (Akeroyd et al., 2007; Damaske, 1971; Sæbø, 2001; Ward, 2001), require the use of specialized playback set-ups (Kirkeby, Nelson & Hamada, 1998a, b; Takeuchi & Nelson, 2002, 2007), and necessarily create a single restricted sweet spot in which the XTC is effective (Takeuchi, Nelson & Hamada, 2001; Ward & Elko, 1999; Xie, 2013, and references therein). Much research effort has been expended on how to relieve the latter constraint and has resulted in potential solutions, of varying degrees of practicality, which include widening the sweet spot through the use of multiple loudspeakers (Bai, Tung & Lee, 2005; Takeuchi & Nelson, 2002; Yang, Gan & Tan, 2003) and/or elevated loudspeakers (Parodi & Rubak, 2010), providing XTC at multiple listening locations through the use of multiple loudspeaker pairs (Bauck & Cooper, 1996; Kim, Deille & Nelson, 2006), and dynamically moving the sweet spot to follow the location of the listener's head by tracking it with optical sensors (Gardner, 1998; Lentz, 2006; Mannerheim, 2008).

The most fundamental challenge in XTC filter design is dealing with the *tonal distortion* (spectral coloration)[5] that XTC filters inherently impose on the sound emitted by the loudspeakers. As we will show in the following sections, the level of tonal distortion depends on the location of the sound source in the sound field, and therefore cannot be corrected through equalization, especially for audio signals containing more than a single sound source. The basic problem of XTC, the fundamental nature of the associated tonal distortion, its main features, its dependencies, and, ultimately, the formulation of a method for the practical design of optimal XTC filters that abate such tonal distortion with minimal degradation of XTC performance are the main subjects of this chapter.

The Problem of XTC-Induced Tonal Distortion

Nature of the Problem

One main difficulty in implementing XTC is to reduce the artifice of crosstalk without adding an artifice of another kind: tonal distortion. Sound waves traveling from two distinct sources to the ears set up an interference pattern in the intervening air space. Depending on the frequency, the distances between an ear and the loudspeakers, the distance between the loudspeakers, and the phase relationship between the left and right components of the recorded stereo signal, the wave interference at that ear of the listener might be constructive, destructive, or complementary (90° out of phase). At the frequencies for which the interference between in-phase recorded signals is destructive at the ears (or, alternatively, the frequencies for which the interference between out-of-phase signals is constructive), XTC control (i.e., signal processing that would cause the waves from the loudspeakers to the contralateral ears to be nulled) would require boosting the amplitude of the emitted waves (Takeuchi & Nelson, 2002).[6] As shown in the section "Benchmark: Perfect Crosstalk Cancellation," in the case of a *perfect* XTC filter (defined as one that theoretically yields, in a free-field or anechoic environment, an infinite XTC level over the entire audio band) for typical listening configurations, these level boosts can easily be in excess of 30 dB, and therefore amount to severe tonal distortion.

Of course, such a "perfect" XTC filter would impose these necessary level boosts *only at the loudspeakers* in such a way that, *at the listener's ears*, not only is the crosstalk cancelled, but the frequency spectrum is also reconstructed perfectly, i.e., with no tonal distortion.

As recognized by Takeuchi and Nelson (2002) and P. A. Nelson and Rose (2005), and as further discussed in the section "Benchmark: Perfect Crosstalk Cancellation," the frequencies at which the level boosts are required correspond to the frequencies at which the system inversion (the mathematical inversion of the system's transfer matrix, which leads to the XTC filter) is ill conditioned. As a result, XTC control becomes highly sensitive to errors at these frequencies, so that even a small error in the alignment of the listener's head in the real world would lead to a significant loss of XTC control at and near these frequencies. Therefore, not only would there be undesired crosstalk at the listener's ears at these frequencies, but also and consequently, the level boosts which must necessarily be imposed at these frequencies would be fully audible, even in the sweet spot, as coloration (tonal distortion).

Takeuchi and Nelson (2002) show that, even in an ideal world where the loudspeakers–listener alignment is perfect, this tonal distortion imposed at the loudspeakers would present three problems: 1) it would be heard by a listener outside the sweet spot, 2) it would cause a relative increase (compared to unprocessed sound playback) in the physical strain on the playback transducers, and 3) it would correspond to a loss in dynamic range. Since even professional audio equipment is seldom designed to have more than a few dB headroom above the levels required to reproduce the full dynamic range of realistic sound pressures (Katz, 2002), in order to avoid clipping in the case of the "perfect" XTC filter defined above, the dynamic range of the program would need to be decreased by more than 30 dB (minus the headroom). This is particularly problematic, for instance, in the case of wide-dynamic-range audio recorded in 16 or 24 bits (see Chapter 2 and references to these early efforts in the Bibliography of this chapter).

Previous Work and Goals of This Chapter

The history of crosstalk cancellation extends back to the seminal work of Bauer, Atal, Hill and Schroeder in the early 1960s (see references to these early efforts in the Bibliography of this chapter) and has since progressed at a faster rate with the advent of digital audio for which XTC can be readily implemented through digital filtering. We shall not attempt to review this history here, nor the various methods of implementing XTC (which range from older techniques applied in the analog domain (Atal et al., 1966), to time-domain signal manipulation algorithms, such as the RACE algorithm (Glasgal, 2007), and FFT-based digital convolution with finite impulse response (FIR) filters (SreenivasaRao, Mahalakshmi & VenkataRao, 2012), and instead focus our discussion on the problem of tonal distortion in XTC filters.

Takeuchi and Nelson (2002) have developed a method that not only yields excellent measured XTC performance (see also Akeroyd et al., 2007; Takeuchi & Nelson, 2007), but also effectively solves the problem of tonal distortion. However, their method, called the "Optimal Source Distribution" (OSD), which is discussed in the section "Benchmark: Perfect Crosstalk Cancellation," requires the use of a minimum of four (but typically six) transducers positioned at various angles around the listener.

The problem of XTC-induced tonal distortion for playback with only two loudspeakers remains compelling due to the simplicity of the two-loudspeaker set-up and its compatibility with existing audio equipment. In this chapter, we study this problem in the context of XTC optimization, which we define as the maximization of XTC performance for a desired tolerable level of tonal distortion or, equivalently, the minimization of tonal distortion for a desired XTC performance.

The ultimate goal of the discussion in this chapter is to describe the design of "optimal XTC filters" (called BACCH filters) that do not suffer from the following drawbacks inherent to regular XTC filters:

- D1: Severe tonal distortion to the sound heard by the listener, even if that listener is sitting in the intended sweet spot.
- D2: Useful XTC levels are reached only at limited frequency ranges of the audio band.
- D3: Severe dynamic range loss when the sound is processed through the XTC filter or processor (while avoiding distortion and/or clipping).

In particular, we use a free-field two-point-source model and address, analytically, the fundamental aspects of tonal distortion control through both constant-parameter (frequency-independent) and frequency-dependent regularization methods. The use of regularization in the design of XTC filters was proposed by Kirkeby and colleagues (1998) to make the inversion of the system transfer matrix better behaved, and has since seen widespread adoption in the field. Specifically, constant-parameter regularization has been employed to control ill-conditioning in the design of HRTF-based XTC filters (e.g., Akeroyd et al., 2007; Kirkeby et al., 1998; Majdak et al., 2013), and frequency-dependent methods have been employed to tame high- and low-frequency amplification due to measured-HRTF inversion (e.g., Kirkeby & Nelson, 1999; Moore et al., 2010) and to control the temporal extent of the XTC filters (e.g., Parodi & Rubak, 2010, 2011a). Regarding the issues of tonal distortion and dynamic range loss, Papadopoulos and Nelson (2010) used constant-parameter regularization to limit the dynamic range loss inflicted by XTC, and Bai et al. (2005) and Bai & Lee (2006a) employed frequency-dependent regularization to impose gain limits on the XTC filters.

In the section "Constant-Parameter Regularization," we show that while the technique of constant-parameter (non-frequency-dependent) regularization may alleviate some of drawback D3, it inherently introduces spectral artifice of its own (specifically, while reducing the amplitude of the spectral peaks in the inverted transfer matrix, constant-parameter regularization results in undesirable narrow-band artifacts at higher frequencies and a roll-off at lower frequencies at the loudspeakers) and does little to alleviate the other two drawbacks (D1 and D2).

A discussion of the fundamental aspects of frequency-dependent regularization in the section "Frequency-Dependent Regularization" will lead us to our ultimate goal: a method for designing "optimal XTC filters" called "BACCH filters." The method relies on calculating the frequency-dependent regularization parameter (FDRP) that results in a flat amplitude versus frequency response at the loudspeakers (as opposed to a flat amplitude versus frequency response at the ears of the listener, as in previous design methods), thus forcing XTC to be effected into the phase domain only and relieving the XTC filter from the drawbacks of audible tonal distortion and dynamic range loss. When the method is used with any effective optimization scheme, it results in XTC filters that yield optimal XTC levels over any desired portion of the audio band, impose no tonal distortion on the processed sound beyond the tonal distortion inherent in the playback hardware and/or loudspeakers, and causes no dynamic range loss. XTC filters designed with this method and used in the system are not only optimal but, due to their being free from drawbacks D1, D2, and D3, allow for a most natural and spectrally transparent 3D audio reproduction of binaural or stereo audio through loudspeakers.

The Fundamental XTC Problem

In this section, we start with the mathematical formulation of the model and the governing transformation matrices. We then define a set of metrics that are useful for evaluating and comparing the tonal distortion and performance of XTC filters, and conclude with the definition and discussion of a benchmark for such comparisons: the perfect XTC filter.

Formulation and Transformation Matrices

In order to render the analysis tractable enough so that fundamental insight is more easily obtained, we make the idealizing assumptions that sound propagation occurs in a free field (with no diffraction or reflection from the head and pinnae of the listener or any other physical objects), and that the loudspeakers radiate like point sources.

In the frequency domain, the air pressure at a free-field point located a distance r from a point source (monopole) radiating a sound wave of frequency ω is given by Morse and Ingard (1986)

$$P(r,i\omega) = \frac{i\omega\rho_0 q}{4\pi}\frac{e^{-ikr}}{r},$$

where ρ_0 is the air density, $k = 2\pi/\lambda = \omega/c_s$ the wavenumber, λ the wavelength, c_s the speed of sound (340.3 m/s), and q the source strength (in units of volume per unit time). It is convenient to define

$$V = \frac{i\omega\rho_0 q}{4\pi},$$

which is the time derivative of $\rho_0 q/4\pi$, the mass flow rate of air from the center of the source.

Therefore, at the left ear of a listener in the symmetric two-source geometry shown in Figure 5.1, the air pressure due to the two sources, under the above-stated assumptions, add up as

$$P_L(i\omega) = \frac{e^{-ikl_1}}{l_1}V_L(i\omega) + \frac{e^{-ikl_2}}{l_2}V_R(i\omega)\,. \tag{5.1}$$

Similarly, at the right ear, we have

$$P_R(i\omega) = \frac{e^{-ikl_2}}{l_2}V_L(i\omega) + \frac{e^{-ikl_1}}{l_1}V_R(i\omega)\,. \tag{5.2}$$

Here, l_1 and l_2 are the path lengths between either source and the ipsilateral and contralateral ears, respectively, as shown in that figure.

In order to maintain a connection with the relevant literature, we adopt the same nomenclature used by Kirkeby et al. (1998a, b), Takeuchi and Nelson (2002), and P. A. Nelson and Rose (2005). Namely, unless otherwise stated, we use uppercase letters for frequency variables, lowercase for time-domain variables, uppercase bold for matrices, and lowercase bold for vectors, and define

$$\Delta l \equiv l_2 - l_1 \quad \text{and} \quad g \equiv l_1 / l_2 \tag{5.3}$$

as the path length difference and path length ratio, respectively. An inspection of the geometry illustrated in Figure 5.1 shows that $0 < g < 1$, and that the path lengths can be expressed as

$$l_1 = \sqrt{l^2 + \left(\frac{\Delta r}{2}\right)^2 - \Delta rl\ sin(\theta)}, \tag{5.4}$$

$$l_2 = \sqrt{l^2 + \left(\frac{\Delta r}{2}\right)^2 - \Delta rl\ sin(\theta)}, \tag{5.5}$$

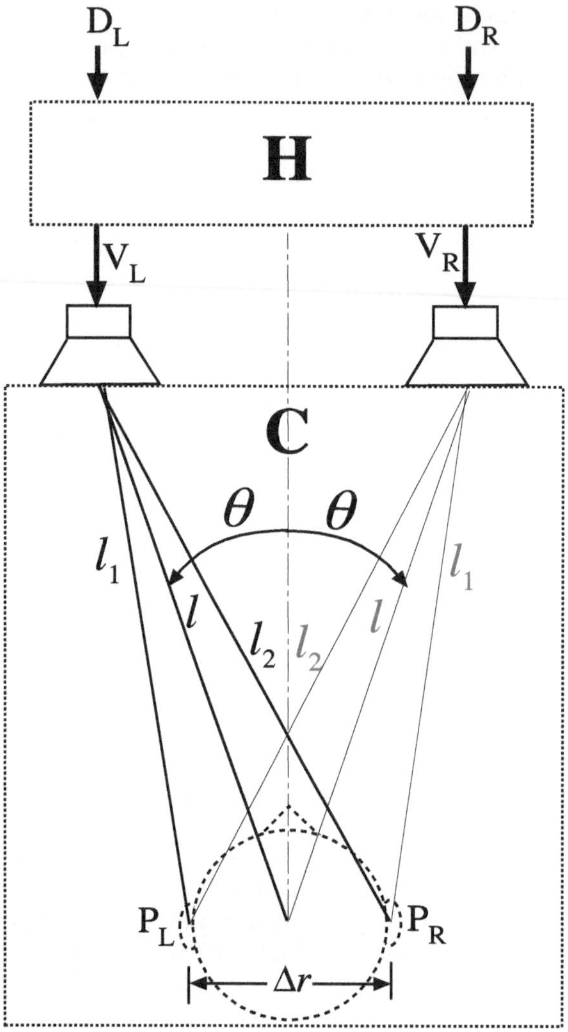

Figure 5.1 Geometry of the free-field two-point-source model. (All symbols are defined in the text.)

where Δr is the effective distance between the entrances of the ear canals, and l is the distance between either source and the interaural mid-point. As defined in Figure 5.1, $2\theta = \Theta$ is the loudspeaker span. Note that for $l \gg \Delta r \sin(\theta)$, as in most loudspeaker-based listening set-ups, we have $g \simeq 1$. Another important parameter is the time delay,

$$\tau_c = \frac{\Delta l}{c_s}, \tag{5.6}$$

defined as the time it takes a sound wave to traverse the path length difference Δl.

Using the above definitions, Equations (5.1) and (5.2) can be re-written in matrix form as

$$\begin{bmatrix} P_L(i\omega) \\ P_R(i\omega) \end{bmatrix} = \alpha \begin{bmatrix} 1 & ge^{-i\omega\tau_c} \\ ge^{-i\omega\tau_c} & 1 \end{bmatrix} \begin{bmatrix} V_L(i\omega) \\ V_R(i\omega) \end{bmatrix}, \tag{5.7}$$

where

$$\alpha = \frac{e^{-i\omega l_1/c_s}}{l_1}. \tag{5.8}$$

In the time domain, α is simply a transmission delay (divided by the constant l_1) that does not affect the shape of the signal. Its role in ensuring causality is discussed in the section "Metrics." The source vector $v = [V_L(i\omega), V_R(i\omega)]^T$ is obtained from the vector of "recorded" signals $d = [D_L(i\omega), D_R(i\omega)]^T$, through the transformation

$$v = Hd, \tag{5.9}$$

where

$$H = \begin{bmatrix} H_{LL}(i\omega) & H_{LR}(i\omega) \\ H_{RL}(i\omega) & H_{RR}(i\omega) \end{bmatrix} \tag{5.10}$$

is the sought 2×2 filter matrix. Therefore, from Equation (5.7), we have

$$p = \alpha CHd, \tag{5.11}$$

where $p = [P_L(i\omega), P_R(i\omega)]^T$ is the vector of pressures at the ears, and C is the system's transfer matrix

$$C \equiv \begin{bmatrix} 1 & ge^{-i\omega\tau_c} \\ ge^{-i\omega\tau_c} & 1 \end{bmatrix}, \tag{5.12}$$

which, like all matrices we deal with here, is symmetric due to the symmetry of the geometry.

In summary, the transformation from the signals d, through the filter matrix H, to the source variables v, then through wave propagation from the sources to the pressures p at the ears of the listener, can be written simply as

$$p = \alpha Rd, \tag{5.13}$$

where we have introduced the performance matrix, R, defined as

$$R = \begin{bmatrix} R_{LL}(i\omega) & R_{LR}(i\omega) \\ R_{RL}(i\omega) & R_{RR}(i\omega) \end{bmatrix} \equiv CH. \tag{5.14}$$

Metrics

We now wish to define a set of metrics by which to judge the tonal distortion and performance of XTC filters. In this context, we note that the diagonal elements of $\boldsymbol{R}$ represent the ipsilateral transmission of the signal to the ears, and the off-diagonal elements represent the undesired contralateral transmission, i.e., the crosstalk.

The responses of the system to a signal fed to only one (either left or right) of the two inputs, as heard at the ears, are called the "side images" of the system (i.e., either $\alpha\boldsymbol{R}\cdot[1,0]^T$ or $\alpha\boldsymbol{R}\cdot[0,1]^T$). We define our first coloration metric as the amplitude spectrum (to a factor α) of the side image at the ipsilateral ear, given by

$$E_{\mathrm{si}_{||}}(\omega) \equiv |R_{LL}(i\omega)| = |R_{RR}(i\omega)|,$$

where the subscripts "si" and "||" stand for "side image" and "ipsilateral ear (with respect to the input signal)," respectively. Similarly, at the contralateral ear to the input signal (subscript "X"), we have the following side-image amplitude spectrum:

$$E_{\mathrm{si}_X}(\omega) \equiv |R_{RL}(i\omega)| = |R_{LR}(i\omega)|.$$

The response of the system to a signal split equally between left and right inputs, as heard at either ear, is called the "center image" of the system (i.e., $\alpha\boldsymbol{R}\cdot[1/2,1/2]^T$). We define another coloration metric as the amplitude spectrum of the center image, given by

$$E_{ci}(\omega) \equiv \left|\frac{R_{LL}(i\omega)+R_{LR}(i\omega)}{2}\right| = \left|\frac{R_{RL}(i\omega)+R_{RR}(i\omega)}{2}\right|,$$

where the subscript "ci" stands for "center image."

Also of importance to our discussions are the frequency responses that would be measured at the sources (loudspeakers). These are denoted by S, and can be obtained from the elements of the filter matrix $\boldsymbol{H}$. They are given using the same subscript convention used above (with "||" and "X" referring to the loudspeakers that are ipsilateral and contralateral to the input signal, respectively) by

$$S_{\mathrm{si}_{||}}(\omega) \equiv |H_{LL}(i\omega)| = |H_{RR}(i\omega)|,$$

$$S_{\mathrm{si}_X}(\omega) \equiv |H_{LR}(i\omega)| = |H_{RL}(i\omega)|,$$

$$S_{\mathrm{ci}}(\omega) \equiv \left|\frac{H_{LL}(i\omega)+H_{LR}(i\omega)}{2}\right| = \left|\frac{H_{RL}(i\omega)+H_{RR}(i\omega)}{2}\right|$$

An intuitive interpretation of the significance of the above metrics is that a signal panned from a single input to both inputs to the system will result in frequency responses going from E_{si} to E_{ci} at the ears, and S_{si} to S_{ci} at the loudspeakers.

Two other tonal distortion metrics are the frequency responses of the system to in-phase and out-of-phase inputs to the system. These two responses are obtained simply from the product of the filter matrix $\boldsymbol{H}$ with the vectors $[1,1]^T$ and $[1,-1]^T$ (or $[-1,1]^T$), respectively, and are given by:

$$S_i(\omega) \equiv |H_{LL}(i\omega) + H_{LR}(i\omega)| = |H_{RL}(i\omega) + H_{RR}(i\omega)|,$$
$$S_o(\omega) \equiv |H_{LL}(i\omega) - H_{LR}(i\omega)| = |H_{RL}(i\omega) - H_{RR}(i\omega)|,$$

where the subscripts "i" and "o" denote the in-phase and out-of-phase responses, respectively. Note that, as defined, S_i is double (i.e., 6 dB above) S_{ci}, as the latter describes a signal of amplitude 1 panned to center (i.e., split equally between L and R inputs), while the former describes two signals of amplitude 1 fed in-phase to the two inputs of the system.

Since a real signal can consist of various components having different phase relationships, it is more useful to combine $S_i(\omega)$ and $S_o(\omega)$ into a single metric, $\hat{S}(\omega)$, which is the *envelope spectrum* that describes the maximum amplitude that could be expected at the loudspeakers, and is given by

$$\hat{S}(\omega) = \max[S_i(\omega), S_o(\omega)].$$

It is relevant to note that $\hat{S}(\omega)$ is equivalent to $\|\boldsymbol{H}\|$, the 2-norm of $\boldsymbol{H}$, and that S_i and S_o are the two singular values, which can be obtained through singular value decomposition of the matrix, as was done by Takeuchi and Nelson (2002).

Finally, an important metric that allows us to evaluate and compare the XTC performance of various filters is $\chi(\omega)$, the crosstalk cancellation spectrum:

$$\chi(\omega) \equiv \frac{|R_{LL}(i\omega)|}{|R_{RL}(i\omega)|} = \frac{|R_{RR}(i\omega)|}{|R_{LR}(i\omega)|} = \frac{E_{\mathrm{si}_{\|}}(\omega)}{E_{\mathrm{si}_{X}}(\omega)}.$$

The above definitions give us a total of eight metrics, ($E_{\mathrm{si}\|}$, $E_{\mathrm{si}X}$, E_{ci}, $S_{\mathrm{si}\|}$, $S_{\mathrm{si}X}$, S_{ci}, $\hat{S}$, and χ), all real functions of frequency, by which to evaluate and compare the tonal distortion and XTC performance of XTC filters.

Benchmark: Perfect Crosstalk Cancellation

A perfect crosstalk cancellation (P-XTC) filter is defined as one that, theoretically, yields infinite crosstalk cancellation at the ears of the listener, for all frequencies.

Crosstalk cancellation, as defined in the section "Background and Motivation," requires that the pressure at each of the two ears be that which would have resulted from the ipsilateral signal alone, namely, in the frequency domain, $P_L = \alpha D_L$ and $P_R = \alpha D_R$, where all quantities are complex functions of frequency. Therefore, in order to achieve perfect cancellation of the crosstalk, Equation (5.13) requires that $\boldsymbol{R} = \boldsymbol{I}$, where $\boldsymbol{I}$ is the identity matrix, and thus, as per the definition of $\boldsymbol{R}$ in Equation (5.14), the P-XTC filter is simply the inverse of the system transfer matrix expressed in Equation (5.12), and can be expressed exactly:

$$\boldsymbol{H}^{[P]} = \boldsymbol{C}^{-1} = \frac{1}{1 - g^2 e^{-2i\omega\tau_c}} \begin{bmatrix} 1 & -ge^{-i\omega\tau_c} \\ -ge^{-i\omega\tau_c} & 1 \end{bmatrix}, \tag{5.15}$$

where the superscript "[P]" denotes perfect XTC. For this filter, the eight metrics we defined above become:

$$
\begin{aligned}
&E^{[P]}_{\mathrm{si}_{\|}} = 1; \quad E^{[P]}_{\mathrm{si}_X} = 0; \quad E^{[P]}_{\mathrm{ci}} = \frac{1}{2} \\
&S^{[P]}_{\mathrm{si}_{\|}}(\omega) = \left|\frac{1}{1-g^2 e^{-2i\omega\tau_c}}\right| \\
&= \frac{1}{\sqrt{g^4 - 2g^2\cos(2\omega\tau_c)+1}}; \\
&S^{[P]}_{\mathrm{si}_X}(\omega) = \left|\frac{-ge^{-i\omega\tau_c}}{1-g^2 e^{-2i\omega\tau_c}}\right| \\
&= \frac{g}{\sqrt{g^4 - 2g^2\cos(2\omega\tau_c)+1}}; \\
&S^{[P]}_{\mathrm{ci}}(\omega) = \frac{1}{2}\left|1-\frac{g}{g+e^{i\omega\tau_c}}\right| \\
&= \frac{1}{2\sqrt{g^2 + 2g\cos(\omega\tau_c)+1}}; \\
&\hat{S}^{[P]}(\omega) = \max\left(\left|1-\frac{g}{g+e^{i\omega\tau_c}}\right|, \left|1+\frac{g}{e^{i\omega\tau_c}-g}\right|\right) \\
&= \max\left(\frac{1}{\sqrt{g^2 + 2g\cos(\omega\tau_c)+1}}\right. \\
&\qquad \left.\frac{1}{\sqrt{g^2 + 2g\cos(\omega\tau_c)+1}}\right);
\end{aligned}
\tag{5.16}
$$

$$
\chi^{[P]}(\omega) = \infty. \tag{5.17}
$$

Therefore, the perfect ($\chi = \infty$) XTC filter gives flat frequency responses at the ears ($E^{[P]}(\omega)$ = constant), but not at the sources. To appreciate the extent of tonal distortion at the loudspeakers, we plot the $S^{[P]}(\omega)$ frequency responses expressed above in Figure 5.2 for a typical value of $g = 0.985$. Throughout this chapter, for the sake of illustration, we complement the non-dimensional plots with dimensional calculations, which are represented by the same curves read in terms of the frequency $f = \omega/2\pi$ on the top axis, for a typical listening geometry characterized by $g = 0.985$ and $\tau_c = 68\ \mu s$ (i.e., 3 samples at the "Red Book" CD sampling rate of 44.1 kHz), which would be the case, for instance, in a set-up with $\Delta r = 15$ cm, $l = 1.6$ m, and $\Theta = 18°$.

The peaks in these spectra occur at frequencies for which the system must boost the amplitude of the signal at the loudspeakers in order to effect XTC at the ears while compensating for the destructive interference at that location. Similarly, minima in the spectra occur when the amplitude must be attenuated.

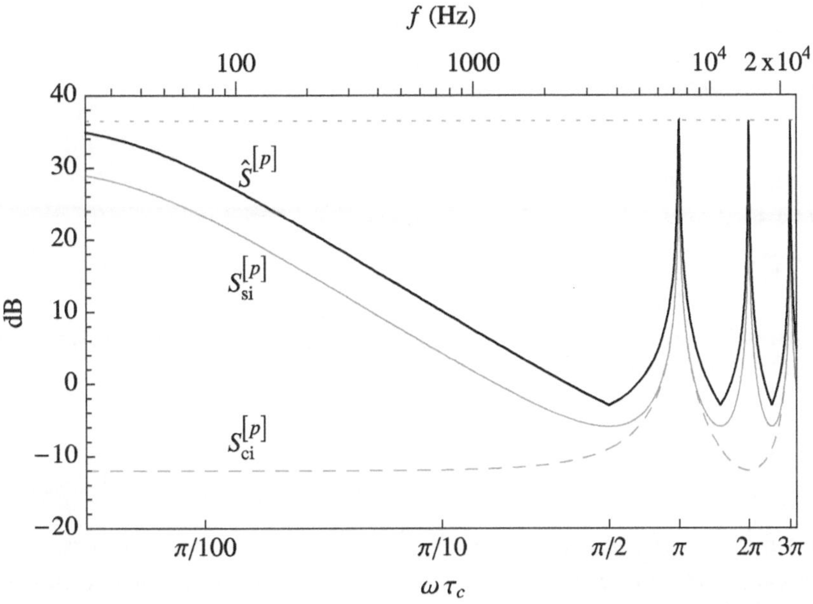

Figure 5.2 Perfect XTC filter frequency responses at the loudspeakers: amplitude envelope (heavy curve), side image (light solid curve), and central image (light dashed curve). The dotted horizontal line marks the envelope ceiling, which for this case (g = 0.985) is 36.5 dB. The non-dimensional frequency $\omega\tau_\chi$ is given on the bottom axis, and the corresponding frequency in Hz, shown on the top axis, is to illustrate a particular (typical) case of τ_c = 3 samples at a sampling rate of 44.1 kHz. (Since $S^{[P]}_{si\parallel} \simeq S^{[P]}_{si\times}$ when $g \simeq 1$, these two spectra are shown as the single curve $S^{[P]}_{si\times}$.)

Using the first and second derivatives (with respect to $\omega\tau_c$) of the above expressions for the various $S^{[P]}(\omega)$ spectra, we find the following amplitudes and frequencies for the associated peaks and minima, denoted by "↑" and "↓" superscripts, respectively:

$$
\begin{aligned}
S^{[P]\uparrow}_{si_{\parallel}} &= \frac{1}{1-g^2} \text{ at } \omega\tau_c = n\pi, \\
S^{[P]\downarrow}_{si_{\parallel}} &= \frac{1}{1-g^2} \text{ at } \omega\tau_c = (2n+1)\frac{\pi}{2}, \\
S^{[P]\uparrow}_{si_{x}} &= \frac{1}{1-g^2} \text{ at } \omega\tau_c = n\pi, \\
S^{[P]\downarrow}_{si_{x}} &= \frac{1}{1-g^2} \text{ at } \omega\tau_c = (2n+1)\frac{\pi}{2}, \\
S^{[P]\uparrow}_{ci} &= \frac{1}{2-2g} \text{ at } \omega\tau_c = (2n+1)\pi,
\end{aligned}
\qquad (5.18)
$$

$$S_{\text{ci}}^{[P]\downarrow} = \frac{1}{2-2g} \text{ at } \omega\tau_c = 2n\pi,$$

$$\hat{S}^{[P]\uparrow} = \frac{1}{1-g} \text{ at } \omega\tau_c = n\pi,$$

$$\hat{S}^{[P]\downarrow} = \frac{1}{\sqrt{1+g^2}} \text{ at } \omega\tau_c = (2n+1)\frac{\pi}{2}, \qquad (5.19)$$

with $n = 0,1,2,3,4,\ldots$

For a typical listening set-up, $g \simeq 1$, say, our reference $g = 0.985$ case shown in Figure 5.2, the envelope peaks (i.e., $\hat{S}^{[P]\uparrow}$) correspond to a boost of

$$20 \log_{10}\left(\frac{1}{1-0.985}\right) = 36.5 \text{ dB}$$

(and the peaks in the other *spectra*, $S_{\text{sill}}^{[P]\uparrow} \simeq S_{\text{six}}^{[P]\uparrow} \simeq S_{\text{ci}}^{[P]\uparrow}$ correspond to boosts of about 30.5 dB). While these boosts have equal frequency widths across the spectrum, when the spectrum is plotted logarithmically (as is appropriate for human sound perception), the low-frequency boost is most prominent in its perceived frequency extent. This "bass boost" has long been recognized as an intrinsic problem in XTC (Kirkeby et al., 1998b; Takeuchi & Nelson, 2002). While the high-frequency peaks could, in principle, be pushed out of the audio range by decreasing τ_c (which, as can be seen from Equations (5.4)–(5.6), is achieved by increasing l and/or decreasing the loudspeaker span Θ, as is done in the so-called Stereo Dipole configuration described by Kirkeby, Nelson and Hamada (1998a, b), where $\Theta = 10°$), the low-frequency boost of the P-XTC filter would remain problematic.

As mentioned in the section "The Problem of XTC-Induced Tonal Distortion," the severe tonal distortion associated with these high-amplitude peaks presents three practical problems: 1) it would be heard by a listener outside the sweet spot, 2) it would cause a relative increase (compared to unprocessed sound playback) in the physical strain on the playback transducers, and 3) it would correspond to a loss in the dynamic range.

These penalties might be a justifiable price to pay if we are guaranteed the infinitely good XTC performance ($\chi = \infty$) and the perfectly flat frequency response ($E^{[P]}(\omega)$ = constant) that the perfect XTC filter promises at the ears of a listener in the sweet spot. However, in practice, these theoretically promised benefits are unachievable due to the solution's sensitivity to unavoidable errors. This problem can best be appreciated by evaluating the condition number of the transfer matrix C.

In matrix inversion problems, the sensitivity of the solution to errors in the system is given by the condition number of the matrix. (For a discussion of the condition number in the context of XTC system errors, see P. A. Nelson and Rose, 2005.) The condition number $\kappa(\text{C})$ of the matrix C is given by

$$\kappa(\text{C}) = \|\text{C}\| \cdot \|\text{C}^{-1}\| = \|\text{C}\| \cdot \|H^{[P]}\|.$$

(It is also, equivalently, the ratio of largest to smallest singular values of the matrix.) Therefore, we have

$$\kappa(C) = \max\left(\sqrt{\frac{2(g^2+1)}{g^2+2g\cos(\omega\tau_c)}-1},\right.$$
$$\left.\sqrt{\frac{2(g^2+1)}{g^2-2g\cos(\omega\tau_c)+1}-1}\right).$$

Using the first and second derivatives of this function, as we did for the previous spectra, we find the following maxima and minima:

$$\kappa^{\uparrow}(C) = \frac{1+g}{1-g} \text{ at } \omega\tau_c = n\pi,$$
$$\kappa^{\downarrow}(C) = 1 \text{ at } \omega\tau_c = (2n+1)\frac{\pi}{2} \quad (5.20)$$

with $n = 0,1,2,3,4,\ldots,$

as was also reported by Ward and Elko (1999) and P. A. Nelson and Rose (2005) in terms of wavelengths, and by Takeuchi and Nelson (2002) in terms of the wave number. First, we note that the maxima and minima in the condition number occur at the same frequencies as those of the amplitude envelope spectrum at the loudspeakers, $\hat{S}^{[P]}$. Second, we note that the minima have a condition number of unity (the lowest possible value), which implies that the filter resulting from the inversion of C is most robust (i.e., least sensitive to errors in the transfer matrix) at the non-dimensional frequencies $\omega\tau_c = \pi/2, 3\pi/2, 5\pi/2, \ldots$. Conversely, the condition number can reach very high values (e.g., $\kappa^{\uparrow}(C) = 132.3$ for our typical case of $g = 0.985$) at the non-dimensional frequencies $\omega\tau_c = 0, \pi, 2\pi, 3\pi, \ldots$. As $g \to 1$, the matrix inversion resulting in the P-XTC filter becomes ill conditioned, or, in other words, infinitely sensitive to errors. The slightest misalignment, for instance, of the listener's head, would thus result in a severe loss in XTC control at the ears (at and near these frequencies) which, in turn, causes the severe tonal distortion in $\hat{S}^{[P]}(\omega)$ to be transmitted to the ears.

We are now in a position to appreciate the prescription proposed and implemented by Takeuchi and Nelson (2002, 2007), which effectively solves both the robustness and tonal distortion problem of the P-XTC filter by ensuring that the system operates always under conditions where $\kappa(C)$ is small. This can be done by allowing the loudspeaker span to be a function of the frequency. More specifically, after noting that typically $l \gg \Delta r$, so that the approximation $\Delta l \simeq \Delta r \sin(\theta)$ holds, and therefore $\omega\tau_c = \omega\Delta l/c_s = 2\pi f\Delta l/c_s$ can be approximated by

$$\omega\tau_c \simeq \frac{2\pi f \Delta \sin(\theta)}{c_s} \text{ for } l \gg \Delta r, \quad (5.21)$$

we can re-write the robustness condition (stated in Equation (5.20)) as

$$\Theta(f) \simeq 2\sin^{-1}\left(\frac{(2n+1)c_s}{4f\Delta r}\right),$$

with $n = 0,1,2,3,4,\ldots$

Since both c_s and Δr are constant, the required loudspeaker span is solely a function of the frequency f. In practice, this prescription, called Optimal Source Distribution (OSD), can be implemented by using a crossover network to distribute adjacent bands of the audio spectrum to pairs of transducers, whose spans are calculated from the above equation so that in each band the condition number does not exceed unity by much, thus insuring robustness and low coloration over the entire audio spectrum. It is clear, however, that this solution is not applicable to the case of a single pair of loudspeakers, which is the focus of our analysis.

We refer the reader interested in the OSD method and XTC errors to Takeuchi and Nelson (2002, 2007) and P. A. Nelson and Rose (2005), and sum up the discussion in this section by stating that, for the case of only two loudspeakers, the perfect XTC filter carries in practice the penalties of over-amplification (and the associated loss of dynamic range) at frequencies where system inversion is ill conditioned, transducer fatigue, and a severe tonal distortion that is heard by listeners inside and outside the sweet spot.

Constant-Parameter Regularization

Regularization methods allow controlling the norm of the approximate solution of an ill-conditioned linear system at the price of some loss in the accuracy of the solution. The control of the norm through regularization can be done subject to an optimization prescription, such as the minimization of a cost function. Hansen (1998) provides a detailed discussion of regularization methods in a general mathematical context, and others (e.g., Bai et al., 2005; Kirkeby & Nelson, 1999; Majdak et al., 2013; Parodi & Rubak, 2010) have demonstrated the use of regularization to control numerical HRTF inversion. We discuss regularization analytically in the context of XTC filter optimization, which we define as the maximization of XTC performance for a desired tolerable level of tonal distortion or, equivalently, the minimization of tonal distortion for a desired minimum XTC performance.

In essence, a nearby solution to the matrix inversion problem is sought:

$$H^{[\beta]} = [C^H C + \beta I]^{-1} C^H, \tag{5.22}$$

where the superscript "H" denotes the conjugate transpose, and β is the regularization parameter which essentially causes a departure from $H^{[P]}$, the exact inverse of $\mathbf{C}$. In this section we take β to be a constant. The pseudoinverse matrix $H^{[\beta]}$ is the regularized filter, and the superscript "$[\beta]$" is used to denote constant-parameter regularization. The regularization stated in Equation (5.22) can be shown to correspond to a minimization of a cost function, $J(i\omega)$, where

$$J(i\omega) = e^H(i\omega)e(i\omega) + \beta v^H(i\omega)v(i\omega), \tag{5.23}$$

and the vector e represents a performance metric that is a measure of the departure from the signal reproduced by the perfect filter (Kirkeby et al., 1998; P. A. Nelson & Elliott, 1993). Physically, then, the first term in the sum constituting the cost function represents a measure of the performance error, and the second term represents an "effort penalty," which is a measure of the power exerted by the loudspeakers. For $\beta > 0$, Equation (5.22) leads to an optimum, which corresponds to the least-squares minimization of the cost function $J(i\omega)$.

Therefore, an increase of the regularization parameter β leads to a minimization of the effort penalty at the expense of a larger performance error, and thus to an abatement of the peaks in the norm of H, i.e., the coloration peaks in the $S(\omega)$ spectra, at the price of a decrease in XTC performance at and near the frequencies where the system is ill conditioned.

Frequency Response

Using the explicit form for **C** given by Equation (5.12), in the last equation above, we find:

$$H^{[\beta]} = \begin{bmatrix} H_{LL}^{[\beta]}(i\omega) & H_{LR}^{[\beta]}(i\omega) \\ H_{RL}^{[\beta]}(i\omega) & H_{RR}^{[\beta]}(i\omega) \end{bmatrix}, \tag{5.24}$$

where

$$\begin{aligned} H_{RR}^{[\beta]}(i\omega) &= H_{RR}^{[\beta]}(i\omega) \\ &= \frac{g^2 e^{4i\omega\tau_c} - (\beta+1)e^{2i\omega\tau_c}}{g^2 e^{4i\omega\tau_c} + g^2 - e^{4i\omega\tau_c}[(g^2+\beta)^2 + 2\beta + 1]}, \end{aligned} \tag{5.25}$$

$$\begin{aligned} H_{LR}^{[\beta]}(i\omega) &= H_{RL}^{[\beta]}(i\omega) \\ &= \frac{g^2 e^{i\omega\tau_c} - g(g^2+\beta)e^{3i\omega\tau_c}}{g^2 e^{4i\omega\tau_c} + g^2 - e^{2i\omega\tau_c}[(g^2+\beta)^2 + 2\beta + 1]}, \end{aligned} \tag{5.26}$$

The eight metric spectra we defined in the section "Metrics" become:

$$\begin{aligned}
E_{\mathrm{si}_{\|}}^{[\beta]}(\omega) &= \frac{g^4 + \beta g^2 - 2g^2\cos(2\omega\tau_c) + \beta + 1}{-2g^2\cos(2\omega\tau_c) + (g^2+\beta)^2 + 2\beta + 1}; \\
E_{\mathrm{si}_X}^{[\beta]}(\omega) &= \frac{2g\beta \cdot |\cos(\omega\tau_c)|}{-2g^2\cos(2\omega\tau_c) + (g^2+\beta)^2 + 2\beta + 1}; \\
E_{\mathrm{ci}}^{[\beta]}(\omega) &= \frac{1}{2} - \frac{\beta}{2[g^2 + 2g\cos(2\omega\tau_c) + \beta + 1]}; \\
S_{\mathrm{si}_{\|}}^{[\beta]}(\omega) &= \frac{\sqrt{g^4 + 2(\beta+1)g^2\cos(2\omega\tau_c) + (\beta+1)^2}}{-2g^2\cos(2\omega\tau_c) + (g^2+\beta)^2 + 2\beta + 1}; \\
S_{\mathrm{si}_X}^{[\beta]}(\omega) &= \frac{g\sqrt{(g^2+\beta)^2 - 2(g^2+\beta)\cos(2\omega\tau_c) + 1}}{-2g^2\cos(2\omega\tau_c) + (g^2+\beta)^2 + 2\beta + 1}; \\
S_{\mathrm{ci}}^{[\beta]}(\omega) &= \frac{\sqrt{g^2 + 2g\cos(\omega\tau_c) + 1}}{2[g^2 + 2g\cos(2\omega\tau_c) + \beta + 1]}; \\
\hat{S}^{[\beta]}(\omega) &= \max\left(\frac{\sqrt{g^2 + 2g\cos(\omega\tau_c) + 1}}{g^2 + 2g\cos(2\omega\tau_c) + \beta + 1} \right.
\end{aligned} \tag{5.27}$$

$$\chi^{[\beta]}(\omega) = \frac{g^4 + \beta g^2 - 2g^2\cos(2\omega\tau_c)+\beta+1}{2g\beta\cdot|\cos(2\omega\tau_c)|}; \tag{5.28}$$

Of course, as $\beta \rightarrow 0$, $H^{[\beta]} \rightarrow H^{[P]}$, and it can be verified that the spectra of the perfect XTC filter are recovered from the expressions above.

The envelope spectrum, $\hat{S}^{[\beta]}(\omega)$, is plotted in Figure 5.3 for three values of β. Two features can be noted in that plot: 1) increasing the regularization parameter attenuates the peaks in the spectrum without affecting the minima, and 2) with increasing β, the spectral maxima split into doublet peaks (two closely spaced peaks).

To get a measure of peak attenuation and the conditions for the formation of doublet peaks, we take the first and second derivatives of $\hat{S}^{[\beta]}(\omega)$ with respect to $\omega\tau_c$ and find the conditions for which the first derivative is nil and the second is negative. These conditions are summarized as follows: If β is below a threshold β^* defined as

$$\beta < \beta^* \equiv (g-1)^2, \tag{5.29}$$

the peaks are singlets and occur at the same non-dimensional frequencies as for the envelope spectrum peaks of the P-XTC filter ($\hat{S}^{[P]\uparrow}$), and have the following amplitude:

$$\hat{S}^{[\beta]\uparrow} = \frac{1-g}{(g-1)^2+\beta} \text{ at } \omega\tau_c = n\pi,$$

with $n = 0,1,2,3,4,\ldots$

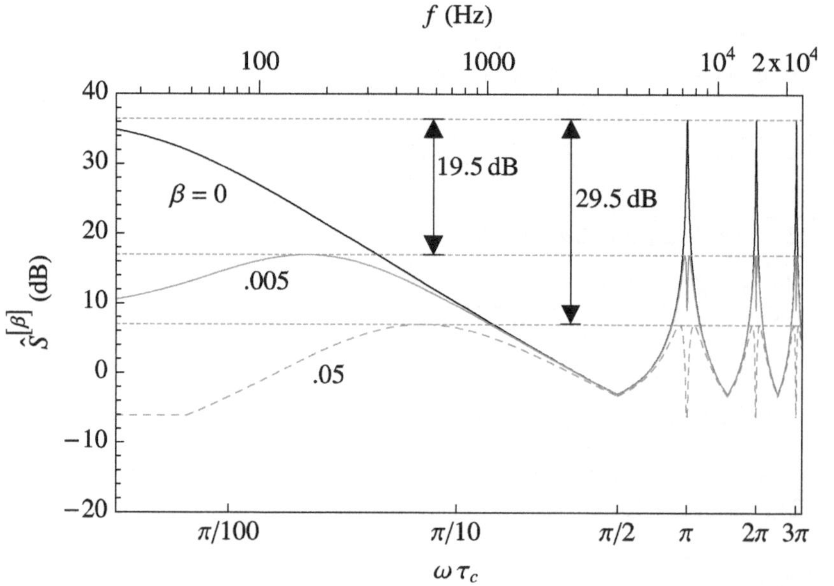

Figure 5.3 Effects of regularization on the envelope spectrum at the loudspeakers, $\hat{S}^{[\beta]}(\omega)$, showing peak attenuation and formation of doublet peaks as β is increased. (Other parameters are the same as for Figure 5.2.)

If the condition

$$\beta^* \le \beta \ll 1 \tag{5.30}$$

is satisfied, the maxima are doublet peaks located at the following non-dimensional frequencies:

$$\omega\tau_c = n\pi \pm \cos^{-1}\left(\frac{g^2 - \beta + 1}{2g}\right) \tag{5.31}$$

with $n = 0,1,2,3,4, \ldots,$

and have an amplitude

$$\hat{S}^{[\beta]\uparrow\uparrow} = \frac{1}{2\sqrt{\beta}}, \tag{5.32}$$

which does not depend on g. (The superscripts "↑" and "↑↑" denote singlet and doublet peaks, respectively.) The attenuation of peaks in the $\hat{S}^{[\beta]}$ spectrum due to regularization can be obtained by dividing the amplitude of the peaks in the P-XTC (i.e., $\beta = 0$) spectrum by that of peaks in the regularized spectrum. For the case of singlet peaks, the attenuation is

$$20\log_{10}\left(\frac{\hat{S}^{[P]\uparrow}}{\hat{S}^{[\beta]\uparrow}}\right) = 20\log_{10}\left[\frac{\beta}{(g-1)^2} + 1\right]\text{dB},$$

and for doublet peaks, it is given by

$$20\log_{10}\left(\frac{\hat{S}^{[P]\uparrow}}{\hat{S}^{[\beta]\uparrow}}\right) = 20\log_{10}\left[\frac{2\sqrt{\beta}}{1-g}\right]\text{dB}.$$

For the typical case of $g = 0.985$ illustrated in Figure 5.3, we have $\beta^* = 2.25 \times 10^{-4}$, and for $\beta = 0.005$ and 0.05, we get doublet peaks that are attenuated (with respect to the peaks in the P-XTC spectrum) by 19.5 and 29.5 dB, respectively, as marked on that plot.

Therefore, increasing the regularization parameter above this (typically low) threshold causes the maxima in the envelope spectrum to split into doublet peaks shifted by a frequency $\Delta(\omega\tau_c) = \cos^{-1}[(g^2-\beta+1)/2g]$ to either side of the peaks in the response of the perfect XTC filter. (For our illustrative case of $g = 0.985$, we have $\beta^* = 2.25 \times 10^{-4}$ and $\Delta(\omega\tau_c) \simeq 0.225$ for $\beta = 0.05$.) Due to the logarithmic nature of frequency perception for humans, these doublet peaks are perceived as narrow-band artifacts at high frequencies (i.e., for $n = 1,2,3, \ldots$), but the first doublet peak centered at $n = 0$ is perceived as a wide-band low-frequency roll-off of typically many dB, as can be clearly seen in Figure 5.3. Therefore, constant-parameter regularization transforms the bass boost of the perfect XTC filter into a bass roll-off.

Since regularization is essentially a deliberate introduction of error into system inversion, we should expect both the XTC spectrum and the frequency responses at the ears to suffer (i.e., depart from their ideal P-XTC filter levels of ∞ and 0 dB, respectively) with increasing β. The effects of constant-parameter regularization on responses at the ears are illustrated in Figure 5.4.

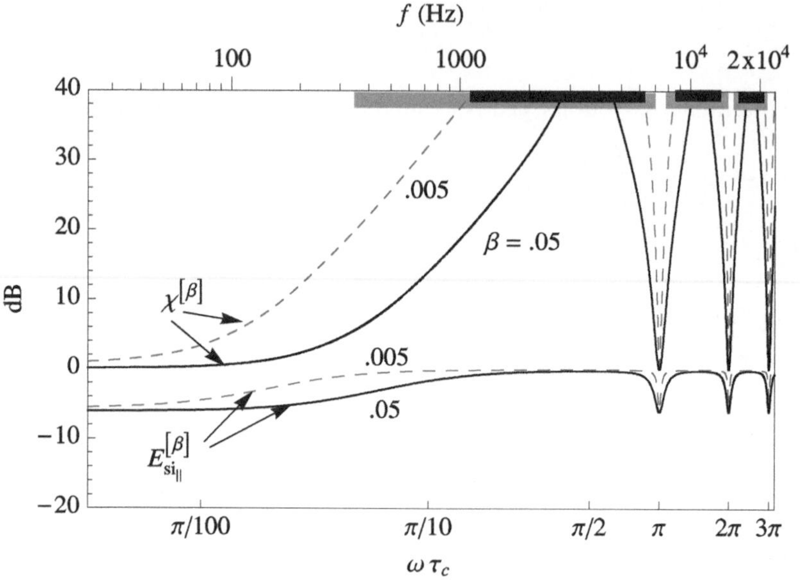

Figure 5.4 Effects of regularization on the crosstalk cancellation spectrum, $\chi^{[\beta]}(\omega)$ (top two curves), and the ipsilateral frequency response at the ear for a side image, $E^{[\beta]}_{\text{si}_{||}}(\omega)$ (bottom two curves). The black horizontal bars on the top axis mark the frequency ranges for which an XTC level of 20 dB or higher is reached with β = 0.05, and the grey bars represent the same for the case of β = 0.005. (Other parameters are the same as for Figure 5.2.)

The black curves in that plot represent the crosstalk cancellation spectra and show that XTC control is lost within frequency bands centered around the frequencies where the system is ill conditioned ($\omega\tau_c = n\pi$ with n = 0,1,2,3,4, . . .) and whose frequency extent widens with increasing regularization. For example, increasing β to 0.05 limits XTC of 20 dB or higher to the frequency ranges marked by black horizontal bars on the top axis of Figure 5.4, with the first range extending only from 1.1 to 6.3 kHz and the second and third ranges located above 8.4 kHz. In many practical applications, such high (20 dB) XTC levels may not be needed or achievable (e.g., because of room reflections and/or HRTF mismatch) and the higher values of β needed to tame the tonal distortion peaks below a required level at the loudspeakers may be tolerated.

The $E^{[\beta]}_{\text{si}_{||}}(\omega)$ responses at the ears, shown as the bottom curves in Figure 5.4, depart only by a few dB from the corresponding P-XTC (i.e., β = 0) filter response (which is a flat curve at 0 dB). More precisely and generally, the maxima and minima of the $E^{[\beta]}_{\text{si}_{||}}(\omega)$ spectrum are given by:

$$E^{[\beta]\uparrow}_{\text{si}_{||}}(\omega) = \frac{g^2+1}{g^2+\beta+1} \text{ at } \omega\tau_c = (2n+1)\frac{\pi}{2};$$

$$E^{[\beta]\downarrow}_{\text{si}_{||}}(\omega) = \frac{g^4+(\beta-2)g^2+\beta+1}{g^4+2(\beta-1)g^2+(\beta+1)^2} \text{ at } \omega\tau_c = n\pi,$$

with n = 0,1,2,3,4, . . .

For the typical ($g = 0.985$) example shown in the figure, we have, for $\beta = 0.05$, and $E_{\mathrm{si}_{||}}^{[\beta]\uparrow} = -0.2\mathrm{dB}$ $E_{\mathrm{si}_{||}}^{[\beta]\uparrow} = -6.1\mathrm{dB}$, showing that even relatively aggressive regularization results in a tonal distortion at the ears that is quite modest compared to the tonal distortion the perfect XTC filter imposes at the loudspeakers.

In sum, we conclude that, while constant-parameter regularization is effective at reducing the amplitude of peaks (including the "bass boost") in the envelope spectrum at the loudspeakers, it typically results in undesirable narrow-band artifacts at higher frequencies and a roll-off of the lower frequencies at the loudspeakers. This non-optimal behavior can be avoided if the regularization parameter is allowed to be a function of the frequency, as we shall see in the section "Frequency-Dependent Regularization."

Before we do so, it is insightful to consider the effects of constant-parameter regularization on the time-domain response of XTC filters.

Impulse Response

We start by making the substitution $z = e^{2i\omega\tau c}$ in Equations (5.25) and (5.26) to get

$$H_{LL}^{[\beta]}(z) = H_{RR}^{[\beta]}(z) = \frac{z^2g^2 - z(\beta+1)}{z^2g^2 + g^2 - z\ [(g^2+\beta)^2 + 2\beta + 1]}, \tag{5.33}$$

$$H_{LR}^{[\beta]}(z) = H_{RL}^{[\beta]}(z) = \frac{z^2[gz^{-1/2} - g(g^2+\beta)z^{1/2}}{z^2g^2 + g^2 - z^2[(g^2+\beta)^2 + 2\beta + 1]}. \tag{5.34}$$

The two expressions above have the same quadratic denominator, which can be factored as

$$z^2g^2 + g^2 - z[(g^2+\beta)^2 + 2\beta + 1] = g^2(z - a_1)(z - a_2),$$

where

$$a_1 = \frac{a - \sqrt{a^2 - 4g^4}}{2g^2}, \quad a_2 = \frac{a - \sqrt{a^2 - 4g^4}}{2g^2}, \tag{5.35}$$

and

$$a = \left(g^2 + \beta\right)^2 + 2\beta + 1. \tag{5.36}$$

We can then re-write Equations (5.33) and (5.34) as

$$H_{LL}^{[\beta]}(z) = H_{RR}^{[\beta]}(z) = \left[z - \frac{(\beta+1)}{g^2}\right] \times \left(\frac{1}{1 - a_1z^{-1}}\right)\left(\frac{1}{z - a_2}\right), \tag{5.37}$$

$$H_{LR}^{[\beta]}(z) = H_{RL}^{[\beta]}(z) = \left[\frac{z^{-1/2} - (g^2+\beta)z^{1/2}}{g^2}\right] \times \left(\frac{1}{1 - a_1z^{-1}}\right)\left(\frac{1}{z - a_2}\right). \tag{5.38}$$

Since $0 < g < 1$, and $\beta \geq 0$, we see from Equations (5.35) and (5.36) that $0 \leq a_1 < 1$ and $a_2 > 1$, and therefore $|a_1 z^{-1}| < 1$ and $a_2 > |z|$. This allows us to express the terms $1/(1 - a_1 z^{-1})$ and $1/(z - a_2)$ in the last two equations as two convergent power series (whose convergence insures that we have a stable filter), and thus write the last two equations as

$$H_{LL}^{[\beta]}(z) = H_{RR}^{[\beta]}(z) = \left[z - \frac{(\beta+1)}{g^2} \right] \times \left(\sum_{m=0}^{\infty} a_1^m z^{-m} \right) \left(\sum_{m=0}^{\infty} - a_2^{-m-1} z^m \right), \tag{5.39}$$

$$H_{LR}^{[\beta]}(z) = H_{RL}^{[\beta]}(z) = \left[\frac{z^{-1/2} - (g^2 + \beta) z^{1/2}}{g^2} \right] \times \left(\sum_{m=0}^{\infty} a_1^m z^{-m} \right) \left(\sum_{m=0}^{\infty} - a_2^{-m-1} z^m \right). \tag{5.40}$$

The filter is now in a form that can be readily transformed into a time-domain filter, $h^{[\beta]}$, represented by

$$h^{[\beta]} = \begin{bmatrix} h_{LL}^{[\beta]}(t) & h_{LR}^{[\beta]}(t) \\ h_{RL}^{[\beta]}(t) & h_{RR}^{[\beta]}(t) \end{bmatrix}. \tag{5.41}$$

We do so by substituting back $e^{2i\omega\tau c}$ for z in Equations (5.39) and (5.40), and taking the inverse Fourier transform (IFT) to get

$$\begin{aligned} h_{LL}^{[\beta]}(t) &= \frac{1}{2\pi} \int_{-\infty}^{\infty} H_{LL}^{[\beta]}(i\omega) e^{i\omega t} d\omega \\ &= h_{RR}^{[\beta]}(t) = \frac{1}{2\pi} \int_{-\infty}^{\infty} H_{RR}^{[\beta]}(i\omega) e^{i\omega t} d\omega \\ &\left[\delta(1 + 2\tau_c) - \frac{\beta+1}{g^2} \delta(t) \right] * \psi(t), \end{aligned} \tag{5.42}$$

$$\begin{aligned} h_{LR}^{[\beta]}(t) &= \frac{1}{2\pi} \int_{-\infty}^{\infty} H_{LR}^{[\beta]}(i\omega) e^{i\omega t} d\omega \\ &= h_{LR}^{[\beta]}(t) = \frac{1}{2\pi} \int_{-\infty}^{\infty} H_{RL}^{[\beta]}(i\omega) e^{i\omega t} d\omega \\ &= \left[\frac{\delta(t - \tau_c)}{g} - \frac{g^2 + \beta}{g^2} \delta(t + \tau_c) \right] * \psi(t), \end{aligned} \tag{5.43}$$

where the asterisk (*) denotes the convolution operation, and $\psi(t)$ is the IFT of the product of the two series appearing in Equations (5.39) and (5.40), and is given by the following convolution of two trains of Dirac delta functions:

$$\psi(t) = \left(\sum_{m=0}^{\infty} a_1^m \delta(t - 2m\tau_c) \right) * \left(\sum_{m=0}^{\infty} -a_2^{-m-1} \delta(t + 2m\tau_c) \right). \tag{5.44}$$

We see that the first train evolves forward in time and the second evolves in reverse time.

The impulse response (IR) represented by Equations (5.42) and (5.43) is plotted in Figure 5.5 for three values of β.

The IR of the perfect XTC filter is shown in the top panel of that figure and consists of two trains of decaying and inter-delayed delta functions of opposite sign. Mathematically, it is the special case of $\beta = 0$, for which Equations (5.37) and (5.38) simplify to

$$H_{LL}^{[P]}(z) = H_{RR}^{[P]}(z) = \frac{1}{1 - a_1 z^{-1}}, \tag{5.45}$$

$$H_{LR}^{[P]}(z) = H_{RL}^{[P]}(z) = \frac{g z^{-1/2}}{1 - a_1 z^{-1}}, \tag{5.46}$$

from which, through the inverse Fourier transform, we recover the IR of the perfect XTC filter derived by Atal et al. (1966):

$$h_{LL}^{[P]}(t) = h_{RR}^{[P]}(t) = \sum_{m=0}^{\infty} a_1^m \delta(t - 2m\tau_c) \tag{5.47}$$

$$h_{LR}^{[P]}(t) = h_{RL}^{[P]}(t) = -g\delta(t - \tau_c) * \sum_{m=0}^{\infty} a_1^m \delta(t - 2m\tau_c) \tag{5.48}$$

where $a_1 = g^2$ (obtained by setting $\beta = 0$ in Equations (5.35) and (5.36)) is the pole of the filter. We see that the perfect XTC IR starts at $t = 0$ with an amplitude of unity and decays to an amplitude $a_1^m = (l_1 / l_2)^{2m}$ after a time $2m\tau_c$.

The physical significance of this impulse response has been discussed by P. Nelson et al. (1997) and Kirkeby et al. (1998b) who, along with Atal et al. (1966) before, recognized the recursive nature of XTC filters. Briefly, a physical appreciation of the perfect XTC IR can be obtained by considering the hypothetical case of a positive pulse whose duration is much smaller than τ_c, fed into only one of the two inputs of the system, say the left input. From Equation (5.9), we see that this pulse, $d_L(t)$, is emitted from the left loudspeaker as a series of positive pulses $d_L(t) * h_{LL}(t)$ (corresponding to the filled circles in the top panel of Figure 5.5) and from the right loudspeaker as a series of negative pulses $d_L(t) * h_{RL}(t)$ (corresponding to the empty circles in the same plot). These two series of pulses are delayed by τ_c with respect to each other so that after the first positive pressure pulse arrives at the left ear, it then reaches the right ear with a slightly smaller amplitude but is cancelled there by a negative pressure pulse of equal amplitude (that was emitted a time l_1/c_s earlier by the right loudspeaker), which in turn is cancelled at the left ear by a positive pressure pulse, and so on. The net result is that only the first pulse is heard and only at the left ear, i.e., with no crosstalk.

The effects of regularization on the XTC IR were recognized by Kirkeby et al. (1998), and can be gleaned from a comparison of the three panels of Figure 5.5. When β is finite, the IR has a "pre-echo" (non-causal) part, i.e., it extends in reverse time ($t < 0$), as shown in Figure 5.5. As can also be seen in that figure and inferred from Equation (5.44), the delta functions in the $t < 0$ and $t > 0$ parts have opposite signs. With increasing regularization, the $t < 0$ part increases in prominence and the IR becomes shorter in temporal extent, which corresponds in the frequency domain to a spectrum with abated peaks.

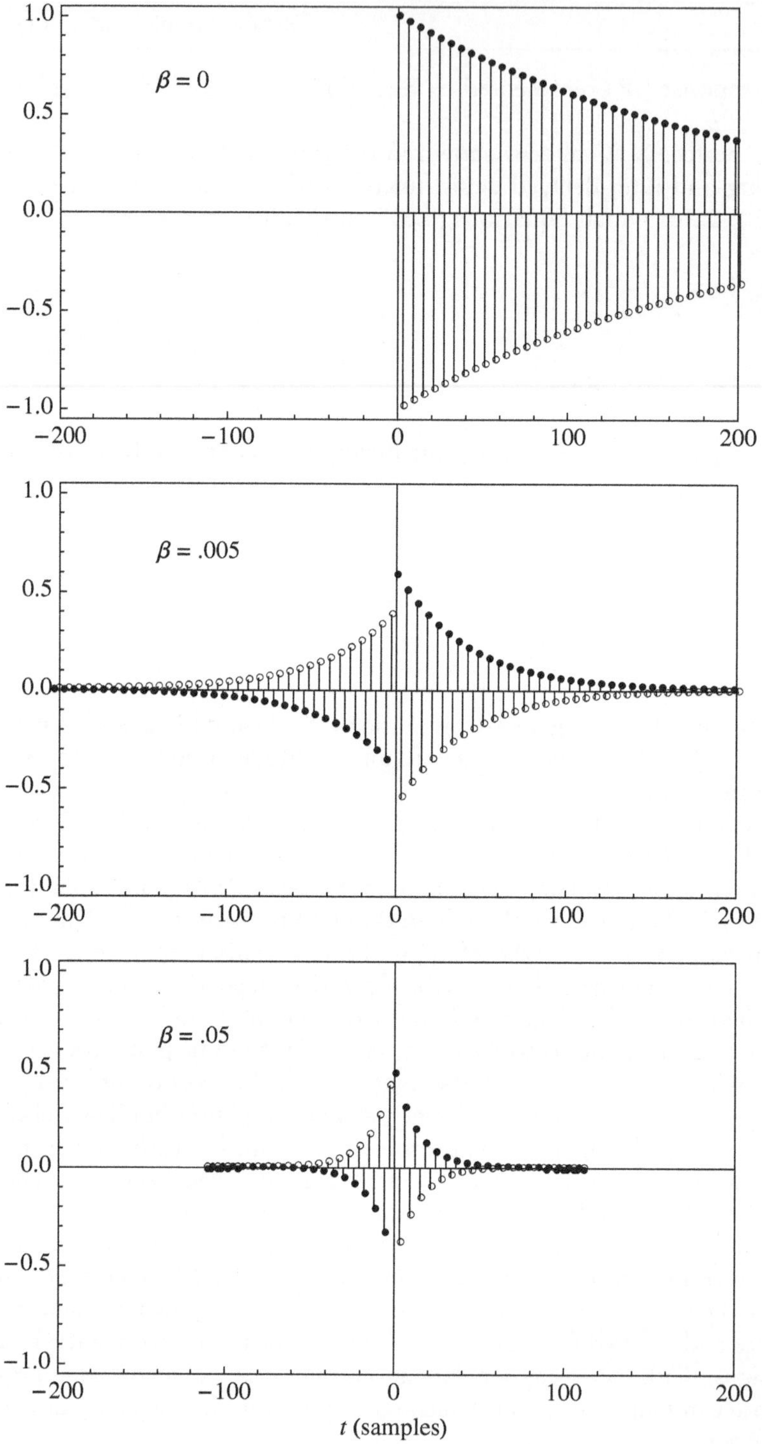

Figure 5.5 Impulse responses $h_{LL}^{[\beta]}(t) = h_{RR}^{[\beta]}(t)$ (filled circles) and $h_{LR}^{[\beta]}(t) = h_{RL}^{[\beta]}(t)$ $h_{RL}^{[\beta]}$ (t) (empty circles) for three values of β. ($g = 0.985$, $\tau_c = 3$ samples)

To insure causality, a time delay must be used to include the $t < 0$ part of the IR. In practice (e.g., when dealing with numerical HRTF inversion), this can be done through a "modeling delay" that accommodates both the non-causal part of the IR and the transmission delay

$$\delta\left(t - \frac{l_1}{c_s}\right)$$

associated with the factor α in Equation (5.8).

The length of a filter having a pole close to the unit circle ($|z| = 1$) is inversely proportional to the distance between that pole and the unit circle (Bellanger, 2000). As β is increased, the poles pull away from the unit circle, as per Equations (5.35) and (5.36), and therefore the length of a finite-β IR is reduced by a factor of

$$\frac{1 - a_1}{1 - g^2}$$

with respect to the length of the perfect XTC IR. This factor (which is based on a_1 since $1 - a_1 < |1 - a_2|$) is accurate as long as $1 - g^2 \ll 1$ and $1 - a_1 \ll 1$.

For instance, in the middle panel of Figure 5.5, we have $\beta = 0.005$ and $g = 0.985$, which give $a_1 = 0.86$ and yield an IR that is about 4.5 times shorter than the perfect XTC IR. (This inverse relationship between regularization and IR length appears to be true in general, as observed by Parodi and Rubak (2010, 2011a) in the case of numerical HRTF inversion via frequency-dependent regularization.)

Frequency-Dependent Regularization

In order to avoid the frequency-domain artifacts discussed in the section "Frequency Response" and illustrated in Figure 5.3, we seek an optimization prescription that would cause the envelope spectrum $\hat{S}(\omega)$ to be flat at a desired level Γ (dB) over the frequency bands where the perfect filter's envelope spectrum exceeds Γ (dB). Outside these bands (i.e., below that level), we apply no regularization. This desired envelope spectrum can be written symbolically as:

$$\hat{S}(\omega) = \begin{cases} \gamma & if\ \hat{S}^{[P]}(\omega) \geq \gamma, \\ \hat{S}^{[P]}(\omega) & if\ \hat{S}^{[P]}(\omega) < \gamma, \end{cases} \tag{5.49}$$

where the P-XTC envelope spectrum, $\hat{S}^{[P]}(\omega)$, is given by Equation (5.16), and

$$\gamma = 10^{\Gamma/20}, \tag{5.50}$$

with Γ given in dB. We take $\Gamma \geq 0$ dB and, since Γ cannot exceed the magnitude of the peaks in the $\hat{S}^{[P]}(\omega)$ spectrum, γ is bounded by the inequalities:

$$1 \leq \gamma \leq \frac{1}{1-g}, \tag{5.51}$$

where the last term is $\hat{S}^{[P]\uparrow}$, given by Equation (5.18).

The frequency-dependent regularization parameter needed to effect the spectral flattening required by Equation (5.49) is obtained by setting $\hat{S}^{[\beta]}(\omega)$, given by Equation (5.27), equal to γ and solving for $\beta(\omega)$, which is now a function of frequency. Since the regularized spectral envelope, $\hat{S}^{[\beta]}(\omega)$, (which is also $\| H^{[\beta]} \|$, the 2 norm of the regularized XTC filter), is the maximum of two functions, we get two solutions for $\beta(\omega)$:

$$\beta_{\rm I}(\omega) = \frac{\sqrt{g^2 - 2g\cos(\omega\tau_c) + 1}}{\gamma} - (g^2 - 2g\cos(\omega\tau_c) + 1), \tag{5.52}$$

$$\beta_{\rm II}(\omega) = \frac{\sqrt{g^2 + 2g\cos(\omega\tau_c) + 1}}{\gamma} - (g^2 + 2g\cos(\omega\tau_c) + 1). \tag{5.53}$$

The first solution, $\beta_{\rm I}(\omega)$, applies for frequency bands where the out-of-phase response of the perfect filter (i.e., the second singular value, which is the second argument of the max function in Equation (5.16)) dominates over the in-phase response (i.e., the first argument of that function):

$$S_o^{[P]} = \frac{1}{\sqrt{g^2 - 2g\cos(\omega\tau_c) + 1}} \geq S_i^{[P]} = \frac{1}{\sqrt{g^2 + 2g\cos(\omega\tau_c) + 1}}. \tag{5.54}$$

Similarly, regularization with $\beta_{\rm II}(\omega)$ applies for frequency bands where $S_i^{[P]} \geq S_o^{[P]}$. Therefore, we must distinguish between three branches of the optimized solution: two regularized branches corresponding to $\beta = \beta_{\rm I}(\omega)$ and $\beta = \beta_{\rm II}(\omega)$, and one non-regularized (perfect-filter) branch corresponding to $\beta = 0$. We refer to these branches by I, II, and P, respectively, and summarize the conditions associated with each as follows:

Branch I: applies where $\hat{S}^{[P]}(\omega) \geq \gamma$ and

$$S_o^{[P]} \geq S_i^{[P]}$$

$$\hat{S}(\omega) = \gamma, \beta = \beta_I(\omega);$$

and requires setting Branch II: applies where $\hat{S}^{[P]}(\omega) \geq \gamma$ and $S_i^{[P]} \geq S_o^{[P]}$

$$\hat{S}(\omega) = \gamma, \beta = \beta_{II}(\omega);$$

and requires setting Branch P: applies where $\hat{S}^{[P]}(\omega) < \gamma$, and requires setting $\hat{S}(\omega) = \hat{S}^{[P]}(\omega)$, $\beta = 0$.

Following this three-branch division, the envelope spectrum at the loudspeakers, $\hat{S}(\omega)$, for the case of frequency-dependent regularization is plotted as the thick black curve in Figure 5.6 for $\Gamma = 7$ dB. This value was chosen because it corresponds to the magnitude of the (doublet) peaks in the $\beta = 0.05$ spectrum (i.e., $\Gamma = 20\log_{10}(1/2\sqrt{\beta})$), which is also plotted (light solid curve) as a reference for the corresponding case of constant-parameter regularization. (We call a spectrum obtained with frequency-dependent regularization and one obtained with constant-parameter regularization "corresponding spectra," if the peaks in $\hat{S}^{[\beta]}(\omega)$, whether singlets or doublets, are equal to γ.)

It is clear from that figure that the low-frequency boost and the high-frequency peaks of the perfect XTC spectrum, which would be transformed into a low-frequency roll-off and narrow-band artifacts, respectively, by constant-parameter regularization, are now flat at the desired maximum coloration level, Γ. The rest of the spectrum, i.e., the frequency bands with amplitude below Γ, is allowed to benefit from the infinite XTC level of the perfect XTC filter and the robustness associated with relatively low condition numbers.

Band Hierarchy

The three-branch prescription therefore splits the audio spectrum into a series of adjacent frequency bands, which we number consecutively starting with Band 1 for the lowest-frequency band. The frequency bounds for each band can be found by setting $S^{[P]}(\omega)$, given by Equation (5.16), equal to γ and solving for $\omega\tau_c$. Doing so results in the following hierarchy of bands and their associated frequency bounds:

- Bands 1, 5, 9, 13, 17, . . ., $4n + 1$ belong to Branch I, and are bounded by

$$2n\pi - \varphi \le \omega\tau_c \le 2n\pi + \varphi; \tag{5.55}$$

- Bands 2, 6, 10, 14, 18, . . ., $4n + 2$ belong to Branch P, and are bounded by

$$2n\pi + \varphi \le \omega\tau_c \le (2n+1)\pi - \varphi; \tag{5.56}$$

- Bands 3, 7, 11, 15, 19, . . ., $4n + 3$ belong to Branch II, and are bounded by

$$(2n+1)\pi - \varphi \le \omega\tau_c \le (2n+1)\pi + \varphi; \tag{5.57}$$

- Bands 4, 8, 12, 16, 20, . . ., $4n + 4$ belong to Branch P, and are bounded by

$$(2n+1)\pi + \varphi \le \omega\tau_c \le (2n+2)\pi - \varphi; \tag{5.58}$$

  where $n = 0, 1, 2, 3, 4, \ldots$ and

$$\phi = \cos^{-1}\left(\frac{(g^2+1)\gamma^2 - 1}{2g\gamma^2}\right). \tag{5.59}$$

For instance, applying this hierarchy to the case of $g = 0.985$, and $\Gamma = 7$ dB (i.e., $\gamma = 10^{7/20} = 2.24$), shown in Figure 5.6, we have the following set of eight consecutive frequency bounds for the seven consecutive bands between $\omega\tau_c = 0$ and 3ϖ: {0, 0.45, 2.69, 3.60, 5.83, 6.74, 8.97, 3ϖ}, which correspond to dimensional frequencies, f (Hz) (with τ_c = 3 samples at 44.1 kHz) given by the set: {0, 1061.5, 6288.5, 8411.5, 13638.5, 15761.5, 20988.5, 22050}, as marked by the vertical lines in Figure 5.6. Bands 1 and 5 belong to Branch I and are regularized with $\beta = \beta_{\mathrm{I}}(\omega)$; Bands 3 and 7 belong to Branch II and are regularized with $\beta = \beta_{\mathrm{II}}(\omega)$; and Bands 2, 4, and 6 belong to Branch P and are not regularized. In general, successive bands, starting from the lowest-frequency one, are mapped to the following succession of branches: I, P, II, P, I, P, II, P, . . .

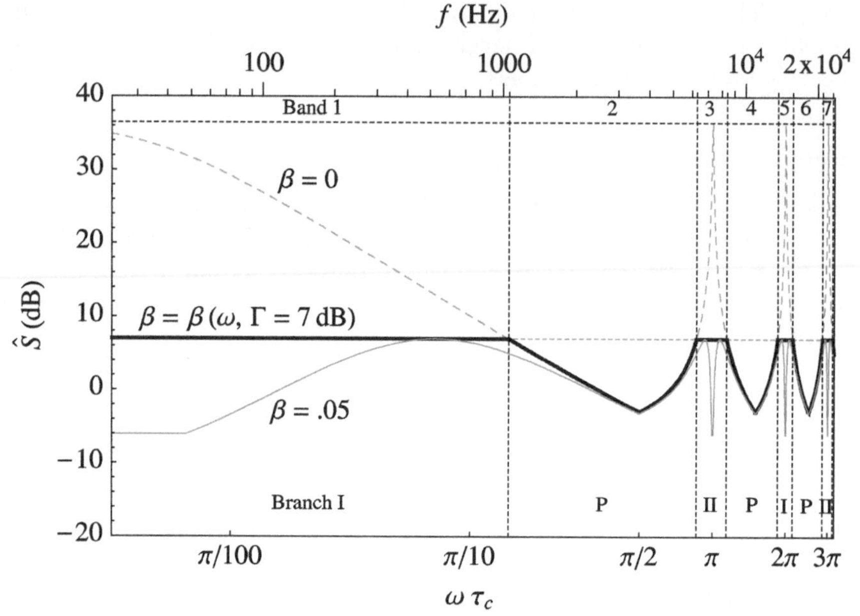

Figure 5.6 Envelope spectrum at the loudspeakers, $\hat{S}(\omega)$, for the case of frequency-dependent regularization with $\Gamma = 7$ dB (thick black curve) and for the corresponding reference case of $\beta = 0.05$ (grey curve). The benchmark case of the perfect XTC filter is also shown (dashed grey curve). The vertical dotted lines show the frequency bounds of the resulting seven bands, which are numbered consecutively at the top of the plot, and labeled with the corresponding branch name at the bottom. (Other parameters are the same as for Figure 5.2.)

Frequency Response

The amplitude envelope of the frequency response at the loudspeakers, given by Equation (5.49), was already shown in Figure 5.6. The other optimized metric spectra can be derived as follows:

$$Y_I^{[O]}(\omega) = Y^{[\beta_I(\omega)]}(\omega), \text{ for Branch–I bands;} \tag{5.60}$$

$$Y_I^{[O]}(\omega) = Y^{[\beta_{II}(\omega)]}(\omega), \text{ for Branch–II bands;} \tag{5.61}$$

$$Y_P^{[O]}(\omega) = Y^{[P]}(\omega), \text{ for Branch–P bands;} \tag{5.62}$$

where $Y(\omega)$ represents any of the eight metric spectra we defined in the section "Metrics," the superscript "[O]" denotes the sought optimized version of that metric spectrum, the subscripts "I," "II," and "P" denote each of the three branches, and the superscripts "$[\beta_I(\omega)]$" and "$[\beta_{II}(\omega)]$" denote regularization following the formulas for the regularized metric spectra in the section "Frequency Response," but with β taken to be frequency-dependent according to Equations (5.52) and (5.53).

For example, following the above hierarchical prescription, and using Equations (5.28), (5.52), (5.53), and (5.17), the optimized crosstalk cancellation spectrum becomes

$$\chi^{[O]}_{\mathrm{I,II}}(\omega) = \mp\frac{\gamma x(b \mp x) \mp b\sqrt{b \mp x}}{|x|(\gamma(b \mp x) - \sqrt{b \mp x})}, \tag{5.63}$$

$$\chi^{[O]}_{\mathrm{P}}(\omega) = \chi^{[P]}(\omega) = \infty, \tag{5.64}$$

where, for compactness, we have used the definitions $x \equiv 2g\cos(\omega\tau_c)$ and $b \equiv g^2 + 1$, and combined both branches into one expression using the double subscripts "I, II" and the double sign ($\pm$ or $\mp$) with the top and bottom signs associated with Branches I and II, respectively. Similarly, the optimized version of the ipsilateral frequency response at the ear for a side image, $E_{\mathrm{si}_{\|}}(\omega)$, becomes

$$E^{[0]}_{\mathrm{si}_{\|\,\mathrm{I,II}}}(\omega) = \pm\frac{\gamma^2 x(b \mp x) \;\pm \gamma b\sqrt{b \mp x}}{(b \mp x) \;\pm\; 2\gamma x\sqrt{b \mp x}} \tag{5.65}$$

$$E^{[0]}_{\mathrm{si}_{\|\,\mathrm{P}}}(\omega) = E^{[P]}_{\mathrm{si}_{\|}}(\omega) = 1 \tag{5.66}$$

These spectra are plotted in Figure 5.7 where it is immediately clear from the $\chi(\omega)$ curves that frequency-dependent regularization yields a significant enhancement of XTC level over that

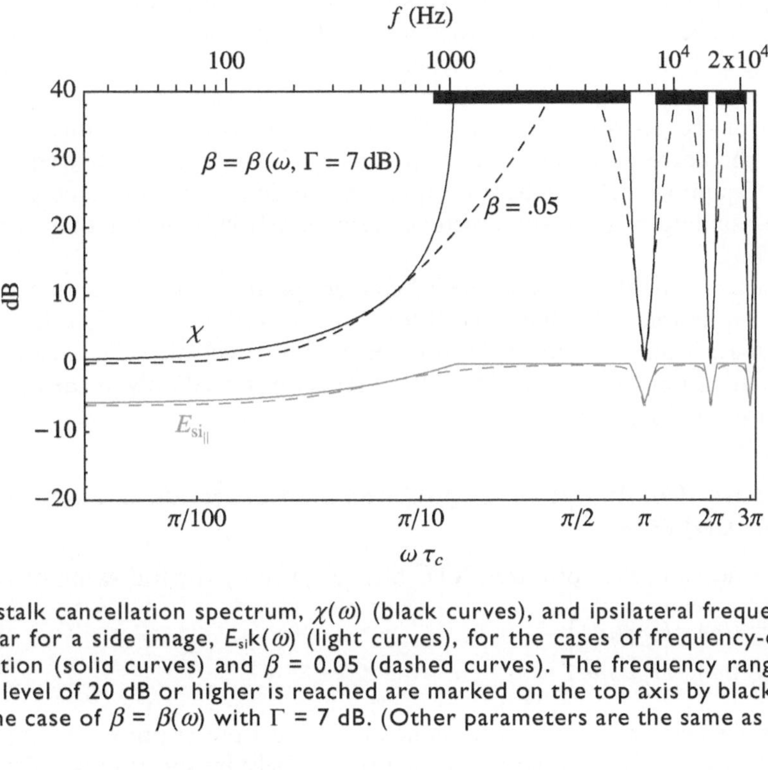

Figure 5.7 Crosstalk cancellation spectrum, $\chi(\omega)$ (black curves), and ipsilateral frequency response at the ear for a side image, $E_{\mathrm{si}_k}(\omega)$ (light curves), for the cases of frequency-dependent regularization (solid curves) and β = 0.05 (dashed curves). The frequency ranges for which an XTC level of 20 dB or higher is reached are marked on the top axis by black horizontal bars for the case of $\beta = \beta(\omega)$ with Γ = 7 dB. (Other parameters are the same as for Figure 5.2.)

obtained with constant-parameter regularization. We can also deduce from this plot that the higher the desired minimum level of XTC, the larger the XTC enhancement over that attained with the corresponding constant-parameter regularization.

Furthermore, this XTC enhancement occurs with minimal penalty to the frequency response at the ears, as can be seen by comparing the $E_{si_{II}}(\omega)$ spectrum with frequency-dependent regularization (solid grey curve) to that with $\beta = 0.05$ (dashed grey curve) in the same figure.

It can be verified through Equations (5.28) and (5.63) that constant-parameter regularization yields an XTC level that is equal to that obtained with the corresponding frequency-dependent regularization only at the discrete frequencies at which the peaks in the corresponding $\hat{S}^{[\beta]}(\omega)$ spectrum are located, i.e., at

$$\omega\tau_c = \begin{cases} n\pi & \text{if} \dfrac{1}{4\gamma^2} < (g-1)^2, \\ n\pi \pm \cos^{-1}\left(\dfrac{g^2-\beta+1}{2g}\right) & \text{if} (g-1)^2 \leq \dfrac{1}{4\gamma^2} \ll 1, \end{cases} \tag{5.67}$$

with $n = 0, 1, 2, 3, 4, \ldots$

(where the inequalities are those conditioning singlet or doublet peaks in the corresponding $\hat{S}^{[\beta]}(\omega)$ spectrum, and are derived from Equations (5.29), (5.30), and (5.32)). At all other frequencies, frequency-dependent regularization yields superior XTC performance to that obtained with constant-parameter regularization. This behavior, which can also be seen graphically in the $\chi(\omega)$ curves of Figure 5.7, is due to the fact that forcing the envelope spectrum to be flat (in bands belonging to Branches I and II) through frequency-dependent regularization clamps the effort penalty term in the cost function (second term in the sum in Equation (5.23)), leading to a minimization of the performance error. This in turn leads to a maximization of XTC level, which exceeds the corresponding constant-parameter XTC level at all frequencies (except at those given by Equation (5.67), where both corresponding $\hat{S}$ spectra reach the same value, γ), since the corresponding constant-parameter envelope, $\hat{S}^{[\beta]}(\omega)$, is lower than (or equal to) γ (as seen in Figure 5.6).

Therefore, we conclude that if we define XTC filter optimization as "the maximization of XTC performance for a desired tolerable level of tonal distortion" as we did earlier, only frequency-dependent regularization leads to an optimal XTC filter over all frequencies, while constant-parameter regularization leads to an XTC filter that is optimized only at the discrete frequencies given by Equation (5.67).

Impulse Response: The Analytical Band-Assembled Crosstalk Cancellation Hierarchy (BACCH) Filter

In the frequency domain, the optimized XTC filter is given by the following matrix:

$$H^{[O]} = \begin{bmatrix} H_{LL}^{[O]}(i\omega) & H_{LR}^{[O]}(i\omega) \\ H_{RL}^{[O]}(i\omega) & H_{RR}^{[O]}(i\omega) \end{bmatrix}, \tag{5.68}$$

whose elements are derived following the same hierarchical prescription (i.e., Equations (5.60)—(5.62)) we used to get the optimized metric spectra, namely by substituting $\beta_I(\omega)$ from Equation

(5.52), $\beta_{\text{II}}(\omega)$ from Equation (5.53), and $\beta = 0$ into each of Equations (5.25) and (5.26) to get the Branch-I, Branch-II, and Branch-P versions, respectively, of the filter's matrix elements. This leads to

$$H^{[O]}_{LL_{\text{I,II}}}(i\omega) = H^{[O]}_{RR_{\text{I,II}}}(i\omega) = \frac{\gamma^2[\pm x - g^2(1+e^{2i\omega\tau_c})] + \gamma\sqrt{b \mp x}}{(b \mp x) \pm 2\gamma x\sqrt{b \mp x}}, \tag{5.69}$$

$$H^{[O]}_{LR_{\text{I,II}}}(i\omega) = H^{[O]}_{RL_{\text{I,II}}}(i\omega) = \frac{\mp\gamma^2[\pm x - g^2(1+e^{2i\omega\tau_c})] + g\gamma e^{i\omega\tau_c}\sqrt{b \mp x}}{(b \mp x) \pm 2\gamma x\sqrt{b \mp x}}, \tag{5.70}$$

$$H^{[O]}_{LL_{\text{P}}}(i\omega) = H^{[O]}_{RR_{\text{P}}}(i\omega) = H^{[P]}_{LL}(i\omega) = H^{[P]}_{RR}(i\omega), \tag{5.71}$$

$$H^{[O]}_{LR}(i\omega) = H^{[O]}_{RL}(i\omega) = H^{[P]}_{LR}(i\omega) = H^{[P]}_{RL}(i\omega), \tag{5.72}$$

where, again, $x \equiv 2g\cos(\omega\tau_c)$ and $b \equiv g^2 + 1$, and we have followed the same subscript and sign conventions used to compact the XTC spectrum in Equation (5.63). Equations (5.71) and (5.72) give the Branch-P elements of the matrix of the optimized filter, which are also the elements of the perfect XTC filter's matrix given by Equations (5.45) and (5.46), whose inverse Fourier transforms had given us the IRs expressed in Equations (5.47) and (5.48). Therefore, we need to derive only the IRs associated with Branches I and II of the optimized filter.

To do so, we follow, albeit through more cumbersome algebra, the same approach we used to obtain the constant-parameter IRs in the section "Metrics"; namely, we seek to factor the frequency-domain representation of each element of the filter matrix into a product of terms, whose IFT can be readily found, or which can be expressed as a convergent series of functions whose IFT can be readily found. The complete IR is then the convolution of the IFTs of all the terms in the factored frequency-domain representation of the filter. The challenge is to carry out the factorization in such a way that all the invoked power series expansions converge over the parameter space of interest.

The derivation is carried out in Appendix A, where we also discuss the convergence of the adopted series expansions. The resulting filter in the time domain is given by the following two IRs:

$$h^{[O]}_{LL_{\text{I,II}}}(t) = h^{[O]}_{RR_{\text{I,II}}}(t) = (\psi_0 + \gamma\psi_1) * \psi_a, \tag{5.73}$$

$$h^{[O]}_{LR_{\text{I,II}}}(t) = h^{[O]}_{RL_{\text{I,II}}}(t) = [\mp\psi_0 + g\gamma\delta(t+\tau_c) * \psi_1] * \psi_a, \tag{5.74}$$

where

$$\psi_a = \pm(\psi_2 * \psi_3) \pm (\psi_1 \mp \psi_4) * \psi_5 * \psi_6(c_1) * \psi_6(c_2),$$

$$\psi_0 = \pm g\gamma^2[\delta(t-\tau_c) + \delta(t+\tau_c)] - g^2\gamma^2[\delta(t) + \delta(t+2\tau_c)],$$

$$\psi_1 = \sum_{m-0}^{\infty} \binom{\frac{1}{2}}{m} (\mp g)^m \left(g^2+1\right)^{\frac{1}{2}-m} \times \sum_{k=0}^{m} \binom{m}{k} \delta\left(t-(2k-m)\tau_c\right),$$

$$\psi_2 = \pm\frac{1}{4g\gamma} \sum_{m-0}^{\infty} \binom{-\frac{1}{2}}{m} 4^{-m} \times \sum_{k=0}^{2m} \binom{2m}{k} (-1)^k \delta\left(t+(2(\mathrm{m}-k)\tau_c\right),$$

$$\psi_3 = \sum_{m-0}^{\infty} \binom{-\frac{1}{2}}{m} (\mp g)^m \left(g^2+1\right)^{-\frac{1}{2}-m} \times \sum_{k=0}^{m} \binom{m}{k} \delta\left(t-(2k-m)\tau_c\right), \tag{5.75}$$

$$\psi_4 = 2g\gamma[\delta(t+\tau_c)+\delta(t+\tau_c)],$$

$$\psi_5 = \pm\frac{1}{(4g\gamma)^3} \sum_{m-0}^{\infty} \binom{-\frac{3}{2}}{m} 4^{-m} \times \sum_{k=0}^{2m} \binom{2m}{k} (-1)^k \delta\left(t+(2(\mathrm{m}-k)\tau_c\right),$$

$$\psi_6(c) = \sum_{m-0}^{\infty} \left(\frac{\pm c}{2g}\right)^p \sum_{m=0}^{\infty} \binom{-\frac{p}{2}}{m} 4^{-m} \times \sum_{k=0}^{2m} \binom{2m}{k} (-1)^k \delta\left(t+(2(\mathrm{m}-k)\tau_c\right),$$

with the constants c_1 and c_2 given by

$$c_1 = \frac{\sqrt{16\gamma^2(g^2+1)+1} \mp 1}{8\gamma^2}, \tag{5.76}$$

$$c_2 = \frac{-\sqrt{16\gamma^2(g^2+1)+1} \mp 1}{8\gamma^2}. \tag{5.77}$$

The impulse responses are valid for values of γ and g that satisfy the condition:

$$\max\left(\frac{\sqrt{5+\sqrt{5}}}{2\sqrt{g^2+1}}, 1\right) \leq \gamma \leq \frac{1}{1-g}, \tag{5.78}$$

which is shown graphically as a region plot in Figure 5.A.1 in Appendix A.

The impulse responses for Branch I and Branch II of this optimal filter are shown in Figure 5.8 for our typical case of $g = 0.985$ and $\tau_c = 3$ samples, and, along with the perfect filter IRs shown in the top panel of Figure 5.5, completely specify the optimal XTC filter.

Compared to the corresponding ($\beta = 0.05$) constant-parameter IRs in the bottom panel of Figure 5.5, the optimal XTC IRs shown in Figure 5.8 are more complex in their structure. Furthermore, each IR consists of a train of deltas that are spaced by τ_c as opposed to the $2\tau_c$ intervals we had for the perfect and constant-parameter filters.

These IRs are difficult to interpret physically because they also include the time response associated, in the frequency domain, with frequency bands where the IR is not valid. This is illustrated in Appendix B, in the bottom panel of Figure 5.B.1, where the envelope spectrum

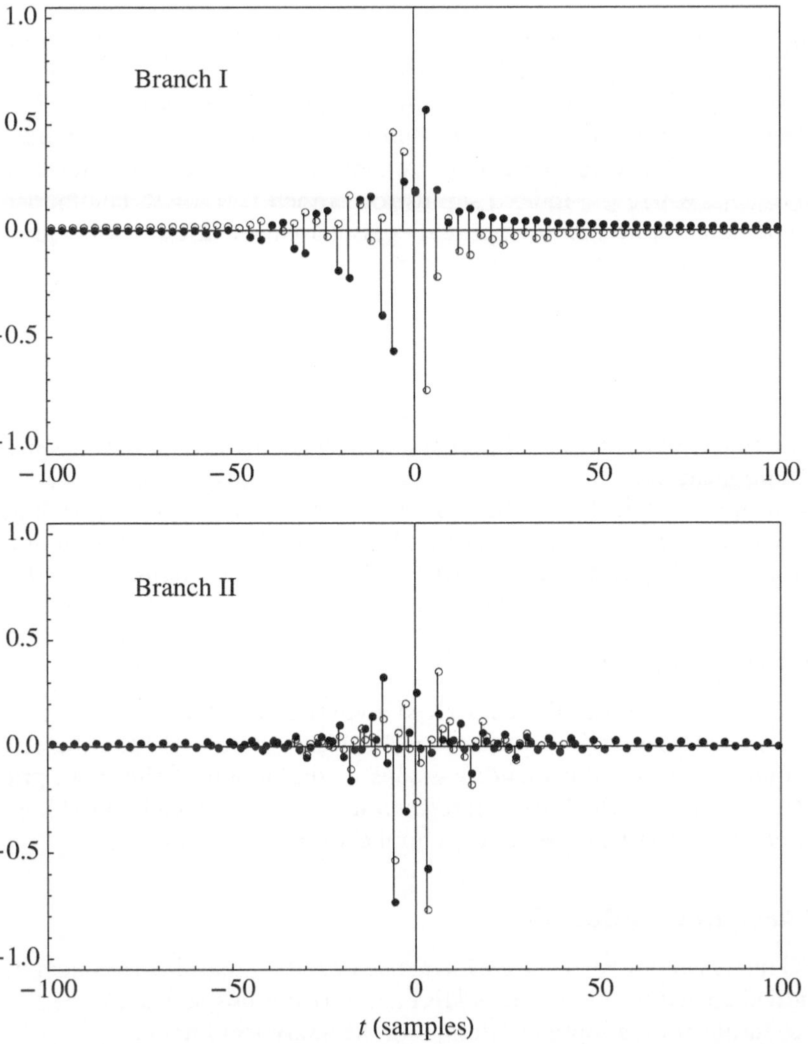

Figure 5.8 Impulse responses of the optimal XTC filter: $h_{LL}^{[O]}(t) = h_{RR}^{[O]}(t)$ (filled circles) and $h_{LR}^{[O]}(t) = h_{RL}^{[O]}(t)$ (empty circles), for Branch I (top panel) and Branch II (bottom panel). (Γ = 7 dB, g = 0.985, and τ_c = 3 samples, as in Figure 5.2.)

obtained from the Fourier transform of the Branch-I optimal IR is compared to the expected flat envelope spectrum, $\hat{S}_{\mathrm{I}}^{[O]}(\omega) = \gamma$. The agreement is excellent only in the bands belonging to the branch for which the IR is intended (which, in the case illustrated in that plot, are the first and fifth bands). In other bands, not only is the IR not valid, but, as discussed in the appendices, its application may lead to singularities associated with the divergence of some of the series that constitute it (see for instance the singularities appearing in the Branch-P bands in Figure 5.B.1).

Therefore, in principle, the application of the optimal filter requires that, prior to XTC filtering, the recorded signal, $[d_{Li}(t), d_{Ri}(t)]^T$, be passed through a crossover filter whose crossover frequencies are set to the band bounds given by the hierarchical prescription in Equations (5.55)–(5.58). The resulting bands are then assembled into three groups (I, II, and P) according to their branch identity. The combined recorded stereo signals in each group can thus be represented by a vector $[d_{Li}(t), d_{Ri}(t)]^T$, where the index i stands for Branch I, II, or P. The loudspeakers source vector, in the time domain, needed for optimal crosstalk cancellation, is then given by the time-domain version of Equation (5.9):

$$\begin{bmatrix} v_L(t) \\ v_R(t) \end{bmatrix} = \sum_i \left(\begin{bmatrix} h^{[O]}_{LL_i}(t) & h^{[O]}_{LR_i}(t) \\ h^{[O]}_{RL_i}(t) & h^{[O]}_{RR_i}(t) \end{bmatrix} * \begin{bmatrix} d_{L_i}(t) \\ d_{R_i}(t) \end{bmatrix} \right), \tag{5.79}$$

where the summation is over the three branches, and the convolution operates in the same fashion as matrix multiplication.

Causality is ensured by calculating the IRs with a "pre-delay," starting back at a time $t < 0$, whose exact temporal extent is not important as long as it allows the inclusion of the salient part of the IR. For the IRs in Figure 5.8, this pre-delay should start at about $t = -100$ samples.

The Analytical BACCH Filter

We refer to the XTC filter whose IR was derived analytically in the previous section as the analytical Band-Assembled Crosstalk Cancellation Hierarchy (BACCH) filter. Before we apply the insight we obtained from our previous discussions to the design of the more practically useful HRTF-based BACCH filters (which we will refer to simply as "BACCH filters"), we discuss some aspects of the analytical BACCH filter and its applications.

The Value of Analytical BACCH Filters

Analytical XTC filters cannot rival the performance of HRTF-based XTC filters because the former are not individualized to the listener's HRTF and real loudspeakers, at best, only approach the point-source idealization adopted in designing the analytical filters.

It has been shown that with non-individualized HRTF-based XTC filters, the practically achievable XTC level seldom exceeds 17 dB over a wide frequency range, and that this mismatch generally leads to a corruption of localization cues (Akeroyd et al., 2007) and an increase in localization errors (Majdak et al., 2013). However, as argued in the section "Background and Motivation," even relatively low levels of XTC can significantly enhance the spatial perception evoked by most binaural and stereo recordings. Consequently, for applications where the precise localization of virtual sound sources is not critical, an optimal analytical XTC filter, even one based on a free-field model (such as the analytical BACCH filter derived in the previous section), can become competitive, especially in situations where it is both calculated for and used with a loudspeaker span that is small enough to diminish the relative importance of head-shadowing effects (which are, of course, not accounted for in a free-field model). In such applications, an optimal analytical XTC filter can offer the following advantages over an HRTF-based XTC filter:

1. The simplicity of using a single (i.e., "universal") filter for all individuals.
2. Shorter filters which incur lower CPU loads on the digital processor.
3. Easy automatic re-calculation of the filter as a function of the changing parameters of the listening configuration.

With this justification for the usefulness of analytically derived BACCH filters, we turn our attention to some practical issues related to their specific design and their application to real listening situations.

Analytical BACCH Filter Design Strategy

Of course, filter design strategies depend on performance requirements (desired maximum tolerable coloration level or minimum XTC level) and the specifics and constraints of the listening configuration (constraints on the listening distance, l, and the loudspeaker span, Θ, and, to some extent, the sound reflection characteristics of the listening room).

One approach to analytical BACCH filter design is to start with the specification of the maximum tolerable coloration level, that is, Γ in dB. For instance, in critical (e.g., audiophile) listening and audio mastering applications, it may be undesirable to have Γ exceed 3–5 dB, while in home-theatre applications, audio (spectral) fidelity may be intentionally compromised with higher values of Γ in exchange for the advantage of having more XTC headroom for reproducing surround effects with the two loudspeakers.

The choice of loudspeaker span is particularly important. In cases where the span angle is constrained to a set value, as for compatibility with the so-called standard stereo triangle (i.e., $\Theta = 60°$), the value of Θ becomes a fixed input to the design process and is used, along with l, to calculate g and τ_c from Equations (5.3)–(5.6). (The inequality in Equation (5.78), which is typically easy to satisfy, must hold for that particular combination of $\gamma = 10^{\Gamma/20}$ and g. If not, one of the input parameters, usually Γ, must be adjusted accordingly before proceeding further with the design.) In cases where Θ is not constrained to a preset value, it becomes a useful variable in the filter design process and can be used to simplify the filter, as discussed in the section "Simplified Implementation" below.

With γ, g, and τ_c specified, one has all the parameters needed to calculate the spectra associated with the optimal XTC filter, as described in the sections "Band Hierarchy" and "Frequency Response," and thus evaluate the various aspects of the filter. (These evaluations are more conveniently done in terms of the dimensional frequency, f, in Hz, by selecting the intended sampling rate.) In particular, a plot of the XTC spectrum according to Equations (5.63) and (5.64) allows the evaluation of the XTC performance of the filter (defined as the frequency extent over which a desired minimum XTC level is reached or exceeded), which, by virtue of the implicit optimization (i.e., minimization of the cost function in Equation (5.23)), is the maximum achievable XTC performance for that particular set of input parameters. If the calculated XTC performance is judged by some empirical standards to exceed that achievable in the intended listening environment (for instance, sound reflections in a reverberant room may limit the achievable XTC to only a few dB over a good part of the audio spectrum), the calculation can be repeated with a lower value of Γ, thus leading to even higher spectral fidelity. Conversely, a lower than desired XTC performance can be amended by raising Γ.

Once the target XTC performance and coloration level are reached, one proceeds to the time domain by calculating the Branch-P IRs from Equations (5.42)–(5.44) (with $\beta = 0$), and the Branch-I and -II IRs from Equations (5.73)–(5.77). The loudspeakers source vector can then be calculated according to Equation (5.79), following the prescription given in the text preceding that equation, i.e., by appropriately convolving the 3-part IRs with the recorded stereo signal after having passed the latter through a multi-band crossover filter whose crossover frequencies are set to the band bounds given in Equations (5.55)–(5.58). The convolution operations can be carried out digitally, and in real-time if desired, using a digital convolution plugin. (Such software plugins often rely on FFT-based algorithms (e.g., Gardner, 1995) for fast convolution and have become readily available in the commercial and public domains for use as IR-based reverberation processors.)

Simplified Implementation

An XTC system consisting of the properly configured crossover filter, the three XTC IR matrices, and the multiple instances of convolution plugins can be considered as a single filter, having stereo inputs and outputs, which acts as a linear operator. Therefore, once assembled, the filter can be "rung" once by a single delta impulse, applied to one of its two inputs, and the recorded stereo output would then represent one of the two columns of the 2×2 IR matrix of the entire filter. Due to the symmetry of the filter, the other column of the IR matrix is obtained by simply flipping the two recorded outputs. This results in a single IR matrix, representing the entire three-branch multi-band filter, and simplifies any future application of Equation (5.79) to a simpler one (with no crossover filtering) in which the summation and indices are foregone.

The Role of Loudspeaker Span

Another important simplification arises in applications where the loudspeaker span, $\Theta = 2\theta$, is not constrained to a preset value, such as the 60° of the standard stereo triangle, and therefore can be a variable in the filter design process. Since τ_c depends on the loudspeaker span, the bounds of the bands can be moved by varying θ. By setting θ equal to a particular value, θ^*, the upper bound of the second band (which belongs to Branch P) can be made to coincide with a cutoff frequency, f_c, above which XTC is psychoacoustically not needed. Such a band-limited optimal XTC filter has the advantage that it requires only a 2-band crossover filter, and its IR consists of only the Branch-I and Branch-P parts, thus leading to significant simplifications in the design and implementation of the filter.

To find an expression for θ^* as a function of f_c, under the typically valid approximations $g \simeq 1$ and $l \gg \Delta r$, we set $\omega\tau_c$ equal to the upper bound of the second band (which, from Equation (5.56), is $\pi - \varphi$), use Equation (5.21), and solve for θ, to get

$$\theta^* \simeq \sin^{-1}\left[\frac{c_s\left(\pi - \cos^{-1}\left[\dfrac{2\gamma^2 - 1}{2\gamma^2}\right]\right)}{2\pi f_c \Delta r}\right]. \tag{5.80}$$

A number of studies have suggested that XTC above a frequency of about 6 kHz is not critical or perhaps even necessary (Bai & Lee, 2007; Gardner, 1998; Majdak et al., 2013). Therefore, we set f_c equal to that value in the above equation, solve for θ^*, design the filter for a loudspeaker span of $2\theta^*$, use a 2-band crossover filter to separate the first two bands, apply the Branch-I and Branch-P parts of the filter to the first and second bands, respectively, and allow the part of the audio spectrum above f_c to bypass the filter. (Of course, to do so would require an additional 2-band crossover at f_c that precedes the one used to apply the XTC filter.)

It is relevant to mention in the context of loudspeaker span that keeping Θ small offers advantages that have been recognized since Kirkeby et al., (1998b) presented their analysis of the "stereo dipole" configuration, which has a span of only 10°. Objective and subjective evaluations of the effects of loudspeaker span in XTC systems have indicated that such a low-Θ configuration gives a larger sweet spot than that obtained with larger loudspeaker spans (Bai & Lee, 2006b; Parodi & Rubak, 2010; Takeuchi et al., 2001). This effect can be attributed to the relative insensitivity of the path length difference, Δl, to head movements when the span is small. On the other hand, the study by Bai and Lee (2006b) favored larger spans partly because increasing the span (while keeping the distance l fixed) lowers the value of g and consequently decreases the magnitude of the coloration peaks as well as the condition numbers. We do however expect, in light of our study of regularization, that an optimal XTC filter in which regularization is used to flatten these peaks and lower the condition numbers, while maintaining good XTC performance, should tip the balance in favor of lower values of Θ. The results of the study by Parodi and Rubak (2010), in which frequency-dependent regularization was employed subject to a 12 dB gain-limit on the XTC filters, seem to suggest that this is indeed the case.

Another argument in favor of small loudspeaker spans is particular to the use of analytical filters based on a free-field model, such as those discussed in this chapter. Since the free-field model ignores the presence of the listener's head, it should be expected that filters based on it perform better when the effects of head shadowing are minimized. This situation can be approached by decreasing the span angle as can be seen, for instance, in Figure 3.13 of Gardner (1998), where the inter-aural transfer function (the ratio of the frequency responses at the two ears) of a typical human head, measured as a function of the azimuthal position of a sound source, is small (about –2 dB) and flat (within 2 dB) for a small horizontal source azimuth ($\theta = 5°$), but increases and becomes less flat with increasing azimuths.

An Example

To illustrate the above design guidelines and discussions, we give the example of a listening situation whose only two design requirements are a distance $l = 1.6$ m and a maximum coloration level of $\Gamma = 7$ dB. From Equation (5.80), with $f_c \simeq 6$ kHz, and $\Delta r = 15$ cm,[7] we get $\theta = 9°$, which we take as half the loudspeaker span. From Equations (5.3)–(5.6), we then find $g = 0.985$ and $\tau_c = 3$ samples at a sampling rate of 44.1 kHz. These are precisely the dimensional and non-dimensional parameters chosen for the calculations that are illustrated in the plots throughout this chapter. The Branch-P and Branch-I IRs are therefore given by those shown in the top panels of Figure 5.5 and Figure 5.8, respectively. The Branch-II IR is not needed as the XTC filter is limited to 6 kHz, which, by design, was made to be the upper bound of the second band (Branch P). The spectra associated with this filter are given by the solid curves in Figure 5.6 and Figure 5.7, with the dimensional frequency read off the top axes of the plots, up to the cutoff frequency of 6 kHz. In particular, we note that the XTC performance (top curve in Figure 5.7) exceeds 20 dB for a wide range of frequencies that extends from the 6 kHz cutoff down to 850 Hz, then drops off with decreasing frequency, reaching 5 dB at 290 Hz.

Individualized BACCH Filters

The BACCH Filter Design Method

Individualizing the BACCH filter to include the particular characteristics of the loudspeakers and the HRTF of the listener can lead to a significant enhancement of the realism of the 3D spatial imaging of binaural audio through loudspeakers.

We now describe the steps (shown schematically in Figure 5.9) of the technique (Choueiri, 2015) for designing such BACCH filters starting from the measured transfer function of a real listener in front of a pair of real loudspeakers.

- The starting point is a 2 × 2 impulse response measurement of the two loudspeakers using a binaural microphone in the ears of the listener. Such a measurement can be obtained through standard IR deconvolution using, for instance, the exponential sine-sweep technique (Farina, 2000, 2007). Each of the 4 impulse responses of this transfer function is FFTed to obtain the system's measured transfer matrix in the frequency domain (i.e., matrix C as in Equation (5.12)).
- In Step 1, the system's measured transfer matrix C is inverted numerically, using zero or a very small constant regularization parameter (large enough to avoid machine inversion problems) to obtain the corresponding perfect XTC filter, $\boldsymbol{H}^{[P]}$.
- In Step 2, the amplitude vs frequency response at the loudspeaker $\hat{S}^{[P]}$ is calculated and its lowest value (in dB) is taken to be Γ^*, then $\gamma^* = 10\Gamma^*/20$ is calculated.
- In Step 3, the frequency-dependent regularization parameter (FDRP), $\beta(\omega)$, that would result in a flat frequency response at the loudspeakers is calculated, so that $\hat{S}^{[\beta]}(\omega)$ = constant $\leq \gamma^*$, thus forcing XTC to be caused by phase effects only.
- In Step 4, the FDRP thus obtained, $\beta(\omega)$, is used to calculate the pseudoinverse of the system's transfer matrix (e.g., according to Equation (5.22)), which yields the sought regularized optimal XTC filter $\boldsymbol{H}^{[\beta]}$ that has a flat frequency response at the loudspeakers. Finally, if needed for applying the resulting filter through a time-base convolution, as is often done in practical XTC implementation, a time domain version (impulse response) of the filter is obtained in the final step by simply taking the inverse Fourier transform of $\boldsymbol{H}[\beta]$.

It should be noted that in Step 3, if the FDRP is calculated so that $\hat{S}^{[\beta]}(\omega)$ = constant $\leq \gamma^*$, the spectral flattening occurs for a side image (i.e., a sound panned to either channel and thus would be perceived by a listener to be located at or near the ipsilateral ear when the XTC level is sufficiently high). However, the same method can be used to flatten the response at the loudspeakers for an image that is not a pure side image by simply requiring that $S^{[\beta]}(\omega)$ = constant $\leq \gamma^*$, where $S^{[\beta]}(\omega)$ is the XTC filter's frequency response for an image of source panned anywhere between the left and right channels. For instance, to flatten for a central image, we set $S_{ci}^{[\beta]}(\omega)$ (given, for instance, by the equation preceding Equation (5.27)) to a constant $\leq \gamma^*$, and proceed with the steps of the method as outlined above. In this context it is relevant to mention that for some

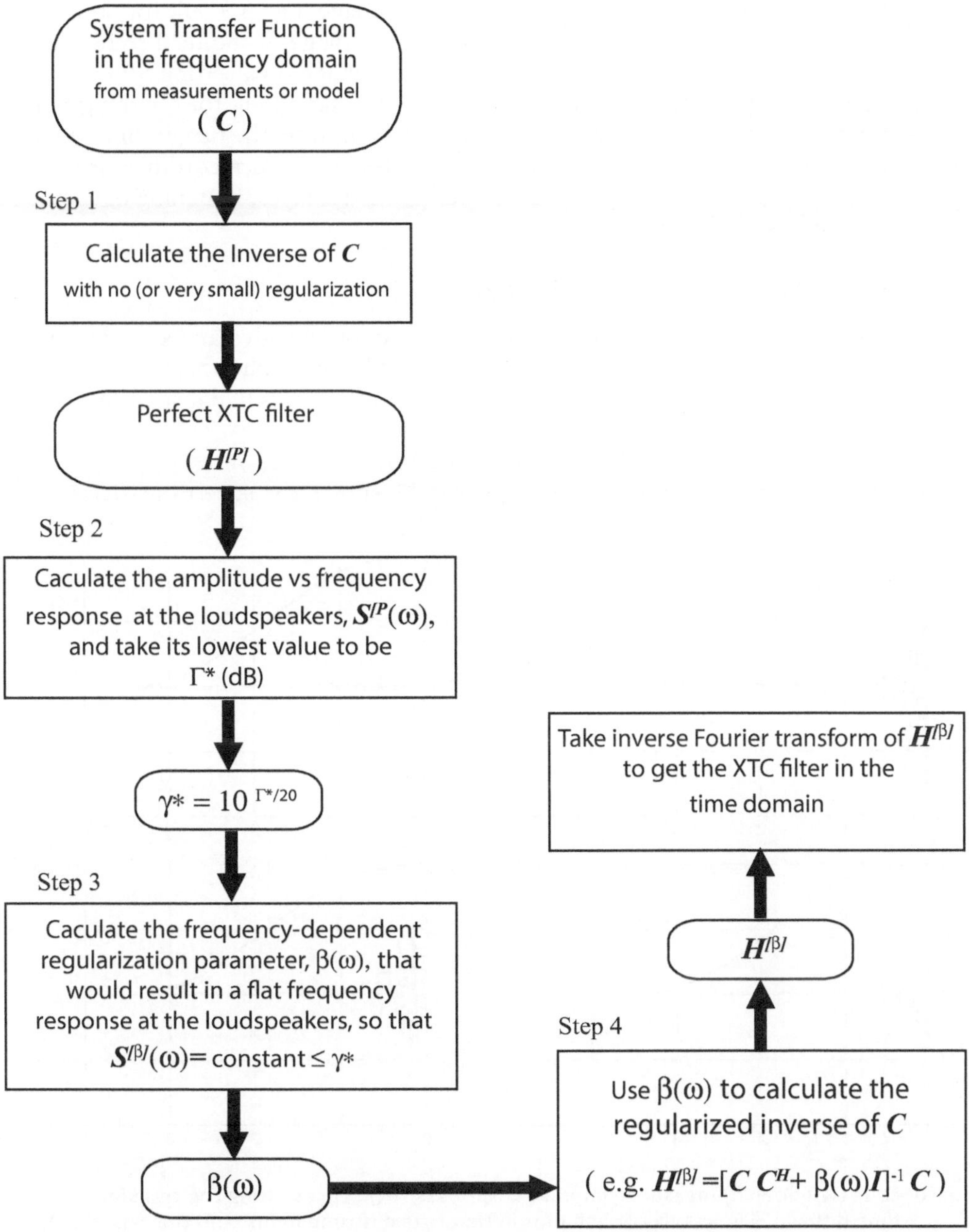

Figure 5.9 Flow chart illustrating the method for designing optimal XTC (i.e., BACCH) filters.

applications, for instance, pop music recording where the lead vocal audio is panned dead center, it might be desirable to flatten the response for a center image, i.e., $S_{ci}^{[\beta]}(\omega)$, (or an image of any other desired panning) in order to avoid coloration of that image. It should also be noted in that context that since $\hat{S}^{[\beta]}(\omega) \geq S^{[\beta]}(\omega)$, only flattening the side image (i.e., setting $S^{[\beta]}(\omega)$ = constant $\leq \gamma^*$) would result in no dynamic range loss. In other words, flattening for anything but the side image would incur a dynamic range loss that must be balanced by the benefit of a reduced tonal distortion for the desired panned image. For instance, for binaural recordings of real acoustic sound fields, which typically contain no dead-center panned images, flattening of the side image is advisable as it incurs no dynamic range loss.

Example Using a Measured Transfer Function

To illustrate the method described in the previous subsection, we give an example based on the transfer function of two loudspeakers in a room measured by microphones placed at the ear canal entrances of a dummy head (Neumann KU-100). The loudspeakers had a span of 60° at the listening position, which was about 2.5 m from each loudspeaker.

Figure 5.10 shows the four (windowed) measured impulse responses (IR) representing the transfer function in the time domain, and Figure 5.11 shows the spectra associated with the

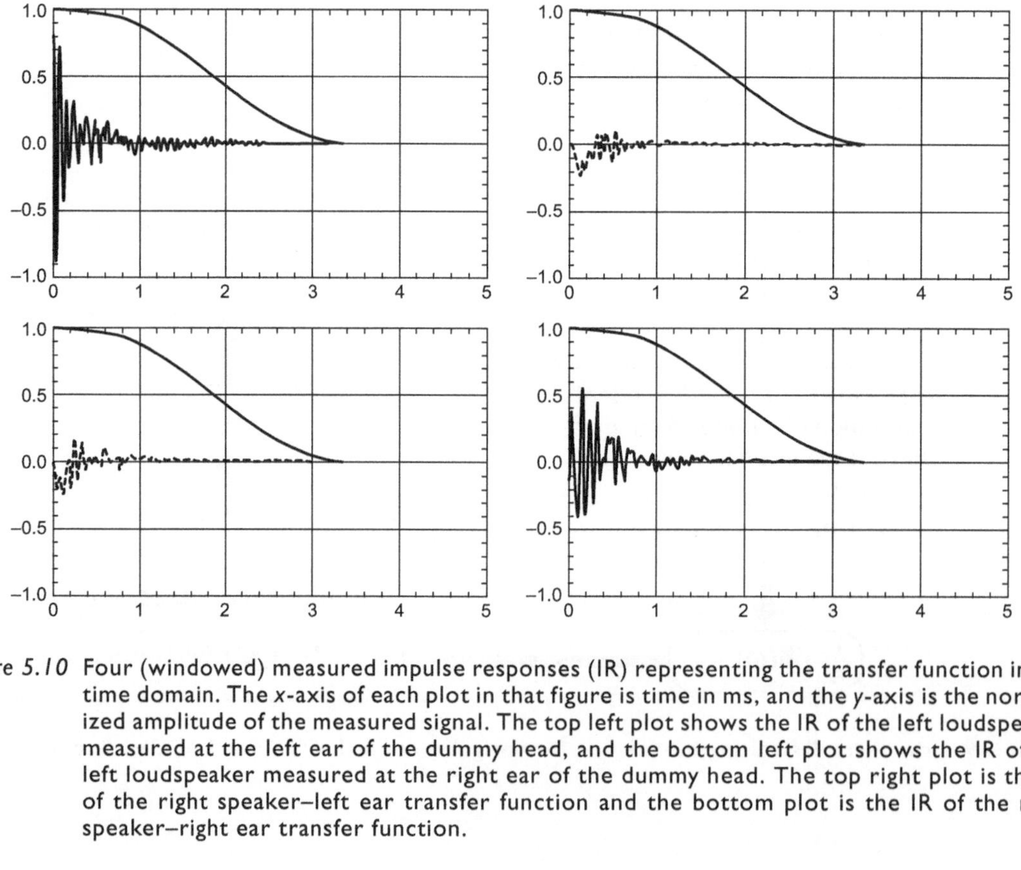

Figure 5.10 Four (windowed) measured impulse responses (IR) representing the transfer function in the time domain. The *x*-axis of each plot in that figure is time in ms, and the *y*-axis is the normalized amplitude of the measured signal. The top left plot shows the IR of the left loudspeaker measured at the left ear of the dummy head, and the bottom left plot shows the IR of the left loudspeaker measured at the right ear of the dummy head. The top right plot is the IR of the right speaker–left ear transfer function and the bottom plot is the IR of the right speaker–right ear transfer function.

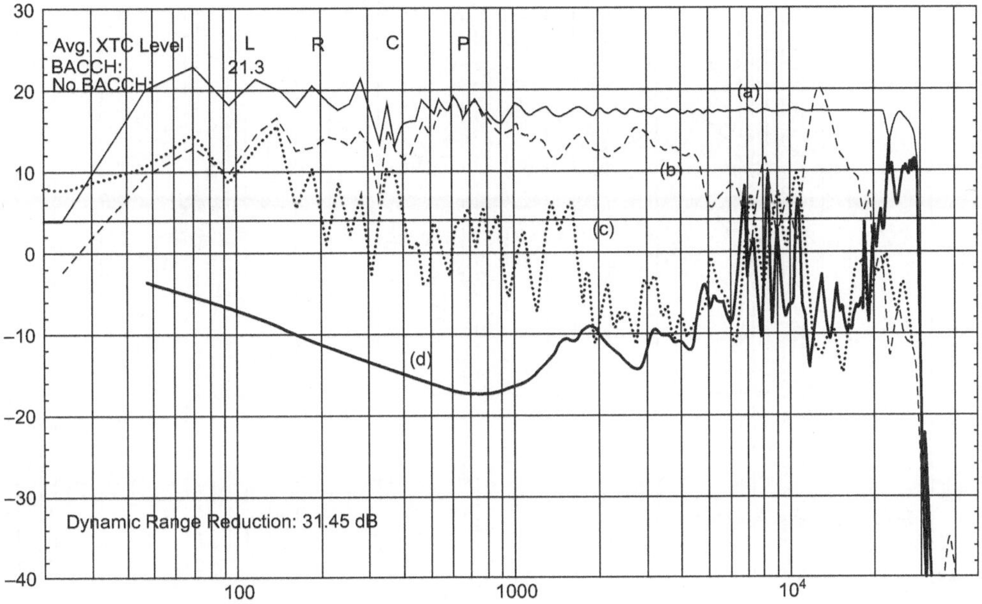

Figure 5.11 Measured spectra associated with the perfect XTC filter for the measured transfer function shown in Figure 5.10. The four curves represent: (a) the frequency response at the left (ipsilateral) ear; (b) the frequency response C_{LL} that corresponds to the left speaker–left ear transfer function; and (c) the frequency response measured at the right (contralateral) ear, E_{si_X}.

perfect XTC filter. The (b) curve in Figure 5.11 is the frequency response C_{LL} that corresponds to the left speaker-left ear transfer function in the frequency domain obtained by panning the test sound completely to the left channel. The ripples in that curve above 5 kHz are due to the HRTF of the head and the left ear pinna. The other curves in Figure 5.11 are the measured frequency responses associated with the perfect XTC filter, that is, an XTC filter obtained by inverting the transfer function with essentially no regularization ($\beta = 10^{-5}$). In particular, the (d) curve is the response at the left loudspeaker, $\hat{S}^{[\beta]}(\omega)$, and shows a dynamic range loss of 31.45 dB (difference between the maximum and minimum in that curve). The (a) curve is the frequency response at the left (ipsilateral) ear, $E_{\mathrm{si\,ll}}$, which, as expected from a perfect XTC filter, is essentially flat over the entire audio band. The faint grey curve labeled (c) is the corresponding frequency response measured at the right (contralateral) ear, E_{six}, and shows significant attenuation with respect to the (c) curve due to XTC. The difference in amplitude between the (a) curve and (c) red curve, linearly averaged over frequencies, is the average XTC level, which for this case is 21.3 dB.

We contrast these curves with those curves in Figure 5.12, which shows the responses due to a filter designed in accordance with the BACCH filter design method.

By design, the curve labeled (d) in that plot, representing $\hat{S}^{[\beta]}(\omega) \geq S^{[\beta]}(\omega)$, the response at the left loudspeaker, is completely flat over the entire audio spectrum. Consequently, the frequency

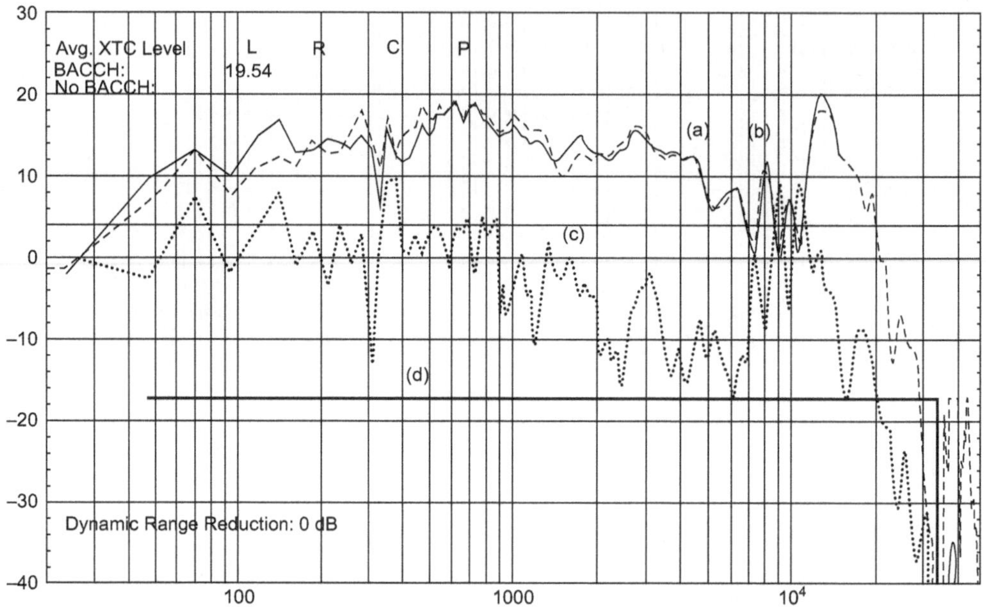

Figure 5.12 Measured spectra associated with the optimal XTC (BACCH) filter for the measured transfer function shown in Figure 5.10. The curves represent the frequency responses defined in the caption of Figure 5.11.

response at the left ear, curve (a), matches very well the corresponding measured system transfer function, C_{LL}, curve (b). Since $\hat{S}^{[\beta]}(\omega) \geq S^{[\beta]}(\omega)$ is flat, there is no dynamic range loss associated with this filter. The average XTC level for this filter (obtained by taking the linear average of the difference between the (a) and (c) curves) is 19.54 dB, which is only 1.76 dB lower than the XTC level obtained with the perfect filter, testifying to the optimal nature of the regularized filter.

In sum, the filter designed with the method described above imposes no audible coloration to the sound of the playback system, has no dynamic range loss, and yields an XTC level that is essentially the same as that of a perfect XTC filter.

Conclusions

Three-dimensional reproduction of binaural audio with two loudspeakers requires cancellation of the crosstalk between the loudspeakers and the contralateral ears of the listener. A perfect XTC filter (i.e., one with infinite crosstalk cancellation) can be easily designed but causes severe tonal distortion to the sound emitted by the loudspeakers due to the ill-conditioned inversion of the system's transfer function.

The coloration produced by the perfect XTC filter consists of peaks in the frequency spectrum that can typically exceed 30 dB and thus strain the playback transducers and significantly reduce the dynamic range of the playback system. Furthermore, the coloration is heard throughout the listening space and, due to extreme sensitivity to errors in the system, it is also heard by the listener in the sweet spot.

Using a free-field two-point-source model, we showed that constant-parameter regularization, which has been used previously to design HRTF-based XTC systems, can lower these peaks, but also produces a bass roll-off and high-frequency artifacts in the filter's frequency response. Furthermore, we demonstrated that constant-parameter regularization does not lead to the optimization of XTC filters across all frequencies, but rather only at discrete, widely spaced frequencies.

Full optimization can be achieved through frequency-dependent regularization and requires the audio spectrum to be divided into a hierarchical set of adjacent frequency bands, each of which belongs to one of three solution branches that make up the complete optimal filter. We derived analytical expressions for the three branches of the filter in terms of series expansions, which we showed are convergent for typical listening situations. The corresponding impulse responses were then obtained analytically and expressed as convolutions of trains of Dirac deltas.

The analytical XTC filters we derived under the simplifying assumptions of a free-field model can be useful in practical situations where individualized HRTF-based XTC filters are either too cumbersome to implement or not needed to attain the XTC levels required for enhancing the spatial fidelity of playback in reflective environments. We described a strategy for designing such optimal filters that meets practical design requirements and we gave an illustrative example for a typical listening configuration.

We concluded with a discussion of a method for designing optimal individualized (HRTF-based) XTC filters (BACCH filters) that impose no audible coloration to the sound of the playback system, have no dynamic range loss, and yield the high XTC level attainable from a perfect XTC filter.

Notes

1 Throughout this chapter, the words "recording" and "signal" are used interchangeably and are meant to also represent a live feed, or the HRTF-encoded signal for the artificial placement of sounds in a virtual acoustic space.

2 Throughout this chapter, the word "level" is meant to represent, generally, a frequency-dependent amplitude.

3 An exception could be made for recordings in which the specific placement of sound images was made with full accounting for crosstalk during playback, e.g., the case of stereo sound fields constructed with pan-potted mono images and monitored over loudspeakers, common in popular music recording.

4 While it has been shown that reliable discrimination of frontal and rear images requires highly controlled playback and individualized XTC systems (Majdak et al., 2013), the larger portion of the *direct* sound content in acoustic recordings, e.g., performed music, is of frontal origin and, with playback through frontal loudspeakers at modest levels of XTC, is largely immune to such localization confusion.

5 We use the terms "spectral coloration" and "tonal distortion" interchangeably.

6 On the other hand, at and near the frequencies for which the interference between in-phase (or out-of-phase) signals is complementary at the ears, XTC control requires slight attenuation instead of boosting (and implies a dynamic range gain, instead of loss). As shown by Takeuchi and Nelson (2002) and P. A. Nelson and Rose (2005), and as is reviewed in the section "Benchmark: Perfect Crosstalk Cancellation," these attenuations are not problematic as they correspond to frequencies where XTC control is most robust.

7 This value for the effective inter-ear separation, $\Delta r = 15$ cm, is justified by the relatively small loudspeaker span, following the guidelines of Takeuchi and Nelson (2002), who reported that good correlation between the peak frequencies in the data calculated using a free-field model, and those measured with the KEMAR dummy head, can be obtained by taking an effective $\Delta r \simeq 13$ cm for low values of θ, and $\Delta r \simeq 25$ cm for large source azimuths. The larger value, which is much larger than the minimum distance between the entrances of the ear canals of the dummy head, reflects the effects of diffraction around the head.

Bibliography

Akeroyd, M. A., Chambers, J., Bullock, D., Palmer, A. R., Summerfield, A. Q., Nelson, P. A., & Gatehouse, S. (2007). The binaural performance of a cross-talk cancellation system with matched or mismatched setup and playback acoustics. *The Journal of the Acoustical Society of America*, *121*(2), 1056–1069. Retrieved from http://scitation.aip.org/content/asa/ journal/jasa/121/2/10.1121/1.2404625 doi: http:// dx.doi.org/ 10.1121/1.2404625

Atal, B., Hill, M., & Schroeder, M. (1966, February 22). *Apparent Sound Source Translator*. Retrieved from www.google.com/patents/ US3236949. US Patent 3,236,949.

Bai, M. R., & Lee, C.-C. (2006a). Development and implementation of cross-talk cancellation system in spatial audio reproduction based on subband filtering. *Journal of Sound and Vibration*, *290*(3–5), 1269–1289. Retrieved from www.sciencedirect.com/science/ article/pii/S0022460X05003421 doi: http://dx.doi.org/10.1016/j.jsv.2005.05.016

Bai, M. R., & Lee, C.-C. (2006b). Objective and subjective analysis of effects of listening angle on crosstalk cancellation in spatial sound reproduction. *The Journal of the Acoustical Society of America*, *120*(4), 1976–1989. Retrieved from http://scitation.aip.org/content/asa/ journal/jasa/120/4/10.1121/1.2257986 doi: http://dx.doi.org/ 10.1121/1.2257986

Bai, M. R., & Lee, C.-C. (2007). Subband approach to bandlimited crosstalk cancellation system in spatial sound reproduction. *EURASIP Journal of Advanced Signal Processing*, *2007*(071948), 1–9.69.

Bai, M. R., Tung, C.-W., & Lee, C.-C. (2005). Optimal design of loudspeaker arrays for robust cross-talk cancellation using the taguchi method and the genetic algorithm. *The Journal of the Acoustical Society of America*, *117*(5), 2802-2813. Retrieved from http://scitation.aip.org/content/asa/ journal/jasa/117/5/10.1121/1.1880852 doi: http://dx.doi.org/ 10.1121/1.1880852

Bauck, J., & Cooper, D. H. (1996). Generalized transaural stereo and applications. *Journal of Audio Engineering Society*, *44*(9), 683–705. Retrieved from http:// www.aes.org/e-lib/browse.cfm?elib=7888

Bauer, B. B. (1961). Stereophonic earphones and binaural loudspeakers. *Journal of Audio Engineering Society*, *9*(2), 148–151. Retrieved from www.aes.org/e-lib/ browse.cfm?elib=471

Bellanger, M. (2000). *Digital Processing of Signals: Theory and Practice*. Chichester, UK: John Wiley & Sons.

Choueiri, E. (2015). *Spectrally Uncolored Optimal Crosstalk Cancellation for Audio Through Loudspeakers*. Retrieved from www.google.com/ patents/WO2012036912A1?cl=en. International Patent Application No. PCT/US2011/050181, Granted November 18, 2015 under Patent No. 2612437.

Cooper, D. H., & Bauck, J. L. (1989). Prospects for transaural recording. *Journal of Audio Engineering Society*, *37*(1-2), 3–19. Retrieved from www.aes.org/e-lib/ browse.cfm?elib=6108

Damaske, P. (1971). Head-related two-channel stereophony with loudspeaker reproduction. *The Journal of the Acoustical Society of America*, *50*(4B), 1109-1115. Retrieved from http://scitation.aip.org/content/ asa/ journal/jasa/50/4B/10.1121/1.1912742 doi: http://dx.doi.org/ 10.1121/1.1912742

Farina, A. (2000). Simultaneous measurement of impulse response and distorsion with a swept-sine technique. *Proceedings of the 108th Audio Engineering Society Convention*. Paris.

Farina, A. (2007). Advancements in impulse response measurements by sine sweeps. *Proceedings of the 122nd Audio Engineering Society Convention*. Vienna.

Gardner, W. G. (1995). Efficient convolution without input-output delay. *Journal of Audio Engineering Society*, *43*(3), 127–136. Retrieved from www.aes.org/ e-lib/browse.cfm?elib=7957

Gardner, W. G. (1998). *3-D Audio Using Loudspeakers*. Boston, MA: Kluwer Academic Publishers.

Glasgal, R. (2007). 360 degrees localization via 4. x RACE processing. *Proceedings of the 12rd Audio Engineering Society Convention*. Vienna.

Hansen, P. C. (1998). *Rank-deficient and Discrete Ill-Posed Problems: Numerical Aspects of Linear Inversion*. Philadelphia, PA: Society for Industrial and Applied Mathematics.

Hugonnet, C., & Walder, P. (1997). *Stereophonic Sound Recording: Theory and Practice*. Chichester, UK: John Wiley & Sons.

Katz, B. (2002). *Mastering Audio: The Art and the Science* (pp. 61–74). Oxford, UK: Focal Press.

Kim, Y., Deille, O., & Nelson, P. (2006). Crosstalk cancellation in virtual acoustic imaging systems for multiple listeners. *Journal of Sound and Vibration*, *297*(1–2), 251–266. Retrieved from www.sciencedirect.com/science/article/pii/S0022460X06002884 doi: http:// dx.doi.org/10.1016/j.jsv.2006.03.042

Kirkeby, O., & Nelson, P. A. (1999). Digital filter design for inversion problems in sound reproduction. *Journal of Audio Engineering Society*, *47*(7-8), 583–595. Retrieved from www.aes.org/e-lib/browse.cfm?elib=12098

Kirkeby, O., Nelson, P. A., & Hamada, H. (1998a). Local sound field reproduction using two closely spaced loudspeakers. *The Journal of the Acoustical Society of America*, *104*(4), 1973–1981. Retrieved from http://scitation.aip.org/content/asa/journal/jasa/ 104/4/10.1121/1.423763 doi: http://dx.doi.org/10.1121/1.423763

Kirkeby, O., Nelson, P. A., & Hamada, H. (1998b). The "stereo dipole": A virtual source imaging system using two closely spaced loudspeakers. *Journal of Audio Engineering Society*, *46*(5), 387–395. Retrieved from www.aes.org/ e-lib/browse.cfm?elib=12148

Kirkeby, O., Nelson, P. A., Hamada, H., & Orduna-Bustamante, F. (1998, March). Fast deconvolution of multichannel systems using regularization. *Speech and Audio Processing, IEEE Transactions On*, *6*(2), 189–194. doi: 10.1109/89.661479

Lentz, T. (2006). Dynamic crosstalk cancellation for binaural synthesis in virtual reality environments. *Journal of Audio Engineering Society*, *54*(4), 283–294. Retrieved from www.aes.org/e-lib/browse.cfm?elib=13677

Majdak, P., Masiero, B., & Fels, J. (2013). Sound localization in individualized and non-individualized crosstalk cancellation systems. *The Journal of the Acoustical Society of America*, *133*(4), 2055–2068. Retrieved from http://scitation.aip.org/content/asa/ journal/jasa/133/4/10.1121/1.4792355 doi: http://dx.doi.org/ 10.1121/1.4792355

Mannerheim, P. V. H. (2008). *Visually Adaptive Virtual Sound Imaging Using Loudspeakers*, Unpublished doctoral dissertation, University of Southampton, Southampton, UK.

Moore, A. H., Tew, A. I., & Nicol, R. (2010). An initial validation of individualized crosstalk cancellation filters for binaural perceptual experiments. *Journal of Audio Engineering Society*, *58*(1-2), 36–45. Retrieved from www.aes.org/ e-lib/browse.cfm?elib=15240

Morse, P. M., & Ingard, K. U. (1986). *Theoretical Acoustics* (pp. 306–312). Princeton, NJ: Princeton University Press.

Nelson, P. A., & Elliott, S. J. (1993). *Active Control of Sound*. London, UK: Academic Press.

Nelson, P., Kirkeby, O., Takeuchi, T., & Hamada, H. (1997). Sound fields for the production of virtual acoustic images. *Journal of Sound and Vibration*, *204*(2), 386–396. Retrieved from www.sciencedirect.com/ science/article/pii/S0022460X97909676 doi: http://dx.doi.org/ 10.1006/jsvi.1997.0967

Nelson, P. A., & Rose, J. F. W. (2005). Errors in two-point sound reproduction. *The Journal of the Acoustical Society of America*, *118*(1), 193–204. Retrieved from http://scitation.aip.org/content/asa/ *BIBLIOGRAPHY*73journal/jasa/118/1/10.1121/1.1928787 doi: http://dx.doi.org/ 10.1121/1.1928787

Nicol, R. (2010). *Binaural Technology* (pp. 30–44). New York, NY: Audio Engineering Society Inc.

Papadopoulos, T., & Nelson, P. A. (2010). Choice of inverse filter design parameters in virtual acoustic imaging systems. *Journal of Audio Engineering Society*, *58*(1-2), 22–35. Retrieved from www.aes.org/e-lib/browse.cfm ?elib=15239

Parodi, Y. L., & Rubak, P. (2010). Objective evaluation of the sweet spot size in spatial sound reproduction using elevated loudspeakers. *The Journal of the Acoustical Society of America*, *128*(3), 1045–1055. Retrieved from http://scitation.aip.org/content/asa/ journal/jasa/128/3/10.1121/1.3467763 doi: http:// dx.doi.org/ 10.1121/1.3467763

Parodi, Y. L., & Rubak, P. (2011a). Analysis of design parameters for crosstalk cancellation filters applied to different loudspeaker configurations. *Journal of Audio Engineering Society*, *59*(5), 304–320. Retrieved from www.aes.org/ e-lib/browse.cfm?elib=15931

Parodi, Y. L., & Rubak, P. (2011b). A subjective evaluation of the minimum channel separation for reproducing binaural signals over loudspeakers. *Journal of Audio Engineering Society*, *59*(7-8), 487–497. Retrieved from www.aes.org/ e-lib/browse.cfm?elib=15974

Sæbø, A. (2001). *Influence of Reflections on Crosstalk Cancelled Playback of Binaural Sound*, Unpublished doctoral dissertation, Norwegian University of Science and Technology, Trondheim, Norway.

SreenivasaRao, C., Mahalakshmi, N., & VenkataRao, D. (2012). Real-time dsp implementation of audio crosstalk cancellation using mixed uniform partitioned convolution. *Signal Processing: An International Journal (SPIJ)*, *6*(4), 118–127. Retrieved from www.scribd.com/ document/299653040/Real-time-DSP-Implementation-of -Audio-Crosstalk-Cancellation-using-Mixed-Uniform -Partitioned-Convolution

Takeuchi, T., & Nelson, P. A. (2002). Optimal source distribution for binaural synthesis over loudspeakers. *The Journal of the Acoustical Society of America*, *112*(6), 2786–2797. Retrieved from http://scitation.aip.org/ content/asa/journal/jasa/112/6/10.1121/1.1513363 doi: http://dx.doi.org/10.1121/1.1513363

Takeuchi, T., & Nelson, P. A. (2007). Subjective and objective evaluation of the optimal source distribution for virtual acoustic imaging. *Journal of Audio Engineering Society*, *55*(11), 981–997. Retrieved from www.aes.org/e-lib/ browse.cfm?elib=14181

Takeuchi, T., Nelson, P. A., & Hamada, H. (2001). Robustness to head misalignment of virtual sound imaging systems. *The Journal of the Acoustical Society of America*, *109*(3), 958–971. Retrieved from http://scitation.aip.org/content/asa/journal/jasa/ 109/3/10.1121/1.1349539 doi: http://dx.doi.org/10.1121/1.1349539

Ward, D. B. (2001). On the performance of acoustic crosstalk cancellation in a reverberant environment. *The Journal of the Acoustical Society of America*, *110*(2), 1195–1198. Retrieved from http://scitation.aip.org/ content/asa/journal/jasa/110/2/10.1121/1.1386635 doi: http://dx.doi.org/10.1121/1.1386635

Ward, D. B., & Elko, G. (1999, May). Effect of loudspeaker position on the robustness of acoustic crosstalk cancellation. *Signal Processing Letters, IEEE*, *6*(5), 106–108. doi: 10.1109/97.755428

Xie, B. (2013). *Head-Related Transfer Function and Virtual Auditory Display* (2nd ed., pp. 283–326). Plantation, FL: J. Ross Publishing.

Yang, J., Gan, W.-S., & Tan, S.-E. (2003). Improved sound separation using three loudspeakers. *Acoustics Research Letters Online*, *4*(2), 47–52. Retrieved from http://scitation.aip.org/content/asa/ journal/arlo/4/2/10.1121/1.1566419 doi: http://dx.doi.org/10 .1121/1.1566419

Appendix A

Derivation of the Optimal XTC Filter

Here we carry out the derivation of Equations (5.73)–(5.75) following the approach outlined in the section "Impulse Response."

We start by factoring the expressions appearing in Equations (5.69) and (5.70), which, we note, have the same denominator, into the following products of terms:

$$H^{[O]}_{LL_{I,II}}(i\omega) = H^{[O]}_{RR_{I,II}}(i\omega) = (\psi_0 + \gamma\psi_1)\psi_a, \tag{A1}$$

$$H^{[O]}_{LR_{I,II}}(i\omega) = H^{[O]}_{LR_{I,II}}(i\omega) = (\mp\psi_0 + g\gamma e^{i\omega\tau_c}\psi_1)\psi_a, \tag{A2}$$

where

$$\psi_0 = \gamma^2\left[\pm x - g^2(1+e^{2i\omega\tau_c})\right], \tag{A3}$$

$$\psi_2 = \sqrt{g^2 \mp x + 1}, \tag{A4}$$

$$\psi_a = \frac{1}{(g^2 \mp x + 1) \pm 2\gamma x\sqrt{g^2 \mp x + 1}}. \tag{A5}$$

The term Ψ_a can be factored as

$$\Psi_a = \pm(\Psi_2 \cdot \Psi_3) \pm (\Psi_1 \mp \Psi_4) \cdot \Psi_5 \cdot \Psi_6(c_1) \cdot \Psi_6(c_2),$$

where

$$\psi_2 = \frac{1}{2\gamma x}, \tag{A6}$$

$$\Psi_3 = \frac{1}{\sqrt{g^2 \mp x + 1}} \tag{A7}$$

$$\Psi_4 = 2\gamma x, \tag{A8}$$

$$\Psi_5 = \frac{1}{8\gamma^3 x^3}, \tag{A9}$$

$$\Psi_6(c) = \frac{1}{1 - cx^{-1}}, \tag{A10}$$

and

$$c_1 = \frac{\sqrt{16\gamma^2(g^2+1)+1} \mp 1}{8\gamma^2}, \tag{A11}$$

$$c_2 = \frac{-\sqrt{16\gamma^2(g^2+1)+1} \mp 1}{8\gamma^2}. \tag{A12}$$

In the time domain, the filter expressed by Equations (A1) and (A2) becomes:

$$\begin{aligned} h^{[O]}_{LL_{I,II}}(t) &= h^{[O]}_{RR_{I,II}}(t) \\ &= (\psi_0 + \gamma\psi_1) * \psi_a, \end{aligned} \tag{A13}$$

$$\begin{aligned} h^{[O]}_{LR_{I,II}}(t) &= h^{[O]}_{RL_{I,II}}(t) \\ &= [\mp\psi_0 + g\gamma\delta(t+\tau_c) * \psi_1] * \psi_a. \end{aligned} \tag{A14}$$

where

$$\psi_a \;=\; \pm(\psi_2 * \psi_3) \;\pm\; (\psi_1 \mp \psi_4) * \psi_5 * \psi_6(c_1) * \psi_6(c_2). \tag{A15}$$

The ψ_i terms are functions of time, and are the IFTs of the Ψ_i terms, which are functions of frequency.

We now seek the IFT of each of the Ψ_i terms given above.

- Ψ_0: The IFT of the expression in Equation (A3) can be readily found by substituting back $2g \cos(\omega\tau_c)$ for x and carrying out the IFT integration:

$$\begin{aligned} \psi_0 &= \frac{1}{2\pi}\int_{-\infty}^{\infty} \gamma^2[\pm 2g\cos(\omega\tau_c) - g^2(1+e^{2i\omega\tau_c})]e^{2i\omega t}\,d\omega \\ &= \pm g\gamma^2[\delta(t-\tau_c) + \delta(t+\tau_c)] - g^2\gamma^2[\delta(t) + \delta(t+2\tau_c)]. \end{aligned} \tag{A16}$$

- Ψ_1: Making the substitution $b \equiv g^2 + 1$ in Equation (A4), we get

$$\Psi_1 = \sqrt{b \mp x}, \tag{A17}$$

which can be expressed as the series expansion

$$\Psi_1 = \sum_{m=0}^{\infty} \binom{\frac{1}{2}}{m} (\mp x)^m b^{\frac{1}{2}-m}, \tag{A18}$$

where we have used the binomial coefficient

$$\binom{k}{m} = \begin{cases} \dfrac{k!}{m!(k-m)!} & \text{if } 0 \leq m \leq k, \\ 0 & \text{if } m < 0 \text{ or } k < m. \end{cases}$$

Since $0 < g < 1$, we have $|x| = 2g|\cos(\omega\tau_c)| < g^2 + 1 = b$, and the series in Equation (A18) always converges. However, as $g \to 1$, $b \to 2$, and when $\omega\tau_c \to 2n\varpi$ with $n = 0,1,2,3,4,\ldots$, $x \to b$ and the series converges slowly. Replacing x and b by their explicit values, we get

$$\Psi_1 = \sum_{m=0}^{\infty} \binom{\frac{1}{2}}{m} 2^m (\mp g)^m (g^2+1)^{\frac{1}{2}-m} \cos^m(\omega\tau_c). \tag{A19}$$

Since $\cos^m(\omega\tau_c)$ can be written as the finite sum

$$\cos^m(\omega\tau_c) = \sum_{k=0}^{m} \binom{m}{k} 2^{-m} e^{-i(2k-m)\omega\tau_c}, \tag{A20}$$

and since the IFT of $e^{-i(2k-m)\omega\tau c}$ is

$$\frac{1}{2\pi}\int_{-\infty}^{\infty} e^{-i(2k-m)\omega\tau_c} e^{i\omega t}\, d\omega = \delta(t-(2k-m)\tau_c),$$

the IFT of Ψ_1 can be expressed as

$$\psi_1 = \sum_{m=0}^{\infty} \binom{\frac{1}{2}}{m} (\mp g)^m (g^2+1)^{\frac{1}{2}-m} \times \sum_{k=0}^{m} \binom{m}{k} \delta(t-(2k-m)\tau_c). \tag{A21}$$

- Ψ_2: Explicitly, Equation (A6) is

$$\Psi_2 = \frac{\sec(\omega\tau_c)}{4g\gamma}.$$

The problem is that the IFT of $\sec(\omega\tau_c)$ cannot be expressed in terms of real delta functions. However, the function $\sec(\omega\tau_c)$ can be expressed as

$$\sec(\omega\tau_c) = \frac{1}{\sqrt{1-\sin^2(\omega\tau_c)}}, \tag{A22}$$

$$\text{if } 2n\pi - \frac{\pi}{2} < \omega\tau_c < 2n\pi + \frac{\pi}{2}$$

$$\text{with } n = 0,1,2,3,4,\ldots$$

Furthermore, we note that since

$$1 \le \gamma \le \frac{1}{1-g} \quad \text{and } 0 < g < 1, \tag{A23}$$

the arguments of the inverse cosine function in Equation (5.59) obeys the condition:

$$0 < \frac{(g^2+1)\gamma^2 - 1}{2g\gamma^2} \le 1 \tag{A24}$$

which leads us to write

$$0 \le \phi < \frac{\pi}{2}. \tag{A25}$$

In light of this expression and Equation (5.55), we conclude that the conditions for the validity of Equation (A22) are always satisfied in Branch-I bands.

Similarly, we find that $\sec(\omega\tau_c)$ can be expressed as $-1/\sqrt{1-\sin^2(\omega\tau_c)}$ for conditions that are always satisfied for Branch-II bands. Therefore, we can write

$$\sec(\omega\tau_c) = \pm\frac{1}{\sqrt{1-\sin^2(\omega\tau_c)}} \tag{A26}$$

for which we wish to use the expansion

$$\frac{1}{\sqrt{1-u}} = \sum_{m=0}^{\infty}\begin{pmatrix} -\frac{1}{2} \\ m \end{pmatrix}(-1)^m u^m. \tag{A27}$$

However, this series converges only for $|u| < 1$. For our particular case, $u = \sin^2(\omega\tau_c)$ and the series diverges at $\omega\tau_c = (2n+1)\pi/2$, with $n = 0,1,2,3,4,\ldots$. From the band division conditions in Equations (5.55) and (5.57) we see that these values of $\omega\tau_c$ are always outside Branch-I and Branch-II bands; therefore, the convergence of the series is assured and this allows us to express Equation (A26) as

$$\sec(\omega\tau_c) = \pm\sum_{m=0}^{\infty}\begin{pmatrix} -\frac{1}{2} \\ m \end{pmatrix}(-1)^m \sin^{2m}(\omega\tau_c). \tag{A28}$$

Since $\sin^{2m}(\omega\tau_c)$ can be written as the finite sum

$$\sec^{2m}(\omega\tau_c) = \sum_{k=0}^{2m}\binom{2m}{k}(-1)^{k+m}4^{-m}e^{2i(m-k)\omega\tau_c}, \tag{A29}$$

and since the IFT of $e^{2i(m-k)\omega\tau c}$ is

$$\frac{1}{2\pi}\int_{-\infty}^{\infty} e^{2i(m-k)\omega\tau_c} e^{i\omega t}\, d\omega = \delta(t + 2(m-k)\tau_c), \tag{A30}$$

the IFT of Ψ_2 can be expressed as

$$\psi_2 = \pm\frac{1}{4g\gamma}\sum_{m=0}^{\infty}\binom{-\frac{1}{2}}{m} 4^{-m} \times \sum_{k=0}^{2m}\binom{2m}{k}(-1)^k \delta(t + 2(m-k)\tau_c) \tag{A31}$$

- Ψ_3: The function $1/\sqrt{b \mp x}$, where, again, $b \equiv g^2 + 1$, has a series expansion in the form of Equation (A18), but with the fraction 1/2 (inside the binomial coefficient and in the exponent of b) replaced by –1/2. Therefore, by analogy to the result expressed in Equation (A21), we have

$$\psi_3 = \sum_{m=0}^{\infty}\binom{-\frac{1}{2}}{m} (\mp g)^m (g^2+1)^{-\frac{1}{2}-m} \times \sum_{k=0}^{m}\binom{m}{k}\delta(t - (2k-m)\tau_c). \tag{A32}$$

which has the same convergence behavior as that of ψ_1.

- Ψ_4: The IFT of $\Psi_4 = 2\gamma x = 4g\gamma\cos(\omega\tau_c)$ is straightforward:

$$\psi_4 = 2g\gamma[\delta(t - \tau_c) + \delta(t + \tau_c)]. \tag{A33}$$

- Ψ_5: Explicitly, Equation (A9) is

$$\Psi_5 = \frac{\sec^3(\omega\tau_c)}{(4g\gamma)^3}, \tag{A34}$$

where, following the same arguments as in the case of Ψ_2, the function $\sec^3(\omega\tau_c)$ can be expanded in a convergent series of the form of that in Equation (A28), but with the fraction –1/2 inside the binomial coefficient replaced by –3/2. Therefore, by analogy to the result expressed in Equation (A31), we have

$$\psi_5 = \pm\frac{1}{(4g\gamma)^3}\sum_{m=0}^{\infty}\binom{-\frac{3}{2}}{m} 4^{-m} \times \sum_{k=0}^{2m}\binom{2m}{k}(-1)^k \delta(t + 2(m-k)\tau_c). \tag{A35}$$

- Ψ_6: Equation (A10) can be written as

$$\Psi_6 = \frac{1}{1 - y(c)}, \tag{A36}$$

where

$$y \equiv \frac{c}{x} = \frac{2}{2g\cos(\omega\tau_c)}, \tag{A37}$$

and c represents either c_1 or c_2, given by Equations (A11) and (A12), respectively. We wish to expand the function in Equation (A36) into the power series

$$\sigma(c) \equiv \sum_{p=0}^{\infty} y^p(c), \tag{A38}$$

but this series converges only for

$$|\, y(c)\,| < 1. \tag{A39}$$

We now show that this convergence condition leads to a restriction on the allowable range of γ and g, but that this restriction does not limit the applicability of the IRs to real listening configurations.

The inequalities in Equation (A25) and the band division conditions in Equations (5.55) and (5.57) imply that $x = 2g\cos(\omega\tau_c)$ *is always positive in Branch-I bands and negative in Branch-II bands. Furthermore, we see from Equations (A11) and (A12) that, under the conditions in Equation (A23),* $c_1 \geq 0$ and $c_2 \leq 0$. Therefore, we have

$$y(c_1) = c_1/x \geq 0 \text{ in Branch-I bands,} \tag{A40}$$
$$y(c_1) = c_1/x \leq 0 \text{ in Branch-II bands,} \tag{A41}$$

and

$$y(c_2) = c_2/x \leq 0 \text{ in Branch-I bands,} \tag{A42}$$
$$y(c_2) = c_2/x \geq 0 \text{ in Branch-II bands.} \tag{A43}$$

If we define $\eta^+(c)$ and $\eta^-(c)$ to be the lowest (between 0 and π) non-dimensional frequencies, $\omega\tau_c$, at which $y(c) = +1$ and $y(c) = -1$, respectively, we can, in light of the expressions above, restate the convergence condition in Equation (A39) as:

$$\sigma(c_1) \text{ converges in Branch-I bands if } \phi \leq \eta^+(c_1) \tag{A44}$$
$$\sigma(c_1) \text{ converges in Branch-II bands if } \eta^-(c_1) \leq \pi - \phi \tag{A45}$$

and

$$\sigma(c_2) \text{ converges in Branch-I bands if } \phi \leq \eta^-(c_2) \tag{A46}$$
$$\sigma(c_2) \text{ converges in Branch-II bands if } \eta^+(c_2) \leq \pi - \phi. \tag{A47}$$

Therefore, for $\sigma(c)$ to converge both in Branch-I and in Branch-II bands, all four inequalities must be satisfied. To express these convergence conditions explicitly (i.e., in terms of conditions on γ

and g), we first set $y(c) = +1$ and $y(c) = -1$, and solve for $\eta^{+}(c)$ and $\eta^{-}(c)$, respectively, to find, for Branch-I bands,

$$\eta^{+}(c_1) = \cos^{-1}\left(\frac{f(g,\gamma)-1}{16g\gamma^2}\right), \quad \text{(A48)}$$

$$\eta^{-}(c_2) = \cos^{-1}\left(\frac{f(g,\gamma)-1}{16g\gamma^2}\right), \quad \text{(A49)}$$

and, for Branch-II bands,

$$\eta^{+}(c_2) = \cos^{-1}\left(\frac{-f(g,\gamma)-1}{16g\gamma^2}\right), \quad \text{(A50)}$$

$$\eta^{-}(c_1) = \cos^{-1}\left(-\frac{f(g,\gamma)+1}{16g\gamma^2}\right), \quad \text{(A51)}$$

where, for compactness, we have used the function $f(g,\gamma)$ defined as

$$f(g,\gamma) \equiv \sqrt{16\gamma^2(g^2+1)+1}.$$

Using these four explicit expressions, along with the definition of φ given by Equation (5.59), we find that the inequalities in Equations (A44) and (A47) lead to the same explicit convergence condition:

$$\frac{f(g,\gamma)+7}{8(g^2+1)\gamma^2} \le 1; \quad \text{(A52)}$$

and the inequalities in Equations (A45) and (A46) lead to

$$\frac{f(g,\gamma)+9}{8(g^2+1)\gamma^2} \le 1. \quad \text{(A53)}$$

Since both of these inequalities need to be satisfied, and since the latter condition is more stringent than the former, we must satisfy the latter. We can finally state the condition for $\sigma(c)$ to converge both in Branch-I and in Branch-II bands explicitly in terms of g and γ:

$$\frac{\sqrt{16(g^2+1)\gamma^2+1}+9}{8(g^2+1)\gamma^2} \le 1. \quad \text{(A54)}$$

This convergence condition is illustrated in the region plot of Figure 5.A.1, where the black-shaded region denotes the values of g and γ for which the convergence condition is violated. It is clear that this restriction only slightly limits the range of allowable γ and g, and is not relevant to real listening geometries, where $g \simeq 1$.

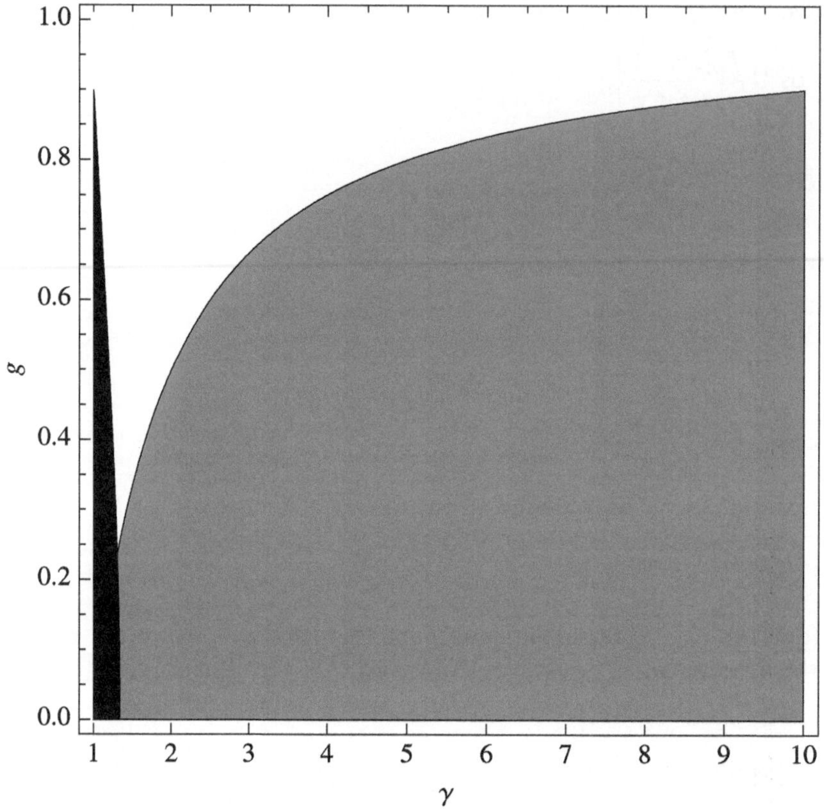

Figure 5.A.1 Region plot showing the allowed values for g and γ (white). The black-shaded region is where the series convergence condition in Equation (A54) is not satisfied, and the grey-shaded region is where the general condition in Equation (5.51) is violated.

Aside from the series convergence condition above, γ must satisfy the general condition given by Equation (5.51) (whose region of violation is shaded in grey in Figure 5.A.1). Therefore, we combine both conditions in the following expression:

$$\max\left(\frac{\sqrt{5+\sqrt{5}}}{2\sqrt{g^2+1}},1\right) \leq \gamma \leq \frac{1}{1-g}, \tag{A55}$$

where the first argument of the max function comes from setting the left-hand side of the convergence condition in Equation (A54) to 1, and solving for γ.

Now that we have found the convergence condition for the series in Equation (A38), we can express Ψ_6 as that series and proceed to find its IFT. Replacing y and x in that series by their explicit values, we write

$$\Psi_6 = \sum_{p=0}^{\infty} \left(\frac{c}{2g} \right)^p \sec^p(\omega\tau_c). \tag{A56}$$

The $\sec^p(\omega\tau_c)$ term can be expanded in a convergent series of the same form as the series in Equation (A28), but with the fraction –1/2 inside the binomial coefficient replaced by $-p/2$, and this leads to:

$$\Psi_6 = \sum_{p=0}^{\infty} \left(\frac{\pm c}{2g} \right)^p \sum_{m=0}^{\infty} \binom{-\frac{p}{2}}{m} (-1)^m \sin^{2m}(\omega\tau_c). \tag{A57}$$

Finally, recalling the finite sum in Equation (A29), and the associated IFT in Equation (A30), we arrive at the sought expression for the IFT of $\Psi_6(c)$:

$$\psi_6 = \sum_{p=0}^{\infty} \left(\frac{\pm c}{2g} \right)^p \sum_{m=0}^{\infty} \binom{-\frac{p}{2}}{m} 4^{-m} \times \sum_{k=0}^{2m} \binom{2m}{k} (-1)^k \delta(t + 2(m-k)\tau_c). \tag{A58}$$

The complete impulse response of the optimal XTC filter is assembled according to Equations (A13)–(A15), and is valid under the condition stated in Equation (A55).

Appendix B

Numerical Verification

The optimal XTC IRs derived in the previous appendix were evaluated for the typical case of $g = 0.985$ and $\Gamma = 7$ dB, and plotted in Figure 5.8. To verify the validity of the IRs and assess the effect of the number of terms in the series expansions, we calculated their Fourier transforms and compared the resulting spectra to those obtained from the frequency-domain expressions of the section "Frequency Response." An example is shown in Figure 5.B.1 for the Branch-I part of the XTC spectrum (top panel) and that of the envelope spectrum (bottom panel).

We found that excellent agreement (within a few tenths of a dB) over all frequencies does not require taking more than the first few (5–10) terms of the infinite series in the expressions for all the ψ functions constituting the IRs, with the exception of ψ_1 and ψ_3, which, due to their slow convergence at and near the frequencies $\omega\tau_c = 2n\varpi$ with $n = 0, 1, 2, 3, 4, \ldots$, require taking a larger number of terms. Approximating the infinite series in the expressions for ψ_1 and ψ_3 by a sum having a finite number of terms causes departures from the correct amplitude spectra at and near these frequencies. Due to the logarithmic frequency scale, the $n = 0$ departure appears as a slight bass roll-off in the first band (seen as the first dot in the first Branch-I band in the bottom panel of Figure 5.B.1), and the $n \geq 1$ departures appear as narrow-band spikes (such as the one appearing as three vertical dots in the fifth band in the same plot). Increasing the number of terms in the series above 1,000 reduces the amplitude of the bass roll-off and pushes it into the subwoofer frequency range, where XTC is not needed, and causes the $n \geq 1$ spikes to diminish in amplitude and frequency extent so as to become inaudible. (The XTC spectrum is more immune from the aforementioned departures, as seen in the top panel, because it is a ratio of left to right spectra.)

A similar analysis of the Branch-II part of the IRs is not shown, as the resulting spectra exhibit the same behavior as that described above.

Acknowledgments

The author wishes to thank Joseph Tylka for his help in checking the manuscript and updating the citations, and J. S. Bach for his Mass in B Minor, whose reproduction in 3D was a main motive for this work.

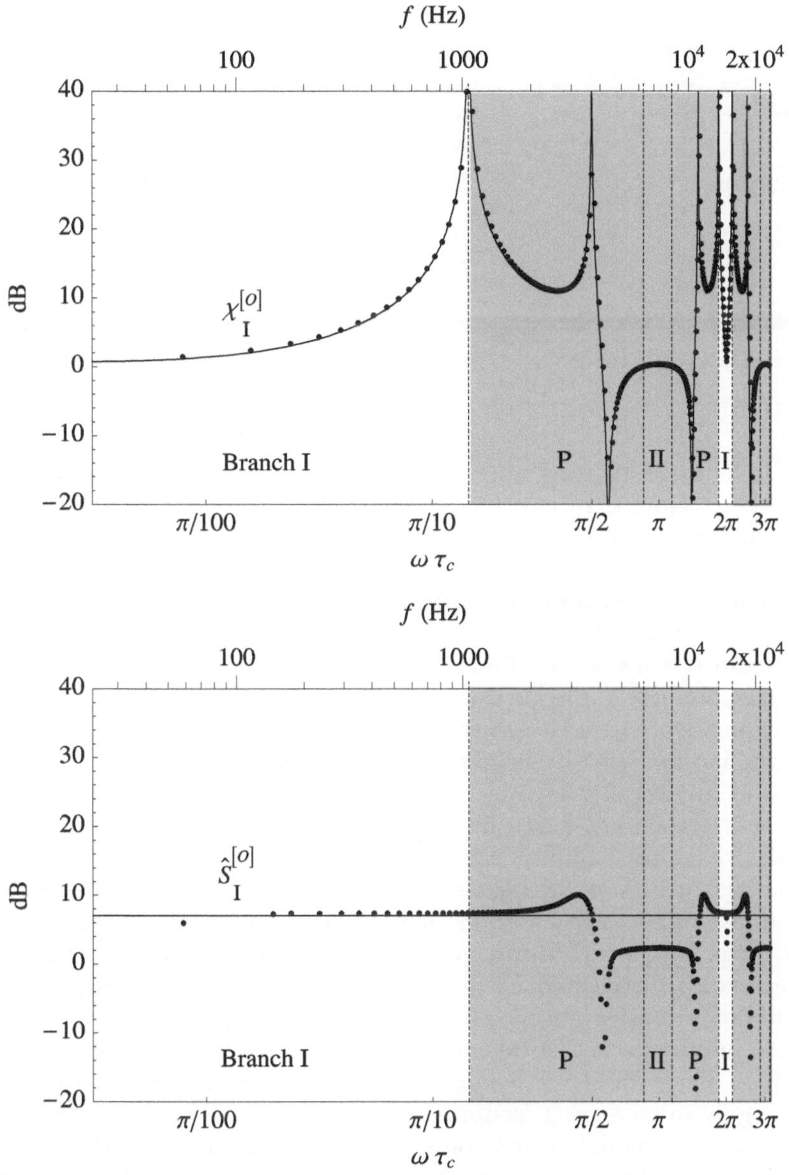

Figure 5.B.1 XTC spectrum of optimal filter for Branch-I bands, $\chi_I^{[O]}(\omega)$, shown in top panel, and the associated envelope spectrum, $\hat{S}_I^{[O]}(\omega)$, shown in bottom panel. The small dots represent the spectra calculated by taking the Fourier transform of the Branch-I part of the IRs derived in Appendix A. (The IRs are shown graphically in the top panel of Figure 5.8.) Only the first 20 terms of the infinite series representing the ψ functions were taken, with the exception of the series for ψ_1 and ψ_3, for which the first 2,500 terms were used. The hard curve in the top panel is the Branch-I XTC spectrum calculated directly from Equation (5.63), and the horizontal line in the bottom panel is the Branch-I envelope spectrum $\hat{S}_I^{[O]}(\omega)$, with Γ = 7 dB. (Other parameters are the same as for Figure 5.2.) Since these spectra are valid only in Branch-I bands, all other bands are shaded in grey. (The vertical dashed lines represent the frequency bounds of the successive bands, and the branch numbers of the first five bands are given in the bottom half of each panel.)

Chapter 6

Surround Sound[1]

Francis Rumsey

Surround sound is a 'catch-all' term often used to describe any form of loudspeaker sound reproduction that involves more than two loudspeakers and attempts to surround the listener with sounds from multiple directions. It started as a marketing term that aimed to imply something more spatially interesting than 2-channel stereo. These days the term may be rather dated, as it has to some extent been superseded or replaced by more generic terms such as *spatial audio*, or trumped by systems termed *3D* or *immersive*. It is still useful, nonetheless, to describe a set of entertainment audio systems and techniques that involve more than two loudspeakers arranged around the listener, but only in the horizontal plane. In this chapter, therefore, the discussion will be restricted to approaches based on conventional stereophonic principles that range from three channels upward, but don't involve height information.

Surround sound loudspeaker layouts and production techniques grew out of the methods used in 2-channel stereophony, based on ideas arising from Bell Labs, Blumlein and others in the 1930s, as described in Chapter 3. Spatialization of sound images in basic stereophony was achieved by the introduction of simple time and/or level differences between channels, created out of the relationships between microphone outputs or by simple panpots. The aim was one of creating an adequate spatial illusion. As more loudspeakers were added to create 'surround sound' these techniques were extended in an attempt to make them work for more loudspeakers, but often still only considering the relationships between pairs of channels, or at most three at a time. As the number of loudspeakers grew from four up to more than ten, the challenges of deciding how to divide the signal between the loudspeakers to create a successful spatial illusion grew, leading to an increasing dichotomy between those systems that were based on some underlying mathematical model of an acoustical sound field and those that simply kept adding loudspeakers and attempting to apply basic stereophonic principles. The former could be considered as the scientist's conception of the way to do surround sound, and is characterized by those methods of sound field reconstruction such as sound field and wave field synthesis. These approaches are covered in Chapters 9 and 10. This chapter is therefore mainly dedicated to the latter group of systems (which could be thought of as the recording engineer's idea of the way to do surround sound) and deals with systems that mainly, but not exclusively, grew out of cinema sound formats. A modern nomenclature was developed that describes formats as '*n*-*m* stereo (or surround)', where *n* represents the number of front channels and *m* the number of surround channels, more about which is said later on in the chapter.

The Evolution of Surround Sound

Cinema Systems

Early work on directional reproduction at Bell Labs in the 1930s involved attempts to approximate the sound wavefront that would result from an infinite number of microphone/loudspeaker channels by using a smaller number of channels, as shown in Figure 6.1. Spaced pressure (omnidirectional) microphones were used, each connected by a single amplifier to the appropriate loudspeaker in a listening room. Steinberg and Snow (1934) found that three channels gave quite convincing results, and that when reducing the number of channels from three to two, central sources appeared to recede towards the rear of the stage and that the width of the reproduced sound stage appeared to be increased. In fact, as Snow later explained, the situations shown in the two diagrams are in fact rather different because the small number of channels do not really recreate the original source wavefront, but depend upon the precedence effect for success. This could be seen as an early break between the concept of sound field synthesis and basic stereophony.

Steinberg and Snow's work was principally intended for large auditorium sound reproduction with wide screen pictures, rather than small rooms or consumer equipment, and in fact most of the early development of surround sound was driven by the need to attract audiences to the cinema. Three front channels were the norm in cinema sound reproduction for many years, partly because of the wide range of seating positions and size of the image. The center channel had the effect of stabilizing the important central image (where the dialog was located) for off-center listeners, and was used increasingly since the Disney film *Fantasia* in 1939.

The *Fantasound* system used multiple operators to do the mixing, and eventually a pilot-tone-based control track alongside the film to manipulate the panning of three sound tracks to a number of loudspeakers around the auditorium. It was an expensive historical experiment into surround sound, in many ways well before its time, and one version even had a 'voice of God' loudspeaker mounted on the ceiling above the listeners, possibly the first example of immersive audio in the business. The development and implications were well described in an article in the *SMPTE Motion Imaging Journal* (Garity & Hawkins, 1941, p. 127), where some critically important observations were made that still have application for entertainment audio today. "Therefore," they said

> we must take large steps forward, rather than small ones, if we are to inveigle the public away from softball games, bowling alleys, nightspots, or rapidly improving radio reproduction. The public has to hear the difference and then be thrilled by it, if our efforts toward the improvement of sound-picture quality are to be reflected at the box-office. Improvements perceptible only through direct A-B comparisons have little box-office value. While dialog is intelligible and music is satisfactory, no one can claim that we have even approached perfect simulation of concert hall or live entertainment. It might be emphasized that perfect simulation of live entertainment is not our objective. Motion picture entertainment can evolve far beyond the inherent limitations of live entertainment.

In saying this they made the crucial point that the purpose of much entertainment audio is just that—entertainment—and does not have to emulate or recreate natural environments. It can

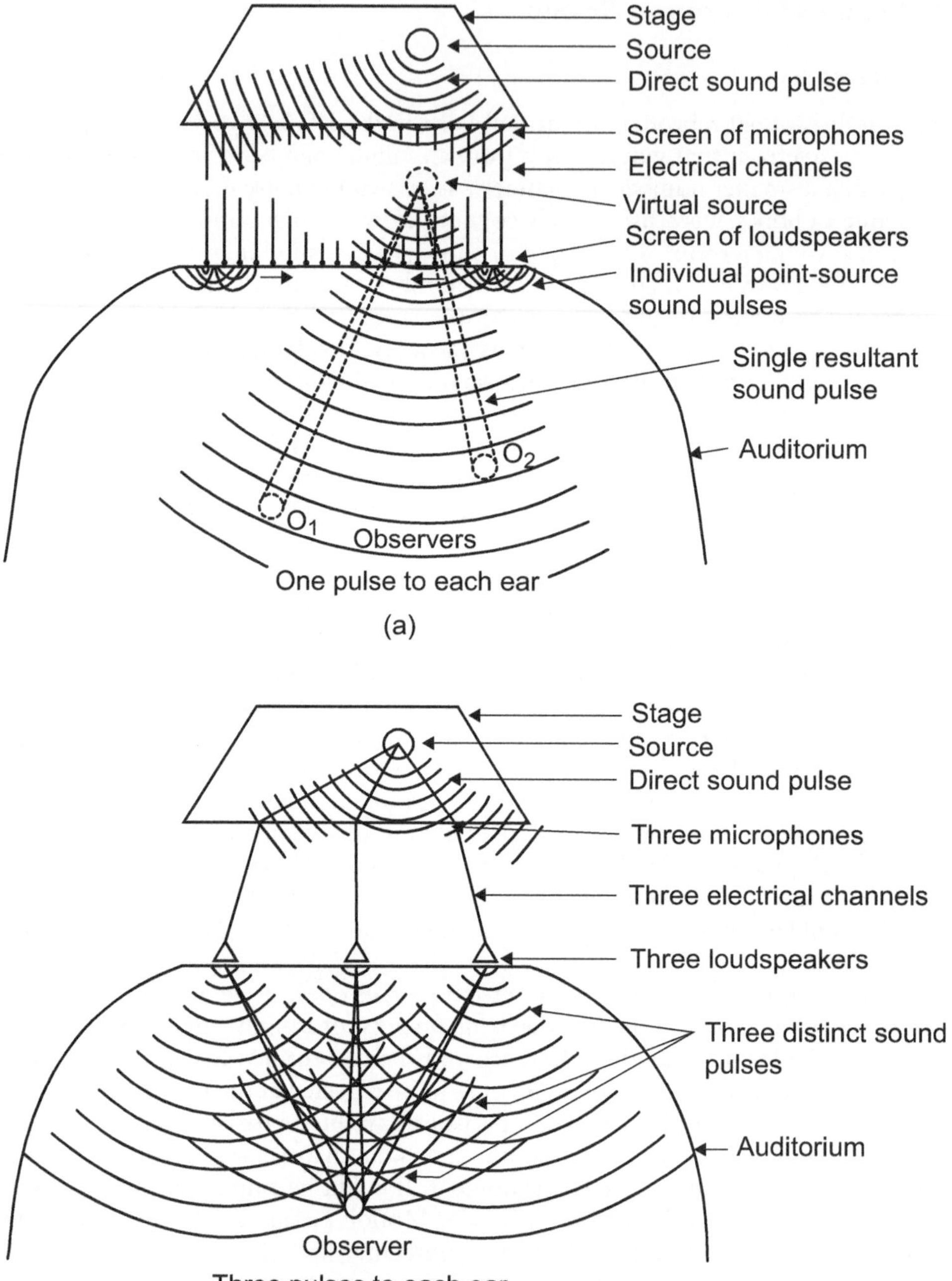

Figure 6.1 Steinberg and Snow's attempt to reduce the number of channels needed to convey a source wavefront to a reproduction environment with appropriate spatial features intact. (a) 'Ideal' arrangement involving a large number of transducers. (b) Compromise arrangement involving only three channels, relying more on the precedence effect.

be hyper-real. *Fantasound*, however, did not last beyond the early war years as it was too time-consuming and costly to set up.

Apart from the unusual stereo effects used in *Fantasia*, cinema sound did not incorporate stereo reproduction until the 1950s. Stereo film sound tracks often employed dialog panned to match the visual scene elements, which was a laborious and time-consuming process. This technique gradually died out in favor of central dialog, accompanied by stereo music and sound effects. During the 1950s Warner Brothers introduced a large screen format with three front channels and a single surround channel, and the 20th Century Fox *Cinemascope* format also used a similar arrangement. *Cinerama* used seven discrete channels of surround sound to accompany wide screen pictures using three projectors, prefiguring the 7.1-channel systems of later years with its five front screen channels and stereo surrounds.

Multichannel stereo formats for the cinema became increasingly popular in the late 1950s and 1960s, culminating in the so-called baby boomer 70 mm format involving multiple front channels, a surround channel and a subwoofer channel to accompany high-quality, wide screen cinema productions. In the early 1970s, Dolby's introduction of Dolby Stereo enabled a 4-channel surround sound signal to be matrix encoded into two optical sound tracks recorded on the same 35mm film as the picture. It was later released in a consumer form called Dolby Surround for home cinema applications. The main problem with analog matrix formats was the difficulty of maintaining adequate channel separation, requiring sophisticated 'steering' circuits in the decoder to direct dominant signal components to the appropriate loudspeakers.

In the 1990s cinema surround sound moved to all-digital sound tracks that typically incorporated either five or seven discrete channels of surround sound plus a sub-bass effects channel. A variety of commercial digital low-bit-rate coding schemes were used to deliver surround sound signals with movie films, such as Dolby Digital, Sony SDDS and Digital Theatre Systems (DTS).

Ambiophony and Similar Techniques

Although surround sound did not appear to be commercially feasible for consumer music reproduction applications during the late 1950s and early 1960s, a number of researchers were experimenting at the time with methods for augmenting conventional reproduction by radiating reverberation signals from separate loudspeakers. This is an interesting precursor of the modern approach that tends to recommend the use of surround channels for the augmentation of conventional frontal stereo with ambience or effects signals. One of the most developed examples in this respect was the 'Ambiophonic' concept developed by Keibs and colleagues in 1960, nicely summarized in a paper by Steinke (1996) and later developed by others (e.g., Glasgal, 1995).

Quadraphonic Sound

Quadraphonic sound represents an attempt to introduce surround sound to the consumer in the 1970s that ultimately failed from a commercial point of view. It can be thought of as a 2–2 surround system in terms of modern nomenclature introduced below. A variety of competing encoding methods, having different degrees of compatibility with each other and with 2-channel stereo, were used to convey four channels of surround sound on 2-channel analogue media such as vinyl LPs (so-called 4–2–4 matrix systems). Unlike Dolby Stereo, quadraphonic sound used no center channel, but was normally configured for a square arrangement of loudspeakers, two at the front and two behind the listener. The 90° angle of the front loudspeakers proved problematic because of lack of compatibility with ideal 2-channel reproduction, and gave poor front images, often with a 'hole in the middle'. A review of some of the issues can be found in Scheiber (1971).

While a number of LP records were issued in various 'quad' formats, the approach failed to capture a sufficiently large part of the consumer imagination to succeed. It seemed that people were unwilling to install the additional loudspeakers required, and there were too many alternative forms of quad encoding for a clear standard to emerge. Also, many people felt that quad encoding compromised the integrity of 2-channel stereo listening (the matrix encoding of the rear channels was supposed to be 2-channel compatible but unwanted side effects could often be heard). Although some efforts were made to release 4-channel recordings on magnetic tape, and some 4-track tape recorders were made for the consumer market, the popularity of magnetic tape as a consumer medium was not sufficient to carry this forward.

The Home Cinema and 5.1-Channel Surround Sound

During the 1990s the development of new consumer audio formats such as DVD, the 'home cinema', and digital sound formats for cinema and broadcasting, gave a new commercial impetus to surround sound. The ITU 5.1-channel configuration became widely adopted for broadcasting and recording applications, as described below, and the discrete channel delivery possibilities of digital transmission and storage formats helped to avoid the former problems of matrix encoding. As will be explained later, this ITU standard did not define anything about the way that sound signals should be represented or coded for surround sound; it simply stated the layout of the loudspeakers. Most other things were open, and there was no 'correct' method of sound field representation or spatial encoding for this standard.

Surround Sound Formats

The principal loudspeaker formats for surround sound that have featured in both professional and consumer environments since the 1990s will be described in this section. Surround sound standards of the type dealt with in this chapter often specify little more than the channel configuration and the way the loudspeakers should be arranged. This leaves the business of how to create or represent a spatial sound field entirely up to the user. Then there is the separate question of how to matrix or encode the channels for delivery to the end user, which is often the domain of commercial technology, and that is introduced in a subsequent section.

In international standards describing stereo loudspeaker configurations, such as ITU-R BS.775-3 (2012), the nomenclature for the configuration is often in the form '*n-m stereo*', where *n* is the number of front channels and *m* is the number of rear or side channels (the latter only being encountered in surround systems). This distinction can be helpful as it reinforces the slightly different role of the surround channels. It was one of the underlying principles of those who designed these layouts that the front channels would fulfill a different role to the rear ones, rather than all being equal. The front left and right channels are often in positions that are compatible with 2-channel stereo, which makes them quite narrowly spaced, then there can be a large gap at the sides where imaging is difficult, followed by rear loudspeakers that are primarily intended for delivering effects and ambience. There is no explicit aim to deliver full 360° imaging with equal accuracy in all directions, although many have ignored this fact and attempted to do just that, suffering from the inevitable problems. Some non-standard approaches have then been adopted for consumer music applications that widen the spacing of the front loudspeakers and fill in the gaps at the sides with additional loudspeakers, and then use sophisticated decoding techniques to render content more effectively.

Another common nomenclature is 'something point something' surround, for example, 5.1 surround, which makes no distinction between front and rear channels but highlights the number of main channels and the number of low frequency effects (LFE) channels that go with them. The 'point one' channel in this case is an LFE channel, and relates mainly to cinema systems where these are commonly used for ground-shaking effects, explosions and the like.

Three-Channel (3–0) Stereo

3–0 stereo forms the basis of the front layout of a lot of surround sound systems. It requires the use of a left (L), center (C) and right (R) channel, the loudspeakers arranged equidistantly across the front sound stage, as shown in Figure 6.2. It has some precedents in historical development, in that the stereophonic system developed by Steinberg and Snow in the 1930s used three channels, as mentioned earlier. Three front channels have also been commonplace in cinema stereo systems, mainly because of the need to cover a wide listening area and because wide screens tend to result in a large distance between left and right loudspeakers. Two channels only became the norm in consumer systems for reasons of economy and convenience, and particularly because it was much more straightforward to cut two channels onto an analog disk than three.

There are various advantages of 3–0 stereo. First, it allows for a somewhat wider front sound stage than 2-channel stereo, if desired, because the center channel acts to 'anchor' the central image and the left and right loudspeakers can be placed further out to the sides (say ± 45°). (Note, though, that in the ITU-R standard the L and R loudspeakers are in fact placed at ± 30°, for compatibility with 2-channel stereo material.) Second, the center loudspeaker enables a wider range of listening positions in many cases, as the image does not collapse quite as readily into the nearest loudspeaker. It also anchors dialog more clearly in the middle of the screen in sound-for-picture applications. Third, the center image does not suffer the same timbral modification as the center image in 2-channel stereo, because it emanates from a real source.

A practical problem with it is that the center loudspeaker position is often very inconvenient. Although in cinema reproduction it can be behind an acoustically transparent screen, in consumer environments, studios and television environments it is almost always just where one wants a television monitor or a window. Consequently the center channel has to be mounted above or below the object in question, and possibly made smaller than the other loudspeakers.

Four-Channel Surround (3–1 Stereo)

The form of stereo called '3–1 stereo' in some international standards, or 'LCRS surround' in some other circles, will briefly be described. Proprietary encoding and decoding technology relating to this format is described later. Although it uses four channels, it is different from

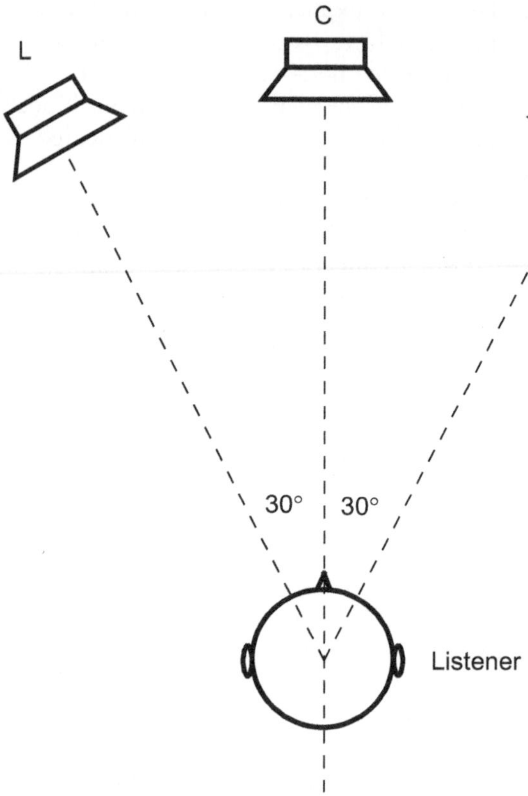

Figure 6.2 Three-channel stereo reproduction usually involves three equally spaced loudspeakers in front of the listener. The angle between the outer loudspeakers is 60° in the ITU standard configuration, for compatibility with 2-channel reproduction, but the existence of a center loudspeaker makes wider spacings feasible if compatibility is sacrificed.

quadraphonics because the loudspeakers are not at 90° intervals, but are configured as three front channels plus one surround.

In the 3–1 approach, an additional 'effects' channel or 'surround' channel is added to the three front channels, routed to a loudspeaker or loudspeakers located behind (and possibly to the sides) of listeners. It was developed first for cinema applications, enabling a greater degree of audience involvement in the viewing/listening experience by providing a channel for 'wrap-around' effects. This development is attributed to 20th Century Fox in the 1950s, along with wide screen Cinemascope viewing, being intended to offer effective competition to the new television entertainment.

There is no specific intention in 3–1 stereo to use the effects channel as a means of enabling 360° image localization. In any case, this would be virtually impossible with most configurations as there is only a single audio channel feeding a larger number of surround loudspeakers, effectively in mono.

Figure 6.3 shows the typical loudspeaker configuration for this format. In the cinema there are usually a large number of surround loudspeakers fed from the single surround channel, in order to cover a wide audience area. This gives rise to a relatively diffuse or distributed reproduction of the effects signal. The surround speakers are sometimes electronically decorrelated to increase the degree of spaciousness or diffuseness of surround effects, in order that they are not specifically localized to the nearest loudspeaker or perceived inside the head.

In consumer systems reproducing 3–1 stereo, the mono surround channel is normally fed to two surround loudspeakers located in similar positions to the 3–2 format described below. The gain of the channel is usually reduced by 3 dB so that the summation of signals from the two

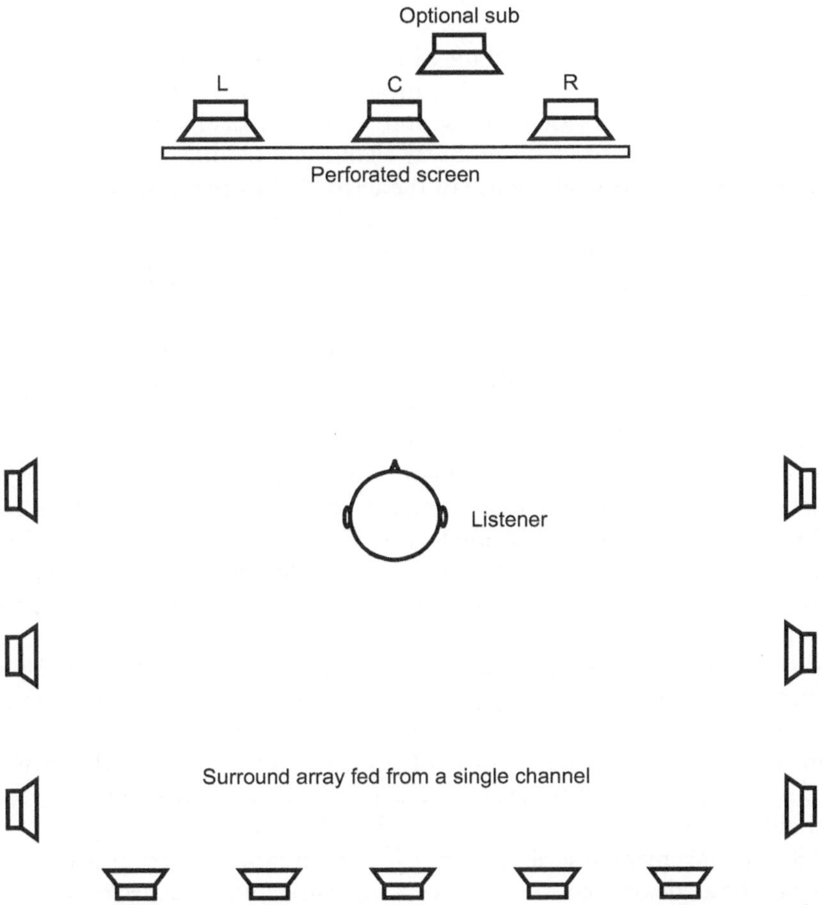

Figure 6.3 3–1 format reproduction uses a single surround channel usually routed (in cinema environments) to an array of loudspeakers to the sides and rear of the listening area. In consumer reproduction the mono surround channel may be reproduced through only two surround loudspeakers, possibly using artificial decorrelation and/or dipole loudspeakers to emulate the more diffused cinema experience.

speakers does not lead to a level mismatch between front and rear. The mono surround channel is the main limitation in this format. Despite the use of multiple loudspeakers to reproduce the surround channel, it is still not possible to create a strong sense of envelopment or spaciousness without using surround signals that are different on both sides of the listener.

The 3–1 audio format was also used in MUSE/Hi-Vision, high-definition laser discs released in Japan in the 1990s. It had 4 channels of digital audio encoded with 32 kHz sampling frequency and 12-bit quantization and multiplexed into MUSE video. MUSE decoders would play out the 4 analog signals into 5 outputs: left, right, center, surround left, surround right, with the surround channels sharing the same signal.

5.1-Channel Surround (3–2 Stereo)

Four-channel systems have the disadvantage of a mono surround channel, and this limitation is removed in the 5.1-channel system, enabling the provision of stereo effects or room ambience to accompany a primarily front-orientated sound stage. The distinction between front channels providing the stereo imaging of sources and rear channels providing ambience or effects explains the insistence in some standards on the use of the term '3–2 stereo' rather than 'five-channel surround'. As introduced above, the '.1' component is a dedicated low-frequency effects (LFE) channel or sub-bass channel, so termed because of its limited bandwidth. Strictly, the international standard nomenclature for 5.1 surround is '3–2–1', the last digit indicating the number of LFE channels.

The loudspeaker layout and channel configuration is shown in Figure 6.4. There can be little doubt that this format's widespread success was the result of a political compromise that aimed to come up with speaker locations that would fulfill a range of functions across cinema, broadcast and music recording applications, not being entirely ideal for any of them, but adequate for most. A display screen is also shown in the figure for sound with picture applications, and there are recommendations concerning the relative size of the screen and the loudspeaker base width. The left and right loudspeakers are located at ± 30° for compatibility with 2-channel stereo reproduction, a necessary compromise that limited the options at the time. The surround loudspeaker locations, at approximately ± 110°, are placed so as to provide a compromise between the need for effects panning behind the listener and the lateral energy important for good envelopment. In this respect they are more like 'side' loudspeakers than rear loudspeakers, and in many installations this is an inconvenient location causing people to mount them nearer the rear than the standard suggests.

In this standard there are normally no loudspeakers directly behind the listener, which can make for creative difficulties, and some commercial variants have attempted to remedy this. The ITU standard allows for additional surround loudspeakers to cover the region around listeners, similar to the 3–1 arrangement described earlier. If these are used then they are expected to be distributed evenly in the angle between ± 60° and ± 150°. Surround loudspeakers should be the same as front loudspeakers where possible, in order that uniform sound quality can be obtained all around. That said, there are arguments for use of dipole loudspeakers in these positions. Dipoles radiate sound in more of a figure-eight pattern and one way of obtaining a diffuse surround impression is to orient these with the nulls of the figure-eight towards the listening position. In this way the listener experiences more reflected than direct sound and this can give the impression of a more spacious ambient sound field that may better emulate the cinema listening experience

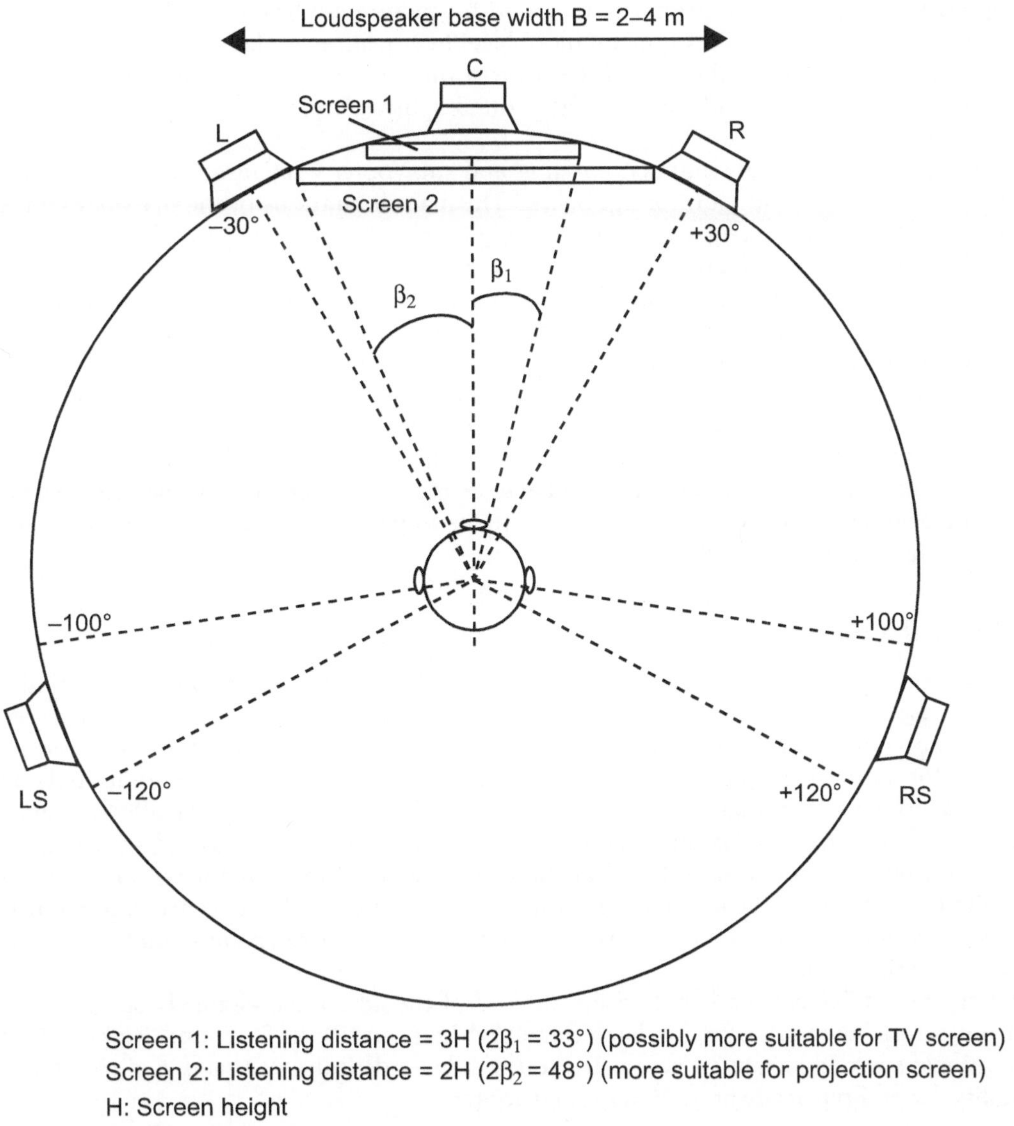

Figure 6.4 3–2 format reproduction according to the ITU-R BS.775 standard uses two independent surround channels routed to one or more loudspeakers per channel.

in small rooms. Dipoles make it correspondingly more difficult to create defined sound images in rear and side positions, though.

The low-frequency effects channel is a separate sub-bass channel with an upper limit extending to a maximum of 120 Hz. It is intended for conveying special low-frequency content that requires greater sound pressure levels and headroom than can be handled by the main channels.

It is not intended for conveying the low-frequency component of the main channel signals, and its application is likely to be primarily in sound-for-picture applications where explosions and other high-level rumbling noises are commonplace. In consumer audio systems, reproduction of the LFE channel is considered optional. Because of this, recordings for general purpose applications should normally be made so that they sound satisfactory even if the LFE channel is not reproduced.

Further discussion of LFE channel handling and subwoofer configuration is contained in the section on surround sound monitoring, below. The main limitations of the 5.1 surround format are first, that it was not designed for accurate 360° phantom imaging capability, as explained above. Second, the front sound stage is narrower than it could be if compatibility with 2–0 reproduction was not a requirement. Third, the center channel can prove problematic for music balancing, as conventional panning laws and coincident microphone techniques are not usually optimized for three loudspeakers. Fourth, the LS and RS loudspeakers are located in a compromise position, leading to a large hole in the potential image behind the listener and making it difficult to find physical locations for the loudspeakers in practical rooms.

Such limitations of the format led to various non-standard uses of the five or six channels available. For example, some used the sixth channel to create a height channel. Others made a pair out of the LFE channel and the center channel so as to feed a pair of front-side loudspeakers, enabling the rear loudspeakers to be farther back.

7.1-Channel Surround

Deriving from wide screen cinema formats, the 7.1-channel configuration normally adds two further loudspeakers to the 5.1-channel configuration, located at center-left (CL) and center-right (CR), as shown in Figure 6.5. This was not a format primarily intended for consumer applications, but for large cinema auditoria where the screen width is such that the additional channels are needed to cover the angles between the loudspeakers satisfactorily for all the seats in the auditorium. Some consumer equipment manufacturers have also implemented a 7-channel mode in their consumer surround decoders, but the recommended locations for the loudspeakers are not quite the same as in the cinema application. The additional channels are used to provide a wider side-front component and allow the rear speakers to be moved round more to the rear than in the 5.1 arrangement.

In some recent 7.1 cinema installations, instead of the additional channels being used across the front screen, they are used for Back Surround Left, and Back Surround Right, leaving the conventional surround channels connected to the side loudspeakers. This is said to offer greater flexibility for audio placement with 3D visual content.

10.2-Channel Surround

Tomlinson Holman developed a 10.2-channel surround sound system, which began the process of bridging the gap to later immersive audio systems that include height. To the basic 5-channel array he added wider side-front loudspeakers and a center-rear channel to 'fill in the holes' in the standard layout. He also added two height channels and a second LFE channel. The second LFE channel was intended to provide lateral separation of decorrelated low bass content to either side

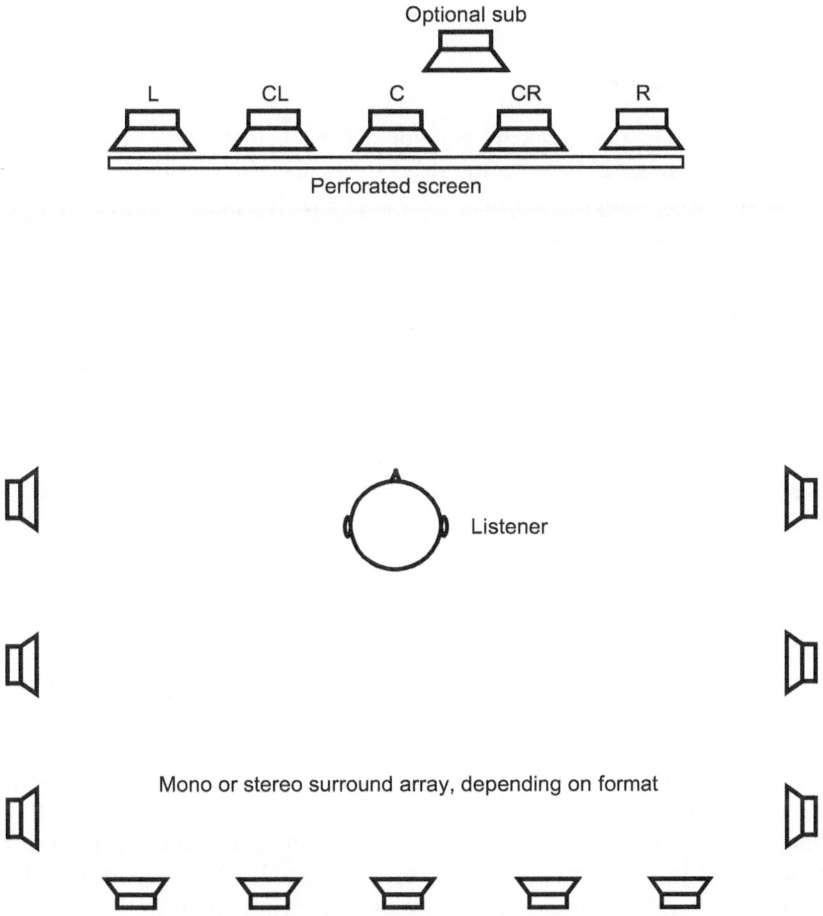

Figure 6.5 Some cinema sound formats for large auditorium reproduction enhance the front imaging accuracy by the addition of two further loudspeakers, center-left and center-right.

of the listening area, to enhance low-frequency spaciousness. (Surround sound with height channels will be discussed in detail in Chapter 7.)

Surround Sound Delivery and Coding

This part of the chapter concerns ways in which surround sound signals can be encoded so as to be carried over analog or digital media. It does not specifically deal with computer file formats, though. Initially, analog surround sound signals often had to be matrixed to get them to the end user, because most delivery media were 2-channel only, although some wide screen cinema formats had individual sound tracks for each channel. These days, surround sound is almost entirely carried in the digital domain, using some form of data reduction coding, although for some high-end applications there may be sufficient bit rate to carry it as linear PCM data.

Matrixed Surround Sound Systems

By matrixing surround signals they can be represented using fewer channels than the source material contains. This can give rise to some side effects and the signals need careful dematrixing, but the approach was used widely for many years. The Dolby Stereo approach is described as an example, but there are alternatives, and a number of other companies made enhanced decoders to improve the decoding of Dolby-matrixed sound tracks.

The original Dolby Stereo system involved a number of different formats for film sound with three to six channels, particularly a 70 mm film format with six discrete tracks of magnetically recorded audio, and a 35 mm format with two optically recorded audio tracks onto which were matrixed four audio channels in the 3–1 configuration. Dolby Surround was introduced in 1982 as a means of emulating the effects of Dolby Stereo in a consumer environment. Essentially the same method of matrix decoding was used, so movies transferred to television formats could be decoded in the home in a similar way to the cinema. Dolby Stereo optical sound tracks for the cinema were Dolby A noise-reduction encoded and decoded, in order to improve the signal-to-noise ratio, but this is not a feature of consumer Dolby Surround.

The Dolby Stereo matrix (see Figure 6.6) is a form of '4–2–4' matrix that encodes the mono surround channel so that it is added out of phase into the left and right channels (+90° in one channel and –90° in the other). The center channel signal is added to left and right in phase. The resulting sum is called L_t/R_t (left total and right total). By doing this the surround signal can be separated from the front signals upon decoding by summing the L_t/R_t signals out of phase (extracting the stereo difference signal), and the center channel can be extracted by summing L_t/R_t in phase. A decoder block diagram for the consumer version (Dolby Surround) is shown in Figure 6.7. In addition to the sum-and-difference-style decoding, the surround channel is subject to an additional delay, band-limiting between 100 Hz and 7 kHz and a modified form of Dolby B noise reduction. The low-pass filtering and the delay are both designed to reduce matrix side effects that could otherwise result in front signals appearing to come from behind. The delay (of the order of 20–30 ms in consumer systems, depending on the distance of the rear speakers) relies on the precedence effect to cause the listener to localize signals according to the first arriving wavefront which will now be from the front rather than the rear of the sound stage. The rear

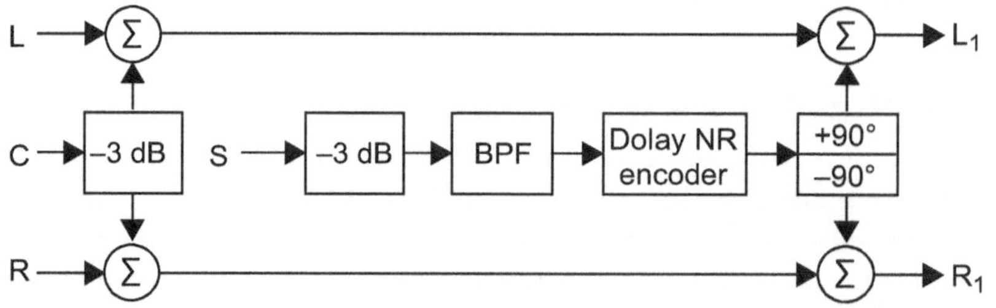

Figure 6.6 Basic components of the Dolby Stereo matrix encoding process.

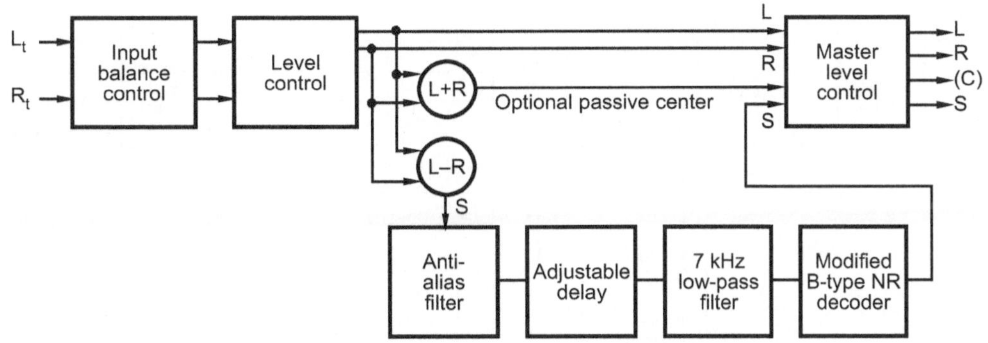

Figure 6.7 Basic components of the passive Dolby surround decoder.

signal then becomes psychoacoustically better separated from the front and localization of primary signals is biased more towards the front. The modified B-type NR reduces surround channel noise and also helps to reduce the effects of decoding errors and interchannel crosstalk, as some distortions introduced between encoding and decoding will be reduced by B-type decoding.

A problem with passive matrix decoding is that the separation between adjacent channels is relatively modest, although the separation of left/right and center/surround remains high. When a signal is panned fully left it will tend to appear only 3 dB down in the center, and also in the surround, for example. Dolby's ProLogic system, based on principles employed in the professional decoder, attempted to resolve this problem by including sophisticated 'steering' mechanisms into the decoder circuit to improve the perceived separation between the channels. A basic block diagram is shown in Figure 6.8. This enables a real center loudspeaker to be employed. Put crudely, ProLogic works by sensing the location of 'dominant' signal components and selectively attenuating channels away from the dominant component. ProLogic 2 added support for full-bandwidth stereo rear channels, with various options that made it more suitable for music programs. It was also claimed to be effective in the up-conversion of unencoded 2-channel material to 5-channel surround.

In 1998 Dolby and Lucasfilm THX joined forces to promote an enhanced surround system that added a center rear channel to the standard 5.1-channel setup. They introduced it, apparently, because of frustrations felt by sound designers for movies in not being able to pan sounds properly to the rear of the listener—the surround effect typically being rather diffuse. This system was christened 'Dolby Digital—Surround EX', and used matrix-style center channel encoding and decoding between the left and right surround channels of a 5.1-channel mix. The loudspeakers at the rear of the auditorium were then driven separately from those on the left and right sides, using the feed from this 'rear-center' channel, as shown in Figure 6.9.

Digital Surround Coding

Dolby Digital or AC-3 encoding (Todd et al., 1994) was developed as a means of delivering 5.1 channel surround to cinemas or the home without the need for analog matrix encoding. It has

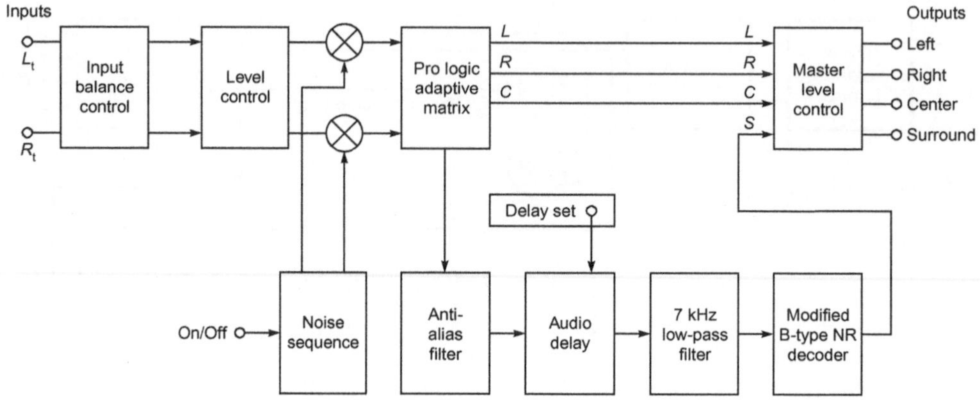

Figure 6.8 Basic components of the active Dolby ProLogic decoder.

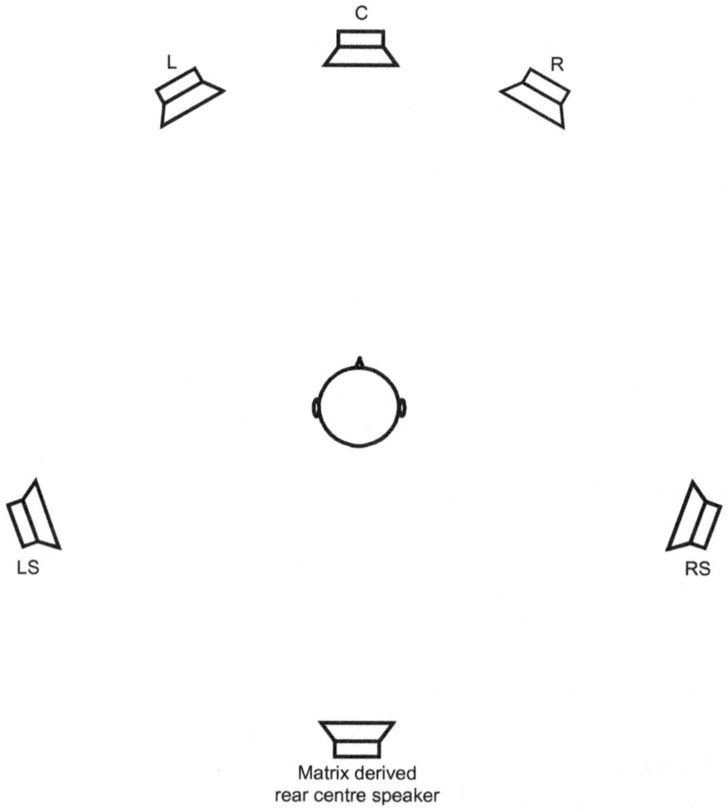

Figure 6.9 Dolby EX adds a center-rear channel fed from a matrix-decoded signal that was originally encoded between left and right surround channels in a manner similar to the conventional Dolby Stereo matrix process.

been used widely for the distribution of digital sound tracks on 35 mm movie films, broadcast and consumer media, on films the data being stored optically in the space between the sprocket holes on the film. The process involves a number of techniques by which the data representing audio from the source channels is transformed into the frequency domain and requantized to a lower resolution, relying on the masking characteristics of the human hearing process to hide the increased quantizing noise that results from this process. A common bit pool is used so that channels requiring higher data rates than others can trade their bit rate requirements provided that the overall total bit rate does not exceed the constant rate specified.

Aside from the representation of surround sound in a compact digital form, Dolby Digital includes a variety of operational features that enhance system flexibility and help adapt replay to a variety of consumer situations. These include dialog normalization ('dialnorm') and the option to include dynamic range control information alongside the audio data for use in environments where background noise prevents the full dynamic range of the source material being heard. Downmix control information can also be carried alongside the audio data in order that a 2-channel version of the surround sound material can be reconstructed in the decoder. As a rule, Dolby Digital data is stored or transmitted with the highest number of channels needed for the end product to be represented and any compatible downmixes are created in the decoder. This differs from some other systems where a 2-channel downmix is carried alongside the surround information.

Dolby Digital Plus is an extension to AC-3 (Dolby Digital), with higher data-rate options and shorter frames if required. It is designed to offer enhanced quality to Dolby Digital, running at data rates up to 6 Mbit/s, although the typical data rate on HD optical disks is said to be between 768 kbit/s and 1.5 Mbit/s. The data stream can be decoded by legacy receivers, which will only decode the Dolby Digital core at up to 640 kbit/s.

Dolby's TrueHD, based on Meridian Lossless Packing (MLP), is a lossless codec resulting in decoded quality that is identical to the studio master. It enables 7.1-channel playback on BD although it has the capacity to support more than 16 channels of audio. Operating at data rates of up to 18 Mbit/s, it supports the Blu-Ray Disk (BD) standard's requirement for eight full-range channels at 96 kHz/24 bits and up to 5.1 channels at 192 kHz/24 bits. An entirely separate artistic stereo mix can be carried if desired.

The original DTS (Digital Theater Systems) 'Coherent Acoustics' system (Smyth et al., 1996) is another digital signal coding format that can be used to deliver surround sound in consumer or professional applications, using low bit rate coding techniques to reduce the data rate of the audio information. The DTS system can accommodate a wide range of bit rates from 32 kbit/s up to 4.096 Mbit/s (somewhat higher than Dolby Digital), with up to eight source channels and with sampling rates up to 192 kHz. Variable bit rate and lossless coding are also optional. Downmixing and dynamic range control options are provided in the system.

DTS currently offers two codecs that can be used for higher-resolution audio on optical disks. Both are backwards compatible with the original DTS Digital Surround decoder because they are based on a lossy core plus extension model. Some other lossless formats take a similar form, for backwards compatibility, whereas others are lossless from the bottom up. DTS-HD High Resolution Audio offers data rates from 2–6 Mbit/s, offering quality that is not identical to the studio master but claimed to be close (it's still a lossy coding format). This version allows for a maximum of 7.1 channels at 96 kHz in a CBR (constant bit-rate) stream. DTS-HD Master Audio operates at data rates up to 24.5 Mbit/s in a variable bit-rate (VBR) stream, offering 7.1 channels at 96 kHz, or 5.1 at 192 kHz. This version is lossless, and therefore bit-for-bit compatible with the original master. The core coding, which works at up to 1509 kbit/s with 6.1 channels, is at a higher bit rate than typical DVD audio data rates, so non-HD players still get a quality increase. This data stream can be routed to legacy AV receivers using a SPDIF connection. A tool is available (Neural Upmix) that enables one to upmix creatively from 5.1 to surround formats with higher numbers of channels. The encoder enables one to set the downmix coefficients from surround to stereo. There is also a QC control tool that enables one to hear the effect of conversion of 5.1 material to different loudspeaker layouts such as non-standard 7.1 speaker positions where there are sides and rears.

Of the MPEG multichannel coding formats (e.g., Bosi et al., 1997), the MPEG-2 BC (backwards compatible) version worked by encoding a matrixed downmix of the surround channels and the center channel into the left and right channels of an MPEG-1 compatible frame structure. Although MPEG-2 BC was originally intended for use with DVD releases in Region 2 countries (primarily Europe), this requirement was dropped in favor of Dolby Digital. MPEG-2 AAC, on the other hand, is a more sophisticated algorithm that codes multichannel audio to create a single bit stream that represents all the channels, in a form that cannot be decoded by an MPEG-1 decoder. Having dropped the requirement for backwards compatibility, the bit rate could be optimized by coding the channels as a group and taking advantage of interchannel redundancy if required. The MPEG-2 AAC system contained contributions from a wide range of different manufacturers, and evolved into MPEG-4. High Definition AAC (HD AAC) has a lossy core accompanied by a lossless extension that enables decoding to provide bit-for-bit compatibility with the original master recording. The AAC core part is compatible with existing decoders in mobile devices such as the iPod and iTunes. It can operate at sampling rates up to 192 kHz and at 24-bit resolution.

The most recent standard related to surround audio coding is MPEG-H (Herre et al., 2014), which is capable of handling a wide range of surround and fully immersive content, as well as ambisonic material and audio objects, rendering to a number of possible loudspeaker layouts. Because of the increasing number of format options, the trend here is away from fixed loudspeaker layouts and towards the idea that material may have to be rendered to whatever is available at the reproduction end of the chain.

Auro 3D and its variants is not mentioned here because it is principally aimed at fully immersive content with height information. The same applies to other coding formats aimed primarily at with-height content.

Spatial Audio Object Coding

The MPEG Spatial Audio Object Coding (SAOC) standard (ISO, 2010) describes a user-controllable rendering method for multiple audio objects based on transmission of a mono or stereo downmix of the object signals. Audio objects are individual signals that can be manipulated independently of other objects under the control of metadata and user interaction. Rendering can be controlled so as to place audio objects in desired positions and at different levels. An increase in level difference and/or repositioning of a speech dialog object, for example, can also improve intelligibility with certain speaker layouts and environments. SAOC encodes Object Level Differences (OLD), Inter-Object Cross Coherences (IOC) and Downmix Channel Level Differences (DCLD) into a parameter bitstream, and so does not discretely encode input audio signals. The

SAOC bitstream is independent of loudspeaker configuration, and a default downmix option ensures backwards compatibility.

Object-based audio representation is also a key feature of the more recent MPEG-H standard (Herre et al., 2014), as well as an increasing number of other immersive audio coding systems.

Parametric Audio Coding

One variant on lossy low-bit rate coding involves the encoding of audio signals in the form of a 'core' signal alongside a sparse stream of 'parameters' that describe one or more features needed to reconstruct an approximation of the original signals. The idea is to code a basic version of the original audio signal, that could be decoded by compatible decoders, and to transmit spatial or spectral enhancements in the form of much lower bit rate 'side' information. MPEG-Surround (Breebaart et al., 2007) transmits a mono or stereo downmix of the original surround, plus side information to enable the surround spatial impression to be approximated upon decoding (see Figure 6.10). The downmix is encoded using a 'legacy' or conventional stereo coder such as MP3. The additional bit rate required for the side information is usually only a few kilobits per second, as opposed to the few hundred that might be needed to transmit the surround information as conventionally coded audio. This enables convincing surround to be transmitted at bit rates as low as 64 kilobits per second.

Time-Frequency Representation

Related to parametric coding of surround sound is time–frequency representation. A technique known as Directional Audio Coding (DirAC) embodies some of the same principles but is not a conventional parametric multichannel encoder like MPEG-Surround (Pulkki & Faller, 2006). Rather it is a method for spatial sound representation that can be applied to arbitrary reproduction scenarios, which can be linked to parametric multichannel audio coding. The authors explain that existing methods for the capture and reproduction of spatial sound fields, such as coincident and spaced microphone arrays, suffer from compromises that limit their ability either to create accurate directional cues or a sufficiently diffuse sound field. DirAC, on the other hand, is designed to represent and render these two components separately.

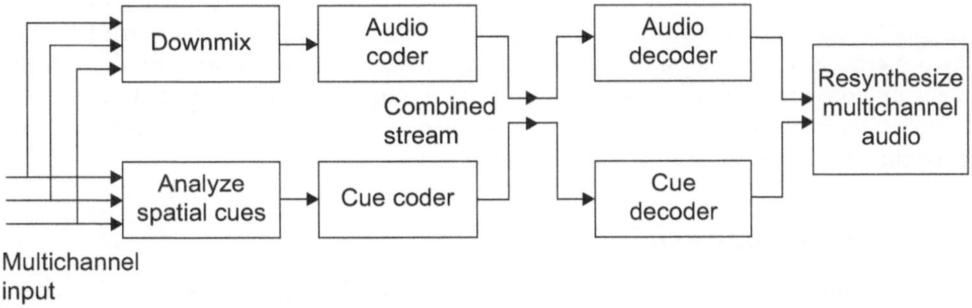

Figure 6.10 Block diagram of MPEG-Surround process showing (a) encoder, and (b) decoder.

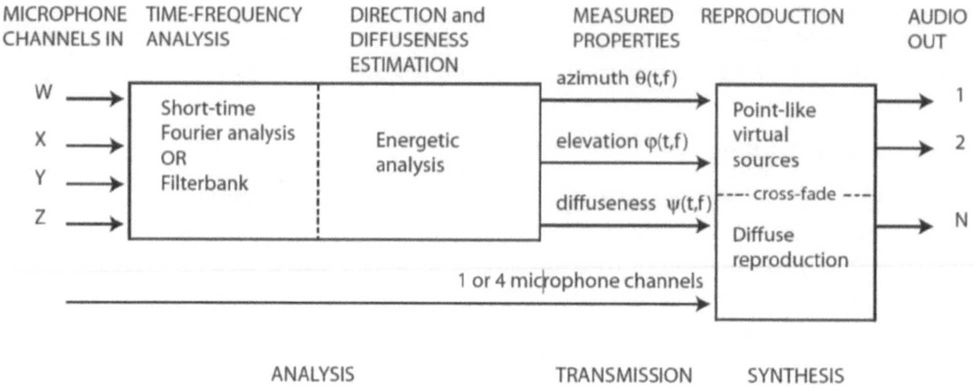

Figure 6.11 Conceptual diagram of DirAC spatial audio encoding system. (Reproduced courtesy of Ville Pulkki and Christof Faller, with permission of the Audio Engineering Society)

The DirAC approach is based on a number of assumptions about the relationship between perceptual parameters and physical cues; namely, that directional arrival of sound will transform into interaural time and level differences (ITD, ILD), that diffuseness will transform into interaural coherence cues, and that timbre depends on the monaural spectrum together with ITD, ILD and coherence information. A final assumption is that these factors together determine the auditory spatial image that the listener perceives. In order to ensure a match between the system's representation and the characteristics of the human auditory process, the captured signals are split into filter bands similar to those of the auditory system, and the temporal resolution of the analysis is similarly defined. In the example presented in their paper, the authors show a flow diagram of the directional audio coding process based on the use of B format ambisonic signals as inputs (see Figure 6.11), which are further described in Chapter 9.

From the microphone signals, instantaneous direction vectors and diffuseness values are derived. The diffuseness values can be averaged over a period of some tens of milliseconds to reduce the rate at which they are transmitted. Upon reproduction the direction vectors in each frequency band are processed so as to render point-like virtual sources, using a panning technique such as VBAP (vector-based amplitude panning) (Pulkki, 1997). Diffuseness is resynthesized by one of two methods, the simplest involving the decorrelation of the transmitted omnidirectional component by convolving it with exponentially decaying white noise bursts having a time constant of 20 ms. By using a different noise signal for each loudspeaker signal, multiple decorrelated versions of the omni component can be generated.

Surround Sound Monitoring

This section is mainly about monitoring environments and configurations for 5.1-channel surround sound, although many of the principles apply in other configurations. The Audio Engineering Society published an information document on this topic (AES, 2001).

Main Loudspeakers

Rooms for surround monitoring should have an even distribution of absorbing and diffusing material. This is so that the rear loudspeakers function in a similar acoustic environment to the front loudspeakers. This is contrary to a number of popular two-channel control room designs that have one highly absorbtive end and the other end more reflective.

In larger rooms designed for film sound mixing, a distributed array of surround loudspeakers is often used, in some cases with decorrelation between them to avoid strong comb filtering effects. In smaller control rooms used for music and broadcasting mixing the space may not exist for such arrays. The ITU standard allows for more than one surround loudspeaker on either side and recommends that they are spaced equally on an arc from 60° to 150° from the front.

It can be difficult to install loudspeaker layouts according to the ITU standard, with equal spacing from the listening position and the surrounds at 110° ± 10°, because of the required width of the space. This often makes it necessary for the room to be laid out 'wide' rather than 'long', and if the room is one that was previously designed for 2-channel stereo the rotation of the axis of symmetry may result in the acoustic treatment being inappropriately distributed. Also the location of doors and windows may make the modification of existing rooms difficult. If building a new room for surround monitoring then it is obviously possible to start from scratch and make the room wide enough to accommodate the surround loudspeakers and absorption in more suitable places.

As a rule, front loudspeakers can be similar to those used for 2-channel stereo, although noting the particular problems with the center loudspeaker described in the next section. Low-directivity front loudspeakers may be desirable when trying to emulate the effect of a film mixing situation in a smaller surround control room. This is because in the large rooms typical of cinema listening the sound balancer is often well beyond the critical distance where direct and reflected sound are equal in level, and using speakers with low directivity helps to emulate this scenario in smaller rooms. Film mixers generally want to hear what the large auditorium audience member would hear, and this means being farther from the loudspeakers than for small room domestic listening or conventional music mixing.

Ideally the center speaker should be of the same type or quality as the rest. It may be possible to use somewhat smaller monitors for the main channels than would be used for 2-channel stereo, handling the low bass by means of a subwoofer or two. This makes it more practical to mount a center loudspeaker behind the mixing console, but its height will often be dictated by a control room window or video monitor. The center loudspeaker should be on the same arc as that bounding the other loudspeaker positions, otherwise the time delay of its direct sound at the listening position will be different from that of the other channels. If the center speaker is closer than the left or right channels, then it should be delayed slightly to put it in the correct place acoustically.

A lot of surround mixing work is carried out in conjunction with pictures, and this presents challenges for the center speaker location. In cinemas the screen is normally acoustically 'transparent' and uses front projection, although this transparency is never complete and usually requires some equalization. In smaller mixing rooms the display is often a flat-screen plasma monitor or a CRT display and these do not allow the same arrangement. With modestly sized solid displays for television purposes it can be possible to put the center loudspeaker underneath the display, with the display raised slightly, or above the display angled down slightly. The presence of a mixing console may dictate which of these is possible, and care should be taken to avoid strong reflections from the center loudspeaker off the console surface. Dolby suggests that if the center loudspeaker has to be offset height-wise it could be turned upside down compared with the left and right channels to make the tweeters line up, as shown in Figure 6.12.

Recommendations for professional setups suggest that the surround loudspeakers should be of the same quality as the front ones. In consumer environments this can be difficult to achieve, and the systems sold at the lower end of the market often incorporate much smaller surround loudspeakers than front. The use of a subwoofer to handle the low bass makes the required volume of the main speakers quite a lot smaller.

The directivity requirements of the surround loudspeakers have been the basis of some disagreement (see for example the exchange between Holman and Zacharov, 2000). The debate centers around the use of the surround loudspeakers to create a diffuse, enveloping sound field—a criterion that tends to favor either decorrelated arrays of direct radiators or dipole surrounds (bidirectional speakers that are typically arranged so that their main axis does not point towards the listener). If the creation of a diffuse, enveloping sound field is the only role for surround loudspeakers, then dipoles can be quite suitable if only two loudspeaker positions are available, particularly in small rooms and for the translation of large auditorium film mixes into smaller spaces. If, on the other hand, attempts are to be made at all-round source localization, direct radiators are probably more suitable.

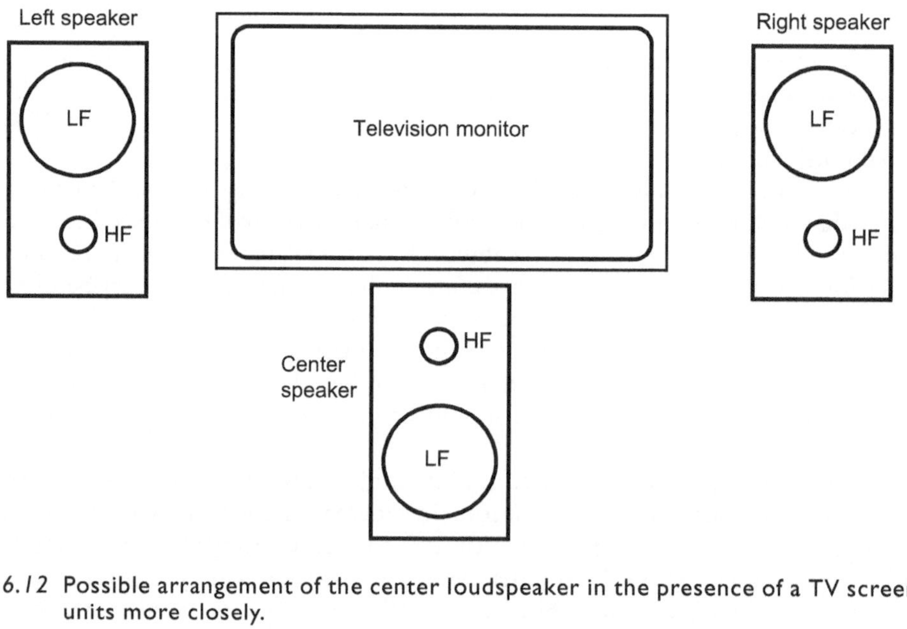

Figure 6.12 Possible arrangement of the center loudspeaker in the presence of a TV screen, aligning HF units more closely.

Subwoofers

Low-frequency interaction between loudspeakers and rooms affects the placement and equalization of subwoofers. In choosing the optimum locations for subwoofers one must remember the basic principle that loudspeakers placed in corners tend to give rise to a noticeable bass boost, and couple well to most room modes. Some subwoofers are designed specifically for placement in particular locations whereas others need to be moved around until the most subjectively satisfactory result is obtained. Some equalization may be needed to obtain a reasonably flat overall frequency response at the listening position. Phase shifts or time-delay controls are sometimes provided to enable some correction of the time relationship of the subwoofer to other loudspeakers, but this will be a compromise with a single unit.

Multiple low-frequency drivers generating decorrelated signals can create a more natural spatial reproduction than monaural low-frequency reproduction from a single driver. Griesinger (1997) proposes that if mono LF content is reproduced it is better done through two units placed to the sides of the listener, driven 90° out of phase, to excite the asymmetrical lateral modes more successfully and improve LF spaciousness.

The LFE channel of a 5.1 surround system should be aligned so that its in-band gain on reproduction is 10 dB higher than that of the other channels. This does not mean that the overall subwoofer output should have its level raised by 10 dB compared with the other channels, as this would incorrectly boost any LF information routed to the subwoofer as a result of bass management (filtering off the LF content of the main channels and sending it to the sub).

It is a common misconception that any sub-bass or subwoofer loudspeaker(s) that may be used on reproduction must be fed directly from the LFE channel in all circumstances. While this may be the case in the cinema, bass management in consumer systems is not specified in the standard and is entirely system-dependent. It is not mandatory to feed low-frequency information to the LFE channel during the recording process, neither is it mandatory to use a subwoofer, indeed it has been suggested that restricting extreme low-frequency information to a monophonic channel may limit the potential for low-frequency spaciousness in balances. In music mixing it is likely to be common to send the majority of full-range LF information to the main channels, in order to retain the stereo separation between them.

In practical systems it may be desirable to use one or more subwoofers to handle the low-frequency content of a mix on reproduction. The benefit of this is that it enables the size of the main loudspeakers to be correspondingly reduced. In such cases crossover systems split the signals between main loudspeakers and subwoofer(s) somewhere between 80 Hz and 160 Hz. In order to allow for reproduction of the LFE channel and/or the low-frequency content from the main channels through subwoofer loudspeakers, a form of bass management akin to that shown in Figure 6.13 is typically employed.

Sound Bars

A brief mention should be made here of 'sound bars' in surround sound monitoring. Although not recommended for professional monitoring (except perhaps to discover what a mix might sound like when replayed using such a system), these are increasingly widely used in consumer

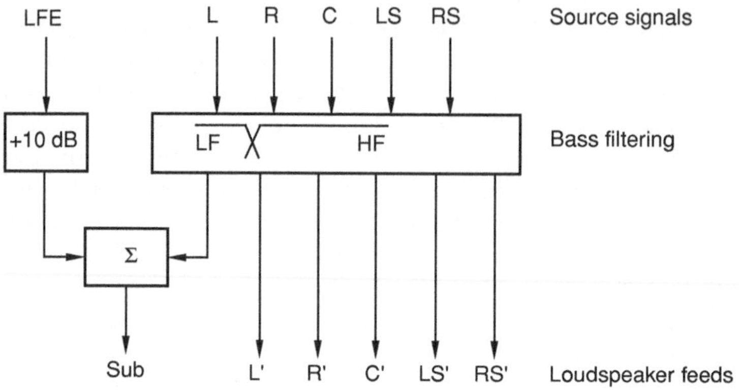

Figure 6.13 Low-frequency management in a 5.1 reproduction system.

reproduction as a compact and convenient alternative to multiple separate loudspeakers. Rear loudspeakers are particularly difficult to locate and wire in homes and other such environments, so the workaround adopted with sound bars is to radiate sound directionally from an array of loudspeakers, usually arranged in the form of a narrow 'bar' that can be mounted above or below a television screen. In this case rear channel content is radiated indirectly so as to bounce off the side and rear walls of the room. The same concept has also been extended to fully immersive reproduction.

Surround Sound Recording Techniques

Many of the concepts used in surround sound recording have at least some basis in conventional 2-channel stereo techniques. However, the challenges are greater, particularly when dealing with the region to the sides of the listener, where there can be a large gap between the loudspeakers and it is hard to deliver convincing phantom images (see Figure 6.14). A similar challenge can exist in the rear sector, where there is again a wide angle between the loudspeakers, at least in the 3–2 configuration. Some of the more recent extensions to the surround channel layout used for the cinema, such as 7.1, have helped to make these 'dead spots' less problematic.

Most 'production' recording techniques make most use of panned monophonic content and artificial effects, whereas a great deal of research and experimentation has gone into the design of surround microphone arrays that attempt the authentic directional pickup of entire acoustic environments. This is an example of the disconnection between purist/academic research and the mainstream of the recording industry. Broadcasting techniques live in a crossover domain between these two extremes, with some use of single-point surround microphones or arrays to capture live events.

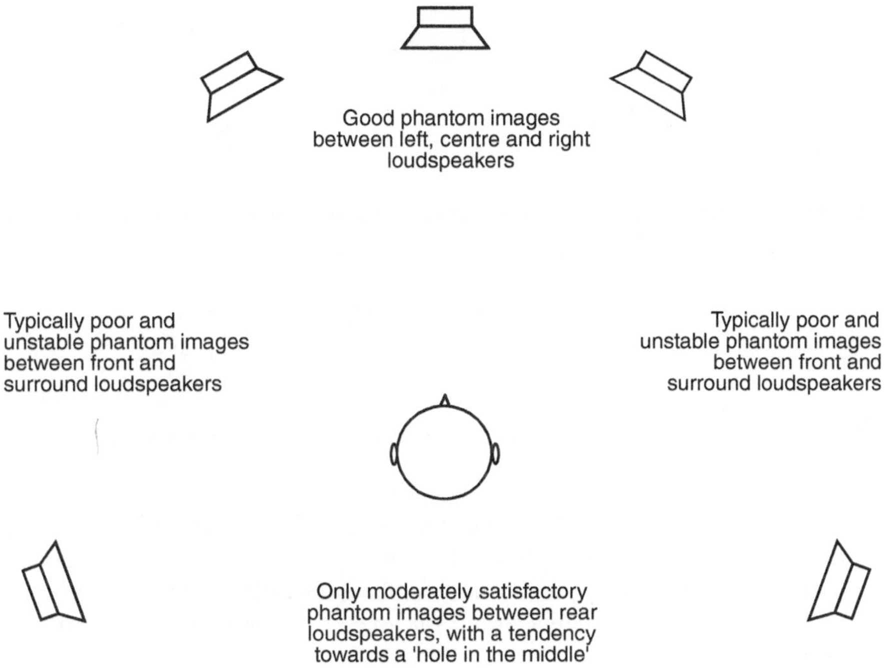

Figure 6.14 Quality of phantom images in a 3–2 surround system.

Microphone Arrays

Surround microphone array techniques split into two main groups: those that are based on a single array of microphones in reasonably close proximity to each other, and those that treat the front and rear channels separately. The former are usually based on an attempt to generate phantom images with different degrees of accuracy around the full 360° in the horizontal plane. The latter usually have a front array providing reasonably accurate phantom images in the front, coupled with a separate means of capturing the ambient sound of the recording space.

Of the first type there are variants on a common theme involving fairly closely spaced microphones (often cardioids) configured in a five-point array. A book by Michael Williams deals with this idea in some detail (Williams, 2004). The basis of most of these arrays is pairwise time—intensity trading, usually treating adjacent microphones as pairs covering a particular sector of the recording angle around the array. The generic layout of such arrays is shown in Figure 6.15. Cardioids or even supercardioids tend to be favored because they offer the increased direct-to-reverberant ratio of sound capture when aimed at the source. The center microphone is typically spaced slightly forward of the L and R microphones, thereby introducing a useful time advance in the center channel for center-front sources. The spacing and angles between the capsules

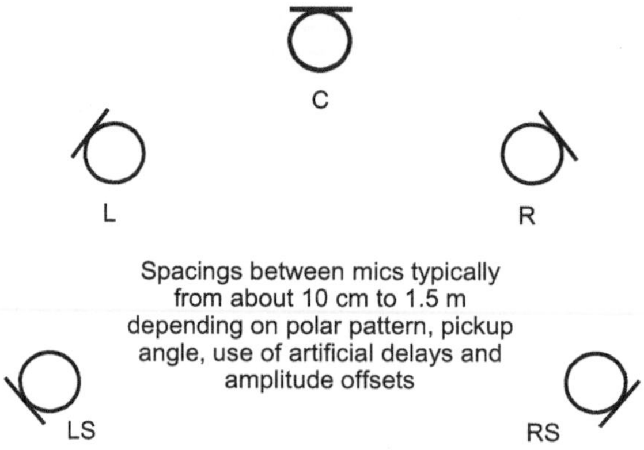

Figure 6.15 Generic layout of 5-channel microphone arrays based on time–amplitude trading.

are typically based on the so-called Williams curves, based on time and amplitude differences required between single pairs of microphones to create phantom sources in particular locations. Some success has also been had by the author's colleagues using omni microphones instead of cardioids, with appropriate adjustments to the spacings. These tend to give better overall sound quality but poorer front imaging.

The second group of techniques treats the stereo imaging of front signals separately from the capture of a natural-sounding spatial reverberation and reflection component. Most do this by adopting a 3-channel variant on a conventional 2-channel technique for the front channels, coupled with a more or less decorrelated combination of microphones in a different location for capturing spatial ambience (sometimes fed just to the surrounds, other times to both front and surrounds). Sometimes the front microphones also contribute to the capture of spatial ambience, depending on the proportion of direct to reflected sound picked up, but the essential point here is that the front and rear microphones are not intentionally configured as an attempt at a 360° imaging array.

Hamasaki of NHK (the Japanese broadcasting company) has proposed an arrangement based on near-coincident cardioids (30 cm) separated by a baffle, as shown in Figure 6.16 (Hamasaki, 2003). Here the center cardioid is placed slightly forward of left and right, and omni outriggers are spaced by about 3 meters. These omnis are low-pass filtered at 250 Hz and mixed with the left and right front signals to improve the LF sound quality. Left and right surround cardioids are spaced about 2–3 meters behind the front cardioids and 3 meters apart. An ambience array is used farther back, consisting of four figure-eight mics facing sideways, spaced by about 1 meter, to capture lateral reflections, fed to the four outer channels. This is placed high in the recording space.

Theile (2000) proposes a front microphone arrangement shown in Figure 6.17. While superficially similar to the front arrays described in the previous section, his arrangement reduces crosstalk between the channels by the use of supercardioid microphones at ± 90° for the left and

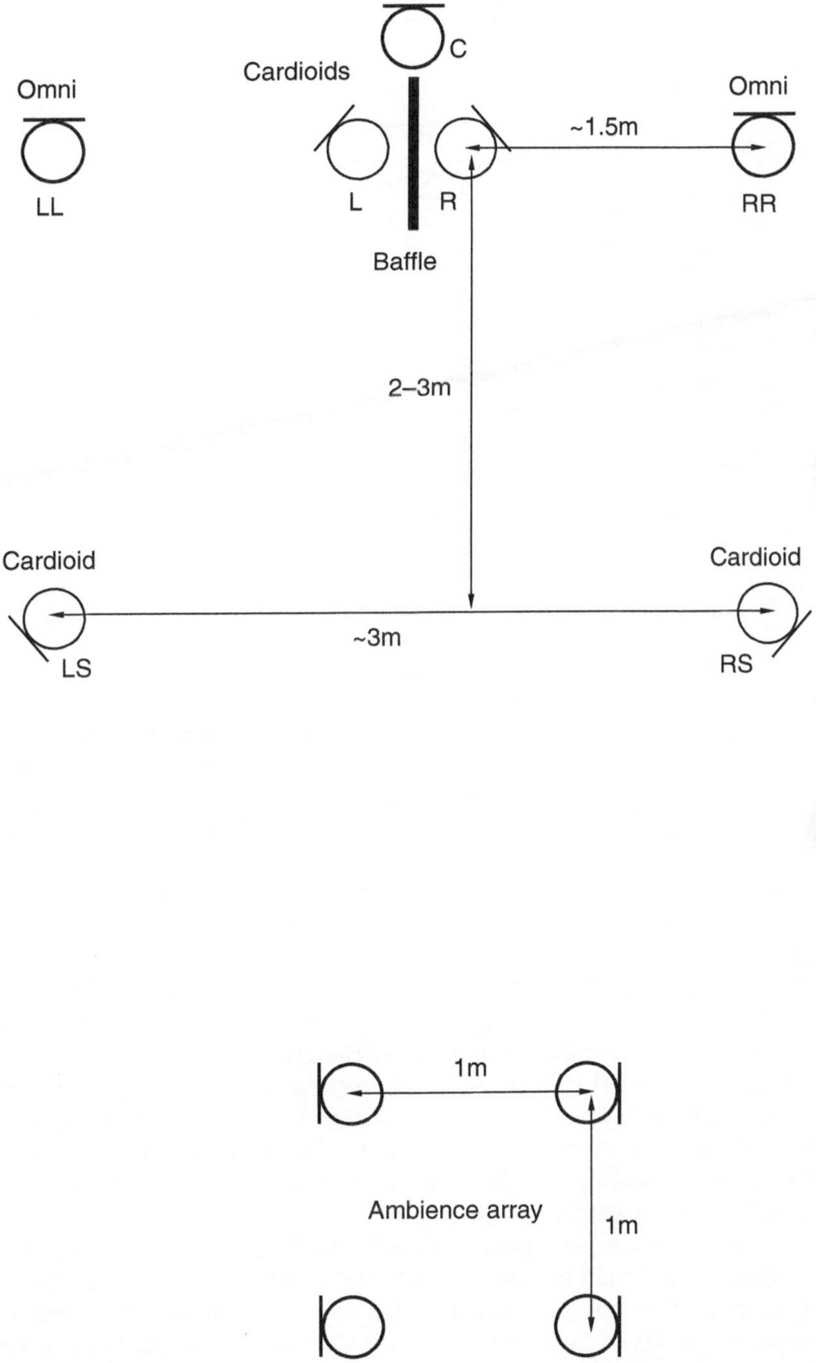

Figure 6.16 A surround technique proposed by Hamasaki (NHK) consisting of a cardioid array, omni outriggers and separate ambience matrix.

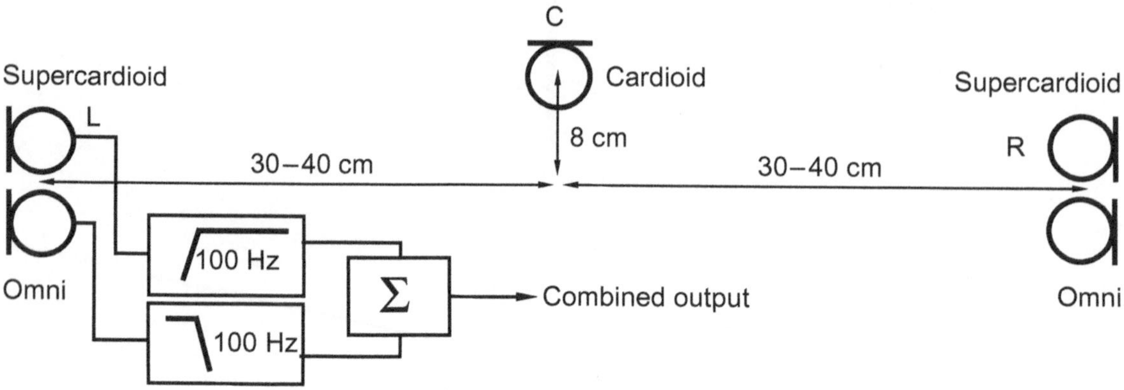

Figure 6.17 Theile's proposed 3-channel array for front pickup using supercardioids for the outer mics, crossed over to omni at LF. The spacing depends on the recording angle (C–R = 40 cm for 90° and 30 cm for 110°).

right channels and a cardioid for the center. (Supercardioids are more directional than cardioids and have the highest direct/reverberant pickup ratio of any first-order directional microphone. They have a smaller rear lobe than hypercardioids.) Theile's rationale behind this proposal is the avoidance of crosstalk between the front segments. He proposes to enhance the LF response of the array by using a hybrid microphone for left and right, which crosses over to omni below 100 Hz, thereby restoring the otherwise poor LF response of supercardioids. The center channel is high-pass filtered above 100 Hz. Furthermore, the response of the supercardioids should be equalized to have a flat response to signals at about 30° to the front of the array (they would normally sound quite colored at this angle). Schoeps has developed a prototype of this array, and it has been christened 'OCT' for 'Optimum Cardioid Triangle'.

For the ambient sound signal, Theile proposes the use of a crossed configuration of microphones, which has been christened the 'IRT cross' or 'atmo-cross'. This is shown in Figure 6.18. The microphones are either cardioids or omnis, and the spacing is chosen according to the degree of correlation desired between the channels. Theile suggests 25 cm for cardioids and about 40 cm for omnis, but says that this is open to experimentation. Small spacings are appropriate for more accurate imaging of reflection sources at the hot spot, whereas larger spacings are appropriate for providing diffuse reverberation over a large listening area. The signals are mixed in to L, R, LS and RS channels, but not the center.

A 'double MS' technique has been proposed by Curt Wittig and others, shown in Figure 6.19. Two mid-side pairs are used, one for the front channels and one for the rear. The center channel can be fed from the front M microphone. (As with any MS technique, the signals from the two microphones have to be added and subtracted using a simple transformer arrangement, mixer configuration, or signal processor, to derive left and right channels.) The rear pair is placed at or just beyond the room's critical distance. S channel gain can be varied to alter the image width in

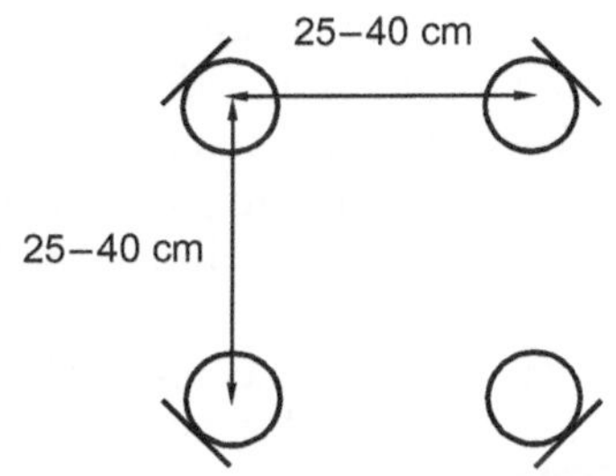

Figure 6.18 The IRT 'atmo-cross' designed for picking up ambient sound for routing to four loudspeaker channels (omitting the center). Mics can be cardioids or omnis (wider spacing for omnis).

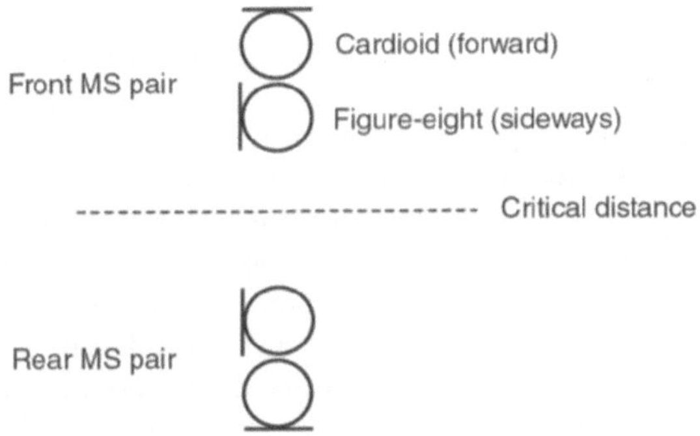

Figure 6.19 Double MS pair arrangement with small spacing between front and rear pair.

either sector, and the M mic's polar pattern can be chosen for the desired directional response (it would typically be a cardioid). Others have suggested using a fifth microphone (a cardioid) in front of the forward MS pair, to feed the center channel, delayed to time align it with the pair. If the front and rear MS pairs are co-located it may be necessary to delay the rear channels somewhat (10–30 ms) so as to reduce perceived spill from front sources into rear channels. In a co-located situation the same figure-eight microphone could be used as the S channel for both front and back pairs.

Spaced omni approaches to surround pickup include those proposed by Erdo Groot of Polyhymnia International, and Richard King of McGill University. Groot developed a largely undocumented array for Polyhymnia's classical recordings that used omnis instead of cardioids, to take advantage of their better sound quality. Using an array of omnis separated by about three meters between left—right and front—back he achieves a spacious result where the rear channels are well integrated with the front. The center mic is placed slightly forward of left and right. It is claimed that placing the rear omnis too far away from the front tree makes the rear sound detached from the front image, so one gets a distinct echo or repeat of the front sound from the rear. In Richard King's design, the spacings are slightly different, but the principle is essentially the same, with 1.3–2.6 m between the front left and right, 2–3 m between the rear left and right, and 3.6–5 m between the front and back. The rear microphones in this case are fitted with 50 mm spherical diffractive attachments (acoustic pressure equalizers or APEs) to modify the high-frequency directivity. The elevation of such arrays above the floor is a matter for experimentation, but typically between 2.5 and 6 m.

In general, the signals from separate ambience microphones fed to the rear loudspeakers may often be made less obtrusive and front–back 'spill' may be reduced by rolling off the high-frequency content of the rear channels. Some additional delay applied to the front channels or the rear channels may also assist in the process of integrating the rear channel ambience. The precise values of delay and equalization can only really be arrived at by experimentation in each situation.

Multi-Microphone Techniques and Panning

Most recording involves the use of spot 'accent' or 'support' microphones in addition to a main microphone technique of some sort. Indeed in many situations the spot microphones may end up at higher levels than the main microphone or there may be no main microphone. Alternatively sources such as recorded effects and synthesized material will be mixed and panned, using panoramic potentiometers, or panpots. Artificial reverberation of some sort is almost always helpful when trying to add spatial enhancement to panned mono sources, and some engineers prefer to use amplitude-panned signals to create a good balance in the front image, plus artificial reflections and reverberation to create a sense of spaciousness and depth.

The panning of signals between more than two loudspeakers presents a number of psychoacoustic problems, particularly with regard to appropriate energy distribution of signals, accuracy of phantom source localization, off-center listening and sound timbre. A number of different solutions have been proposed, some rather sophisticated, in addition to the relatively crude pairwise approach used in much film sound, but the simplicity and relative success of amplitude panning still seems to make it the most popular solution in practical applications.

English inventor Michael Gerzon came up with some criteria for a good panning law for surround sound (Gerzon, 1992c, p. 2):

> The aim of a good panpot law is to take monophonic sounds, and to give each one amplitude gains, one for each loudspeaker, dependent on the intended illusory directional localization of that sound, such that the resulting reproduced sound provides a convincing and sharp phantom illusory image. Such a good panpot law should provide a smoothly continuous range of image directions for any direction between those of the two outermost loudspeakers, with no "bunching" of images close to any one direction or "holes" in which the illusory imaging is very poor.

Pairwise amplitude panning involves adjusting the relative amplitudes between a pair of adjacent loudspeakers so as to create a phantom image at some point between them. This has been extended to three front channels and is also sometimes used for panning between side loudspeakers (e.g., L and LS) and rear loudspeakers. The typical sine/cosine panning law devised by Blumlein for 2-channel stereo is often simply extended to more loudspeakers. Most such panners are

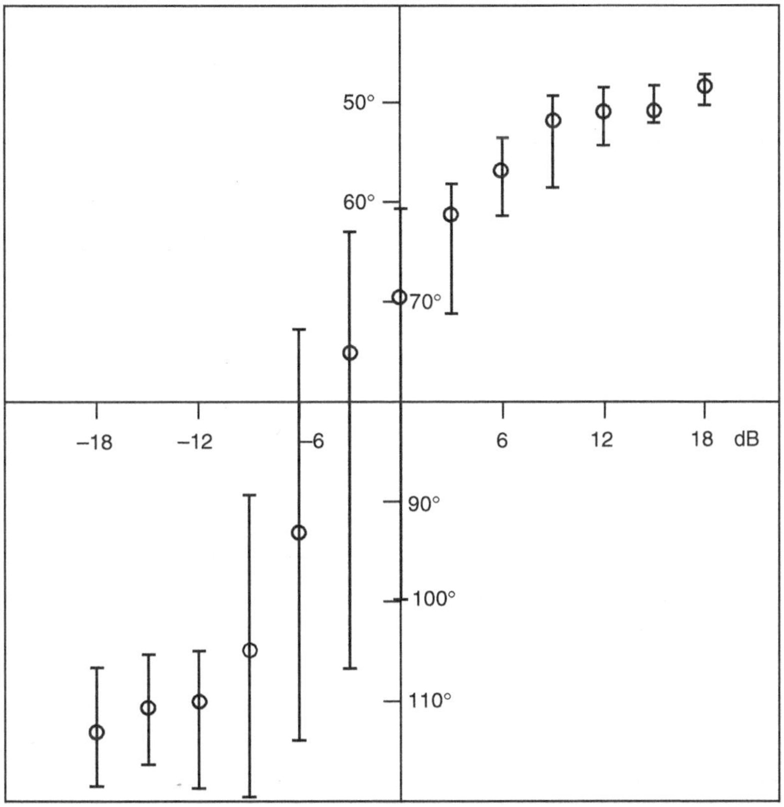

Figure 6.20 Perceived location of phantom image versus interchannel level difference between side loudspeakers centered on 80° offset from front-center, showing error bars. The forward loudspeaker is at 50° and the rear at 110°. It can be seen that the greatest uncertainty is in the middle of the range and that the image jumps rapidly from front to back. There is also more uncertainty towards the rear than the front (Theile & Plenge, 1977).

constructed so as to ensure constant power as sources are panned to different combinations of loudspeakers, so that the approximate loudness of signals remains constant.

Panning using amplitude or time differences between widely spaced side loudspeakers is not particularly successful at creating accurate phantom images. Side images tend not to move linearly as they are panned and tend to jump quickly from front to back. Data from Theile and Plenge (1977) illustrating this is shown in Figure 6.20. Spectral differences resulting from differing HRTFs of front and rear sound tend to result in sources appearing to be spectrally split or 'smeared' when panned to the sides.

In some mixing consoles designed for surround work, particularly in the film domain, separate panners are provided for L–C–R, LS–RS, and front-to-rear surround. Combinations of positions of these amplitude panners enable sounds to be moved to various locations, but some more successfully than others. For example, sounds panned so that some energy is emanating from all loudspeakers (say, panned centrally on all three pots) tend to sound diffuse for center listeners, and in the nearest loudspeaker for those sitting off-center. Joystick panners combine these amplitude relationships under the control of a single lever that enables a sound to be 'placed' dynamically anywhere in the surround sound field. Moving effects made possible by these joysticks are often unconvincing and need to be used with experience and care.

Research undertaken by Jim West (1999) at the University of Miami showed that, despite the limitations of constant power 'pairwise' panning, it proved to offer reasonably stable images for center and off-center listening positions, for moving and stationary sources, compared with some other more esoteric algorithms. Front–back confusion was noticed in some cases, for sources panned behind the listener. In Martin et al.'s (1999) subjective tests of image focus using different panning laws it was found that conventional pairwise constant-power panning provided the most focused images, followed by a relatively simple polarity-restricted cosine law and a second-order ambisonic law. These tests were conducted at the hot spot only, and the authors subsequently concluded that the polarity-restricted cosine law appeared to create fewer unwanted side effects than the constant power law (such as changes in perceived distance to the source).

The amplitude panning concept was extended to a general model that can be used with combinations of loudspeakers in arbitrary locations, known as vector-based amplitude panning or VBAP (Pulkki, 1997). This approach enables amplitude differences between two or three loudspeakers to be used for the panning of sources. Borß (2014) describes a novel alternative to VBAP as a way of rendering phantom sources to immersive loudspeaker arrays. VBAP is basically an extension to the tangent panning law, based on amplitude differences between loudspeakers in a triad, and is widely used because it is simple and effective. However, there are some limitations under certain circumstances and Borß proposes a system that uses symmetric panning gains for symmetric loudspeaker setups, using *N*-wise panning defined using polygons. This scheme uses a larger number of loudspeakers and seems to stabilize the position and trajectory of phantom sources. However, it also introduces a slightly greater bass boost and slightly more spread images. The author christens the approach 'Edge Fading Amplitude Panning' (EFAP). The method is based on many of the same principles as VBAP, in that it uses minimal computing resources, is still based only on amplitude panning, and has power normalized gains. It aims to offer smooth transition of gains between the speakers, and tries to avoid panning where summing localization principles don't work, where there are large angles between speakers.

Mixing Aesthetics

How to use the center channel in mixes has aroused controversy. Some engineers strongly protest using it, claiming that the center channel is a distraction and a nuisance, and that they can manage very well without it, while others are equally strongly convinced of its merits. The psychoacoustical advantages of using a center channel have been mentioned earlier, but existence of this channel complicates panning laws and microphone techniques, as well as makes conversion between formats more difficult.

Some classical engineers find that simultaneous surround and 2-channel recordings of the same session are made easier by adopting 4-channel rather than 5-channel recording techniques, but this may be more a matter of familiarity than anything else. For many situations a separate mix and different microphones will be required for 2-channel and 5-channel versions of a recording.

In multitrack recording using panned mono sources, the panning law chosen to derive the feed to the center channel will have an important effect on the psychoacoustic result. Numerous studies have highlighted the timbral differences between real and phantom center images, which leads to the conclusion that the equalization of a source sent to a hard center would ideally be different from that used on a source mixed to a phantom center. Vocals, for example, panned so as only to emanate from the center loudspeaker may sound constricted spatially compared with a phantom image created between left and right loudspeakers, as the center loudspeaker is a true source with a fixed location. Some 'bleed' into the left and right channels is sometimes considered desirable, in order to 'defocus' the image, or alternatively stereo reverberation can be used on the signal.

The technique of spreading mono panned sources into other channels is often referred to as a 'divergence' or 'focus' control, and can be extended to the surround channels as well, using a variety of different laws to split the energy between the channels. Holman (1999) advises against the indiscriminate use of divergence controls as they can cause sounds to be increasingly localized into the nearest loudspeaker for off-center listeners.

Surround channels are in most cases best reserved for mix components that are not to be clearly or accurately localized, unless very close to loudspeaker positions. In film sound the concept of a surround 'loudspeaker position' is somewhat alien in any case, as there are usually numerous surround loudspeakers connected together. In mixing music for consumer applications it may be possible to treat the surround loudspeakers as point sources, although they may not be accurately localized by listeners.

Upmixing and Downmixing

Content can be converted from one spatial format to another, using a matrix or algorithm of some kind, but this can come with compromises in both spatial and timbral quality. In upmixing an attempt is made to generate surround sound with more channels than exist in the source material, whereas in downmixing the aim is to create fewer channels.

Many upmixing algorithms, using 2-channel stereo as a source, extract some of the ambience contained in the difference information between the L and R channels and use it to drive the rear channels, often with quite sophisticated directional steering to enhance the stereo separation of the rear channels. Sometimes a proportion of the front sound is placed in the rear channels to increase envelopment, with suitable delay and filtering to prevent front sounds being pulled towards the rear. Experiments by the author found that the level of signal extracted by such algorithms to the center and rear channels was strongly related to the sum and difference components of the 2-channel signal (Rumsey, 1998).

Surround matrix decoding algorithms, such as described earlier, may be used for this purpose, although some are optimized better than others for dealing with 2-channel material that has not previously been matrix-encoded. Often a separate collection of settings is needed for upmixing unencoded 2-channel stereo to surround than is used for decoding matrix-encoded surround.

There are also a number of algorithms used in home cinema and surround systems that add 'effects' to conventional stereo in order to create a surround impression. Rather than extract existing components from the stereo sound to feed the rear channels they add reverberation on top of any already present, using effects called 'Hall' or 'Jazz Club' or some other such description. These alter the acoustic characteristics of the original recording quite strongly.

Subjective experiments carried out by the author on a range of such upmixing algorithms found that the majority of 2-channel material suffered from a degradation of the front image quality when converted to 5-channel reproduction (Rumsey, 1999). This either took the form of a narrower image, a change in perceived depth or a loss of focus. On the other hand, the overall spatial impression was often improved, although listeners differed quite strongly in their liking of the spatial impression created (some claiming it sounded artificial or phasy).

Faller et al. (2013) propose that upmixing is often based on an unrealistically simple model that separates direct and ambient sound. For example, if a model assumes that there is the same ambient signal power in all channels, some discrete 5.1 mixes can't be represented very well because there are different ambient signals (with different power) in the front and rear channels. They go on to show how a cascade of 2-channel upmixes to surround, called a ring upmix, can be used to generate channels for more loudspeakers with full support for 360° panning and high channel separation. The 'ring' referred to in this case is the ring of loudspeakers that defines the input format, and the aim is to extend this to more channels by adding loudspeakers between the original channels, as shown in Figure 6.21. In each case the aim is to take a 2-channel original

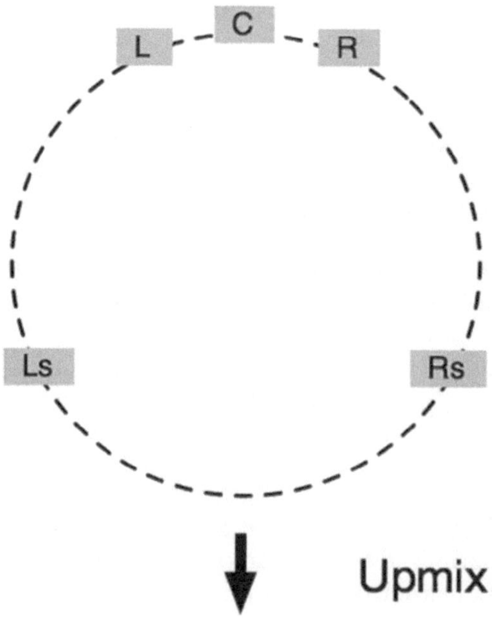

Figure 6.21 The 5.1 'ring' is extended to more channels by adding loudspeakers between the original channels. (Figures 6.19 and 6.20 courtesy of Faller et al., 2013. Reproduced by permission of the Audio Engineering Society.)

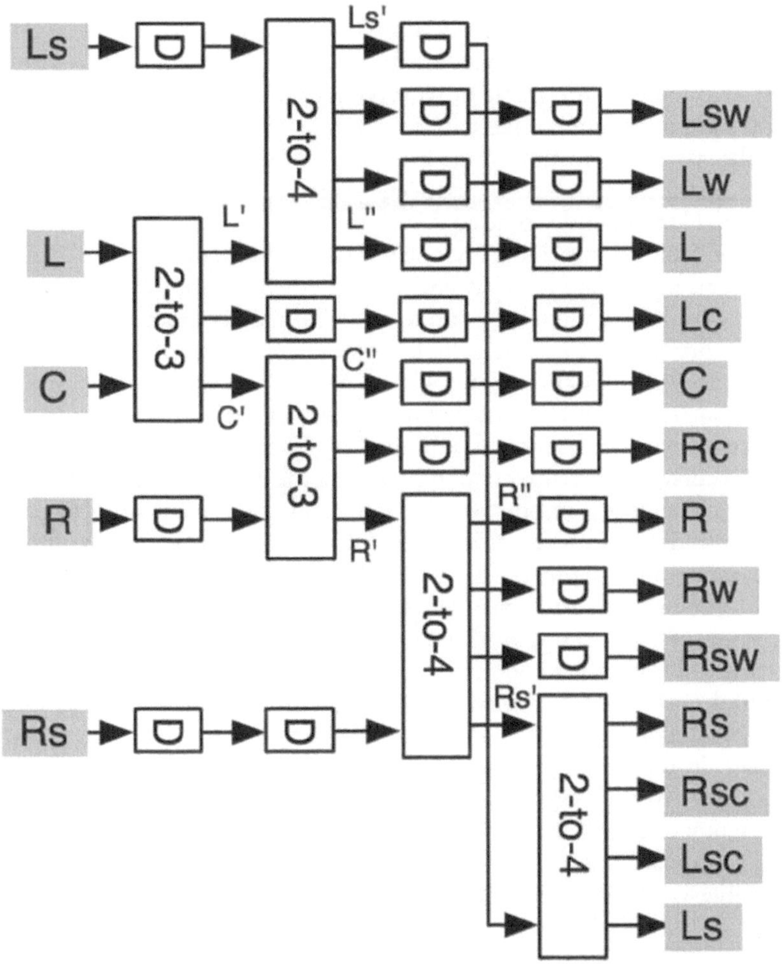

Figure 6.22 A cascaded series of 2-channel upmixes can be used to fill in the gaps in the 'ring' shown in Figure 6.19. The D symbols are delays to match those of the processes in the other arms of the algorithm to ensure that all the samples get to the output at the same time.

pair and reproduce it over *N* loudspeakers, ideally with the same sound and sound stage. As shown in Figure 6.22, the 5-to-13 upmix example in Figure 6.21 can be implemented with a cascade of 2-channel upmixes that include a number of delays (D) to compensate for the fact that some upmixes will have run through fewer stages than others. The total delay of each channel through the system should then be the same. The authors also show an alternative that uses frequency-domain processing to avoid the build-up of such delays. One application of this idea described in the paper is in the IOSONO 3D sound system, where a multichannel ring upmix is employed to render standard content such as 2.0, 5.1, 7.1 and so forth over the large number of loudspeakers used in the WFS (wave field synthesis) layouts concerned.

Making separate mixes for every format can be extremely time-consuming, and this has led to the need for semi-automatic or automatic downmixing of multichannel mixes. The total amount of reverberant sound in multichannel mixes can be different to that in 2-channel mixes, though. This is partly because the spatial separation of the loudspeakers enables one to concentrate on the front image separately from the all-round reverberation, whereas in 2-channel stereo all the reverberation comes from the front. Consequently some control is required over the downmix coefficients and possibly the phase relationships between the channels, for optimal control over a 2-channel downmix.

Downmix equations are given in ITU-R BS.775, intended principally for broadcasting applications where a 'compatible' 2-channel version of a 5-channel program needs to be created. These are relatively basic approaches to mixing the LS and RS channels into L and R, respectively, and the center equally into front left and right, all at –3 dB with respect to the gain of the front channels. Recognizing that this may not be appropriate for all program material the recommendation allows for alternative coefficients of 0 dB and –6 dB to be used. Formulae for other format conversions are also given. Experiments conducted at the BBC Research Department suggested that there was little consistency among listeners concerning the most suitable coefficients for different types of surround program material, with listeners preferring widely differing amounts of surround channel mixed into the front. It is possible that this was due to listeners having control over the downmix themselves, and that in cases where there was little energy in the surround channels a wide range of settings might have been considered acceptable. Averaged across all program types a setting of between –3 and –6 dB appeared to be preferred, but with a wide variance.

In Dolby Digital decoders the downmix coefficients can be varied by the originator of the program at the post-production or mastering stage and included as side information in the Dolby Digital data stream. In this way the downmix can be optimized for the current program conditions and does not have to stay the same throughout the program. Listeners can choose to ignore the producer's downmix control if they choose, creating a custom version that they prefer.

Gerzon (1992a) proposed that, in order to preserve stereo width and make the downmix hierarchically compatible with other formats, an alternative downmix formula from 5–2 channels should be used:

$$L_0 = 0.8536L + 0.5C \quad 0.1464R + 0.3536k(LS+RS) + 0.3536k2(LS-RS)$$
$$R_0 = -0.1464L + 0.5C + 0.8536R + 0.3536k(LS+RS) - 0.3536k2(LS-RS)$$

where k is between 0.5 and 0.7071 (–6 and –3 dB) and $k2$ = between 1.4142k and 1.4142 (–3 to +3 dB).

The result of this matrix is that the front stereo image in the 2-channel version is given increased width compared with the ITU downmix proposal, and that the rear difference gain component $k2$ has the effect of making rear sounds reproduce somewhat wider than front sounds.

He suggests that this would be generally desirable because rear sounds are generally 'atmosphere' and the increased width would improve the 'spatial' quality of such atmosphere and help separate it from front stage sounds. Based on the above equation he proposes that values of $k = 0.5$ and $k2 = 1.1314$ work quite well, making the folded-down rear stage wider than the front stage and the rear channels between 3.5 and 6 dB lower in level than the front.

Perceptual Evaluation

There are many perceptual dimensions or attributes making up human judgments about sound quality. These may be arranged in a hierarchy, with an integrative judgment of quality at the top, and judgments of individual descriptive attributes at the bottom (see Figure 6.23). According to Letowski's model (1989) this 'tree' may be divided broadly into spatial and timbral attributes, the spatial attributes referring to the three-dimensional features of sounds such as their location, width and distance, and the timbral attributes referring to aspects of sound color. Effects of non-linear distortion and noise are also sometimes put in the timbral group. The higher one goes up the tree, the more one is usually talking about the acceptability or suitability of the sound for some purpose and in relation to some frame of reference, whereas at the lower levels one may be able to evaluate the attributes concerned in value-free terms. In other words a high-level judgment of quality is an integrative evaluation that takes into account all of the lower-level attributes and weighs up their contribution. The nature of the reference, the context and the definition of the task govern the way in which the listener decides which aspects of the sound should be taken into consideration.

Although researchers have tended to concentrate on analyzing the ability of surround sound systems to create optimally localized phantom images and to reconstruct original wavefronts accurately, other subjective factors such as image depth, width and envelopment relate strongly to subjective preference in entertainment audio applications. These factors are much harder to define and measure, but they appear nonetheless to be quite important determinants of overall quality. Mason (1999) proposed a hierarchy of spatial attributes for use in perceptual evaluation (see Figure 6.24) and the author has published an extensive review of terminology and schema for evaluation in Rumsey (2002).

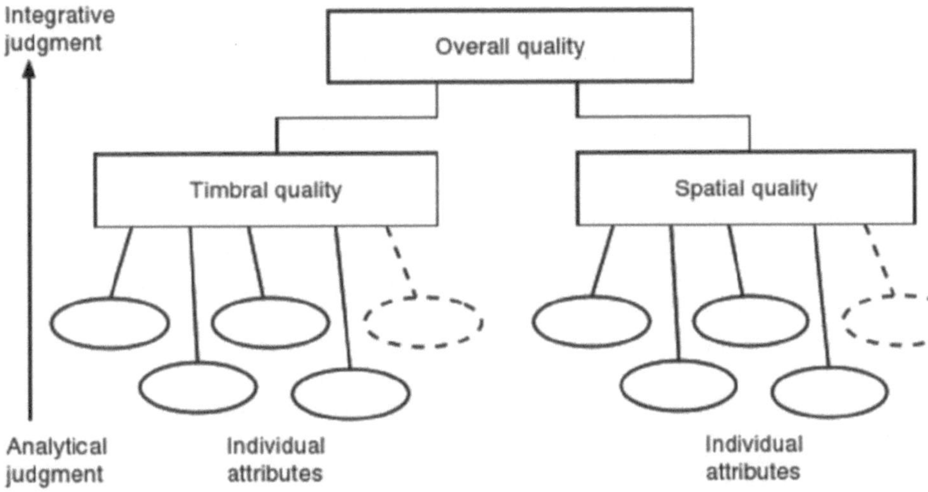

Figure 6.23 Overall sound quality can be subdivided into timbral and spatial quality domains. Each of these consists of contributions from a number of relevant low-level attributes.

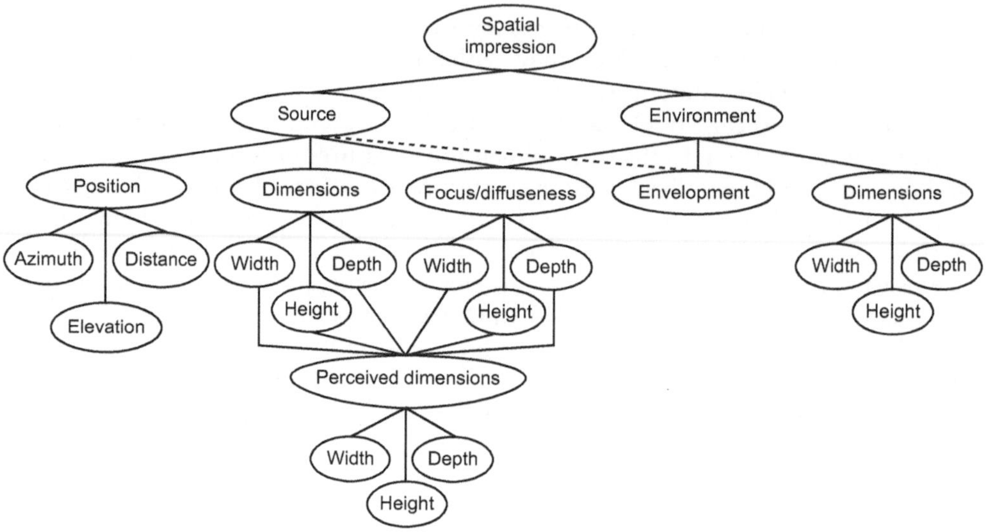

Figure 6.24 Proposed hierarchy of spatial attributes for use in subjective analysis (Mason, 1999).

If true correspondence of all source locations were possible (or indeed desirable) between recording environment and reproducing environment, in all three dimensions and for all listening positions, then it might be reasonable to suppose that ability of a surround sound system to create accurate phantom images of all sources (including reflections) would be the only requirement for fidelity. Since true identity is rarely possible or desirable, some means of creating and controlling adequate illusions of the most important subjective cues for consumer enjoyment could be held as the primary aim of recording and reproducing techniques. This attitude is particularly relevant for entertainment audio applications, but might not be the right one to take for flight simulators, for example.

An interesting conclusion of work by the author and his colleagues on the factors affecting overall quality judgments of surround sound was that timbral fidelity is considerably more important than spatial fidelity (Rumsey et al., 2005b). In other words, listeners care more about the tonal or sound color features of reproduction than they do about the spatial ones. Naïve listeners (ordinary consumers) hardly notice aspects of stereophonic source location, and are more affected by the immersive effect of surround channels; it is only trained listeners that seem to appreciate accurate phantom imaging. It was also found that the overlap between spatial and timbral domains cannot be ignored, as each affects the perception of the other (Conetta et al., 2014a).

One of the few examples of spatial subjective quality tests carried out during a previous intense period of interest in surround sound reproduction is the work of Nakayama et al. (1971). The subjective factors they identified as important in explaining listener quality ratings were interpreted as (a) 'depth of image sources', (b) 'fullness', (c) 'clearness'. An examination of their results suggests that 'fullness' is very similar to what others have called 'envelopment', as it is heavily loaded for reproductions involving more loudspeakers to the sides and rear of the listener, and weak for 2-channel frontal stereo. 'Fullness' was most important, followed by 'depth of sources', followed by 'clearness'. The authors' concluding remarks are still relevant today with regard to the problem of assessing recorded material that does not conform to 'natural' acoustic layouts of sources and reverberation.

> Needless to say, the present study is concerned with the multichannel reproduction of music played only in front of the listeners, and proves to be mainly concerned with extending the ambience effect. . . . In other types of four-channel reproduction the localizations of image sources are not limited to the front. With regard to the subjective effects of these other types of reproduction, many further problems, those mainly belonging to the realm of art, are to be expected. The optimization of these might require considerably more time to be spent in trial, analysis and study.
>
> (Nakayama et al., 1971, p. 750)

No one has really solved this problem yet, as the mixing of surround sound for entertainment purposes is really an art and not a science.

In studies of the perceived effects of spatial sound reproduction, it is sometimes useful to distinguish between judgments and sentiments (Nunally and Bernstein, 1994). Judgments are human responses or perceptions essentially free of personal opinion or emotional response and can be externally verified (such as the response to questions like 'how long is this piece of string?' or indeed 'what is the location of this sound source?'). Sentiments are preference-related or linked to some sort of emotional response, and cannot be externally verified. Obvious examples are 'like/dislike' and 'good/bad' forms of judgment.

In experiments designed to determine how subjects described spatial phenomena in reproduced sound systems, including surround sound, Berg and Rumsey (2006) separated descriptive attributes or constructs from emotional and evaluative ones. Descriptive features could then be analyzed separately from emotional responses, and relationships established between them in an attempt to determine what spatial features were most closely related to positive emotional responses. In this experiment it seemed that high levels of envelopment and room impression created by surround sound, rather than accurate imaging of sources, were the descriptive features most closely related to positive emotional responses.

Predictive Models of Surround Sound Quality

It may be possible to arrive at an overall prediction of quality or listener preference by some weighted combination of ratings of individual low-level attributes. However, such weightings are strongly context- and task-dependent. Listening tests are time-consuming and resource-intensive, so there is a strong motivation to develop perceptual models that aim to predict the human response to different aspects of sound quality. Relationships are established between metrics of the signals or sound field and the results of listening tests. A typical perceptual model for sound quality is calibrated in a similar way to that shown in Figure 6.25. Audio signals, usually consisting of a set of reference and impaired versions of chosen program items,

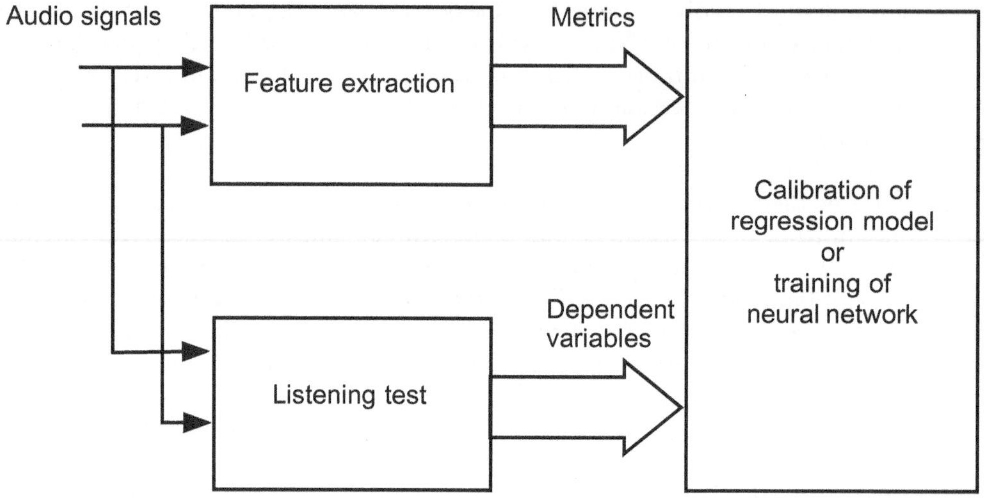

Figure 6.25 A typical approach used in the calibration of sound quality prediction models.

are scaled in standard listening tests to generate a database of 'subjective' grades. In parallel with this a set of audio features is defined and measured, leading to a set of metrics representing perceptually relevant aspects of the audio signals concerned. These are sometimes termed 'objective metrics'. The statistical model or neural network is then calibrated or trained based on these data so as to make a more or less accurate prediction of the quality ratings given by the listeners.

The author and his colleagues developed one such model for surround sound quality evaluation, which was able to predict the quality ratings of trained listeners with reasonably high accuracy, based on measurements made with probe signals (Conetta et al., 2014b). This was reasonably generalizable to a range of entertainment audio content types, but evidence suggested the need for adaptation to different mixing styles.

Note

1 Parts of this chapter are drawn from material appearing in *Spatial Audio* (Rumsey, 2001) and *Sound and Recording*, 7th ed. (Rumsey & McCormick, 2014), and are used by permission of Focal Press.

References

AES. (2001). *Multichannel Surround Sound Systems and Operations: Technical Document AESTD1001.1.01–10*. Audio Engineering Society, New York.

Berg, J., & Rumsey, F. (2006). Identification of quality attributes of spatial audio by repertory grid technique. *Journal of Audio Engineering Society*, *54*(5), 365–379.

Borß, C. (2014). A polygon-based panning method for 3D loudspeaker setups. *Presented at the AES 137th Convention, Los Angeles, USA, 9–12 October*. Paper 9106. Audio Engineering Society.

Bosi, M., Brandenburg, K., Quackenbush, S., Fielder, L., Akagiri, K., Fuchs, H., & Dietz, M. (1997). ISO/IEC MPEG-2 advanced audio coding. *Journal of Audio Engineering Society*, *45*(10), 789–812.

Breebaart, J., Hotho, G., Koppens, J., Schuijers, E., Oomen, W., & Van de Par, S. (2007). Background, concept, and architecture for the recent MPEG surround standard on multichannel audio compression. *Journal of Audio Engineering Society*, *55*(5), 331–351.

Conetta, R., Brookes, T., Rumsey, F., Zielinski, S., Dewhirst, M., Jackson, P., Bech, S., Meares, D., & George, S. (2014a). Spatial audio quality perception (Part 1): Impact of commonly encountered processes. *Journal of Audio Engineering Society*, *62*(12), 831–846.

Conetta, R., Brookes, T., Rumsey, F., Zielinski, S., Dewhirst, M., Jackson, P., Bech, S., Meares, D., & George, S. (2014b). Spatial audio quality perception (Part 2): A linear regression model. *Journal of Audio Engineering Society*, *62*(12), 847–860.

Faller, C., Altmann, L., Levison, J., & Schmidt, M. (2013). A multi-channel ring upmix. *Presented at the 134th AES Convention, Rome, May 4–7*. Paper 8908. Audio Engineering Society.

Garity, W., & Hawkins, J. (1941). Fantasound. *SMPTE Motion Imaging Journal*, *37*(8), 127–146.

Gerzon, M. (1992a). Compatibility of and conversion between multispeaker systems. *Presented at 93rd AES Convention, San Francisco, 1–4 October*. Preprint 3405. Audio Engineering Society.

Gerzon, M. (1992b). Optimum reproduction matrices for multispeaker stereo. *Journal of Audio Engineering Society*, *40*(7-8), 571–589.

Gerzon, M. (1992c). Panpot laws for multispeaker stereo. *Presented at 92nd AES Convention, Vienna*. Preprint 3309. Audio Engineering Society

Glasgal, R. (1995). Ambiophonics: The synthesis of concert hall sound fields in the home. *Presented at the 99th AES Convention, New York, October 6–9*. Preprint 4113. Audio Engineering Society.

Griesinger, D. (1997). Spatial impression and envelopment in small rooms. *Presented at AES 103rd Convention, New York, September 26–29*. Preprint 4638. Audio Engineering Society

Hamasaki, K. (2003). Multichannel recording techniques for reproducing adequate spatial impression. *Proceedings of the AES 24th International Conference: Multichannel Audio, The New Reality*. Paper 27. Audio Engineering Society

Herre, J., Hilpert, J., Kuntz, A., & Plogsties, J. (2014). MPEG-H Audio-The new standard for universal spatial/3d audio coding. *Journal of Audio Engineering Society*, *62*(12), 821–830.

Hertz, B. (1981). 100 years with stereo: The beginning. *Journal of Audio Engineering Society*, *29*(5), 368–372.

Holman, T. (1999). *5.1 Surround Sound: Up and Running*. Oxford and Boston: Focal Press.

Holman, T., & Zacharov, N. (2000). Comments on "subjective appraisal of loudspeaker directivity for multichannel reproduction" (in Letters to the Editor). *Journal of Audio Engineering Society*, *48*(4), 314–321.

ISO. (2010). *ISO/IEC 23003-2—Information technology—MPEG audio technologies—Part 2:Spatial Audio Object Coding (SAOC)*. International Standards Organization.

ITU-R. (2012). *BS. 775-3 (2012) Multichannel Stereophonic sound System with and without Accompanying Picture*. International Telecommunications Union.

Letowski, T. (1989). Sound quality assessment: Cardinal concepts. *Presented at the 87th Audio Engineering Society Convention, New York*. Preprint 2825.

Martin, G., Woszczyk, W., Corey, J., & Quesnel, R. (1999). Controlling phantom image focus in a multichannel reproduction system. *Presented at 107th AES Convention, New York, 24–27 September*. Preprint 4996. Audio Engineering Society.

Mason, R. (1999). *Personal communication.*

Nakayama, T., Miura, T., Kosaka, O., Okamoto, M., & Shiga, T. (1971). Subjective assessment of multichannel reproduction. *Journal of Audio Engineering Society*, *19*(9), 744–751.

Nunally, J., & Bernstein, I. (1994). *Psychometric Theory* (3rd ed.). New York and London: McGraw-Hill.

Pulkki, V. (1997). Virtual sound source positioning using vector base amplitude panning. *Journal of Audio Engineering Society*, *45*(6), 456–466.

Pulkki, V., & Faller, C. (2006). Directional Audio Coding: Filterbank and STFT-based Design *Presented at the AES 120th Convention, Paris, May 20–23*. Paper 6658. Audio Engineering Society.

Rumsey, F. (1998). Synthesized multichannel signal levels versus the M-S ratios of 2-channel programme items. *Presented at 104th AES Convention, Amsterdam, 16–19 May*. Preprint 4653. Audio Engineering Society.

Rumsey, F. (1999) Controlled subjective assessments of 2-to-5 channel surround sound processing algorithms. *J. Audio Eng. Soc., 47*(7/8), pp. 563–582.

Rumsey, F. (2001). *Spatial Audio*. Oxford and Boston: Focal Press.

Rumsey, F. (2002). Spatial quality evaluation for reproduced sound: Terminology, meaning and a scene-based paradigm. *Journal of Audio Engineering Society, 50*(9), 651–666.

Rumsey, F., & McCormick, T. (2014). *Sound and Recording: Applications and Theory* (7th ed.). Oxford and Boston: Focal Press.

Rumsey, F., Zielinski, S., Kassier, R. & Bech, S. (2005a) Relationships between experienced listener ratings of multichannel audio quality and naïve listener preferences. *Journal of Acoustical Society of America, 117*(6), 3832–3840.

Rumsey, F., Zielinski, S., Kassier, R., & Bech, S. (2005b). On the relative importance of spatial and timbral fidelities in judgments of degraded multichannel audio quality. *Journal of Acoustical Society of America, 118*(2), 968–977.

Scheiber, P. (1971). Suggested performance requirements for compatible four-channel recording. *Journal of Audio Engineering Society, 19*(8), 647–650.

Smyth, S., Smith, W. P., Smyth, M. H. C., Yan, M., & Jung, T. (1996). DTS coherent acoustics: Delivering high quality multichannel sound to the consumer. *Presented at 100th AES Convention, Copenhagen, 11–14 May*. Workshop 4a-3.

Steinberg, J., & Snow, W. (1934). Auditory perspectives-physical factors. *Stereophonic Techniques*, 3–7. Audio Engineering Society.

Steinke, G. (1996). Surround sound—the new phase. An overview. *Presented at the 100th AES Convention, Copenhagen, May 11–14*. Preprint 4286. Audio Engineering Society.

Theile, G. (2000). Multichannel Natural Recording Based on Psychoacoustic Principles. *Presented at the AES 108th Convention, Paris, France, 19–22 February*. Paper 5156. Audio Engineering Society.

Theile, G., & Plenge, G. (1977). Localization of lateral phantom images. *Journal of Audio Engineering Society, 25*(4), 196–200.

Todd, C., Davidson, G. A., Davis, M. F., Fielder, L. D., Link, B. D., & Vernon, S. (1994). Flexible perceptual coding for audio transmission and storage. *Presented at 96th AES Convention*. Preprint 3796.

West, J. (1999). *Five-channel panning laws: An analytical and experimental comparison*. Master's thesis, University of Miami, Florida.

Williams, M. (2004). *Microphone Arrays for Stereo and Multichannel Sound Recordings*. Milano: Editrice Il Rostro.

Chapter 7

Height Channels

Sungyoung Kim

Listeners in real spaces are immersed in acoustic information that arrives from all directions. Capturing and reproducing the vertical aspect of an acoustic environment is essential for the complete rendering of immersive audio. To deliver periphonic information, several multichannel sound capture and reproduction systems have been developed that include dedicated height channels. By providing vertically oriented sound images, systems with height channels deliver a more convincing, natural, and immersive three-dimensional (3D) sound environment. This chapter will introduce loudspeaker configurations, microphone techniques, and the perceptual characteristics that are associated with height channels.

Background

Architectural Acoustics

Because an acoustic space alters the timbral and spatial characteristics of an original sound, architectural acousticians have long endeavored to construct a space within which a performed sound field can satisfy listeners. Even a musician or priest in prehistoric times knew that radiated sound in a cave increased the perceived loudness and also the reverential fear of a god (Blesser & Salter, 2006). As music has systematically evolved, both musicians and audiences have found that acoustic conditions dramatically alter the appreciation of performed music.

As Forsyth (1985) wrote in his book Mozart realized the effect of acoustical conditions on his operas. He stated in a letter to his wife:

> . . . By the way, you have no idea how charming the music sounds when you hear it from a box close to the orchestra—it sounds much better than from the gallery. . . .
>
> (p. 94)

Many composers, from Johann Sebastian Bach to Richard Wagner, often considered the acoustics of certain venues and composed music specifically for those acoustical conditions (Blesser & Salter, 2006). A given acoustical condition has been treated as an instrument whereby a composer (and a performer) could express musical creativity effectively. A musical space has historically been a significant factor for composers.

Therefore, architects have worked diligently to create a space that can assist both musicians and audiences in apprehending the music. Over this long process, they have acquired empirical knowledge on the importance of acoustic information coming from above and its influence on performance. For instance, Beranek (2008) showed that the ceiling type and the height of a space could change the initial time-delay gap (ITDG)—defined as the difference between the arrival time of a direct sound and the arrival time of the first reflected sound. This ITDG value is highly correlated with subjective attributes including "clarity" and "intimacy." Therefore, acoustical information modulated by the ceiling type and height can influence the perceived clarity and intimacy of a sound field within a space. Beranek introduced an exemplary project at the Koussevitzky Music Shed in Tanglewood, Massachusetts, which clearly demonstrated the importance of ceiling reflections (Beranek, 2007).

> The hall's sound was unclear, termed "muddy," and the music appeared "remote"—sort of like in a barn. Our 1959 solution was to recommend a canopy over the orchestra and the front part of the audience. This canopy, 50% open, is an array of large triangles that look like enormous bats flying wing-to-wing. Half the sound can get through the between-bat openings to the upper spaces so the reverberation is the same as before, but the other half is reflected down to the audience area, lending clarity and "intimacy" to the sound.
>
> (p. 128)

A more recent example (Miyazaki, 2010) also showed the importance of the ceiling in architectural acoustics. The space inside the new Yamaha Ginza Hall has a relatively narrow width yet a tall height. The narrow width of the hall created strong lateral reflections, which generated an unnaturally wide sound image, due to strong early reflections. To solve this problem, Mr. Miyazaki, the designer, used multi-angle reflective panels in the lateral walls to increase lateral diffusion thus decreasing the direct lateral reflections. However, the amount of acoustical energy required for full spaciousness was not enough, which led him to add a swelling surface and movable reflectors (as illustrated in Figure 7.1). The reflectors reinforce direct sounds by adding reflections in front of a listener to provide a clear instrument sound and adjust the balance between the frontal and lateral acoustic energy for proper impression of spaciousness. In addition, the height of the reflectors can be used to control the perceived size (or extent) of sound source image.

Multi-Loudspeaker Reproduction of Immersive Audio

5.1 surround sound systems can successfully create the illusion of being in another space. However, this experience is limited to the horizontal plane. Researchers have investigated new electro-acoustic systems that vertically extend the spatial impression and provide listeners with a 3D listening experience. These systems generally fall into one of three categories: a binaural system, a sound-field synthesis system, or a (discrete) channel-based system. This chapter will focus on the discrete channel-based systems. A discrete channel-based system captures or reproduces a number of independent signals through the designated channels. For example, mono, stereo, and 5.1 surround sound are discrete channel-based systems. Newer systems use *audio objects*, and are not channel-dependent. Hybrid systems, such as Dolby ATMOS, combine channel-based and object-based approaches. Since the focus of this chapter is on discrete channel-based systems,

Figure 7.1 An interior view of Yamaha Hall (Tokyo, Japan). The ceiling has a swelling surface and movable reflectors to provide the performers and audiences with a pleasingly diffusive and spacious sound field.

readers are advised to refer to other chapters of this book to find information about alternative periphonic systems (including sound-field synthesis in Chapter 9). The following section will present fundamental psychoacoustic concepts that relate to the perception of elevated sound sources.

Fundamental Psychoacoustics of Height-Channel Perception

Directional Band

In the horizontal plane of a sound field, localization of a sound source is heavily dependent on binaural cues such as the Interaural Level Difference (ILD) and the Interaural Time Difference (ITD). However, the contribution of these interaural differences is less critical in determining the perceived direction and spatial information of an elevated sound source than that of a horizontal sound source. Elevated perception is mainly dependent on the spectral modification caused by the shoulders, head, and pinna reflections, especially when a sound source is located in the median plane.

Blauert (Blauert, 1996) has investigated this phenomenon and found that specific spectral bands are boosted and cut according to sound source positions, and named them "directional bands." For example, a region around 8 kHz was associated with an overhead source position. Similarly, Hebrank and Wright (1974) identified narrower bands corresponding to certain positions and found that an overhead position was associated with a 1/4-octave peak between 7 kHz and 9 kHz. Recently, Wallis and Lee (2015) investigated various physical factors including source bandwidth, duration, and loudspeaker position, and their influences on perceived location. The results showed that "the 1/3-octave band bursts tends to agree with Blauert's findings for 1, 4, and 8 kHz" (Wallis and Lee, 2015, p. 6) and the aforementioned factors–bandwidth, duration, and loudspeaker position–affected the perceived localization of an elevated sound source (see Figure 7.2). While all three studies had subtle differences in findings, they all confirmed the existence of the directional bands and that a 1/3-octave band at 8 kHz is associated with an overhead impression in the median plane.

When height signals are radiated, the directional bands influence perceived elevation. For example, if one allocates a high-hat cymbal signal in a front height (FH) loudspeaker, it may be heard

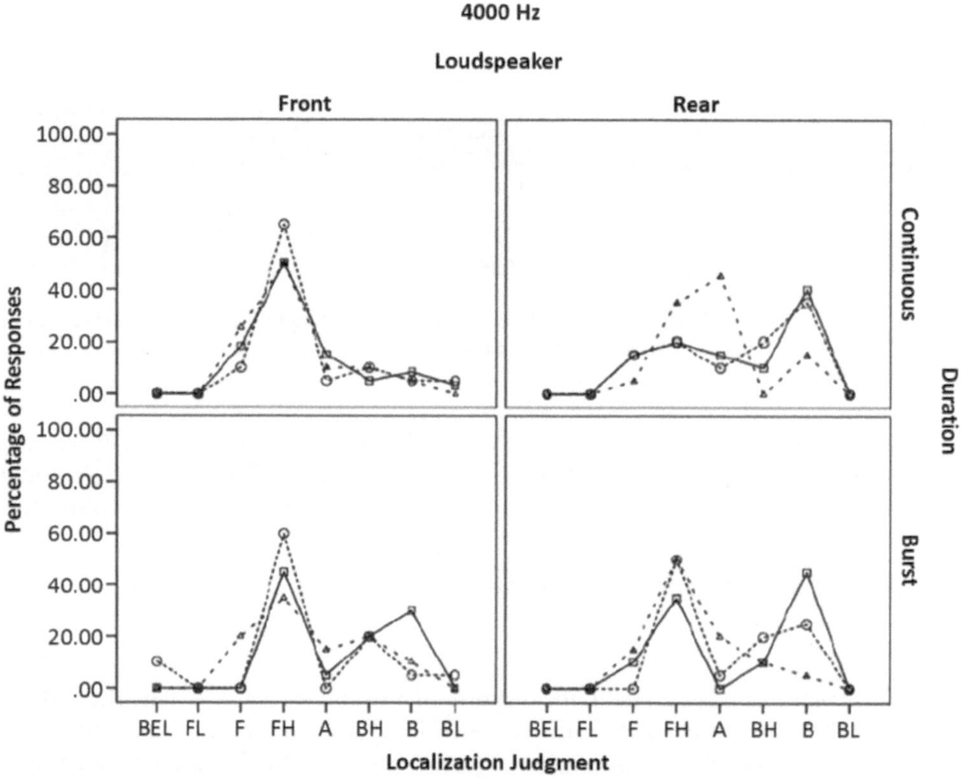

Figure 7.2 The percentage of listeners' responses for the 8 kHz localization (courtesy of Wallis & Lee, 2015). The abscissa indicates eight regions of the median plane. For instance, A refers to *above* the listener's head position—i.e., the overhead position and FH refers to *front high*. The results show that this region provided listeners with elevated localization regardless of the duration of stimuli. Most of the responses were either front high (FH), above (A), or *back high* (BH).

as more elevated than its intended target elevation due to spectral dominance in the directional band for elevation. Therefore, the entire recording and reproduction process should carefully consider the possible influence of the spectral content and appropriate distribution through the upper hemisphere. Another application of the directional bands is to render a virtually elevated sound image. Directional bands can be used to virtually render an elevated sound image. Modifying spectral content in the directional bands can elevate the perceived image without using a physical height loudspeaker.

Vertical Localization of a Phantom Image

The ability to create a phantom sound source image between two loudspeakers is one of the key benefits of using a channel-based system, but is it the same for a vertical phantom image between horizontal (middle) and height (upper) loudspeakers? If so, what would be the required inter-channel relation to attain a stable vertical phantom image? To answer these questions, researchers have conducted various psychoacoustic experiments.

Lee (2011) investigated the inter-channel level-difference (ICLD) relation between a middle and upper loudspeaker in the median plane and found that an ICLD of less than 6 to 7 dB produced only horizontal localization for a height channel signal within a 10-ms inter-channel time difference (ICTD) range. This implies that a level difference higher than 7 dB in an upper loudspeaker could elevate the perceived image. Theile and Wittek (2011) also reported that a 10-dB ICLD was required between two layers to shift a sound image from one layer to another. The study also reported that when the two layers have a smaller ICLD relationship, the auditory image does not appear as a phantom source and was unstable due to propagation-delay differences especially when the height loudspeakers are positioned at a ±45° azimuth angle. Further, this unstable phantom image could deteriorate the overall program image due to a strong inter-channel crosstalk.

Barbour conducted a controlled experiment of sound source localization in an upper hemisphere working in conjunction with an ITU-R 5-channel audio system (Barbour, 2003). As the title of study implies, the goal was to investigate the influence of the loudspeaker position perception on elevated phantom sound images. The results show that listeners generally perceived a stable height phantom image along the frontal plane as compared to the median plane. Thus, he suggested the use of two height loudspeakers at an azimuth of ±90° from the center position and an elevation of +60° to achieve effective localization and height envelopment. Kim, Ikeda and Takahashi (2009) also showed that perceived elevations monotonically matched vertical loudspeaker positions particularly when located at the frontal plane.

These results indicate that an elevated image location is not as precise as that of a frontal horizontal phantom image, although vertical localization can be rendered using two loudspeakers. Williams (2013) proposed to use three loudspeakers positioned in an isosceles triangle configuration and asserted that this formation provided precise localization. This isosceles configuration became the foundation for the Williams height microphone array discussed later in this chapter.

Multichannel Reproduction Systems With Height Channels

Based on listening experiments and known psychoacoustic principles relating to elevated sound sources, various research institutions have proposed new loudspeaker configurations incorporating height channels to provide listeners with a vertically extended sound field. A comprehensive summary of the loudspeaker arrangement and configuration for multichannel audio reproduction can be found in the ITU-R BS.2051–0 specification (ITU, 2014). The recommendation illustrates both the positional and directional configurations of loudspeakers using three layers: upper, middle, and bottom. The middle layer indicates the horizontal (near ear level) plane while the upper and bottom layers indicate the elevated and ground level planes, respectively. In addition, the recommendation denotes each configuration using the numbers of loudspeakers in each layer. Thus a "U+M+B" configuration indicates a loudspeaker layout using U loudspeakers in the upper layer, M loudspeakers in the middle layer, and B loudspeakers in the bottom layer. The conventional 5-channel system (ITU-R BS. 775) can be denoted as a 0+5+0 configuration.

Various configurations have been used to capture and reproduce sound recordings with height channel(s). The TELARC recording label proposed the 1+5+0 configurations to capture the height information in a recording venue and deliver it via the sixth channel of a Super Audio CD (SACD) or DVD-Audio (DVD-A) instead of the subwoofer channel (LFE signal) (e.g., Tchaikovsky, 1880). A variation of this format is 2+4+0 that utilizes two height channels (the MDG and Divox recording labels in Germany and Switzerland, respectively (Sunier, 2008) as illustrated in Figure 7.3.

2+5+0 / 2+7+0 / 2+8+0 (THX 10.2)

These configurations utilize two loudspeakers in the upper layer coupled with either five, seven, or eight loudspeakers in the middle layer as illustrated in Figure 7.4. The loudspeakers in the middle layer increase the horizontal resolution while the two upper loudspeakers add *presence* to the entire sound image. The position of the upper loudspeakers is usually above the front left and right loudspeakers. Dolby ProLogic IIz recommends a similar configuration for playback systems with two height channels (Dolby Laboratories, 2015). The use of two front-height loudspeakers has

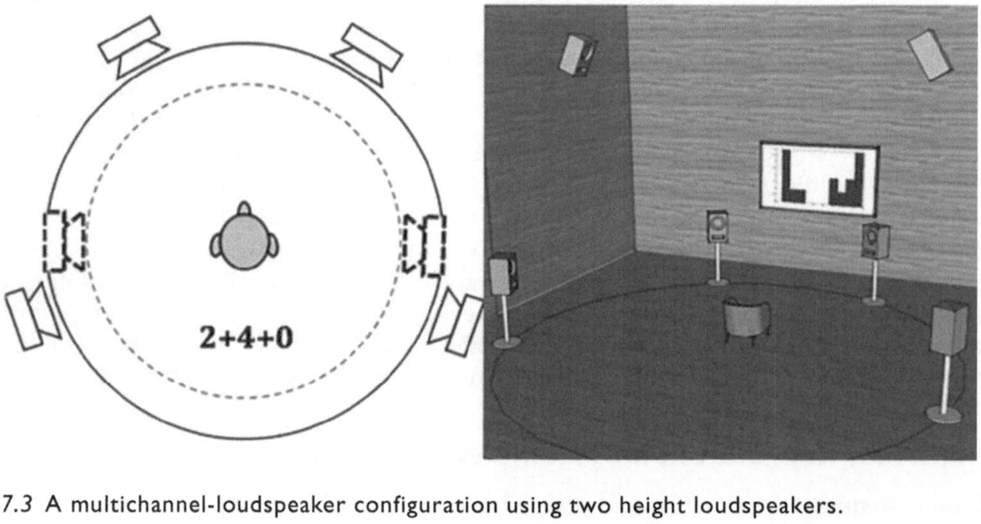

Figure 7.3 A multichannel-loudspeaker configuration using two height loudspeakers.

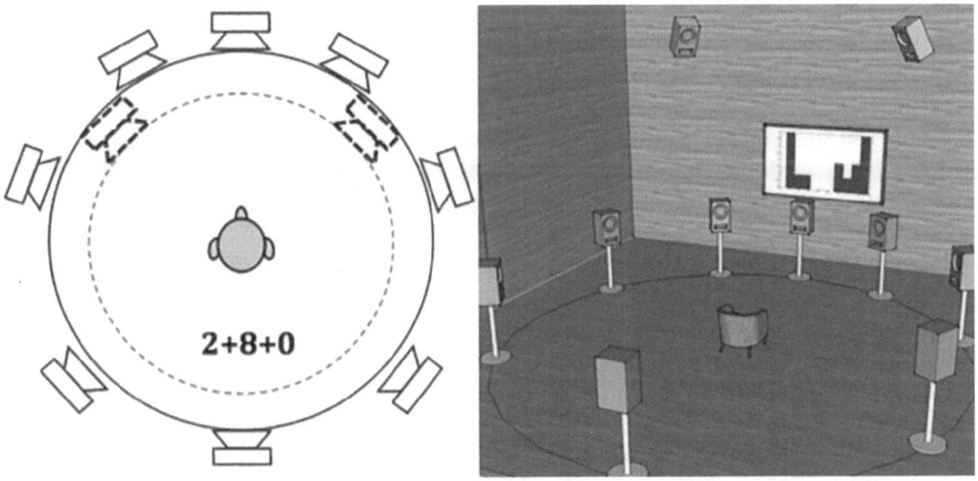

Figure 7.4 Illustrations of a 2+8+0 configuration also known as the THX 10.2 format.

been thoroughly investigated by Kamekawa et al. (2011) in the context of 3D sound integration. The authors reported that when the height-channel content had been appropriately captured and reproduced through two upper loudspeakers, it created a natural-sounding perception of depth that complemented the accompanying 3D video more suitably than horizontal-only loudspeaker reproduction systems. In particular, the 2+8+0 configuration has been proposed by Tomlinson Holman with two subwoofers, which is better known as the THX 10.2 system (Holman, 2007).

While the frontal position of two height channels is popular (due to its function to enhance the presence of sound images), it should be noted that Barbour's previous results demonstrated a possible benefit from two loudspeakers located in the frontal plane. Kim et al. (2013) subsequently conducted a subjective comparison of various positions of two upper loudspeakers and found that listeners preferred the upper loudspeaker position at the frontal plane because it provided them with a more continuous (and homogenous) envelopment of the overall sound field.

Kim, Lee and Pulkki (2010) conducted a study to determine how many height loudspeakers would be required. The authors compared 0, 2, 3, and 4 upper-layer loudspeaker configurations to a 9-upper loudspeakers configuration. When asked to evaluate localization and spaciousness, the listeners preferred the 9-upper loudspeaker configuration (9+12+3). The 3-upper loudspeaker configuration (at ±35°–45° azimuth range with 30°–45° elevation range, and 180° azimuth with 45°–90°) provided the listeners with the plausible *directional quality* as well, similar to the 4-upper loudspeaker configuration. The results imply that three or four loudspeakers are required to render convincing and pleasant sound images.

4+5+0 / 5+5+0 / 6+5+0 (AURO-3D)

In contrast to previous configurations with few height loudspeakers, the following three configurations are based on the concept of *vertical extension* of the current surround layout (0+5+0;

ITU-R BS. 775) and incorporate four or more height loudspeakers. The upper loudspeakers are positioned directly above the middle loudspeakers to emphasize "compatibility with the existing standards and formats" (Van Baelen, 2010, p. 196) in 3D-audio reproduction, especially in the new digital cinema market. A 4+5+0 (9-channel version) does not use a center-height loudspeaker (top panel of Figure 7.5), while a 6+5+0 uses an additional loudspeaker at the overhead listening position (bottom panel of Figure 7.5). To fully utilize this overhead loudspeaker's feature, it is sometimes located above the upper layer. This center-overhead loudspeaker is known as the "voice-of-the-god" (VOG) loudspeaker and allows a sound designer to produce a true semi-hemispherical sound field that has three height layers.

These configurations were proposed by AURO Technologies. Wilfried van Baelen presented the AURO-3D concept in 2006 (Hamasaki et al., 2006). AURO Technologies, based in Belgium at Galaxy Studios, pioneered the development of high-quality compact digital audio data storage solutions for the delivery of multichannel audio to home theaters and digital cinemas (including but not limited to the AURO-Codec). The proprietary AURO-MATIC upmixing system can render AURO-3D sound fields from conventional stereo and/or surround sound sources as well as the discrete channel contents produced for the AURO-3D configuration.

Theile and Wittek (2011) found three benefits of the AURO-3D configuration: an appropriate spatial diffusion by reproducing early reflections in the entire upper half space; enhanced envelopment, spatial impression, and depth; and effective stereo imaging in the upper layer (similar to the middle layer).

9+10+3 (NHK 22.2)

Known as the 22.2 Multichannel Sound System, this configuration utilizes 9 loudspeakers in the upper layer, 10 loudspeakers in the middle layer, and 3 loudspeakers in the bottom layer as illustrated in Figure 7.6. The azimuth angles of height loudspeakers are 0°, ±45–60°, ±90°, ±110–135°, and 180°, and the vertical angle is + 30–45°. The overhead loudspeaker is located directly above the listening position. The azimuth angles of horizontal loudspeakers are 0°, ±22.5–30°, ±45–60°, ±90°, ±110–135°, and 180°. The azimuth angles of bottom loudspeakers are 0° and ±30–45°, and the vertical angle is 15–25°. Two subwoofers are located at ±30–90° azimuth on the floor (Hamasaki, 2011).

Originally proposed by the Science and Technology Research Laboratories (STRL) of Nippon Hōsō Kyōkai (NHK) in 2003, the system was designed to provide ultra-high definition (UHD) video viewers with a highly precise, natural, and fully immersive listening experience. The UHD video system features 4,000 scanning lines providing a viewing angle of up to 100° (Hamasaki et al., 2007). The accompanying audio system thus needed to control auditory images over this wide viewing angle. The 22.2 sound system has five middle, three upper, and three bottom loudspeakers to generate auditory images on screen. In addition, 11 loudspeakers are used to extend the integration of the immersive auditory images off screen. A total of nine loudspeakers are in the upper layer to generate precise localization in the elevated sound field. As previously discussed, a phantom sound image in the elevated field yields less precise localization when compared to the horizontal localization. Thus, more loudspeakers are useful in achieving more reliable 3D (periphonic) localization of sound sources. While many have questioned the relative advantages

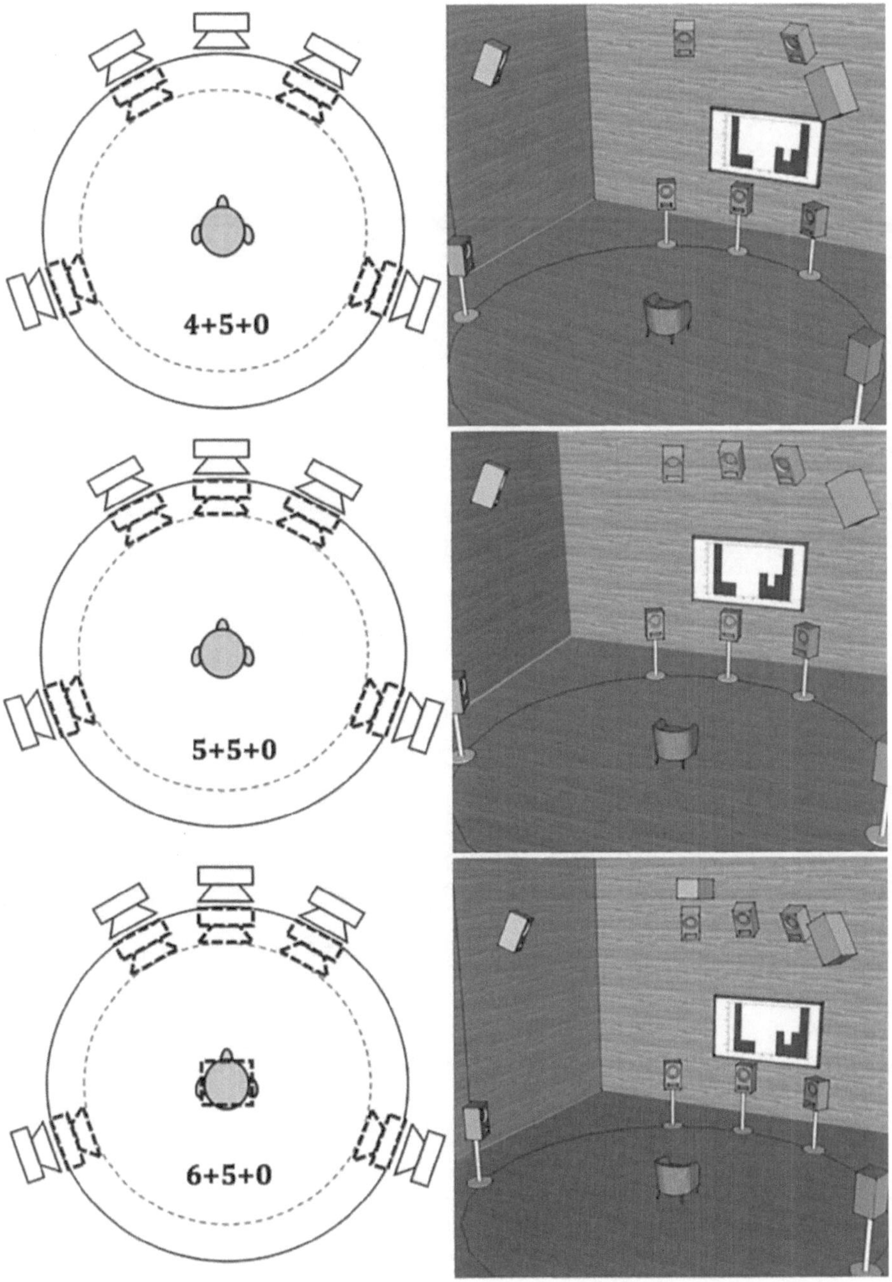

Figure 7.5 Multichannel-loudspeaker configurations using four, five, or six height loudspeakers (AURO-3D).

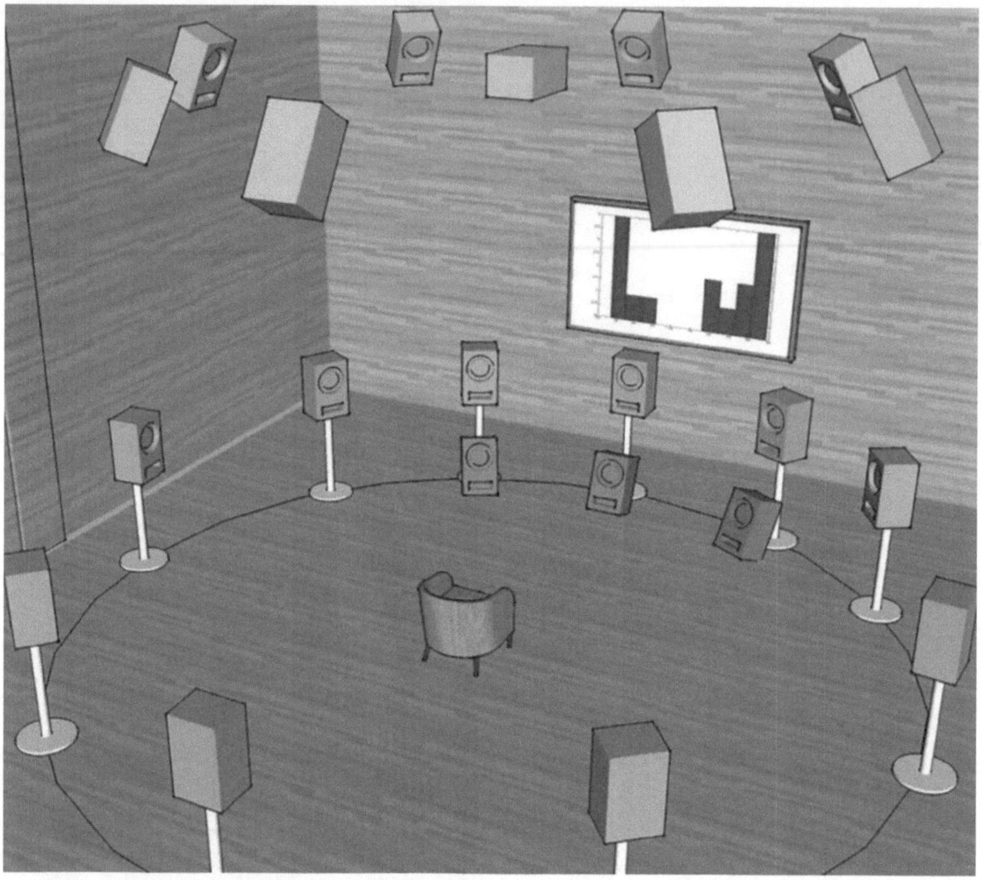

Figure 7.6 Illustration of a spherical arrangement of the 9+10+3 configuration (also known as the NHK 22.2 Multichannel Audio System).

of using as many as 9 height loudspeakers, subjective listening test results support the perceptual merits of this configuration.

Hamasaki et al. (2007) conducted a comparative study on subjective characteristics among 2– (0+2+0), 5.1– (0+5+0 with one subwoofer), and 22.2-channel (9+10+3 with two subwoofers) sound systems proving the sound for an ultra-high definition (UHD) video image. Participants evaluated the 22.2-channel system higher than the 2-channel system for all tested attributes (except loudness) and higher than the 5.1-channel system on the following six attributes: "front/rear" (discrimination), "up/down"(discrimination), "movement," "direction," "reverberant," and "envelopment." The results imply that the 22.2 system provided better localization and spaciousness than the conventional sound system. As the title of the paper indicates, the study focused on the "listening area" of various systems and investigated various sonic characteristics relating to the listener distance from the scene. While the results of the stereo and 5.1 system tests were inconsistent, the 22.2 system produced a consistent impression over various listening positions. This indicates that the 22.2 system can deliver a homogeneous listening experience for more viewers than the other two systems. The extended listening area is a distinct feature associated with a multichannel-reproduced sound field with height channels.

In the following paper (Hamasaki, 2011), NHK summarizes five required aspects of the 9+10+3 configuration, including:

- Integrity: the localization of an audio image anywhere on the screen;
- Periphony: the reproduction of sound coming from all directions surrounding the viewing position;
- Presence: the reproduction of a natural, high-quality 3D acoustic space;
- Compatibility: with existing multichannel configurations;
- Usability: the capability to support a live recording and live broadcasting.

This configuration is being discussed by various organizations as a standard for a future broadcasting audio format including the SMPTE standard: SMPTE ST 2036–2–2008, "Ultra High Definition Television1—Audio Characteristics and Audio Channel Mapping for Program Production."

The author had an exclusive chance to interview an NHK mixing engineer, Kensuke Irie, who recorded the NASA space shuttle launch for this configuration. He pointed out that the use of the top overhead and rear-height loudspeaker enabled him to render a distinct sound field that is more realistic and convincing than other conventional formats. In addition, he asserted that a very common misunderstanding of a 22.2 sound field is that it generates simply a wider sound field than a conventional 5.1 configuration. Precise localization and enhanced envelopment are two key aspects that Mr. Irie considers vital when recording and mixing for a 22.2 production.

Virtual Height Speaker

While the benefit of having height loudspeakers is clear, a consumer may find it challenging to purchase and install additional loudspeakers in the ceiling and walls. Even for a 5.1 surround sound system, the additional loudspeakers has prevented many consumers from experiencing multichannel audio. Consumer electronic companies have competitively developed practical solutions to deliver multichannel audio to consumers through compact systems. Solutions include binaural reproduction over loudspeakers with the use of crosstalk cancellation (see Chapter 5), and the use of sound bars (see Chapter 6). Recently, Kim, Ikeda and Martens (2014) proposed a method to use a conventional five-channel layout to create two upper virtual loudspeakers using crosstalk cancellation. This approach processes the HRTF information mainly in the center and rear loudspeakers and does not process the front left and right channel. Since the front two channels are designated for use in the delivery of salient auditory information in most applications, the proposed method is of practical benefit for sound engineers. Lee, Son and Kim (2010) also proposed a similar approach by combining vector-base amplitude panning (VBAP) with spectral cues in order to create lateralized virtual sound by using the 0+9+0 configuration.

While the virtual approaches have certain limitation to render full immersive sound fields, these allow consumers to experience vertically extended sound fields with relative ease and assist content developers in preparing 3D audio programs for the future.

Significance of Height Configuration

As previously introduced in the section "2+5+0 / 2+7+0 / 2+8+0 (THX 10.2)," Kim, Lee and Pulkki (2010) conducted a listening test that compared various height loudspeaker formats. They asked listeners to report on perceived directional attributes and the overall sound quality of the systems. The directional attributes indicated the perceived fidelity of moving sound sources in the upper layer, and the overall quality was judged on the perceived pleasantness of timbral and spatial qualities of reproduced sound. The results indicated that height configuration had a significant impact on the listeners' ratings for both qualities. The study also showed the importance of the front-to-back distribution of height loudspeakers. The results show that a minimum of 3 or 4 height channels are required for consumer immersive audio systems to suitably deliver a wide range of program material over a wide listening area.

In another relevant study (Kim, King & Kamekawa, 2015), the authors compared various configurations of 4-channel height loudspeakers in the context of a 4+5+0 configuration. Eight configurations of four height loudspeakers were compared. The subjects ranked them on spatial quality and described the perceived characteristics of the configurations.

The authors placed 12 height loudspeakers with elevation of +30° as illustrated in Figure 7.7 and selected four loudspeakers at a time. For example, one configuration used the height

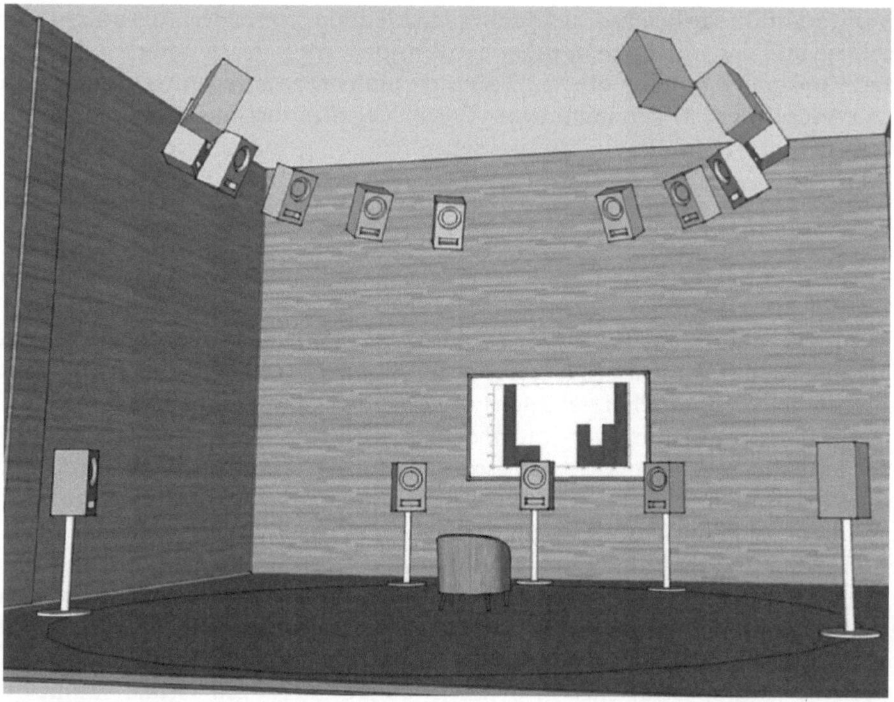

Figure 7.7 Seventeen loudspeakers used in the comparative study (Kim, King & Kamekawa, 2015) to determine the significance of a height-loudspeaker configuration. Five loudspeakers were placed following the ITU-R BS.775 and four out of 12 height loudspeakers were used to deliver height information, which produced a total of eight variations.

loudspeakers at ±30° and ±90° azimuth and +30° elevation, while another used the loudspeakers at ±70° and ±130° azimuth and +30° elevation. In other words, the elevation was constant and only the azimuths were changed. The results indicated that height-loudspeaker positioning has a greater influence on sound quality than the content (specifically, reverberation type and/or musical selection) of height-channel signals. Furthermore, the results were similar for two listener groups in Canada and the U.S.A. A subsequent analysis of the descriptors revealed that listeners chose one 4-channel height configuration (among many variations) that generated "frontal" and "narrow" images to have a higher spatial quality. In other words, the listeners chose a 4-channel height configuration that delivered the sound field with a strong presence of overall auditory scene over a configuration giving increased overall spaciousness. The result implies that the arrangement of height loudspeakers (even the same number) could influence (affect) the perceived characteristics of a reproduced sound field, which subsequently influences the overall sound quality judgment.

Recording With Height Channels

This section introduces the design criteria and psychoacoustic characteristics of recently tested multichannel microphone techniques, created to render an immersive sound field with height channels. In general, a multichannel microphone technique defines an array of microphones that are configured to capture an acoustical event. Directional information is typically captured as interchannel intensity differences, interchannel time differences, or both. As previously stated, height channels can deliver enhanced spatial impressions such as periphonic localization, immersion, and envelopment. A height channel microphone technique should be designed to effectively capture inter-microphone acoustic parameters that can maximize these important attributes, leading to a perception of immersion.

Williams's MAGIC Array

Michael Williams has significantly contributed to the development of 2-channel stereo and surround sound multichannel microphone arrays that capture optimal time-and-intensity differences (for example, using cardioid microphones; Williams & Du, 2000). Williams (2012) later extended the array vertically and proposed a two-layer array, asserting that the height layer should not conflict with the horizontal layer for the localization cues. The proposed microphone array consists of a middle layer that uses four super-cardioid microphones (45°, 135°, 225°, and 315° with distance of 35 cm, also known as "Hypocardioid 'Williams's cross' ") and an upper layer using three bidirectional microphones (0°, 90°, and 270°). The upper layer's microphones face the ceiling so that direct sound is picked up as little as possible. Williams named this array the 7-channel Hypocardioid M.A.G.I.C. array (Multichannel Arrays Generating Inter-format Compatibility).

In the following study, Williams (2013) found that height loudspeakers located directly above the middle layer speakers could not produce convincing elevation. Instead he found that three loudspeakers in an isosceles-triangle form could generate a reliable vertical elevation by allowing diagonal level control between the upper and middle layers. With this observation, Williams proposed a 12-channel microphone array for a complete rendering of 3D sound fields. The array consists of the "Hypocardioid 'Williams's cross' " (a square of four super-cardioid microphones) for the middle layer and four upward-facing bidirectional microphones (0°, 90°, 180°, and 270° separated by distance of 1.5 m) for the upper layer. The array optionally uses an additional four satellite microphones in the middle layer (0°, 90°, 180°, and 270° with a distance of 2.5 m) to enhance compatibility with conventional 5.1 surround sound configurations and to create an isosceles-triangle configuration for precise control of image elevation over a wide range of directions. The configuration is illustrated in Figure 7.8.

OCT-9

With regard to multichannel microphone techniques, there is a concern on whether or not there is a detrimental sonic effect due to interchannel crosstalk (ICC) fed from an adjacent microphone. Proponents, who claim an audible artifact of this ICC, assert that a reasonably balanced localization could only be possible when the intensity of ICC is minimal.

According to this hypothesis, with a relatively large amount of ICC, triple phantom images corresponding to the three horizontal front channels arise due to the similarity between signals.

Theile (2001) investigated this issue in depth and claimed that ICC should be minimized to provide a continuous localization between two channels and to prevent sound coloration. He found that the crosstalk problem was significantly reduced by using two super-cardioid microphones facing the sides. Theile subsequently proposed a multichannel microphone array that incorporates two super-cardioid microphones, one for the left channel and one for the right channel, and an additional cardioid microphone for the center. This array is called an *Optimized*

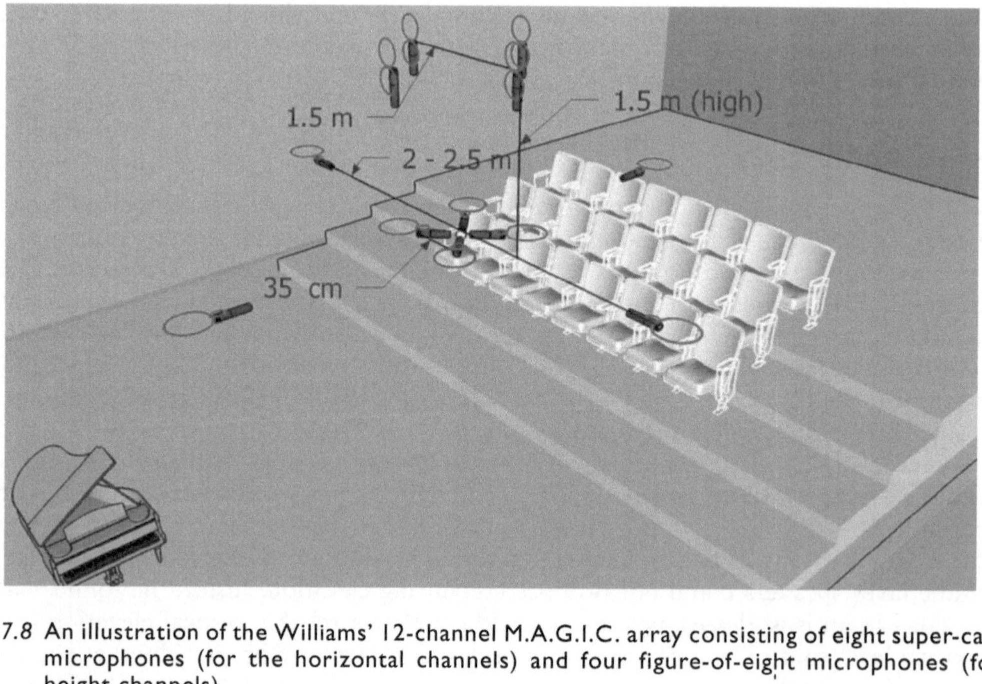

Figure 7.8 An illustration of the Williams' 12-channel M.A.G.I.C. array consisting of eight super-cardioid microphones (for the horizontal channels) and four figure-of-eight microphones (for the height channels).

Cardioid Triangle (OCT) because it is optimized to provide *directional stability* without decreasing the stereophonic quality.

The OCT technique was extended to a 5-channel version called *OCT-Surround* with two additional cardioid microphones facing the rear. Theile designed the surround version and achieved the same aforementioned optimization criteria through the following two conditions: (1) there is sufficient suppression of the direct sound in the left and right surround channel, and (2) there is an effective pickup of early reflections from two lateral walls and the rear wall. This surround array has been further extended to capture 3D immersive sound with the use of four super-cardioid microphones pointing upward (Theile & Wittek, 2011) about 1 m above the four middle layer microphones (as illustrated in Figure 7.9). The main principle of operation for the nine-microphone version of the array remains the same: the optimization of directional stability and stereophonic quality. A user of an OCT microphone array should carefully consider Theile's (2001) caveat that the performance of any OCT microphone array depends heavily on the appropriate choice of the distance and height from the sound source.

2L-Cube

Morten Lindberg from the Norwegian recording label 2L has proposed a microphone array that incorporates eight omni-pressure microphones called 2L-Cube as illustrated in Figure 7.10. Inspired from Decca- or Mercury-tree, 2L-Cube is a direct match between microphones and four corner loudspeakers in both the middle and upper layers, which can be used for the four height channels (such as an Auro-3D 4+5+0 configuration). Lindberg and Shores (2006) asserted

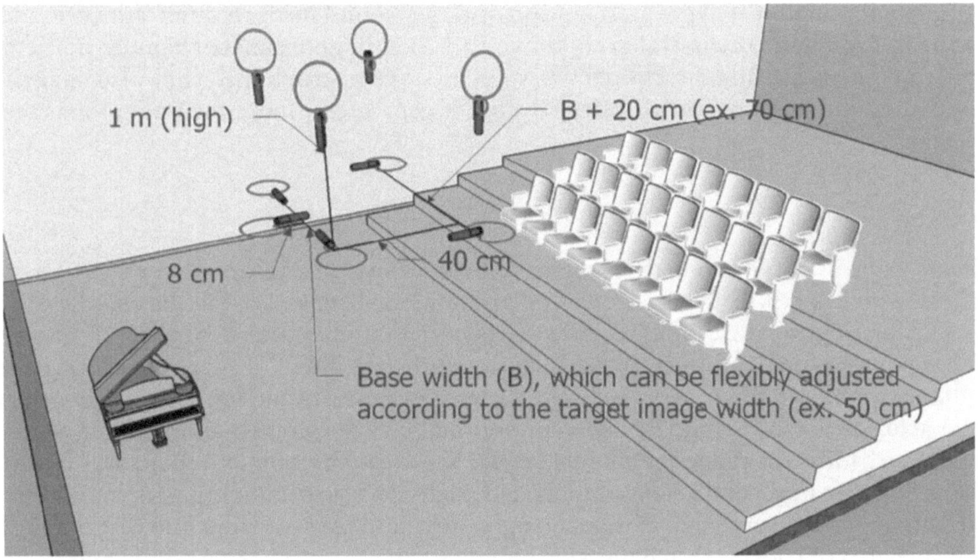

Figure 7.9 An illustration of the 9-channel OCT-9 microphone array consisting of five microphones for the horizontal channels: three cardioid, two super-cardioid, and four super-cardioid microphones for the height channels.

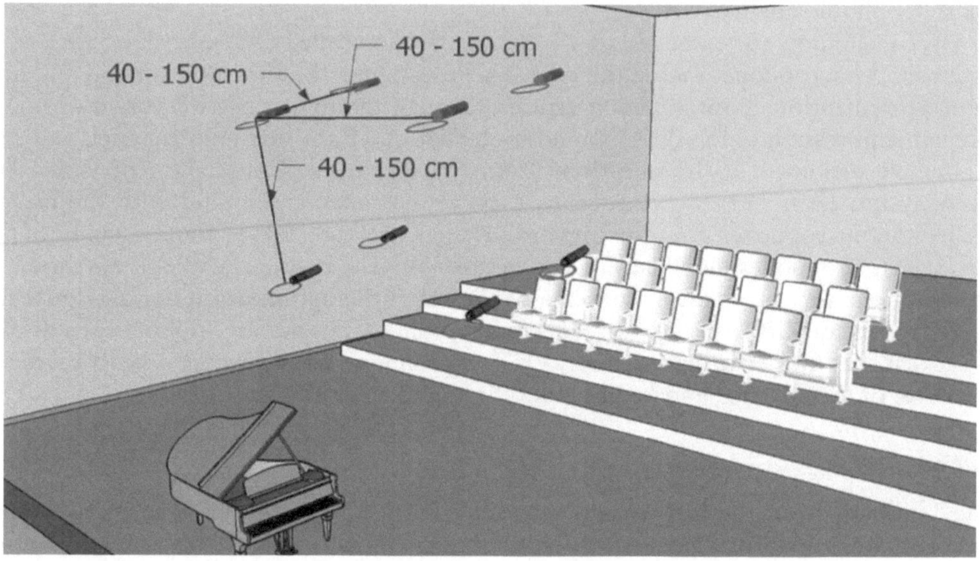

Figure 7.10 An illustration of the eight-channel L2-Cube microphone array.

that the array captures the time of arrival, SPL, and on-axis high-frequency texture of sound appropriately for an immersive representation of target sound fields. A cubic dimension can vary from 150 cm for a large orchestral array down to 40 cm in an intimate chamber music context. Each microphone is an omnidirectional pressure-sensitive microphone. They also suggested that a larger diaphragm microphone is preferred since it generates a more focused on-axis texture of sonic image.

Twins Cube

The Twins Cube and Twins Square array have been proposed and heavily used by Gregor Zielinsky. These arrays utilize a special *Twin* microphone by Sennheiser that has a pair of cardioid capsules facing opposite directions, each with its own discrete output.[1] Since each output of the microphone can be fed to a pre-amplifier independently, the directional characteristic of the twin microphone can be determined remotely live or later in post-production. The Twins Square consists of a spaced pair in the middle layer, an optional center channel, and another equal pair in the upper layer (directly above the middle layer). Since each twin microphone has two outputs, one Twins Square can generate eight outputs: left, right, left-surround, right-surround, upper-left, upper-right, upper-left-surround, and upper-right-surround. As two front and rear diaphragms of each capsule are coincident, the chances of sounds jetting from the front to the rear in playback are minimized.

The Twins Cube system extends the Twins Square by having a second *square* behind the front square thus forming a cube, imitating a playback loudspeaker arrangement with height channels.

An eight-microphone Twins Cube array is illustrated in Figure 7.11. Using all eight microphones in a cubic introduces a time delay between the front and rear. A user may control this time delay optimally to enhance perceived spaciousness. Zielinsky (2015) recommends that one can also use wide cardioids or cardioids if twins are not available, but it is important to use the same mic type for all front 4 or 5 channels, in order to keep the integrity of sound. For the Cube, once a polar pattern has been determined for each microphone, a single output can be assigned to each corresponding loudspeaker.

Coincident Z-Microphone Technique

Based on Ambisonic and middle-side (MS) recording technology, Paul Geluso has proposed a *Z-microphone* technique to record multichannel height information. Using this technique, a vertically oriented bidirectional (figure-of-eight) microphone is paired with a horizontally oriented microphone to create a coincident *middle-Z* (MZ) pair (see Figure 7.12). Since the Z microphone can be paired with virtually any microphone, multiple MZ pairs can exist within stereo and surround sound microphone techniques (Geluso, 2012). Using a standard MS decoder, the vertical pick-up angle for the MZ pair can be determined remotely or in post-production to create effective height channels.

Bowles Array

The *Bowles array* is the microphone array with height channels proposed by David Bowles. It has a surround (horizontal) array, consisting of four omnidirectional microphones and one

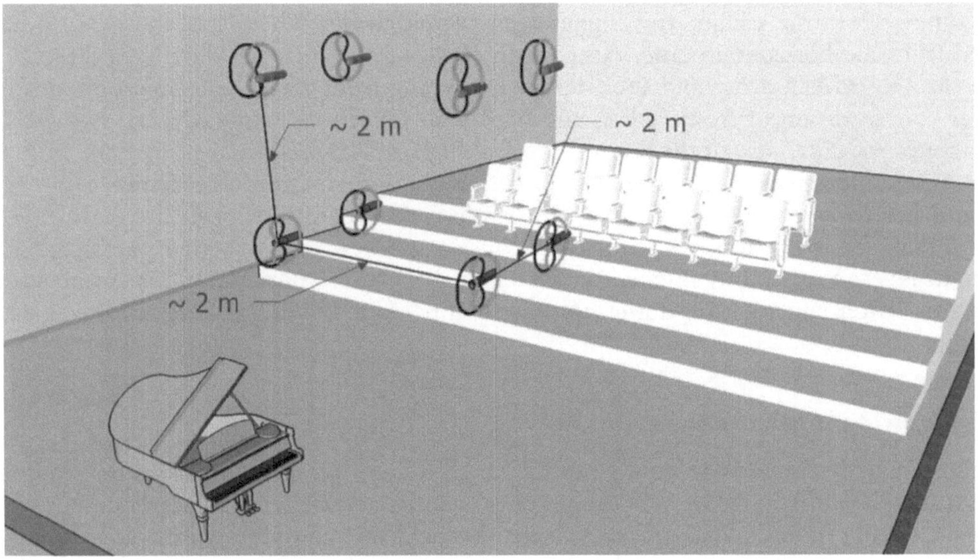

Figure 7.11 An illustration of the eight-channel Twins Cube array.

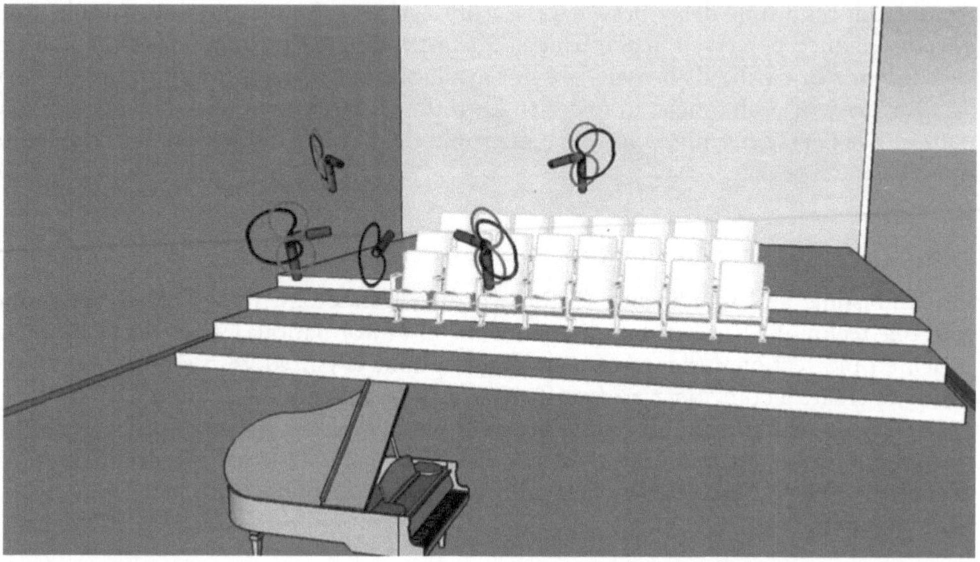

Figure 7.12 An illustration of a spaced MZ array incorporating four vertically oriented bidirectional microphones paired with horizontally oriented cardioid microphones.

uni-directional microphone for the center channel, and a height array of four super-cardioid microphones as illustrated in Figure 7.13. The height array was designed to capture sound reflections coming from the ceiling and higher areas of sidewalls. Therefore, the array points 30° upward from the horizontal plane, rather than pointing the microphone's axes directly above, like in the M.A.G.I.C. array and OCT-9. This way, the front two height microphones pick up a higher concentration of front ceiling and high wall reflections. Similarly, the two rear height microphones pick up more of the rear ceiling and high wall reflections.

Like other microphone arrays that capture height information, the distance between the main and height layers will vary. Also, the amount of spacing between the main and height layers is dependent on how resonant the acoustic space is, and whether the *sweet spot* for the height layer is limited by pitched or barrel-vaulted roofs (Bowles, 2015). Some additional flanking or center super-cardioid microphones can be added, when needed, for large ensemble recordings.

NHK's Coincident Microphone

The recordings for the 9+10+3 configuration (22.2 system) require numerous microphones. As a broadcasting company, NHK saw the practical challenge of having a large number of microphones in a field recording and proposed a new coincident microphone array for the configuration (Ono et al., 2013). The microphone units are coupled with acoustic baffles to achieve "a constant and narrow beam width" (p. 2) and to eventually reduce or eliminate crosstalk. The

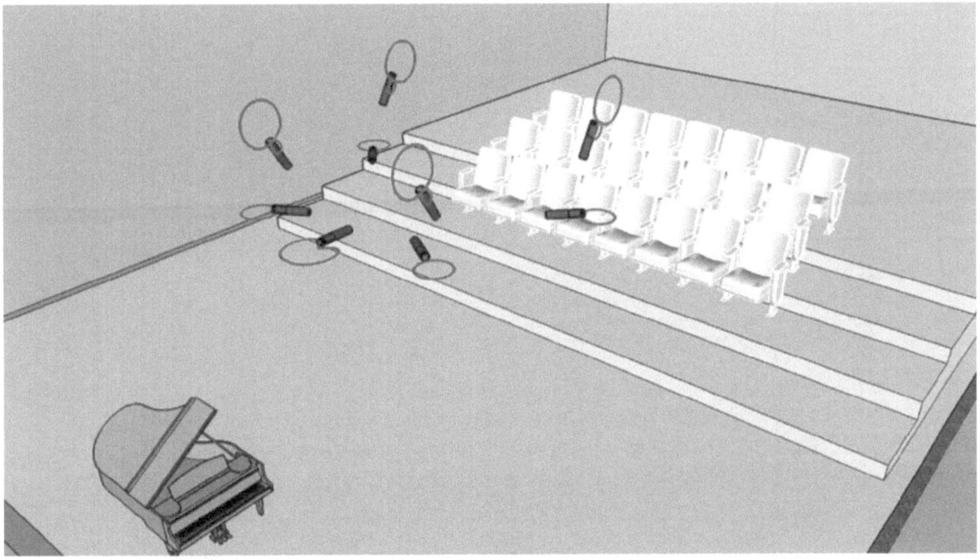

Figure 7.13 An illustration of the Bowles array incorporating the center cardioid microphone.

diameter of the sphere is 45 cm and is partitioned into 8 horizontal parts and 3 vertical parts using the baffles. The weak directivity of the low-frequency range was improved by using inverse filtering to cancel crosstalk from non-targeted directions.

Influence of Upper Microphone-Layer Spacing

Lee and Gribben (2014) investigated the effect of upper microphone-layer spacing from the horizontal layer in the context of a classical music recording. They placed the two layers of microphones in a concert hall and changed the height of the upper layer. The middle layer comprised five cardioid microphones and the upper layer comprised four cardioid microphones facing the ceiling, which is similar to the upper array of the OCT-9. The height of the upper layer varied from 0 m (coincident to the middle layer's left-, right-, left-surround, and right-surround microphone positions), to 0.5 m, 1 m, and 1.5 m. Listeners compared the recordings of four different heights in 9-channel reproduced music from a number of sound sources, and evaluated two attributes: overall spatial impression and preference. The results showed that for most attributes, no significant difference associated with the height of the upper microphone layer was heard, while the 0 m spacing (coincident position) tended to be rated as slightly better than, or similar to, the other heights depending on the type of sound sources (see Figure 7.14). In other words, the study found that the inter-layer distance really had little effect on the listeners' spatial impression and preference.

After correlating the subjective responses with objective measurements (including signal energy level, inter-channel level differences [ICLDs], and more), the authors found that the 0 m distance

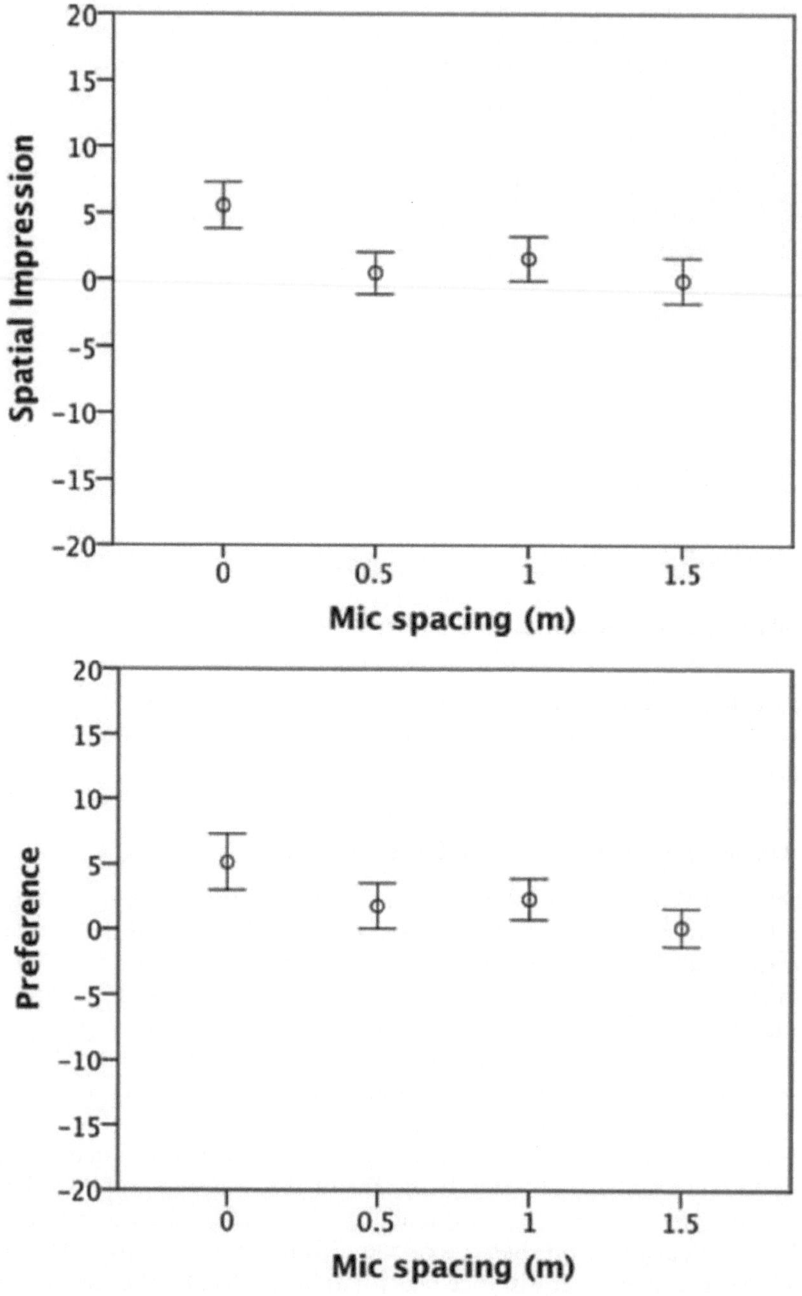

Figure 7.14 Mean values and associated 95% confidence intervals of listeners' preference on spatial impression (top panel) and overall preference (bottom panel) when microphone distance varied (courtesy of Lee & Gribben, 2014).

did not cause significant spectral modification while the other distances created a comb-filtering effect for the listeners, which might lead them to give high ratings for the 0 m coincident array. Therefore, in certain recording conditions, a "vertically coincident" microphone array for height channels might be beneficial due to less comb-filtering caused by a time delay between the middle and upper layers.

Concluding Remarks on Multichannel Microphone Arrays With Height

Although each height microphone technique introduced above has a unique configuration, a skilled audio engineer does not simply place microphones according to a given specification, but rather employs personal discretion based on experience, aural training, and observation. If all audio engineers had the same aesthetic goal, each recording would likely sound the same, or at least similar with respect to acoustical conditions and techniques. Some audio engineers may put more value in *precise* auditory images while others in *spaciousness*. Indeed, compromises due to practical considerations exist as well. The purpose of this subchapter is not to simply quantitatively or qualitatively rank various microphone arrays, but rather to present the design concept of each so that an engineer can choose, adapt, and optimize a recording system (adjust the array's inter-channel relations) based on the acoustics of the venue and the musical content in each case (Kim et al., 2006).

Conclusion

Height channels can help create and improve the immersive listening experience. To take full advantage of the systems discussed in this chapter, height sound must be used with careful attention to acoustic and electroacoustic principles as well as artistic creativity. This chapter introduces various height-channel configurations and techniques used in height sound capture. The configurations include 2 height loudspeaker systems (2+2+2 and THX 10.2); 4, 5, and 6 height loudspeaker systems (AURO-3D 9.1, 10.1, or 11.1); and a 9 height loudspeaker system (NHK 22.2). Research indicates that at least three or four height loudspeakers are required to deliver an enhanced spatial quality that is better than the conventional 5-channel (horizontal only) configuration.

New microphone arrays with height have been proposed to recreate a holistic sound field. The height microphone arrays discussed in this chapter were based on two main approaches: a coincident approach (Z-microphone and NHK), and a vertically spaced approach (M.A.G.I.G., OCT-9, 2L-Cube, Twins Cube, and the Bowles array).

Discrete channel-based methods for height sound reproduction face the challenge of how and where to install the height loudspeakers. The position of the height loudspeakers has a significant influence on perceived spatial attributes. In order to overcome this challenge, researchers proposed a series of methods that can reconstruct the height information, as originally intended by the sound designer or producer, for virtually any loudspeaker configuration. This method is called *object-based* audio. The introduction of the Dolby ATMOS system (a hybrid method that utilizes channel- and object-based audio) has brought object-based audio to the consumer market. The following chapter will introduce and explain the principles and application of object-based audio technology for immersive sound systems.

Note

1 The concept of a "*Twin*" microphone has been originally developed by Pearl Microphones in Sweden using two back-to-back cardioid rectangular-shape condenser capsules (in their TL4 model).

References

Barbour, J. (2003). Elevation perception: Phantom images in the vertical hemi-sphere. *Proceedings of Audio Engineering Society 24th International Conference on Multichannel Audio*. Banff, Canada.

Beranek, L. L. (2007). Seeking concert hall acoustics. *IEEE Signal Processing Magazine*, September, 126–130.

Beranek, L. L. (2008). Concert hall acoustics-2008. *Journal of Audio Engineering Society*, *56*(7-8), 532-544.

Blauert, J. (1996). *Spatial Hearing: The Psychophysics of Human Sound Localization*. Massachusetts, USA: The MIT Press.

Blesser, B., & Salter, L.-R (2006). *Spaces Speaks, Are You Listening? Experiencing Aural Architecture*. Massachusetts, USA: The MIT Press.

Bowles, D. (2015). *Personal communication with Paul Geluso*.

Dolby Laboratories (2015). Dolby Pro Logic Iiz. Retrieved from www.dolby.com/us/en/technologies/dolby-pro-logic-iiz.html, Accessed on June.

Forsyth, M. (1985). *Buildings for Music: The Architect, the Musician, and the Listener from the 17th century to the Present Day*. Massachusetts, USA: The MIT Press.

Geluso, P. (2012). Capturing height: The addition of Z microphones to stereo and surround microphone arrays. *Proceedings of Audio Engineering Society 132nd International Convention*. Budapest, Hungary, AES.

Hamasaki, K. (2011). 22.2 Multichannel audio format standardization activity. *Broadcasting Technology*, *45*(2), 14–19, NHK STRL.

Hamasaki, K., Burgert, M. W., Laborie, A., Levison, J., Holman, T., van Baelen, W., & Woszczyk, W. (2006). *W9-Surround Recording and Reproduction with Height*. [Workshop] Audio Engineering Social 121st International Convention. San Francisco, USA.

Hamasaki, K., Nishiguchi, T., Okumura, R., & Nakayama, Y. (2007). Wide listening area with exceptional spatial sound quality of a 22.2 multichannel sound system. *Proceedings of Audio Engineering Society 132nd International Convention*. Vienna, Austria, AES.

Hebrank, J., & Wright, D. (1974). Spectral cues used in the localization of sound sources on the median plane. *Journal of Acoustical Society of America*, *56*(6), 1829–1834.

Holman, T. (2007). *5.1 Surround Sound, Up and Running*. Oxford, UK: Focal Press.

ITU-Recommendation BS.2051–0. (2014). *Advanced Sound System for Programme Production*. International Telecommunications Union Radiocommunication Assembly, Geneva, Switzerland.

Kamekawa, T., Marui, A., Date, T., & Enatsu, M. (2011). Evaluation of spatial impression comparing 2ch stereo, 5ch surround, and 7ch surround with height channels for 3D imagery. *Proceedings of Audio Engineering Society 130th International Convention*. London, UK, AES.

Kim, S., de Francisco, M., Walker, K., Marui, A., & Martens, L. W. (2006). Listener preferences in multichannel audio: Examining the influence of musical selection on surround microphone technique. *Proceedings of Audio Engineering Society 28th International Conference*. Piteå, Sweden.

Kim, S., Ikeda, M., & Martens, W. L. (2014). Reproducing Virtually elevated sound via a conventional home-theater audio system. *Journal of Audio Engineering Society*, *62*(5), 337–344.

Kim, S., Ikeda, M., & Takahashi, A. (2009). Investigating listeners' localization of virtually elevated sound sources. *Proceedings of Audio Engineering Society 40th International Conference on Spatial Audio*. Tokyo, Japan

Kim, S., King, R., & Kamekawa, T. (2015). A cross-cultural comparison of salient perceptual characteristics of height-channels for a virtual auditory environment. *Virtual Reality, Springer*, 19(3), 149–160.

Kim, S., Ko, D., Nagendra, A., & Woszczyk, W. (2013). Subjective evaluation of multichannel sound with surround-height channels. *Proceedings of Audio Engineering Society 135th International Convention*. NY, USA, AES.

Kim, S., Lee, Y. W., & Pulkki, V. (2010). New 10.2-channel vertical surround system (10.2-VSS); comparison study of perceived audio quality in various multichannel sound systems with height loudspeakers. *Proceedings of Audio Engineering Society 129th International Convention*. San Francisco, USA, AES.

Lee, H. (2011). The Relationship between Interchannel Time and Level Differences in Vertical Sound Localization and Masking. *Proceedings of Audio Engineering Society 131st International Convention*. New York, USA, AES.

Lee, H., & Gribben, C. (2014). Effect of vertical microphone layer spacing for a 3D microphone array. *Journal of Audio Engineering Society*, 62(12), 870–884.

Lee, K., Son, C., & Kim, D. (2010). immersive virtual sound for beyond 5.1 channel audio. *Proceedings of Audio Engineering Society 128th International Convention*. London, UK, AES.

Lindberg, M., & Shores, D. (2006). *SD14—Recording music in immersive audio [spatial audio demo]. Audio Engineering Society 138th International Convention*. Warsaw, Poland.

Miyazaki, H. (2010). The acoustical design of the New Yamaha Hall. *Proceedings of the International Symposium on Room Acoustics*. Melbourne, Australia, ISRA.

Ono, K., Nishiguchi, T., Matsui, K., & Hamasaki, K. (2013). Portable spherical microphone for Super Hi-Vision 22.2 multichannel audio. *Proceedings of Audio Engineering Society 135th International Convention*. New York, USA, AES.

Sunier, J. (2008, April). First Special Feature on 2+2+2 Multichannel Discs. *Audiophile Audition*. Retrieved from http://audaud.com/2008/04/first-special-feature-on-222multichannel-discs/

Tchaikovsky, P. I. (1880). *1812 Overture*. Conducted by Erich Kunzel with Cincinnati Pops Orchestra [CD], TELARC Classic (2001).

Theile, G. (2001). Natural 5.1 music recording based on psychoacoustic principles. *Proceedings of Audio Engineering Society 28th International Conference on Surround Sound*. Elmau, Germany.

Theile, G., & Wittek, H. (2011). Principles in surround recordings with height. *Proceedings of Audio Engineering Society 130th International Convention*. London, UK, AES.

Van Baelen, W. (2010). Challenges for spatial audio formats in the near future. *26Tonmeistertagung VDT International Convention*. Leipzig, Germany.

Wallis, R., & Lee, H. (2015). Directional bands revisited. *Proceedings of Audio Engineering Society 138th International Convention*. Warsaw, Poland, AES.

Williams, M. (2012). Microphone Array Design for localization with elevation cues. *Proceedings of Audio Engineering Society 132nd International Convention*. Budapest, Hungary, AES.

Williams, M. (2013). The psychoacoustic testing of the 3D multiformat microphone array design, and the basic isosceles triangle structure of the array and the loudspeaker reproduction configuration. *Proceedings of Audio Engineering Society 133rd International Convention*. Rome, Italy, AES.

Williams, M., & Du, G. L. (2000). Multichannel microphone array design. *Proceedings of Audio Engineering Society 108th International Convention*. Paris, France, AES.

Zielinsky, G. (2015). *Personal communication with Paul Geluso*.

Chapter 8

Object-Based Audio

Nicolas Tsingos

Introduction

Audio content production or interactive rendering is traditionally based on the manipulation of sound objects. We define sound objects as audio waveforms (audio elements) and associated parameters (metadata) that embody the artistic intent by specifying the translation from the audio elements to loudspeaker signals. Sound objects generally use monophonic audio tracks that have been recorded or synthesized through a process of sound design. These sound elements can be further manipulated, e.g., in a digital audio workstation (DAW), so as to be positioned in a horizontal plane around the listener, or in more recent systems in full three-dimensional (3D) space (see Chapter 6), using positional metadata. An audio object can therefore be thought of as a "track" in a DAW. Similarly, interactive audio engines found in video games or simulators are also manipulating sound objects—generally point source emitters—as the building blocks for complex dynamic soundscapes. In this case, they can incorporate very rich sets of metadata determining their behavior.

This process of positioning sound elements in space has been in use since the early 1940s with the introduction of the *FANTASOUND* system (Garity & Hawkins, 1941) and later evolved in the now-common 5.1 and 7.1 surround sound systems (see Chapter 2). Until recently, due to technical limitations of the various delivery media, these objects were pre-mixed into a small number of *speaker feeds* or *channels* that can be directly played back on matching loudspeaker layouts, without requiring further processing.

Recently, with the transition to digital cinema, overall increase in available bandwidth and advances in parametric audio coding, a number of approaches have been proposed (Robinson, Tsingos & Mehta, 2012) to transport a number of the original objects used during production to be rendered inside the playback environment (movie theaters, home theaters, mobile devices, etc.). As opposed to a pre-mixed speaker feed output targeting a single playback configuration, object-based audio delivery preserves a higher spatial resolution and artistic intent all the way to the playback endpoint. This provides more adaptability and the opportunity to deliver richer and more immersive audio experiences within each environment. More generally, object-based audio production and delivery enables:

- Enhanced immersion, adding height and flexible rendering across speaker layouts and environments;
- Enhanced personalization, allowing consumers to tailor the content to their preferences;

- Enhanced adaptability, ensuring that content is optimized across a wider range of playback devices;
- Enhanced accessibility, with improved multiple language support, improved video description and dialogue enhancement;
- Efficient production workflows and future-proofing of content, by deriving current or future deliverables from a single object-based master mix.

In this chapter, we offer an in-depth look at object-audio production, delivery and rendering across cinema, broadcasting and interactive applications (e.g., gaming). We first review how audio objects can be represented spatially and how the associated spatial metadata is used to render the objects to loudspeakers. In particular, we will cover in detail some common object panning algorithms and their tradeoffs. We further describe several application-specific sets of metadata, beyond spatial representation, enabling interactivity and fine-grain control of artistic intent.

One of the challenges of object-based representations is the added complexity of manipulating, encoding and transmitting a potentially large number of audio elements, compared to legacy stereophonic or multichannel techniques. We review advances in object-domain spatial coding, where large sets of objects can be converted into smaller, more convenient, sets while preserving the original perceptual intent. We also cover specific audio object parametric coding strategies for low bit-rate delivery, e.g. to the home, and provide some insights on where audio object delivery brings the most improvement compared to channel-based delivery.

Making the best out of an object-based workflow requires improved capture techniques in particular in live production environments. Building on previous chapters, we discuss how some new tools and conversion techniques can be used to complement traditional microphone techniques to capture sets of objects for both immersive and interactive applications.

Finally, we focus on two industry-wide topics: extending loudness metering and control to object-based presentations as well as standardization for interchange and delivery of object-based content.

Spatial Representation and Rendering of Audio Objects

Coordinate Systems and Frame of Reference

In order to specify locations in a space, a frame of reference is required (Klatzky, 1998). There are many ways to classify reference frames, but one fundamental consideration is the distinction between allocentric (or environmental) and egocentric (observer) reference (Figure 8.1). An egocentric frame of reference encodes object location relative to the position (location and orientation) of the observer or "self." An allocentric frame of reference encodes object location using reference locations and directions relative to other objects in the environment. An egocentric reference is commonly used for the study and description of perception; the underlying physiological and neurological processes of acquisition and coding most directly relate to the egocentric reference. An allocentric reference is better suited for scene description that is independent of a single observer position, and when the relationship between elements in the environment is of interest. For interactive rendering, video game applications programming interfaces (APIs)

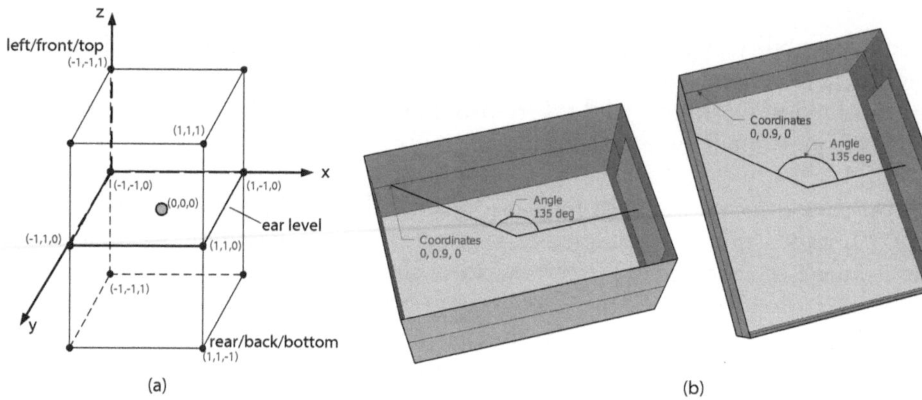

(a) (b)

Figure 8.1 (a) Example allocentric unit coordinate system to describe room-relative object-positions. For production mixing, the position of objects as well as loudspeakers in the playback environment can be expressed in a Cartesian unit room-space with $(x, y, z) \in [-1, 1] \times [-1, 1] \times [-1, 1]$. (b) The same angle in an egocentric frame of reference can translate to different locations relative to the listening room and the screen.

generally express the positions of audio objects as allocentric world-space Cartesian coordinates. The coordinates of the objects may be converted to a listener/egocentric frame of reference at rendering time depending on the player's position, in particular if a single perspective has to be rendered (e.g., on headphones).

When choosing the frame of reference for audio mixing in post-production, the following issues should be taken into consideration: (1) How to best capture artistic intent, (2) How to best preserve and reproduce artistic intent in a variety of listening environments (known as translation), (3) The tools and man-machine interface used to capture artistic intent and (4) Consistent behavior across a wide audience area.

To understand how to best capture and translate artistic intent, one must consider what spatial relationships the mixing engineer is intending to create and preserve. In general, mixing engineers tend to think and mix in allocentric terms, and panning tools are laid out with an allocentric frame—the screen, the room walls—and they expect sounds to be rendered that way: this sound should be on screen, this sound should be off screen, 1/4th of the way from the left to the right wall, and so forth. Movements are also defined in relation to the playback environment e.g., a fly-over from the center of the screen, up across the ceiling and ending at the center of the back wall. Using an egocentric frame of reference can result in an object on the side wall of an elongated mixing room ending up on the back wall of a more square exhibition space. If the egocentric audio framework is 3D, i.e., includes distance, azimuth and elevation, then an object on the rear wall of a small mix stage would end up well within the audience area of a large exhibition auditorium (Figure 8.1(b)). Using an allocentric reference, for every listening position, and for any screen size, the sound can be described at the same relative position on the screen, e.g., 1/3rd left of the middle of the screen. This allows the relationships to be captured and optimally reproduced in the wide range of room sizes and shapes that exist in exhibition.

Modern (surround-sound era) cinema sound uses an allocentric frame of reference (Figure 8.1 (a)). The reference points are the nominal location of loudspeakers (e.g., L, C, R) or loudspeaker zones (e.g., left-side surround array, right-top surround array). These locations have a known and consistent mapping to the important features of the cinema environment: the screen, the audience, the room. These locations also have a known and consistent mapping to authoring tools: left/right fader, joystick position, and the GUI. In this way, when a mix control is full-left and full-front it is understood that sound will be reproduced by a loudspeaker that is nominally located at the left edge of the screen. All location metadata are generated and decoded using this reference. This applies to both objects and channels. Only by using the same frame of reference for both objects and channels can we ensure that the spatial relationship between objects and channels is preserved.

Another particular example is illustrated in Figure 8.2 where characters on screen follow an off-screen sound event with their eyes. The corresponding sound object is perceived as incoming from different directions at each seat but this is consistent with the position of this sound in the room as perceived by both the audience and the characters on screen.

Rendering Approaches

A fundamental operation in spatial sound content creation tools is audio rendering. Audio rendering algorithms map a monophonic audio signal to a set of loudspeakers to generate the perception of an auditory event at an intended source location in space. Such algorithms have long been a key component of channel-based surround sound program creation (Rumsey, 2001; Begault & Rumsey, 2004) and are required to play back object-based content.

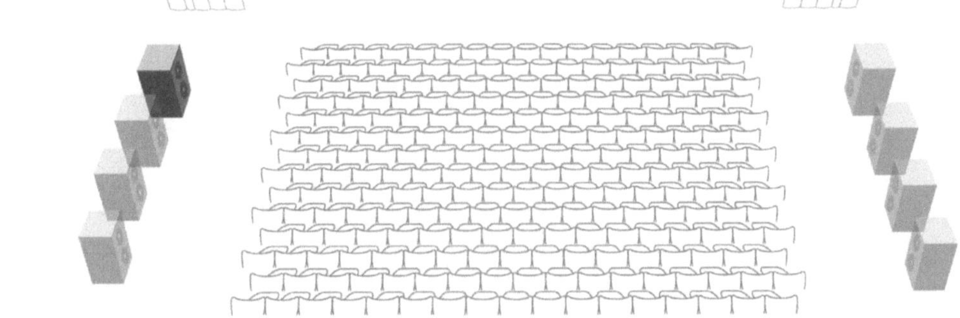

Figure 8.2 Benefits of an allocentric object representation and rendering for audio-visual coherence in a large listening environment such as a movie theater. The audience and characters on screen appear to share the same space.

Some rendering algorithms, such as wave field synthesis (de Vries, 2009) or higher-order ambisonics (Furness, 1990), attempt to recreate a physically based sound field in the listening area. Such techniques will be covered in Chapters 9 and 10 of this book. Other physically based rendering techniques, such as binaural rendering, can also be used specifically for headphone playback of audio objects (see Chapter 4).

Alternatively, object renderers approximate higher-level perceptual cues, such as interaural time/level differences for the desired source position. Most algorithms currently used in professional audio production attempt to recreate suitable cues by amplitude panning (Lossius, 2009; Dickins, 1999; Pulkki, 1997). A normalized gain vector [Gi] ($1 < i < n$), where $\Sigma_i G_i^2 = 1$, is computed and assigned to the source signal for each of the n loudspeakers in use. The object signal $s(t)$ is therefore reproduced by each loudspeaker as $G_i(x, y, z).s(t)$ creating suitable localization cues for a phantom sound source indicated by the object (x, y, z) coordinates. For instance, Figure 8.3 illustrates how different speakers may be used among various 2D rendering (panning) algorithms to simulate an object's perceived position in the playback environment.

For moving objects, the gains are traditionally evaluated over small time-frames and interpolated, either directly or using an overlap-add reconstruction of the output audio signal. Depending on the rendering system, object position updates, gain computations and audio processing can be performed synchronously or asynchronously. Audio renderers traditionally re-sample the incoming object coordinate updates (e.g., originally at 30 Hz) to a fixed audio frame rate (e.g., 100 Hz) at which the evaluation of the panning gains is performed. The gains are then further interpolated on a per sample basis, matching the audio processing rate (e.g., 48 kHz) (Tsingos & Gascuel, 1997; Tsingos, 2001).

While most renderers compute wide-band gain values for efficiency, Pulkki et al. (1999) and Laitinen et al. (2014) conducted several studies of the frequency-dependent localization and loudness bias of directional panning algorithms and demonstrated the additional benefits of frequency-dependent panning gains.

Directional, Vector-Based Panning

Directional pairwise panning (Figure 8.3(a)) is a commonly used strategy that solely relies on the directional vector from a reference position (generally the sweet spot or center of the room) to

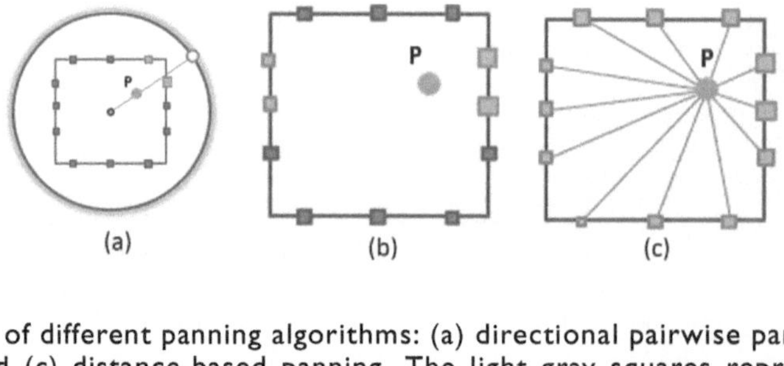

Figure 8.3 Illustration of different panning algorithms: (a) directional pairwise panning, (b) *dual-balance* panning and (c) distance-based panning. The light gray squares represent active speakers used to render the object P (dark squares are inactive speakers). The size of each square is roughly indicative of the amplitude of the signal sent to each speaker.

the desired object position. The pair of speakers bracketing the relevant directional vector is used to place (render) that object's position in space during playback. A well-documented extension of directional panning to 3D loudspeaker layouts is *Vector-Based Amplitude Panning* (VBAP) (Pulkki, 1997), which uses triplets of speakers (see Figure 8.4) to render a sound with a desired 3D direction of incidence to the listener.

A set of speaker triplets can be obtained by triangulating the convex hull of the loudspeaker array, e.g., using a Delaunay triangulation algorithm that provides triangle meshes adapted to the specific geometry of the reproduction loudspeaker setup (Barber, Dobkin & Huhdanpaa, 1996). For a given object position **p** and sweetspot **O**, a single triplet of speakers is selected by intersecting the corresponding direction vector **d** = **p-O/||p-O||** with the triangulated convex hull (Figure 8.4). The direction **d** can be expressed as a function of the unit directions **l1, l2, l3** of the 3 corresponding speakers as: $\mathbf{d} = g_1\mathbf{l1} + g_2\mathbf{l2} + g_3\mathbf{l3}$. The vector of gains for each loudspeaker $\mathbf{G} = [g_1\ g_2\ g_3]$ can therefore be obtained as:

$$\mathbf{G} = \mathbf{d}^T\mathbf{L}^{-1},$$

where **L** is the 3 × 3 matrix of the loudspeaker direction vectors.

Several proprietary extensions, e.g., in the MPEG-H standard, have been developed over a generic triangulation and VBAP algorithm to improve its performance, in particular for arbitrary loudspeaker setups (Herre et al., 2015).

First, triangulation algorithms have been designed such that they yield a left-right and front-back symmetric division of the loudspeaker convex hull into triangles. This prevents asymmetric rendering of symmetrically placed sound objects. To solve this problem, it is also possible to extend the VBAP approach by using a hybrid mesh composed of quadrilaterals and triangles (or in general *n*-gons). Once the intersection of the object direction vector with the mesh has been determined, the panning gains can then be computed as a function of the barycentric coordinates of the intersection point in the triangle or generalized barycentric coordinates in the polygon (Warren et al., 2007).

VBAP also requires that the convex hull of the speakers cover the entire sphere (or upper hemisphere) of directions. In order to prevent uneven source movements and to avoid the need

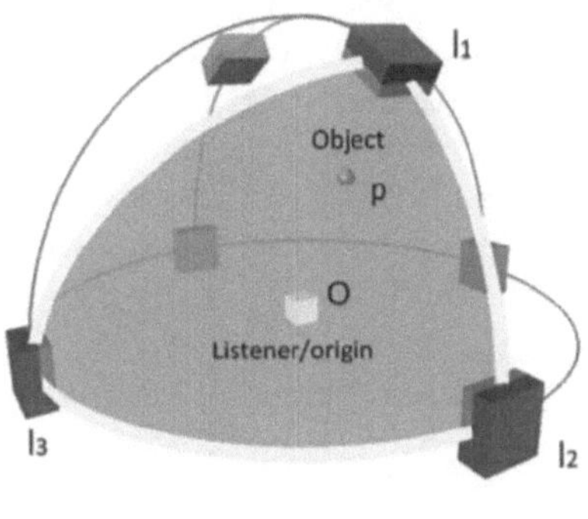

Figure 8.4 Vector-Based Amplitude Panning (VBAP) uses triplets of speakers to render a sound object (P) with a desired direction of incidence to the listener (O). The triplet of loudspeakers l1, l2, l3 brackets the desired direction of incidence (OP).

for clipping object coordinates, some systems include virtual loudspeakers in the target setup in regions not covered by physical loudspeakers. During rendering, VBAP is applied to a loudspeaker setup extended by the virtual loudspeakers. The obtained signals for the virtual loudspeakers are then downmixed to the actual physical loudspeakers. The downmixing gains for mapping virtual to physical loudspeakers are derived by distributing the virtual loudspeakers energy equally to the neighboring loudspeakers (i.e., as defined by the edges in the triangulation). One prominent use case for the added imaginary loudspeakers are reproduction layouts that only consist of loudspeakers in the horizontal plane: In this case, an imaginary loudspeaker is added at the zenith position above the center of the listening area, resulting in smooth perceived movements e.g., for fly-over sound objects (Herre et al., 2015).

Because vector-based panning only uses the direction of the source relative to a reference position, it cannot differentiate among object sources at different positions along the same direction vector. Moreover, some 3D implementations may constrain the rendered objects to the surface of a unit sphere and thus would not necessarily allow an object to cross inside the room without going "up and over." Directional panning solutions can also create sharp speaker transitions as objects approach the center of the room, where a small movement of an object's position would not always translate into a small variation in loudspeaker gains [Gi]. Solving these issues generally requires some form of distance-based blending, where the panning algorithm transitions from the original direction-based behavior to e.g., firing all speakers equally, as the object approaches the origin of the frame of reference (i.e., the sweet spot or center of the room).

Position-Based Panning

Alternative panning approaches relying on position of the objects, rather than direction, provide solutions to the above issues.

The "dual-balance" panning algorithm is the most common approach used in 5.1- or 7.1-channel surround productions today (Figure 8.3(b)). This approach uses the left or right and front or back pan-pot controls widely used for surround panning. As a result, dual-balance panning generally operates on a set of 4 speakers bracketing the desired 2D object position.

Extending to 3D (e.g., when using a vertical layer of speakers above the listener) yields a layered "triple-balance" panner. It generates 3 sets of 1-dimensional (1D) gains corresponding to left or right, front or back, and top or bottom balance values. These values can then be multiplied to obtain the final loudspeaker gains: $G_i(x, y, z) = Gx_i(x) \times Gy_i(y) \times Gz_i(z)$. This approach is fully continuous for objects panned across the room in either 2D or 3D and makes it easier to precisely control how and when speakers on the base or elevation layers are to be used.

The following background and supporting equations provide a representative example of how a simple panning algorithm would be implemented using a simple sine or cosine law. Other panning laws are also possible.

An indicative 1D (stereo) rendering could be derived as follows, using an audio objects x coordinate in [−1, 1]:

$$G_{left} = cos\left(\frac{x+1}{2} * \frac{\pi}{2}\right)$$

$$G_{right} = sin\left(\frac{x+1}{2} * \frac{\pi}{2}\right)$$

An indicative 3/2/0 example (3 front channels: left, center, right, 2 surround channels: left surround, and right surround, 0 LFE channels) would be, with 2D rendering using (x, y) coordinates in $[-1, 1] \times [-1, 1]$, as follows:

- Set all gains to 0.0.
- Using x, compute left and right pan-pot values for front and back speakers as in the previous stereo example (where *ls* is left-surround, *rs* is right-surround, *l* is left, *c* is center, and *r* is right):

$$G(ls) = \cos\left(\frac{x+1}{2} * \frac{\pi}{2}\right)$$

$$G(rs) = \cos\left(\frac{x+1}{2} * \frac{\pi}{2}\right)$$

$$if\,(x \leq\; 0.0)$$

$$G_1 = \; cos(-x\; \pi / 2)$$
$$G_c = \; sin(-x\; \pi / 2)$$

else

$$G_c = \; \cos(x\; \pi / 2)$$
$$G_r = \; \sin(x\; \pi / 2)$$

- Using y, compute front/back pan-pot values and combine with previous left/right gains:

$$f = \cos\left(\frac{y+1}{2} * \frac{\pi}{2}\right) b = \sin\left(\frac{y+1}{2} * \frac{\pi}{2}\right)$$

$$G_l{}^* = f; G_r{}^* = f; G_c{}^* = f;$$
$$G_{ls}{}^* = b; G_{rs}{}^* = b$$

- Normalize power to 1.0 by dividing all gains G_i by $\sqrt{}\,(\Sigma_i\, G_i^2)$. Following the same principle, these examples can be easily extended using a third dimension for elevation (height).

In contrast to the directional and balance-based approaches, distance-based panning (Lossius et al., 2009; Kostadinov, 2010) (Figure 8.3(c)) uses the relative distance from the desired 2D or 3D object location p to each speaker L_i in use to determine the panning gains:

$$G_i(\mathrm{p}) = \frac{1}{\varepsilon + \left(\|L_i - p\|\right)^a},$$

where a is a distance exponent (typical values being $a = 1$ or, preferably, $a = 2$) and ε is a "spatial blur" coefficient that controls how much an object can be rendered by a single speaker only.

As a result, this approach generally uses all available speakers rather than a limited subset, which leads to smoother object pans but has the tradeoff of being prone to timbral artifacts. In addition, as the number of objects increases, even a small leakage to all speakers can lead to an overall mix sounding less discrete.

However, one advantage of this approach is that it does not require an underlying topological structure (e.g., a loudspeaker mesh) and therefore provides ultimate flexibility in terms of supported speaker layouts with a very straightforward implementation.

A generalized solution to object panning over arbitrary speaker layouts is to determine an optimal set of (non-negative) gains $[\mathbf{G}](G_i \geq 0)$ so that the weighted sum of loudspeaker positions (or directions) yields the desired object position (resp. direction) (Dickins et al., 1999). This can be solved through a least-square approach. As multiple solutions are possible, additional regularization terms are generally required to ensure smoothly varying gains for moving objects, for instance, by enforcing the gains to be as small as possible. An optimization approach is likely to be more computationally intensive than solutions that explicitly pre-select subsets of speakers within a given topological structure e.g., a triangle mesh. However, it is more generic in terms of supported speaker layouts and control of image focus and sweet spot robustness.

Tradeoffs of Different Amplitude Panning Strategies

The design of object panning/rendering algorithms ultimately must balance tradeoffs among timbral fidelity, spatial accuracy, smoothness and sensitivity to listener placement in the listening environment, all of which can affect how an object at a given position in space is perceived by listeners.

For instance, Kostadinov, Reiss and Mladenov (2010) compared source localization with DBAP and VBAP and found the two approaches to perform comparably. However, no evaluation of the source timbral fidelity was conducted.

Different rendering approaches may have a significant impact on how object trajectories are perceived in the playback environment. Figure 8.5 illustrates results from an experiment comparing the three panning algorithms of Figure 8.3 to produce 2D pans across a 250-seat movie theater outfitted with a 25-loudspeaker system (Tsingos et al., 2014). Taken together, these results suggest that rendering strategies critically depend on listener distance from the center of the room (i.e., the origin of the egocentric frame of reference for directional panning), with dual-balance panning performing well near the center of the room and direction-based panning achieving good results at far distances, especially near or outside of the room boundary.

Point Objects and Wide Objects

Most object audio renderers include a control of perceptual object size that helps mixers create the impression of spatially extended sound sources. Perceptual object size can be implemented using a combination of spreading the object across multiple neighboring speakers and decorrelating the resulting speaker feeds to prevent the creation of a phantom panned image (Figure 8.6). The set of gains for the spread object is computed by spatial integration, summing the gains for a number of elementary point sources covering a given 2D area or 3D volume (Figure 8.7). For a review of decorrelation and size algorithms, we refer the reader to Potar and Burnett (2004).

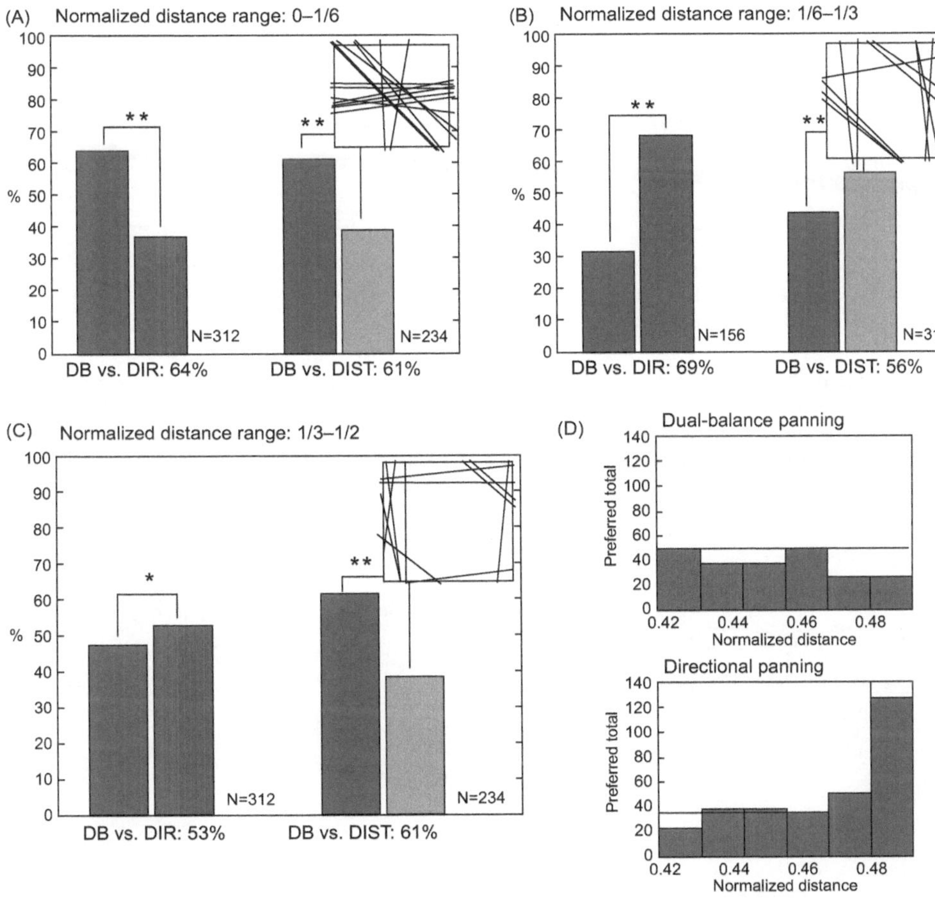

Figure 8.5 Preference judgments of three panning algorithms (directional panning (DIR), dual-balance (DB) and distance-based panning (DIST)) for different linear trajectories (shown in insets). Results were clustered into three groups of trajectories with increasing normalized distance from the center of the room. Corresponding trajectories are shown in the inserts. The bottom-right plot is a detailed breakdown view of the bottom-left plot. *N* indicates the total number of averaged results across all items and listeners.

Different solutions are used to model the spreading of an object. Following an egocentric frame of reference, some interfaces and renderers model spreading as an angular spread in azimuth, elevation and possibly distance (Figure 8.7(a)). For instance, the spread algorithm in MPEG-H 3D Audio is based on Multiple Direction Amplitude Panning (MDAP) (Herre et al., 2015; Pulkki, 1997). Other rendering approaches model spread as a 3D box-shaped volume but can also limit the spreading to the walls if the original object is positioned on the wall boundary (Figure 8.7(b) and (c)) (Robinson et al., 2012).

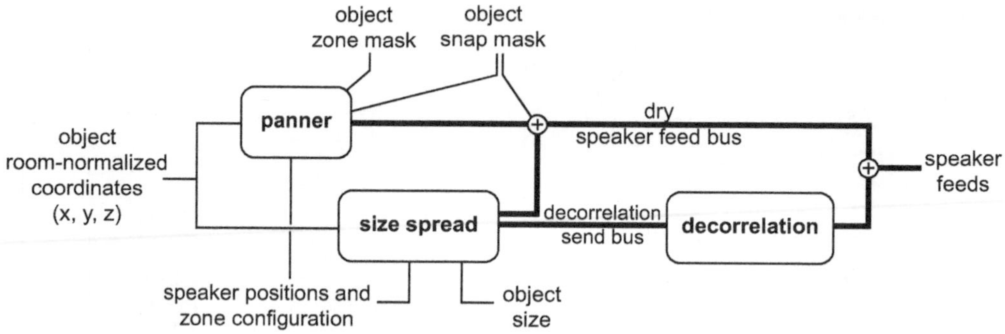

Figure 8.6 Example high-level view of an object renderer with spread and decorrelation processing. The spatially extended component of the object is rendered to a *N*-channel decorrelation bus feeding *N* decorrelators, one for each of the *N* loudspeakers in use for playback. Additional metadata controls can be used to fine-tune which speakers should be used to render the object.

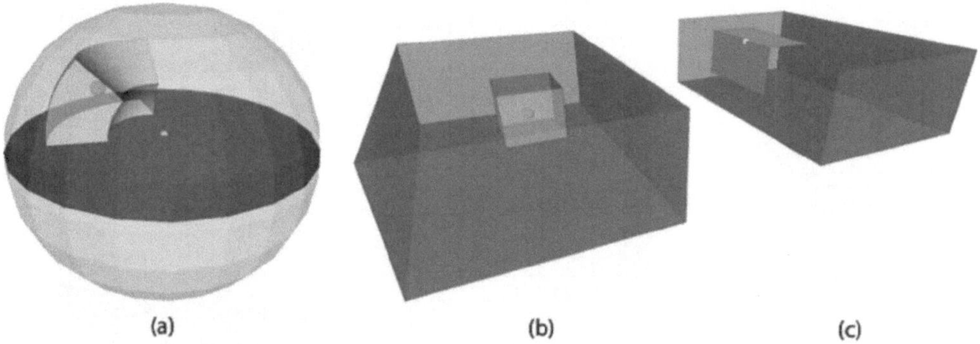

Figure 8.7 Illustration of object size control. In Figure 8.7(a), size is controlled using egocentric parameters (azimuth and elevation angular spreading and distance spreading) relative to a listening position. In Figures 8.7(b) and (c), size is controlled relative to the room, e.g., as a box-shaped region (a) or a 2D surface similar to a "torchlight" (c) as objects approach the room boundaries. A unique parameter in [0,1] can effectively model isotropic object size, where 0 is a point source and 1 covers the entire room or wall surface.

Advanced Metadata and Applications of Object-Based Representations

Artistic Controls for Object Rendering in Cinema Audio

While the use of a consistent core audio rendering technique is desirable, it cannot be assumed that a given rendering technique will always deliver consistent, aesthetically pleasing, results across different playback environments. For instance, cinema mixing engineers commonly remix the same soundtrack for different channel-based formats, such as 7.1/5.1 or stereo, to achieve their desired artistic goals in each configuration. With several hundred audio tracks competing for audibility, maintaining the discreteness of the mix and finding a place for all the key elements is a challenge that all cinema mixing engineers face and that requires mixing rules that are deliberately inconsistent with a physical model or a direct re-rendering across different speaker configurations.

Recent cinema mixing formats (Robinson et al., 2012; Robinson & Tsingos, 2001) introduce additional object-level controls such as loudspeaker-zone metadata, which are used to dynamically reconfigure the object renderer to "mask out" certain loudspeakers (see Figure 8.6 and Figure 8.8(a)). This guarantees that no loudspeaker belonging to the masked zones will be used for rendering the object. Typical zone masks used in production include *no sides*, *no back*, *screen only*, *room only* and *elevation on/off*. In this section, we review how they are used by mixing engineers to control and improve object panning over a large audience, or to render overhead objects for speaker layouts without ceiling speakers, and how they affect rendering of perceived object size. In addition, we illustrate the use of an additional *snap-to-speaker* rendering metadata to control the discreteness of the rendering as well as downmixing of proscenium objects (i.e., objects in the room that are close to the screen).

Usage of Speaker Zone Masks

A main application of speaker zone masks is to help the mixing engineer achieve a tight control of which speakers are used to render each object in order to maximize the desired perceptual effect. For instance, the *no sides* mask guarantees that no loudspeaker on the side wall of the room (see Figure 8.8(a)) will be used. This creates more stable screen-to-back fly-throughs across a wider audience. If the side speakers are used to render such trajectories, they will become audible for the

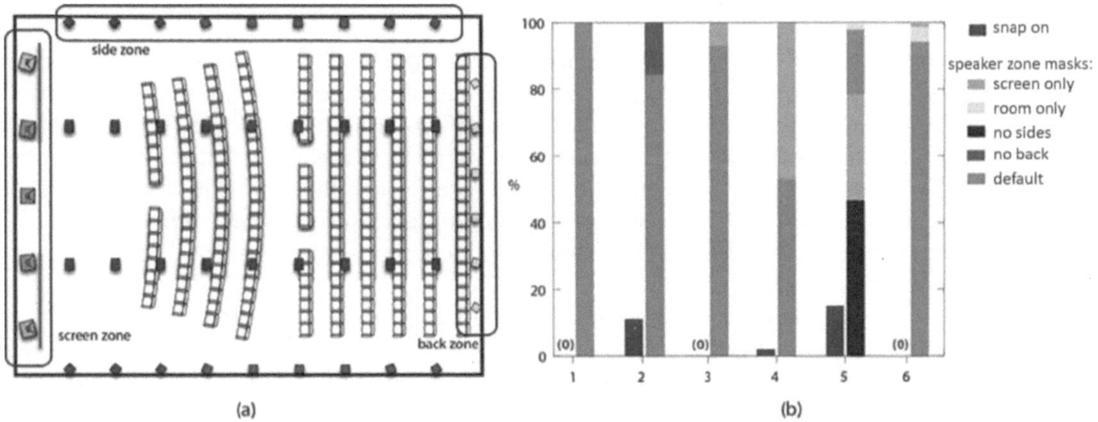

Figure 8.8 (a) Example loudspeaker zones in a theatrical playback environment. (b) Distribution of aesthetic metadata usage (snap-to-speaker and loudspeaker zone masks) for six recent Dolby Atmos movie clips (1–6) expressed as a percentage of the total non-silent audio frames across all objects. While the first clip uses a default behavior without snap or zone masks, clip #5 makes extensive use of both speaker snapping and zone masks.

seats nearest to the side walls and these seats will perceive a distorted trajectory "sliding" along the walls rather than crossing the room.

Another key application of zone masks is to fine tune how overhead objects must be rendered in a situation where no ceiling speakers are available. Depending on the object and whether it is directly tied to an on-screen element, a mixing engineer can choose, e.g., to use the *screen only* or *room only* mask to render this object, in which case it will be rendered only using screen speakers or using surround speakers, respectively, when no overhead speakers are present. Overhead music objects, for instance, are often authored with a *screen only* mask. This is directly achieved as a function of the object elevation coordinate (z). If $z = 1$ the object is rendered fully in the overhead speaker zone using all available overhead speakers. When no ceiling speakers are available, the object is projected on the base plane for rendering (i.e., $z = 0$) and the zone masks apply.

Speaker zone masks provide an effective means to further control which speakers can be used as part of the process in order to optimize the discreteness of the mix. For instance, a wide object can be rendered only in the 2D plane by using the *elevation off* mask. To avoid adding more energy to screen channels, which could compromise dialogue intelligibility, the *room only* mask is also used.

Panned vs. Discrete Sources

Another useful aesthetic control parameter is the *snap-to-speaker* mode. The mixing engineer can select this mode to indicate that consistent reproduction of timbre is more important that consistent reproduction of position. When this mode is enabled, the object renderer does not perform phantom panning to locate the desired sound image. Rather, it renders the object entirely from the single loudspeaker nearest to the intended object location. Reproduction for a single loudspeaker creates a pinpoint, timbrally neutral source and can be used to contrast key effects in the mix, in particular with respect to more diffuse elements such as those rendered directly using the channel-based representation and cinema arrays.

The snap-to-speaker mode can also be combined with zone exclusion masks and size control although size is generally forced to zero (i.e., no spreading) as the goal is mainly to create a sharp point source. To achieve better consistency when the snap mode is turned on and off, the "nearest" speaker is generally chosen as the one that would receive the largest energy if the source was phantom panned (i.e., snap off). To avoid snapping to a speaker too far from the original intended object location (which could happen in sparse speaker configurations), a *snap-release threshold* is provided. If the snapped position is farther from the intended position by more than this distance threshold, the renderer reverts to panning.

For film soundtracks, a key use of the snapped parameter is to create near-screen/wide pairs of objects along the side walls of the cinema by using objects which snap to the proscenium speakers. This is particularly useful for music elements, e.g., to extend the orchestra beyond the screen. When re-rendered to sparser speaker configurations (e.g., legacy 5.1 or 7.1), these elements will be automatically snapped to left/right screen channels. Another use of the snap metadata is to create "virtual channels," for instance to re-position the outputs of legacy multichannel reverberation plug-ins in 3D without risk of "double panning."

Figure 8.8(b) shows aesthetic overall metadata usage in six recent movie clips. As can be seen, the default rendering behavior, based only on object position, is used most of the time while mixing object-based content. However, in a significant number of cases, the baseline rendering does not give the best result and additional artistic input, overriding the default behavior, is beneficial.

Cinematic Virtual Reality and Headphone Playback

An emerging type of cinematic content targets virtual reality (VR) endpoints. VR endpoints leverage head-mounted displays (HMDs) and headphone rendering to deliver an immersive full 360-degree stereoscopic experience to the user. Rendering of spatialized sound is paramount for this use case and object-based descriptions are well suited to create auditory scenes with the required high spatial resolution. However, additional rendering metadata can also be authored and delivered on a per-object basis to help the mixing engineers fine tune the headphone experience. For instance, it can be beneficial to control the rendering algorithm depending on the type of content. A stereo music ambiance could then be rendered as is, without binaural spatial processing and without head tracking (i.e., head-relative) while sound effects can be spatially rendered and head-tracked (i.e., world-relative). Mixing engineers also commonly choose to render percussive sounds (e.g., drums) in stereo to avoid transient smearing due to binaural filtering. Endpoint-specific metadata enables the obtained VR headphone mix to be delivered and adapted for loudspeakers playback, for instance combining an HMD for video playback and a home-theater system for audio.

Interactivity and Personalization in Broadcasting

Traditionally, live-broadcast mixing engineers take all microphone feeds from a live event and mix them down to a high-quality 5.1-channel or stereo program. Production mixing engineers also create different submixes (e.g., ambience or effects, music and dialogue) before mixing down to a final program. These submixes make up the audio *beds* (i.e., traditional multichannel ambiences) and objects that are sent discretely to the playback device, where they can be personalized and rendered. For example, an ambience or effects submix could be represented by a 5.1-channel audio bed, and each of the multiple dialogue submixes (e.g., for multiple commentators or languages) could be represented by an audio object. As a result, existing microphones and microphone plans can be used to create customizable object-based mixes for live broadcast events. Additional microphones (e.g., for capturing height or close-proximity crowd sounds) can be added to provide a more immersive audio experience.

A channel-based live audio production today is composed of a number of individual elements, which may include the following:

- Crowd sound (diffuse) constructed from a number of sources;
- Spot sounds (e.g., ball kick or basketball bounce);
- Off-screen dialogue (e.g., commentary or announcer);
- On-screen dialogue (e.g., studio links and rangefinder camera interviews);
- Audio effects associated with on-screen graphic transitions;
- Pre-recorded audio and video (A/V) playback material (e.g., a highlights package or replay elements from the event);
- Synthesized fill sounds (e.g., a helicopter, garage sounds in a pit lane, or crowd fill).

An object-based audio production is made up of the same elements. The difference, however, is that some of these elements are not blended into the mix during production but instead are sent to the receiver as a number of different audio presentations (made from one or more sub-streams). Then, the selected presentation is rendered in the receiver to the final speaker configuration (Mann et al., 2013).

An additional layer of metadata defines the personalization aspects of the audio program. These personalization metadata serve two purposes: to define a set of unique audio presentations that a consumer could select and to define dependencies (i.e., constraints) between the audio elements (objects) that make up the unique presentation to ensure that personalization always sounds optimal (Riedmiller et al., 2015).

Presentation Metadata

Producers and sound mixing engineers can define multiple audio presentations for a program to allow users to switch easily among several optimally predefined audio configurations. For example, a sound mixing engineer for a sports event could define a default sound mix for general audiences, biased sound mixes for supporters of each team that emphasize their crowd and favorite commentators, and a commentator-free mix. The defined presentations depend on the content genre (e.g., sport or drama) and differ among subgenres (from sport to sport or form to form). Presentation metadata define the details that create these different sound experiences. An audio presentation specifies which object elements or groups should be active, along with their position and their volume level. Defining a default audio presentation ensures that audio is always output for a given program. Presentation metadata can also provide conditional rendering instructions that specify different audio object placements or volumes for different speaker configurations. For example, a dialogue objects playback gain may be specified at a higher level when reproduced on a mobile device as opposed to an A/V receiver. Each object or audio bed may be assigned a category, such as dialogue or music. This category information can be used later either by the production chain to perform further processing or by the playback device to enable specific behavior. For example, categorizing an object as dialogue would allow the playback device to manipulate the level of the dialogue object with respect to the ambience. This categorization can also be used to help in prioritizing objects during playback to conditionally enable ducking of non-prioritized elements and thus enhance intelligibility. Presentation metadata can also identify the program, along with other aspects (e.g., which sports genre or which teams are playing), that could be used to automatically recall personalization details when similar programs are played. For example, if a consumer personalizes a viewing experience to always pick a radio commentary for a baseball game, the playback device can remember this genre-based personalization and select the radio commentary for subsequent baseball games. The presentation metadata also contains unique identifiers for the program, each presentation and each sound element. This allows user interfaces on the playback device to associate user interface elements to each aspect of the personalized program. Presentation metadata typically does not vary temporally, on a frame-by-frame basis. However, they may occasionally change throughout the course of a program. For example, the number of presentations available may be different during live game play but may change during a halftime presentation as an example.

Interactive Metadata

Users may want complete control over personalizing an audio program. To ensure that every customization results in an optimal sounding mix, the content creator or broadcaster can provide interactive metadata defining a set of rendering rules to be used for personalization only. Interactive metadata can specify object parameter minimum or maximum values; inter-object mutual exclusion; inter-object position, volume or ducking rules; and overall mix rules. The interactive metadata are typically leveraged by a consumer user interface to prevent the creation of a non-ideal rendering (mix). For example, if both English and Spanish dialogue objects are present in the audio stream, the interactive metadata would prevent a user from enabling both objects simultaneously. The interactive metadata can be represented using a mixgraph structure (Figure 8.9).

Video Games and Simulation

The definition of audio object in a game engine generally extends far beyond what can be found for theatrical or broadcasting applications. It includes specific sound source modeling (e.g., directivity) as well as propagation metadata (e.g., distance attenuation models and reverberation parameters). For a review of audio rendering for games and simulation, we refer the reader to (Savioja et al., 1999; Funkhouser, Jot & Tsingos, 2002). This is to be contrasted to post-production applications where such effects are generally pre-baked in the objects' audio essence and therefore are not considered part of the object metadata further transmitted downstream e.g., for theatrical or home playback. In addition, audio objects in game engines can comprise multiple elements each tied to the game logic through specific control parameters, therefore defining very

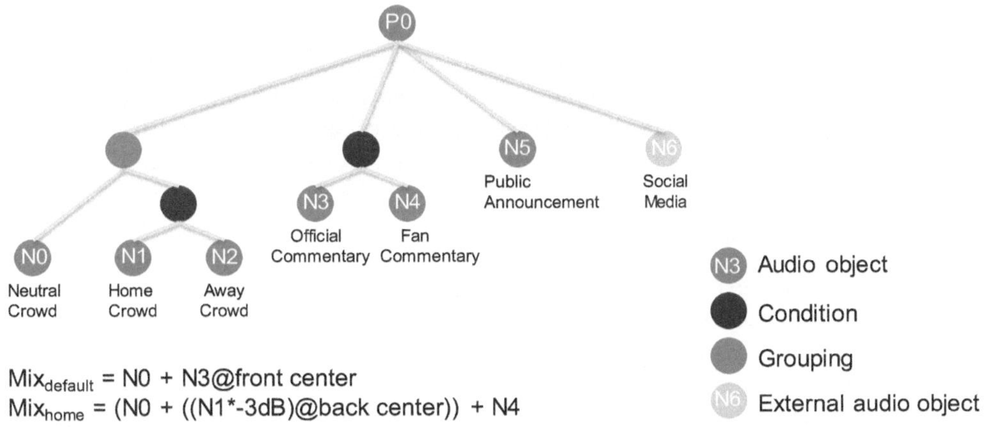

Figure 8.9 Mixing graph for personalized or interactive object-based presentations in the context of a sports broadcast. Default mix presets (e.g., Mix*default* or Mix*Home* can be defined from the leaf objects present in the graph. Constraints can be expressed as specific control nodes that prevent users to generate undesirable results. For instance, conditional nodes express an "either/or" constraint.

complex mixing graphs including scripted behavior e.g., enabling live concatenative synthesis (Roads, 1996).

Several standards and APIs have been historically developed to render object-based content for games. For instance, *DirectSound* (Bargen & Donelly, 1998), *OpenAL* (2000) were introduced to perform rendering of point source objects in 3D space and also included both standardized (IASIG level 2.0, 2016) and proprietary extensions (EAX, 2004) enabling the rendering of environmental occlusion or reverberation effects. Some environmental rendering extensions were also part of the MPEG4 audio BIFS description format (Jot, Ray & Dahl, 1998). Today, video game developers either rely on middleware solutions (FMOD, WWISE) or proprietary audio engines for which specific plug-ins can be created and the flexibility of which extends far beyond what could be realistically captured with a limited set of metadata. Such engines combine real-time object rendering with dynamic busing and audio effect plug-in architectures similar to post-production DAWs or consoles.

However, similar to production environments, game engines traditionally render a channel-based audio output. Object-based audio output, in a way that abstracts the rendering configuration, is highly desirable, in particular as new loudspeaker systems are introduced that support rendering of height for instance. In that sense, the notion of object-based audio output matches the one used in post-production applications. In recent gaming consoles such as Sony's *PS4* or Microsoft's *Xbox One*, this functionality is enabled at the system level and supports plug-ins for encoding of object-based information into proprietary formats supported by external A/V receivers.

Managing Complexity of Object-Based Content

Whether in post-production or for interactive gaming, an audio mix can comprise hundreds of simultaneous elements. It is often desirable to compress the information of the resulting auditory scene in a way independent of the chosen reproduction technique to enable more efficient delivery, rendering or post-processing. Several perceptual auditory properties may be exploited in order to simplify the rendering of a complex object-based scene with limited impact on the overall perceived audio quality. The general approach is to re-organize the sound scene by (1) sorting its components by their relative importance and (2) reducing its spatial complexity.

Prioritizing and Culling of Objects

A first approach to manage the complexity of an object-based audio scene is to prioritize objects and discard the least important ones. This solution builds upon prior work from the field of perceptual audio coding that exploits auditory masking. When a large number of sources are present in the environment, it is very unlikely that all will be audible due to masking occurring in the human auditory system (Moore, 1997). This masking mechanism has been successfully exploited in perceptual audio coding (PAC), such as the well-known MPEG I Layer 3 (mp3) standard (Painter & Spanias, 2000) and several efficient computational models have been developed in this field. This approach is also linked to the illusion of continuity phenomenon (Kelly & Tew, 2002), although current works do not generally include explicit models for this effect. This phenomenon is implicitly used together with masking to discard entire frames of original audio content without perceived artifacts or holes in the resulting mixtures (Gallo, Lemaitre & Tsingos, 2005).

Evaluating all possible solutions to the optimization problem required for optimal rendering of a sound scene would be computationally intractable. An alternative is to use greedy approaches, which first require estimating the relative importance of each source in order to get a good starting point. A key aspect is also to be able to dynamically adapt to the content. Several metrics can be used for this purpose such as energy, loudness or saliency (Kayser et al., 2005). Recent studies have compared some of these metrics showing that they might achieve different results depending on the nature of the signal (speech, music, ambient sounds). Loudness has been found to be generally leading to better results while energy is a good compromise between complexity and quality (Gallo et al., 2005).

Following these principles, recent game audio engines implement "High-Dynamic Range" audio by dynamically adapting the dynamic range window to the current loudness of the mix. This offers the opportunity to simultaneously control the dynamic range of the overall mix while culling the objects of weaker relative level, freeing up computational resources (Frostbite, 2009).

Spatial Coding

Limitations of the human spatial hearing e.g., as measured through perceivable distance and angular thresholds (Begault, 1994) can also be exploited for faster rendering independently of the subsequent signal processing operations. Studies have also shown that our auditory localization is strongly affected in multi-source environments. Localization performance decreases with increasing number of competing sources (Brungart, Simpson & Kordik, 2005) showing various effects such as pushing effect (the source localization is repelled from the masker) or pulling effects (the source localization is attracted by the masker), which depend on the time and frequency overlapping between the concurrent sources (Best et al., 2005). Thus, spatial simplification could be performed even more aggressively as the complexity of the scene, in particular the number of sound sources, grows. However, if interaction is possible at rendering time, for instance the ability for the listener to navigate the scene or some objects to be emphasized, this approach may lead to artifacts unless the specific elements can be prioritized accordingly. Another challenge is spatial coding of specific object rendering metadata (see the section "Advanced Metadata and Applications of Object-Based Representations"). In this case, spatial coding approaches must be extended to preserve objects with different metadata into different groups.

If the reproduction format is known in advance, one straightforward approach to reducing spatial complexity is simply to render some objects to an intermediate smaller set of channels, at the expense of downstream flexibility. This approach is often used in interactive game engines to feed "effect-send" buses (e.g., to multichannel reverberators) with a premix of the objects in the scene.

For applications where the reproduction format or target device is not set in advance, the coding/simplification of the spatial information must be performed in scene-space, ideally directly on the original objects themselves. One solution is to convert objects to a fixed set of spatial basis functions to obtain a new alternative representation that does not depend on the original number of objects. For instance, Ambisonics (Malham & Myatt, 1995; Daniel & Moreau, 2004) uses a spherical harmonics decomposition of the incoming sound pressure at the listening point (we refer the reader to Chapter 9 for more information on Ambisonics and high-order Ambisonics solutions).

However, it is generally desirable to preserve the object-based nature of the scene by converting the original set of objects into a reduced set of perceptually equivalent objects (Figure 8.10). This solution, which we refer to as *spatial coding*, can be divided into three categories:

- Fixed clustering approaches operate in object-space and explicitly group neighboring sound sources belonging to the same cone of directions (Herder, 1999), or using fixed or hierarchical grid structures (Wand & Straßer, 2004).
- Dynamic per-object clustering: The clustering proposed by Sibbald (2001) is an object-based method aiming at progressively rendering objects formed of complex aggregates of elementary sources. Sound sources related to an object, or an area, are grouped according to their distance to the listener. In the near field, secondary sound sources are created and dynamically uncorrelated in order to improve the spatial sensation. In the far field, sources are clustered together, accelerating the spatial rendering. The drawback of the method is that the clustering is evaluated on a per-object basis and does not consider interactions between all the elements of the scene.
- Dynamic global clustering: Dynamic source clustering methods (Tsingos, Gallo & Drettakis, 2004) based on both the geometry of the scene and the signals emitted by each source have also been proposed. This is especially useful for scenes where sound objects are frequently changing in time, varying their shape, energy as well as location. These algorithms flexibly allocate the required number of clusters; thus clusters are not wasted where they are not needed. Dynamic clustering can be greedily derived from e.g., the Hochbaum-Shmoys heuristic. The cost-function used for clustering combines instantaneous loudness (Moore, Glasberg & Baer, 1997) and inter-object distance (allocentric grouping) or distance and incidence direction to the listener (egocentric grouping). An equivalent signal for each object cluster is then computed as a mixture of the signals of the clustered sound sources. In Tsingos et al. (2004), each original object is assigned to a single cluster but it is also possible to

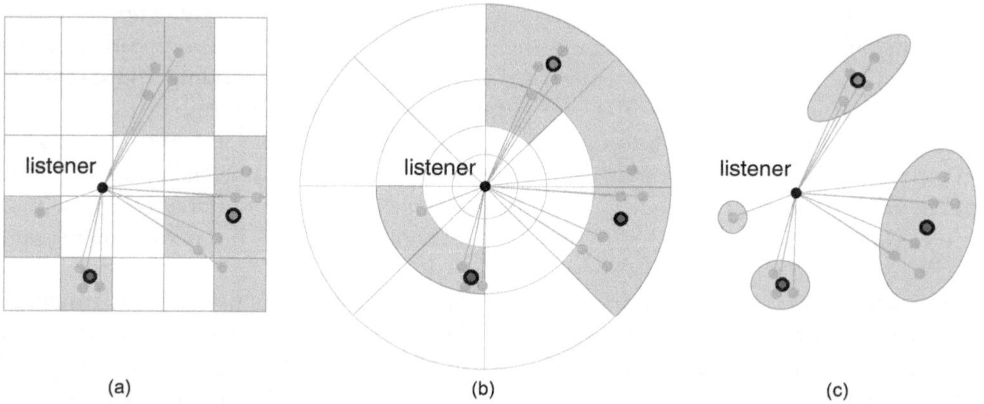

Figure 8.10 Three examples for spatial object grouping: (a) allocentric grouping using a regular grid structure leading to 10 object-clusters, (b) egocentric grouping using a progressively refined structure for objects near the listener (6 object-clusters) and (c) using an egocentric adaptive clustering approach (4 object-clusters).

re-distribute objects to multiple clusters based on their relative position. A representative loudness-weighted centroid is used to spatialize the cluster according to the desired reproduction setup.

In the context of interactive simulations and gaming, adaptive clustering techniques have been shown to preserve very good rendering quality and minimal impact on localization-task performance, even with a small number of clusters (Tsingos et al., 2004). More recently, similar approaches have been successfully used to translate complex object-based theatrical audio mixes comprising dozens of concurrent objects into more compact versions for interchange and home delivery (Riedmiller et al., 2015).

Audio Object Coding

Complexity-reduction approaches, such as spatial coding, can dramatically reduce the bandwidth required for storing or transmitting object-based programs. However, the resulting program might still require several megabits per second (Mbps) for transmission which is incompatible with home delivery via streaming where the target bitrates are typically a few hundred kilobits per second (kbps) (e.g., 192 to 640 kbps for current 5.1/7.1 Dolby Digital Plus). Delivering object-based content at these bitrates requires additional coding tools.

Independent Coding of Objects

For applications requiring full interactivity and manipulation of the audio objects (such as freely adjusting their level), independent coding of the individual objects is the preferred solution. This is typically used in video games, where the objects' audio essence is generally separately encoded as monophonic files that can be efficiently decoded by the hardware (e.g., using AAC, mp3 or WMA codecs). For closed ecosystems, any codec whether lossless or lossy could in theory be used. However, mono coding of objects introduces several challenges to deliver linear production content. First, it is not directly backwards compatible with common legacy playback formats such as stereo or 5.1 surround and requires decoding and rendering the full set of objects to produce legacy channel-based output. Second, for movie playback or applications that do not require full interactivity, maintaining perfect separation of the individual objects is not required and could be exploited for more efficient compression.

To enable backwards compatibility with a typical channel-based surround format, Figure 8.11(a) illustrates a possible workflow where the objects' signals are individually coded either with a lossy or lossless approach and can be canceled out from a core backwards compatible rendering (e.g., 5.1 or 7.1 surround channels) at playback time to be individually re-rendered. Both core and individual objects are simultaneously transmitted. If a lossy algorithm is used, the objects must ideally first be encoded and decoded before rendering to the core set of channels so that coding artifacts can be entirely canceled out during playback. The cost of decoding the full set of objects can be high, as all objects must be decoded and rendered twice (once to be canceled out from the main core channels and another for the actual playback). However, the approach lends itself well to scalable decoding where objects can be extracted in priority order, the least important ones remaining in the channel core if the decoder becomes computationally constrained. It

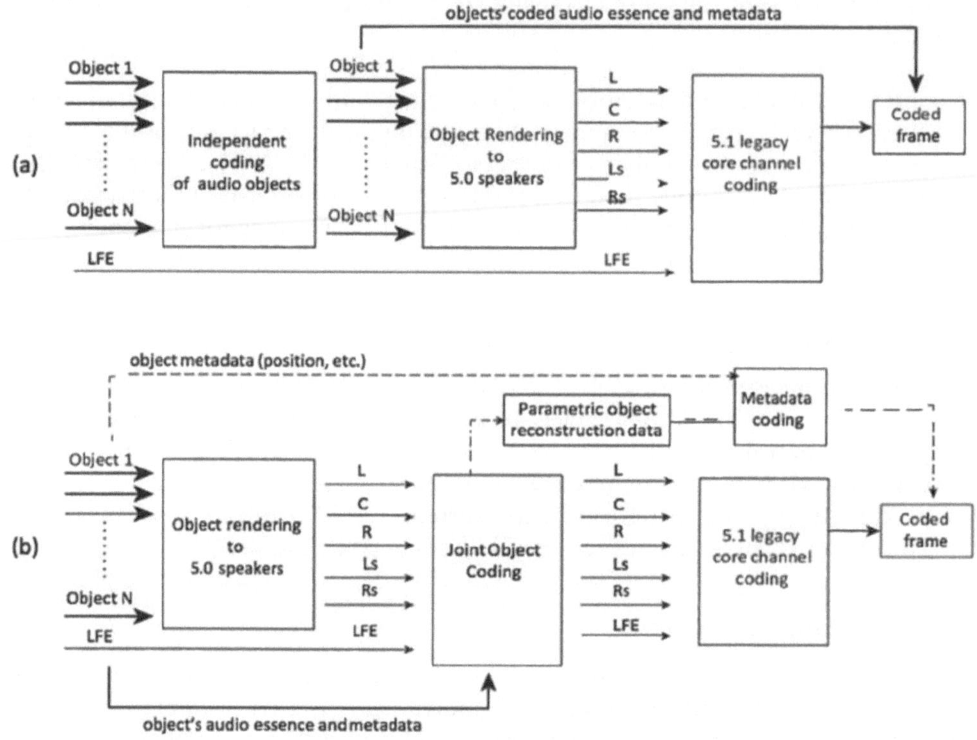

Figure 8.11 Two options for backwards compatible coding of audio objects. (a) Both objects and a backwards compatible channel-based rendering are transmitted. Objects can be progressively canceled in the channel-based core mix to be individually re-rendered at playback time. (b) Objects are parametrically encoded and reconstructed from the backwards compatible channel-based core mix. This is more bit-efficient but requires all objects to be decoded for playback.

is also well suited to carrying a residual channel-based ambience or "bed" in addition to supplemental objects. One challenge, however, is to handle dynamic object priorities as time-varying extraction of objects from the core channels could lead to audio artifacts.

Parametric, Joint Coding of Audio Objects

Parametric joint audio object coding e.g., MPEG-SAOC (Herre & Disch, 2007), overcomes these drawbacks and can achieve further coding efficiency. Such techniques are designed to transmit a number N of audio objects in an audio signal that comprises K downmix channels, where $K < N$ and K is typically two, five or seven channels following standard channel-based layouts. For instance, the downmix core signal can be obtained by rendering the original objects to a canonical stereo or 5.1 speaker configuration. This core signal can then be coded using traditional perceptual coding techniques (Figure 8.11(b)). Together with this backwards compatible downmix signal core, object reconstruction metadata (typically in time-frequency tiles) is transmitted through a dedicated bitstream to the decoder side.

Although the object reconstruction data grows linearly with the amount of objects, it only adds a small overhead to the bitrate required to code the channel-based downmix itself.

A benefit of the approach is that the core downmix can be directly decoded and played back for legacy channel-based systems at no additional cost over legacy coding techniques. Some control over the core downmix is also possible, giving the opportunity to mixing engineers to fine tune the presentation for these important legacy use cases.

For applications where direct backwards compatibility is not a strong requirement, it should be noted that the coded core does not have to match typical channel-based playback configurations. In fact, the core could itself be object-based, thus creating a hierarchical object-based presentation (e.g., where 15 objects could be reconstructed from 4). Applying spatial coding techniques recursively is a solution to build such hierarchical object-based presentations. This approach typically leads to better reconstruction of the original objects. The smaller core set of objects is more efficient to decode and can still be flexibly rendered to different speaker layouts. It can typically be used for efficient virtualization on a mobile device while maintaining battery life.

Capturing Audio Objects

A Good Object Is a Clean Object

In traditional production workflows or interactive rendering systems, sound objects are generally attached to monophonic audio elements. As a single monophonic element cannot reproduce the full spatial characteristics of natural soundscapes, an object-based program is generally assembled from multiple mono, stereophonic and possible 3D spatial captures. Preferably, the monaural audio essence must be individually recorded, ensuring the cleanest possible capture for maximum flexibility. This approach gives the user the freedom to control and position each object and freely navigate throughout the resulting auditory scene. A clean capture also avoids noise buildup as many objects get mixed to form the final auditory scene. In some cases, it is valuable to apply more aggressive noise-removal approaches on the individual objects and introduce a unique global room tone or background ambience to mask possible remaining artifacts in the final mix. In broadcast applications, crosstalk between non-coincident microphones used as audio essence for objects can lead to combing artifacts during re-rendering to smaller speaker layouts (e.g., stereo). A solution to this problem is to introduce decorrelation processing on the acquired signals or modify the microphone layout to capture more diverse signals. Traditional directional microphone techniques (e.g., using close miking or shotgun microphones (Rayburn, 2012; Viers, 2012) can be used to achieve the desired separation. Steerable microphone arrays, either linear or spherical (Meyer & Elko, 2004), also offer a similar type of control with the added benefit of extracting specific components at post-production time that can be further turned into objects. Finally, the combination of directional microphones and orientation-trackers, e.g., in mobile devices such as phones, can enable simultaneous recording of audio essence and orientation data that can directly serve as positional information for object rendering (Tsingos et al., 2016).

The work of Cengarle et al. (Cengarle et al., 2010) proposes to automate the production of audio for sports broadcasting by dynamically blending the contribution of spot microphones around a soccer pitch based on a desired point of interest (Figure 8.12). The point of interest can thus become an audio object with an associated signal, i.e., the weighted sum of the microphones closest to the point of interest and a position. This point-of-interest object can be manually tracked by the user or could be automatically tracked in a live video feed. The gains associated with each microphone can be sent as level automation to the mixing console to provide additional manual control options to the mixer.

From Spatial Capture to Objects

Spatial sound recording techniques which encode directional components of the sound field (Merimaa, 2002; Meyer & Elko, 2004; Soundfield, 2016) can also be directly used to acquire and play back real-world auditory environments as a whole. Converting such recordings to more structured object-based representations is an emerging problem.

A number of solutions have been proposed to extract parametric, object-like, representations from real-world coincident or non-coincident recordings. For instance, Directional audio coding (DirAc) (Pulkki, 2006; Vilkamo, Looki and Pulkki, 2009) reconstructs direction of arrival information for different frequency sub-bands, typically from B-format recordings or using MEMS acoustic velocity probes or small arrays of omnidirectional microphones (Ahonen, 2013). This process of converting spatial recordings into more parametric object-like formats also enables improved, more discrete, rendering. For instance, a B-format recording converted into an object-like format with DirAc outperforms a traditional 1st-order decoding (Vilkamo, Lokki & Pulkki, 2009). Another advantage of such an approach is that it lends itself to more efficient coding and distribution as a downmix or intermediate rendering (e.g., mono or stereo) of the original recordings. It can be more efficiently transmitted but can still be parametrically decoded to a richer spatial representation using the estimated spatial metadata. Recently, it has been shown that such approaches can generate compelling 3D soundscapes including elevation information from widely available XY-stereo microphones (Tsingos et al., 2016).

Similar approaches have proposed a reconstruction extended to 3D position using non-coincident recordings (Gallo et al., 2007; Gallo & Tsingos, 2007). Time-differences of arrival

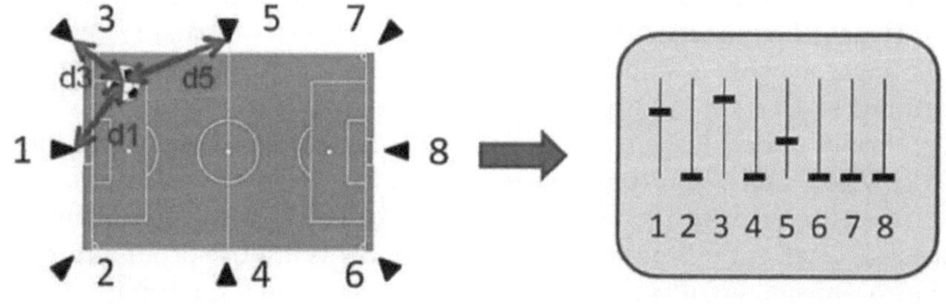

Figure 8.12 A point-of-interest audio object constructed from multiple non-coincident directional recording around a soccer pitch (from Cengarle et al., 2010). The mixing levels of each microphone can be a function of the distance (*di*) of the point of interest to each microphone *i*.

between the non-coincident microphones can be used to reconstruct the 3D position of the different time-frequency tiles in the convex hull of the microphone array. For a moving listener across the microphone hull, a spatial rendering can be reproduced from the estimated position. A combination of the signals of the microphones closest to the listening point can be used as a good representative signal for each time-frequency tile/object. The different sub-band signals and their estimated direction or position can be considered audio objects and be imported into object-audio production workflows to be mixed with other elements or manipulated. For instance, Gallo et al. (2007) demonstrates the ability to move some of the original elements or dynamically occlude them by introducing virtual obstacles in a virtual reality or gaming context.

Tracking of the acoustic intensity and spatial metadata in general can also enable virtual beamforming or blind source separation (BSS) (Yu, Hu & Xu, 2014). Such approaches, however, cannot fully achieve fine-grain separation of semantic objects (e.g., perfectly separate a dog bark from a car engine) and as a result cannot enable full control over the captured auditory scene without introducing audible artifacts. Other categories of BSS algorithms that attempt to segregate sources more finely do not directly rely upon spatial information but more on the assumed statistical independence of the signals themselves, e.g., by performing independent component analysis (Yu et al., 2014). They generally require the number of sources to be known in advance or obtained through other algorithms as a pre-processing step.

Tradeoffs of Object-Based Representations

The previous sections have explored the core components of object-based content creation, encoding and rendering. In this section, we review different tradeoffs that can guide the decision to create and deliver object-based content as opposed to alternative channel-based programs (e.g., loudspeaker feeds or B-format).

Rendering Flexibility and Scalability

Object-based delivery is an obvious choice for high-quality content distribution to high-end playback environment such as movie theaters that use large speaker configurations (e.g., typically 40 and up to 64 in some Dolby Atmos theatrical installations). Enabling spatial audio reproduction over a large number of discrete speakers has a number of benefits over traditional cinema playback where speakers are grouped into arrays: better timbral characteristic, sharper point sources, better A/V and spatial coherence for a large listening area. Objects also offer exhibitors the ability to differentiate or adapt the quality offering to the different markets. Delivering a multitude of different channel-based renderings to match the different room configurations would create a distribution nightmare and is not desirable. Object-based audio production lets content creators streamline the process of creating high-end spatial audio presentations including height information, while being able to re-render in a single pass multiple legacy stereo, 5.1 and 7.1 deliverables or stems for subsequent tuning or internationalization.

For home applications, where the number of loudspeakers will be in general much smaller or the playback will occur on headphones, one can wonder if the same benefits of objects hold in terms of spatial representation. Figure 8.13 shows the result of a preference test conducted on headphones with listeners comparing channel-based virtualization (i.e., using a few discrete

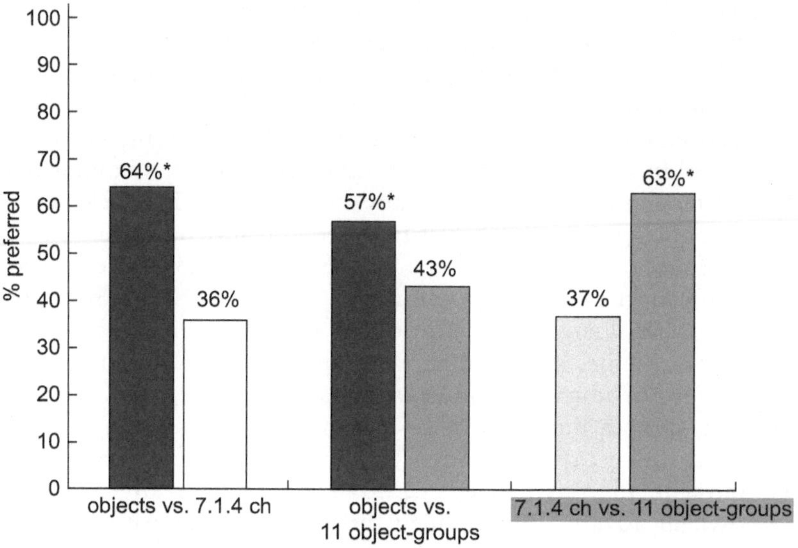

Figure 8.13 Preference test for channel-based vs. object-based virtualization. A set of 7.1.4 channel presentations rendered from original object-based content were virtualized over headphones and compared to a virtualization of all original objects as well as 11 object-groups (see previous Spatial Coding section). The test was conducted with 10 listeners and 13 short clips. The asterisks * indicate statistically significant results.

Head-Related Transfer Functions [HRTFs] for each channel) and native object-based virtualization (using a high-resolution continuous HRTF model). As can be seen, an object-based presentation can be preferred to a channel-based virtualization, assuming a high-resolution or continuous HRTF model is used for rendering.

Coding Efficiency and Transmission

In the theatrical context, object-based content requires on average much less audio tracks to reproduce the soundtrack at any single time-instant than delivering a large number of pre-rendered channels. This is due to object-based content being, in general, temporally sparse. It can therefore be efficiently and losslessly coded, leading to relatively small file sizes compared to channel-based linear PCM (LPCM). Peak bitrates can still be high, however, with up to 128 simultaneous objects in some theatrical formats (Robinson et al., 2012), but this is rare in practice and does not strongly impact average file sizes. On average, a losslessly coded Dolby Atmos digital cinema package may vary in size from 3 gigabytes (Gb) to 10Gb, with a relative size of 0.6× to 1.5× the uncompressed 24-bit LPCM 7.1 deliverables.

In the context of low-bitrate lossy delivery to the home, the quality of object-based playback tends to decrease faster due to the extra cost of sending object metadata and the slightly reduced coding efficiency. This tradeoff also has to be balanced against the improved quality brought by object-based delivery for other endpoints, as illustrated in Figure 8.13.

Finally, for applications requiring some amount of interactivity or conditional rendering, such as the broadcast applications discussed above, object-based coding offers a more bandwidth-efficient way to deliver separate elements that would otherwise require multiple fully separated multichannel stems. This also allows a more efficient and higher-quality processing for playback-time enhancement of dialogue tracks by representing spatialized dialogue as one or few objects. For emerging applications such as cinematic virtual reality, the artistic control of the headphone rendering (e.g., stereo vs. binaural or headtracked vs. non-headtracked sounds) is also made more bit-efficient through the use of objects as opposed to e.g., delivering multiple B-format stems.

Object-Based Loudness Estimation and Control

As object-based audio representations move from production or interactive gaming into more dual-ended transmission workflows such as broadcast delivery, ensuring consistent loudness estimation and control becomes paramount as it ties to legal regulations in many geographies.

International recommendations, such as ITU-R BS.1770 (2015), provide methods to estimate loudness for channel-based content and are widely used throughout the audio industry.

Two issues become of interest in the context of object-based audio production: 1) extending current standards to meter and correct loudness of object-based content, and 2) ensuring that playback of such content has consistent loudness regardless of the rendering configuration.

ITU-R BS.1770 has recently been revised to be able to measure any arbitrary channel-based configuration (see Figure 8.14). This extension to an arbitrary number of channels could form the basis for object loudness measurement, where the loudness is derived from summing the frequency-weighted energies of all the objects monaural audio essence. Similar to current channel-based workflows, this extension would be used to measure and correct object-based programs prior to transmission and would ensure that multiple programs (whether channel-based or object-based) are mixed and delivered at consistent levels. On the playback side, it is indeed desirable that levels do not drastically vary from one rendering configuration to another. As seen above, a widely accepted requirement of object rendering is to be energy preserving, which is different from loudness preservation.

Figure 8.15(a) shows a comparison of loudness measurements for several object-based audio clips being rendered to different output formats from stereo to an immersive 7.1.4 channel layout. The measurements were done using the recently revised version of ITU-R BS.1770 as shown in Figure 8.14. As can be observed, an energy-preserving renderer provides a good baseline for loudness preservation across different rendering configurations (within 2.5 dB from the object-based measurement). It can be expected that the measured loudness generally increases as the number of playback channels decreases, being maximum for stereo. This is caused by having more objects summing up electrically into fewer output channels, therefore departing from the model that different speakers' outputs sum acoustically in a way closer to an energy-preserving model.

A solution to achieve better timbral and loudness preservation would be to perform frequency-dependent rendering where the renderer transitions from energy-preserving panning to amplitude-preserving panning as the frequency decreases (Laitinen et al., 2014). However, this would require a significant increase in rendering cost. Alternatively, the renderer can incorporate level trimming, possibly controlled via metadata that can help achieve both better aesthetic results and

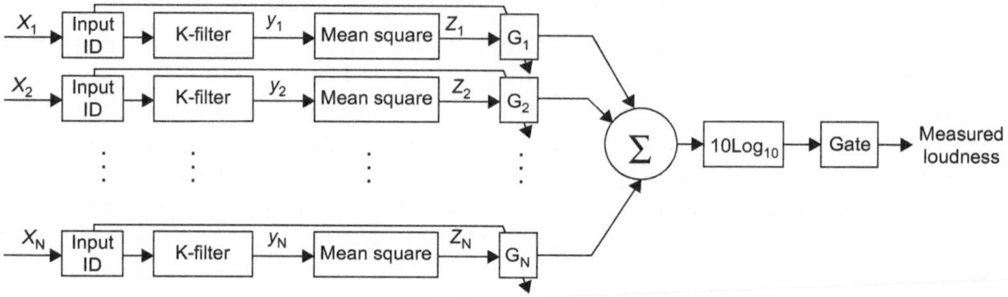

Figure 8.14 Revision of ITU-R BS.1770–4 loudness measurement to include an arbitrary number of channels. This solution could also be used as a basis to measure loudness of object-based content.

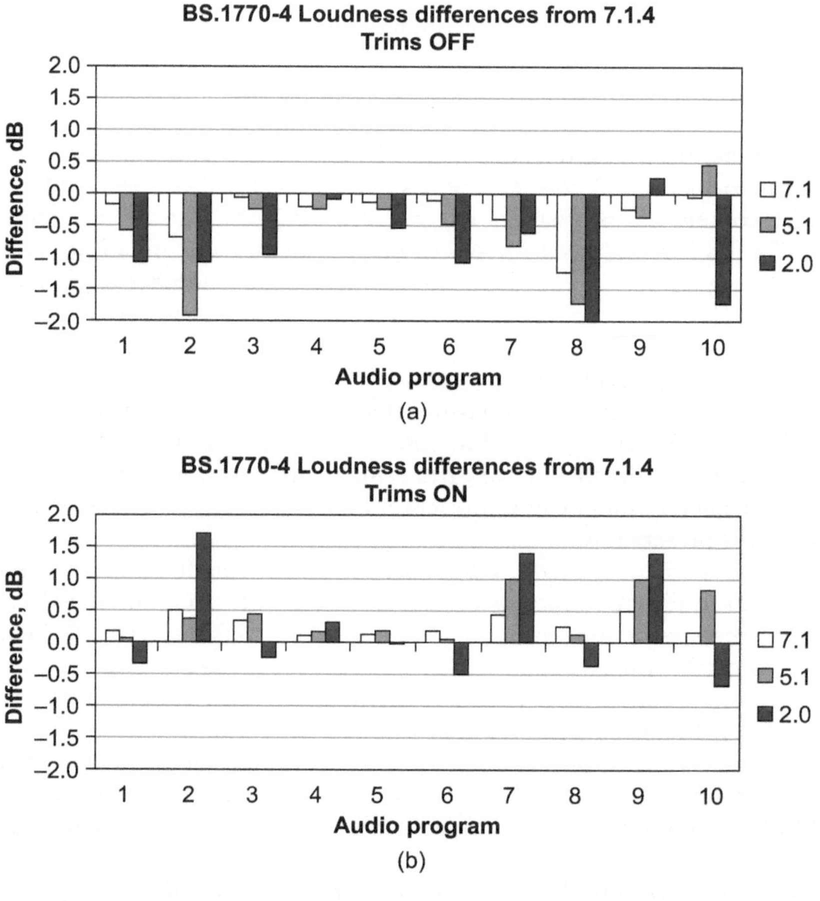

Figure 8.15 (a) ITU-R BS.1770 Loudness differences of a 7.1, 5.1 and stereophonic rendering of an object-based presentation compared to a rendering to an immersive 7.1.4 channel configuration. As expected, loudness differences tend to increase as the number of target channels decreases due to more coherent signal summation. (b) An example where the rendering includes configuration-dependent level trimming for each object, which lets the mixer achieve desired aesthetic goals and also reduces large loudness discrepancies.

loudness preservation, in a way similar to surround downmix coefficients in current channel-based codecs. Figure 8.15(b) illustrates such a mechanism where object-based level trims (a per-object gain that depends on each object's (x,y,z) coordinates as well as the number of playback speakers) are used to alter the core energy-preserving rendering. The trims are applied to the object's audio essence prior to rendering. Such an approach can lead to a reduced loudness discrepancy without increasing the rendering complexity.

Object-Based Program Interchange and Delivery

The spatial coding process described in the section above enables the interchange of countless combinations of spatial object-groups that simplify the enablement of immersive and personalized experiences. Figure 8.16 illustrates example spatial object-groups along with channel-based group program elements envisioned for immersive and/or personalized program interchange (including consumer delivery) represented in blocks A–G. All of the blocks A–G can be generated from a single object-based audio master. In each of the blocks (AC) in Figure 8.16, *N* indicates the number of spatial object-groups, each carrying a monophonic audio signal and metadata. This number is flexible to address a wide range of workflow capabilities and compatibility needs. This approach allows users to tradeoff spatial resolution to meet their business and/or operational needs.

The audio essence and associated metadata for each of the program building blocks can be carried e.g., in the newly defined ITU recommendation for the broadcast WAVE file format and audio definition model (BWAV/ADM) (ITU-R BS 2076–0, 2015; ITU-R BS 2088–0, 2015) The audio essence can be either as encoded as linear PCM or in a mezzanine compressed format such as Dolby E with extensible metadata delivery format (EMDF) extensions (ESTI TS 102 366

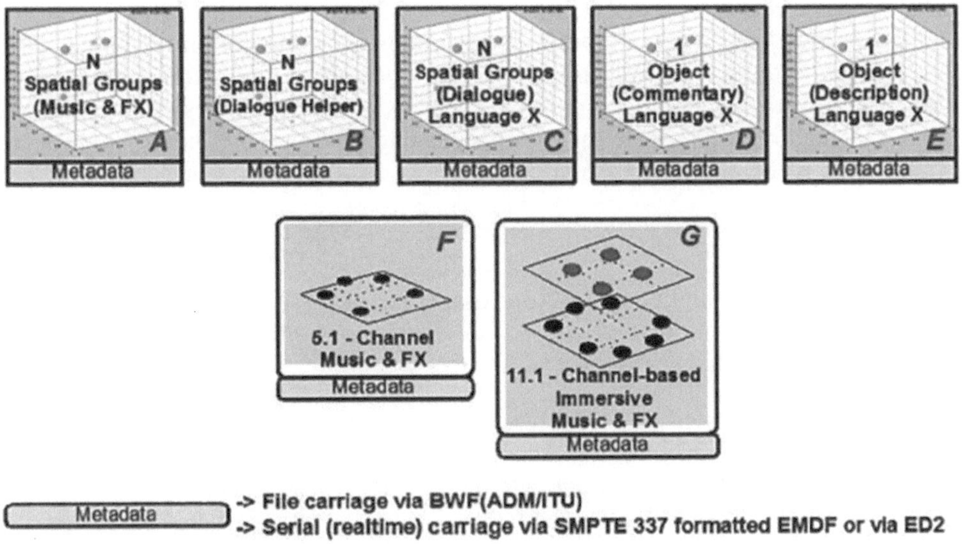

Figure 8.16 Example building-blocks for complex object-based or hybrid channel and object-based immersive programs.

Annex H, 2014). For live applications, the EMDF metadata can also be encoded over AES digital audio interfaces using the SMPTE 337 standard (2015). In addition to formats suitable for traditional audio production, efforts have also been ongoing to standardize richer file formats, such as Interactive XMF (2008) that could be used for interchange and playback in the context of video games and interactive applications in the future.

In addition to the standardization of new professional formats for interchanging audio objects, new commercial formats have also been recently introduced for the home delivery of object-based content. These formats leverage current generation codecs (e.g., Dolby Digital Plus, Dolby TrueHD or DTS Master Audio) using a backwards-compatible core and extension layers. Object-based audio is also an essential component of next-generation codecs such as Dolby AC-4 and MPEG-H, which are now part of the next generation ATSC 3.0, ETSI (2015) and DVB broadcasting standards. These new codecs dramatically broaden the reach of object-based representations and their applications into consumer devices.

Conclusion

Historically limited to interactive gaming or production, object-based audio is now extending throughout the audio industry and all the way to consumers thanks to new tools, codecs and standards. Object-based audio mixing and delivery concepts strongly resonate with the content creation community. Objects are fast becoming the preferred approach to mix and deliver soundtracks in cinemas (9 out of 10 Oscar-nominated movies for sound mixing and editing in 2016 were mixed in an object-based format) and is also making fast progress in the broadcast industry. New game engines and consoles include object-based audio output to address increasingly flexible playback configurations in the home, which may include 3D speaker layouts or virtualized ceiling speakers. Audio objects are also making their first foray into music production and delivery and live DJ-ing and will be critical to emerging applications such as cinematic virtual reality or augmented reality as they become more accessible to the consumers.

More importantly, the growing adoption of object-based production, interchange and delivery throughout the audio industry brings unprecedented future-proofing and evolution capabilities. As rendering or capture techniques improve, the new and upcoming standardized audio and metadata infrastructures will enable these improvements to be readily heard by wide audiences at much reduced cost for the content creators.

Acknowledgments

The author thanks Jeff Riedmiller, Scott Norcross, Charles Robinson, Sripal Mehta, Dan Darcy and Poppy Crum for their input and contributions to this work, as well as the greater sound technology team at Dolby Laboratories.

References

Ahonen, J. (2013). *Microphone Front-ends for Spatial Sound Analysis and Synthesis with Directional Audio Coding*, Doctoral Thesis, Department of Signal Processing and Acoustics, Aalto University.

Barber, C. B., Dobkin, D. P., & Huhdanpaa, H. (1996). The quickhull algorithm for convex hulls. *ACM Transactions on Mathematical Software*, 22(4), 469–483.

Bargen, B., & Donelly, P. (1998). *Inside Direct X*. Microsoft Press.
Begault, D. R. (1994). *3D Sound for Virtual Reality and Multimedia*. Academic Press Professional.
Begault, D. R., & Rumsey, F. (2004). *An Anthology of Articles on Spatial Sound Techniques: Part 2—Multichannel Audio Technologies*. New York: Audio Engineering Society.
Best, V., van Schaik, A., Jin, C., & Carlile, S. (2005). Auditory spatial perception with sources overlapping in frequency and time. *Acta Acustica united with Acustica*, *91*(8), 421–428.
Brungart, D. S., Simpson, B. D., & Kordik, A. J. (2005). Localization in the presence of multiple simultaneous sounds. *Acta Acustica United with Acustica*, *91*(9), 471–479.
Cengarle, G., Mateos, T., Olaiz, N., & Arumı, P. (2010). A new technology for the assisted mixing of sport events: Application to live football broadcasting. *Proceedings of the 128th Audio Engineering Society Convention*, Barcelona, Spain.
Daniel, J., & Moreau, S. (2004). Further study of sound field coding with higher order Ambisonics. *Proceedings of the 116th Audio Engineering Society Convention*, Berlin, Germany.
de Vries, D. (2009). *Wave Field Synthesis*. AES monograph.
Dickins, G., Flax, M., McKeag, A., & McGrath, D. (1999). Optimal 3D-speaker panning. *Proceedings of the 16th AES International Conference, Spatial Sound Reproduction*. Rovaniemi, Finland.
EAX. (2004). Environmental Audio Extensions 4.0, Creativeqc. Retrieved from www.soundblaster.com/eaudio.
ETSI. (2014). Digital Audio Compression (ac-3, enhanced ac-3) Standard, ESTI TS 102 366 Annex H.
ETSI. (2015). Digital Audio Compression (ac-4) Standard part 2: Immersive and Personalized Audio, ESTI TS 103 190–2.
Faller, C., Favrot, A., Langen, C., Tournery, C., & Wittek, H. (2010). Digitally enhanced shotgun microphone with increased directivity. *Proceedings of the 129th Audio Engineering Society Convention*, San Francisco, USA.
FMOD Music and Sound Effects System. Retrieved from www.fmod.org.
Funkhouser, T., Jot, J. M., & Tsingos, N. (2002). *Sounds Good to Me! Computational Sound for Graphics, vr, and Interactive Systems*. Siggraph 2002 course #45, 2002.
Furness, R. K. (1990). Ambisonics—An overview. *Proceedings of the 8th Audio Engineering Society Conference*. Washington, DC.
Gallo, E., Lemaitre, G., & Tsingos, N. (2005). Prioritizing signals for selective real- time audio processing. *Proceedings of International Conference on Auditory Display (ICAD)*. Limerick, Ireland.
Gallo, E., & Tsingos, N. (2007). Extracting and re-rendering structured auditory scenes from field recordings. *Proceedings of the 30th Audio Engineering Society International Conference on Intelligent Audio Environments*, Saariselka, Finland.
Gallo, E., Tsingos, N., & Lemaitre, G. (2007). 3D-Audio matting, post-editing and re-rendering from field recordings. *EURASIP Journal on Applied Signal Processing, Special Issue on Spatial Sound and Virtual Acoustics*.
Garity, W. E., & Hawkins, J. N.A. (1941). Fantasound. *Journal of the Society of Motion Picture Engineers*, 37.
Herder, J. (1999). Optimization of sound spatialization resource management through clustering. *The Journal of Three Dimensional Images, 3D-Forum Society*, *13*(3), 59–65.
Herre, J., & Disch, S. (2007). New concepts in parametric coding of spatial audio: From SAC to SAOC. *Proceedings of the IEEE International Conference on Multimedia and Expo*, Beijing, China.
Herre, J., Hilpert, J., Kuntz, A., & Plogsties, J. (2015). MPEG-H 3D audio-The new standard for coding of immersive spatial audio. *IEEE Journal of Selected Topics in Signal Processing*, *9*(5).
High Dynamic Range Audio in the Frostbite Game Engine. (2009). Retrieved from www.frostbite.com/2009/04/how-hdr-audio-makes-battlefield-bad-company-goboom/
Interactive Audio Special Interest Group (IASIG), 3D Audio Working Group. (2016). Retrieved from www.iasig.org/.

Interactive XMF File Format. (2008). Retrieved from www.iasig.org/wg/ixwg. Format for non-pcm audio and data in an aes3 serial digital audio interface (2015). SMPTE ST 337:2015, 1–17.

International Telecom: Union. *Method for the subjective assessment of intermediate quality level of coding systems*. Recommendation ITU-R BS.1534–1, 2001–2003.

International Telecom: Union. *Advanced Sound System for Programme Production*. Recommendation ITU-R BS.2051–0, 2014.

ITU-R. Algorithms to measure audio programme loudness and true-peak audio level, ITU-R BS 1770–4. 2015.

ITU-R. Audio definition model, ITU-R BS 2076–0. 2015.

ITU-R. Long-form file format for the international exchange of audio programme materials with metadata, ITU-R BS 2088–0. 2015.

Jot, J. M., Ray, L., & Dahl, L. (1998). *Extension of Audio BIFS: Interfaces and Models Integrating Geometrical and Perceptual Paradigms for the Environmental Spatialization of Audio*. ISO Standard SO/IEC JTC1/SC29/WG11 M.

Kayser, C., Petkov, C., Lippert, M., & Logothetis, N. K. (2005). Mechanisms for allocating auditory attention: An auditory saliency map. *Current Biology*, *15*, 1943–1947.

Kelly, M. C., & Tew, A. I. (2002). The continuity illusion in virtual auditory space. *Proceedings of the 112th Audio Engineering Society Convention*. Munich, Germany.

Klatzky, R. L. (1998). *Allocentric and Egocentric Spatial Representations: Definitions, Distinctions, and Interconnections: Spatial Cognition*. Berlin Heidelberg: Springer-Verlag.

Kostadinov, D., Reiss, J., & Mladenov, V. (2010). Evaluation of distance based amplitude panning for spatial audio. *Proceedings of ICASSP2010*, 285–288.

Laitinen, M.-V., Vilkamo, J., Jussila, K., Politis, A., & Pulkki, V. (2014). Gain normalization in amplitude panning as a function of frequency and room reverberance. *Proceedings of the 55th International Audio Engineering Society Conference on Spatial Audio*.

Lossius, T., Baltazar, P., & de la Hogue, T. (2009). DBAP—distance-based amplitude panning. *Proceedings of the International Conference on Computer Music (ICMC)*. Montreal, Canada.

Malham, D. G., & Myatt, A. (1995). 3D sound spatialization using ambisonic techniques. *Computer Music Journal*, *19*(4), 58–70.

Mann, M., Churnside, A., Bonney, A., & Melchior, F. (2013). Object-based audio applied to football broadcasts. *Proceedings of the 2013 ACM International Workshop on Immersive Media Experiences, ImmersiveMe '13*, 13–16. New York, NY.

Merimaa, J. (2002). Applications of a 3D microphone array. *Proceedings of the 112th Audio Engineering Society Convention,* Munich, Germany.

Meyer, J., & Elko, G. (2004). Spherical microphone arrays for 3D sound recording. In Yiteng (Arden) Huang & Jacob Benesty (Eds.), *Audio Signal Processing for Next-Generation Multimedia Communication Systems* (Chap. 2). Boston: Kluwer Academic Publisher.

Moore, B. C. J. (1997). *An Introduction to the Psychology of Hearing* (4th ed.). San Diego, CA: Academic Press.

Moore, B. C. J., Glasberg, B., & Baer, T. (1997). A model for the prediction of thresholds, loudness and partial loudness. *Journal of the Audio Engineering Society*, *45*(4), 224–240. Retrieved from http://hearing.psychol.cam.ac.uk/Demos/demos.html

OpenAL: An Open Source 3D Sound Library. (2000). Retrieved from www.openal.org.

Painter, E. M., & Spanias, A. S.(2000). Perceptual coding of digital audio. *Proceedings of the IEEE*, *88*(4), 451–515.

Potard, G., & Burnett, I. (2004). Decorrelation techniques for the rendering of apparent source width in 3D audio displays. *Proceedings of the 7th International Conference on Digital Audio Effects (DAFX'04)*. Naples, Italy.

Pulkki, V. (1997). Virtual sound source positioning using vector base amplitude panning. *Journal of the Audio Engineering Society, 45*(6), 456–466.

Pulkki, V. (1999). Uniform spreading of amplitude panned virtual sources. *Proceedings of the IEEE Workshop on Applications of Signal Processing to Audio and Acoustics*. New Paltz, NY.

Pulkki, V. (2006). Directional audio coding in spatial sound reproduction and stereo upmixing. *Proceedings of the 28th AES International Conference*. Pitea, Sweden.

Pulkki, V., Karjalainen, M., & Valimaki, V. (1999). Localization, coloration, and enhancement of amplitude-panned virtual sources. *Proceedings of the 16th AES International Conference on Spatial Sound Reproduction*. Rovaniemi, Finland.

Rayburn, R. A. (2012). *Eargle's Microphone Book*. Amsterdam: Focal Press.

Riedmiller, J., Mehta, S., Tsingos, N., & Boon, P. (2015). Immersive and personalized audio: A practical system for enabling interchange, distribution, and delivery of next-generation audio experiences. *Motion Imaging Journal, SMPTE, 124*(5), 1–23.

Roads, C. (1996). *The Computer Music Tutorial*. Cambridge: MIT Press.

Robinson, C., & Tsingos, N. (2001). Cinematic sound scene description and rendering control. *Annual Technical Conference Exhibition, SMPTE 2014*, 1–14.

Robinson, C., Tsingos, N., & Mehta, S. (2012). Scalable format and tools to extend the possibilities of cinema audio. *SMPTE Motion Imaging Journal, 121*(8).

Rumsey, F. (2001). *Spatial Audio*. US: Taylor & Francis.

Savioja, L., Huopaniemi, J., Lokki, T., & Vaananen, R. (1999). Creating interactive virtual acoustic environments. *Journal of the Audio Engineering Society, 47*(9), 675–705.

Sibbald, A. (2001). *MacroFX Algorithm: White Paper*. Retrieved from www.sensaura.co.uk/whitepapers

Soundfield (2016). Soundfield microphones. Retrieved from www.soundfield.com.

Tsingos, N. (2001). *Artifact-free Asynchronous Geometry-based Audio Rendering*. ICASSP'2001, Salt Lake City, USA.

Tsingos, N., Gallo, E., & Drettakis, G. (2004). Perceptual audio rendering of complex virtual environments: ACM Transactions on Graphics. *Proceedings of SIGGRAPH 2004*.

Tsingos, N., & Gascuel, J.-D. (1997). Soundtracks for computer animation: Sound rendering in dynamic environments with occlusions. *Proceedings of Graphics Interface'97*, pp. 9–16.

Tsingos, N., Govindaraju, P., Zhou, C., & Nadkarni, A. (2016). XY-stereo capture and upconversion for virtual reality. *AES International Conference on Augmented and Virtual Reality*. Los Angeles.

Tsingos, N., Robinson, C., Darcy, D., & Crum, P. (2014). Evaluation of panning algorithms for theatrical applications. *Proceedings of the 2nd International Conference on Spatial Audio (ICSA)*. Erlangen, Germany.

Viers, R. (2012). *The Location Sound Bible*. Studio City, CA: Michael Wiese Productions.

Vilkamo, J., Lokki, T. & Pulkki, V. (2009). Directional audio coding: Virtual microphone based synthesis and subjective evaluation. *Journal of Audio Engineering Society, 57*(9), 709–724.

Vries, D. de. (2009). *Wave Field Synthesis*. AES monograph.

Wand, M., & Straßer, W. (2004). Multi-resolution sound rendering. *Symposium on Point-Based Graphics*, Zurich, Switzerland.

Warren, J., Schaefer, S., Hirani, A., & Desbrun, M. (2007). Barycentric coordinates for convex sets. *Advances in Computational Mathematics, 27*(3), 319–338.

WWISE by Audiokinetic. Retrieved from www.audiokinetic.com/products/wwise.

Yu, X., Hu, D., & Xu, J. (eds.) (2014). *Blind Source Separation: Theory and Applications*. Singapore: Wiley.

Chapter 9

Sound Field

Rozenn Nicol

Introduction

In general, stereophony and multichannel surround sound can be defined as *loudspeaker* and *listener-centric* channel-based methods, wherein sound reproduction is based on a specific set of audio channels associated for a given loudspeaker setup. Using these systems, each channel is contributing to a focused sound image for a listener located in the sweet spot. As opposed to stereophony and multichannel surround sound, the sound field approach is based on a non-speaker-centric physical representation of the sound waves.

The term *sound field* refers to the capture, reproduction and description of *sound waves*. This is in contrast to binaural, stereo or surround sound systems, where the objective is to create perceived *object(s)*, or *auditory event(s)*. This directional information is interpreted as spatial properties by the auditory system. With the sound field approach, the properties that are controlled to create or to reproduce sounds are the *physical* properties of sound waves, whereas using binaural, stereo or surround sound techniques, the properties under control are at the perceptual level. Sound field properties are linked to all of the acoustic phenomena encountered by the sound wave from its point of origin to its point of observation.

Free-field propagation is the simplest case where the sound wave propagates in a straight line (at least for atmospheric propagation over a reasonably short distance). The source directivity and the distance of propagation, which causes a delay of the arrival time and a decrease of amplitude, are the most critical properties. These properties define the *direct wave*. Inside a room, or in the presence of any obstacle, the sound wave is affected by acoustic reflections, diffusion and scattering and/or diffraction, which result in the addition of a set of modified and delayed copies of the direct wave. Thus, a sound field is the superposition of all these components (i.e. direct wave, reflected wave, diffuse wave, diffracted wave, etc.). These components can be captured together by taking the impulse response of a room. Each component is characterized by several parameters including arrival time, frequency content and incidence angle. These parameters describe the acoustic and geometric properties of the sound source(s) and its environment.

This chapter will be divided into five parts. First, general ideas about the sound field approach and its development, starting from coincident stereo microphone recording techniques introduced by Blumlein to Ambisonics and High Order Ambisonics (HOA), will be presented. Then capture methods, recording formats and reproduction of sound fields will be discussed in further detail. The physical and mathematical tools in connection of sound fields will be given last as further insight.

Development of the Sound Field

Starting Point: The X/Y and M/S Techniques

The initial development of the sound field approach can be traced to the pioneering work of Blumlein on X/Y and the M/S techniques for stereophonic recording (Blumlein, 1931). The X/Y pair is composed of two directional microphones, which are arranged perpendicularly. The M/S pair is composed of one microphone facing forward (the "Mid" or M signal) and one figure-of-eight (i.e. bidirectional) microphone (the "Side" or S signal) along the left/right axis. In each case, the two microphones are theoretically coincident, but in practice one is put above the other. Since a cardioid microphone can be seen as the combination of one pressure microphone (i.e. omnidirectional) and one figure-of-eight microphone, it is easily shown that the X/Y pair of cardioid microphones and an M/S configuration (with a cardioid for the M microphone) are equivalent (Hibbing, 1989).

The Mid microphone picks up the frontal components of the sound field, whereas the Side microphone gets the lateral components. By convention the left components have a positive phase, and the right ones a negative phase. Thus the two microphones achieve a kind of amplitude and phase encoding of the direction of the sound source. At the reproduction step, this spatial encoding is used to restitute the location of all the components of the sound field. Since the left components picked up by the S microphone are in phase with the frontal components, summing the outputs of the M and S microphones leads to extract the left part of the sound field. On the contrary, since the right components are in opposite phase with the frontal ones, they are eliminated. In the same way, if the difference of the M and S outputs is computed instead of their sum, the right components are extracted and the left ones eliminated. The left (L) and right (R) signals for a stereophonic reproduction are thus derived from the M and S signals by the following matrix process:

$$\begin{aligned} L &= M + S \\ R &= M - S \end{aligned} \tag{9.1}$$

This equation summarizes how spatial information (namely left/right separation) can be extracted from the M/S signals. The M/S pair can be understood as a first level of sound field recording, which is restricted to the horizontal plane. For a full discussion on the X/Y and M/S techniques, see Chapter 3.

Ambisonics

Based on the previous work of Cooper, Shiga and Bruck, Michael Gerzon presented a system that would treat all directional sounds equally, representing both horizontal and vertical sounds. Gerzon (1973) believed that this could be achieved by recording values of sound pressure on

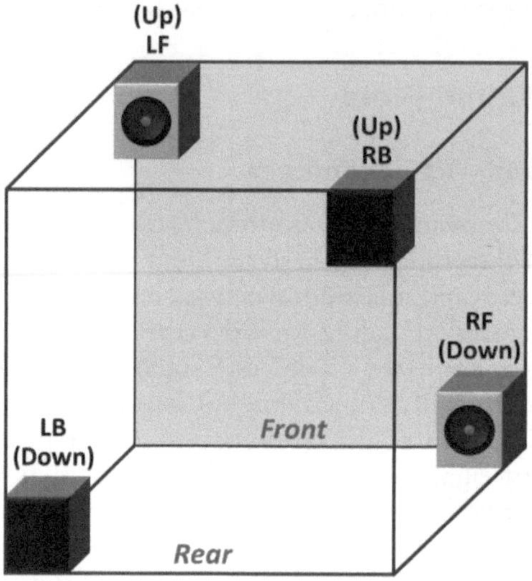

Figure 9.1 Tetrahedral loudspeaker layout embedded in a cube (from Gerzon, 1973).

the surface of a sphere equally from all directions. Realizing the practical limitations of recording, broadcast and playback systems of the day, Gerzon suggested that a minimal number of channels and speakers could be used to fulfill the basic psychoacoustic requirements to perceive both horizontal and vertical sound by placing the listener in a space bounded by loudspeakers defined in a cube (see Figure 9.1). He acknowledged that by increasing the number of spherical harmonics, the directional resolution of the system is improved; however, he desired a system that was practical. Gerzon later introduced Ambisonics Technology as an alternative to channel-based stereo and surround systems. He defined a 4-channel system (using signals X, Y, Z and W discussed later in this chapter) that was compatible with stereo and quadraphonic surround playback systems yet could achieve, in his words, a "full spherical portrayal of directionality" (Gerzon, 1985, p. 859).

First Order Sound Field Capture

Acoustic pressure and particle velocity are two physical variables that are essential to fully describe a sound wave. The former is recorded by a pressure microphone, the latter typically by a ribbon figure-of-eight microphone. Sound field recording techniques lead to a full and accurate representation of sound waves.[1] The first consequence is that the input signals of the loudspeakers are not always the discrete output signals of the microphones. A processing step (for example using an M/S matrix or an Ambisonics decoder) is required to extract the loudspeaker signals from the microphone signals. A second consequence is that sound reproduction is not limited to one single point, but can be extended to a large area, the boundary of which is determined by the accuracy of the sound field information recorded. Another important difference between sound field and conventional multichannel surround sound is that for sound field systems, all directions are equally considered without any frontal focus.

The tetrahedral microphone was proposed by Craven and Gerzon (Craven & Gerzon, 1977; Farrar, 1979a,b). It is composed of four cardioid[2] microphones arranged in a tetrahedron (see Figure 9.2). The fundamental objective of the tetrahedral microphone is to acquire the components (W, X, Y, Z), which represent the 0th and 1st order components of the Spherical Harmonics expansion of the sound field. The most intuitive way to record these signals is to use one omnidirectional microphone (for the 0th order component W) and three figure-of-eight microphones (for the 1st order components X, Y and Z), which are pointed respectively along the x-, y- and z-axis. Four cardioid microphones arranged in a tetrahedron can be used instead, especially when the coincident setup of four microphones (three figure-of-eights and one omnidirectional) is not practical. The tetrahedral microphone (Craven & Gerzon, 1977) provides an elegant solution. It should be noted that the tetrahedral microphone does not directly deliver the components (W, X, Y, Z). A matrixing process (expressed by Equation 9.2) is required to derive these latter components from its output signals (LF, RF, LB, RB). This step is commonly referred to as the *encoding*.

Similar to the X/Y pair, each cardioid microphone can be decomposed into one omnidirectional and one figure-of-eight capsule. By appropriate recombination of the resulting elements, it is shown that the tetrahedral microphone is equivalent to one omnidirectional capsule (W component) and three figure-of-eight capsules (X, Y and Z components), each one oriented along respectively the x-axis, the y-axis and the z-axis (Farrar, 1979a), see Figure 9.3. All the capsules are theoretically coincident. The four cardioid microphones are referred to as LF (for Left Front), RF (for Right Front), LB (for Left Back) and RB (for Right Back), and the components (W, X, Y, Z) are expressed as a linear sum of the signals (LF, RF, LB, RB), as follows:

$$\begin{aligned} W &= LF + LB + RF + RB \\ X &= LF - LB + RF - RB \\ Y &= LF + LB - RF - RB \\ Z &= LF - LB - RF + RB \end{aligned} \tag{9.2}$$

The components (W, X, Y, Z) allow us to reformulate the spatial analysis performed by the tetrahedral microphone. The W component corresponds to an omnidirectional recording of the sound field, and is a kind of “0th-order” analysis of spatial information. The X, Y and Z components provide respectively front/back, left/right and top/bottom separation. In the same way, the X/Y and M/S setups can be recomposed as the combination of one omnidirectional capsule and two figure-of-eight capsules, each one oriented along respectively the x-axis and the

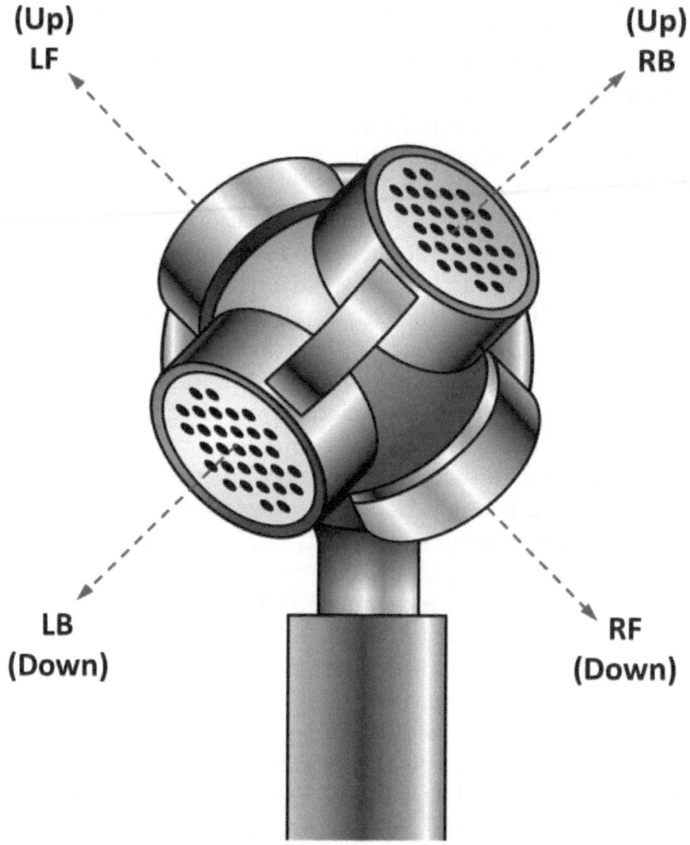

Figure 9.2 Illustration of the tetrahedral microphone.

y-axis. Therefore the X/Y and M/S can be really considered as the 2D restriction of sound field recording.

As previously highlighted, a sound field has some specific features. First, it is composed of many components (i.e. direct and reflected waves), which all have temporal, frequency and spatial properties. Second, a sound field is defined intrinsically over an extensive area. As mentioned earlier, a minimum of two microphones is required to extract some spatial differential. More generally, a sound field can be captured by spatial sampling using a microphone array. However, this solution raises several problems. Analogous to time sampling, the minimum sampling rate is determined by the maximal spatial frequency contained in the sound field. For time sampling, Shannon theorem recommends to take at least two samples per period. In the same way, for

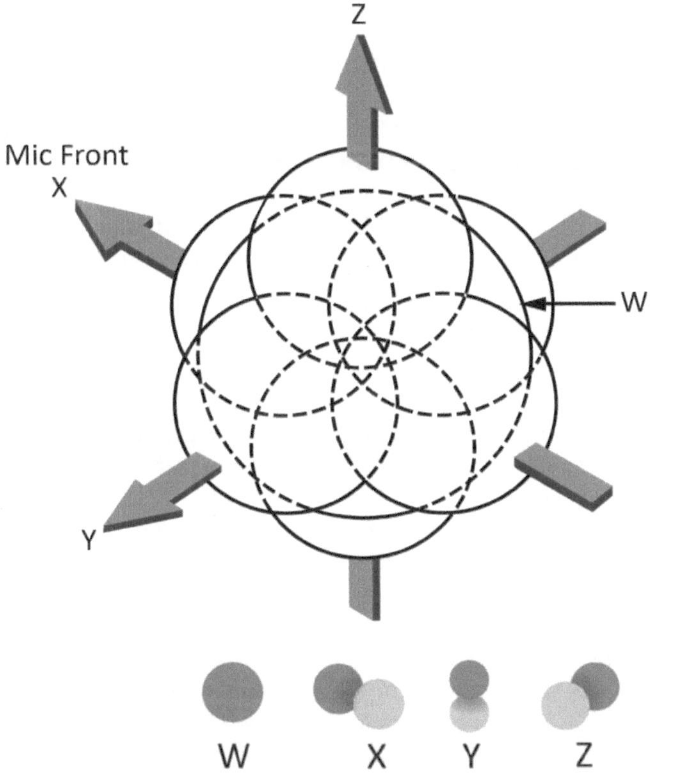

Figure 9.3 Illustration of the components (W, X, Y, Z).

spatial sampling, two samples are needed per wavelength. For instance, at a frequency of 17 kHz, assuming that the speed of sound is 340 m/s, the wavelength will be 2 cm. Thus, to properly sample frequencies as high as 17 kHz, a minimum spacing of 1 cm is required. Consequently, to cover an extensive area will require a huge number of microphones. For these reasons, ideal spatial sampling of a sound field is hardly ever feasible.

A promising alternative is sparse sampling by an irregular array of microphones. Methods of source separation and source localization are then required to extract all the sound components and the associated spatial information from the microphone signals (Gallo, Tsingos & Lemaitre, 2007). Current research indicates that the required post-processing is likely to cause audible artifacts. Most of the time, the audio quality of these systems does not meet the requirements of commercial sound engineers, but the related signal processing methods are constantly being improved.

As introduced in the previous section, the tetrahedral microphone, which is essentially a 3D extension of X/Y and M/S microphone techniques, offers an effective and practical solution to record a sound field. Its concept can be explained in several ways. Intuitively, since the tetrahedral microphone is composed of four cardioid microphones regularly distributed over a sphere, the direction of each sound component is encoded by the differences in amplitude between the four capsules. In a more physical way, it is shown that the microphone outputs can be recombined to form the (W, X, Y, Z) signals, which correspond to the first components of the Spherical Harmonics expansion of the sound field (see Equations 9.39 and 9.41). As it will be explained in the next section, with Higher Order Ambisonics (HOA), this concept is generalized to spherical arrays composed of a higher number of microphones, leading to the extraction of the Spherical Harmonics components of higher orders.

First Order Sound Field Reproduction

In the case of the M/S technique, we have seen that a matrix process (see Equation 9.1) is required to obtain the loudspeaker signals. A similar processing is needed for Ambisonics reproduction. The loudspeaker inputs are obtained as a weighted sum of the components (W, X, Y, Z). This matrix process is called *decoding* in Ambisonics terminology. Because Ambisonics is not a channel-centric reproduction system, it is not constrained to one single and standardized setup of loudspeakers. On the contrary, various loudspeaker layouts can be used, both in terms of the number and the position of the loudspeakers. In addition, for each configuration, there are several ways to decode the components (W, X, Y, Z) into the loudspeaker signals. This flexibility of sound field reproduction is a major advantage of Ambisonics. Ambisonics signals are even compatible with monophonic or stereophonic reproduction. The W component is equivalent to a monophonic recording of the sound field and is therefore useful for monophonic reproduction.

For stereophonic reproduction, there are several ways to matrix the signals (W, X, Y, Z) into *plausible* stereophonic signals. One solution is to derive the signals (M_v, S_v) corresponding to a virtual M/S pair:

$$\begin{aligned} M_v &= \frac{W}{\sqrt{2}} + X \\ S_v &= Y \end{aligned} \tag{9.3}$$

Then Equation 9.1 is used to compute stereophonic signals like for a real M/S recording. In the same way, a virtual X/Y recording can be simulated as:

$$\begin{aligned} X_v &= \frac{X + Y}{\sqrt{2}} \\ Y_v &= \frac{X - Y}{\sqrt{2}} \end{aligned} \tag{9.4}$$

The signals (X_v, Y_v) directly feed the left and right loudspeakers.

For explicit Ambisonics reproduction, the decoding is less intuitive and depends on the loudspeaker layout. Even though the number and the position of the loudspeaker can be freely chosen, some rules (mainly common sense) must be satisfied. First, since spatial information is represented by four signals (i.e. the components W, X, Y, Z), at least four loudspeakers are required (as shown in Figure 9.1). Second, the more regular the layout is, the simpler (and more robust) the decoding is. To reproduce the full 3D spatial information, a 3D array of loudspeakers is needed. A regular layout means that the loudspeakers are arranged as the vertices of a regular polyhedron, which correspond to a regular sampling of a sphere. The simplest example is a cube. It is also possible to focus on spatial information in the horizontal plane, which leads to 2D Ambisonics reproduction, and for which only the (W, X, Y) components are needed. In that case, a regular layout corresponds to a circular array of equidistant loudspeakers. The simplest example is a square.

To illustrate the decoding process, we will examine the following two examples (see Figure 9.4): 2D reproduction over a square setup (i.e. four loudspeakers arranged in a square centered around the listener) and 3D reproduction over a cube setup (i.e. eight loudspeakers arranged as the vertices of a cube). The general idea of Ambisonics reproduction is that the spatial information is remapped over the loudspeaker array, in order that the sum of the contributions of all the loudspeakers properly reconstruct the sound field in the listening area. The input signal of each loudspeaker is therefore a weighted sum of the components (W, X, Y, Z). The weights are defined as a function of the loudspeaker position. Several methods have been proposed to compute the decoding matrix (Gerzon, 1992; Daniel, 2001), which derives the loudspeaker signals from the components (W, X, Y, Z). Each method corresponds to specific properties of the sound field reconstruction. These aspects will be detailed later. Here we will present only one method, which is called basic decoding and, in which the weights are simply the spatial coordinates of the loudspeakers. The loudspeaker located in the direction (φ_l, θ_l) is thus fed by the signal:

$$L_l = \frac{1}{N_L}\left(\frac{W}{\sqrt{2}} + X \cos\phi_l \cos\theta_l + Y \sin\phi_l \cos\theta_l + Z \sin\theta_l\right) \tag{9.5}$$

For a square setup, the four loudspeaker signals are obtained as:

$$\begin{aligned} L_{RF} &= \frac{1}{\sqrt{2}}(W + X - Y) \\ L_{LF} &= \frac{1}{\sqrt{2}}(W + X + Y) \\ L_{LB} &= \frac{1}{\sqrt{2}}(W - X + Y) \\ L_{RB} &= \frac{1}{\sqrt{2}}(W - X - Y) \end{aligned} \tag{9.6}$$

where the signals L_{RF}, L_{LF}, L_{LB}, L_{RB} refer respectively to the right front, the left front, the left back and the right back loudspeakers. In the same way, for a cube setup, the eight loudspeakers are fed by:

$$\begin{aligned}
L_{RFU} &= \frac{W}{\sqrt{2}} + \frac{1}{2}(X - Y) + \frac{Z}{\sqrt{2}} \\
L_{LFU} &= \frac{W}{\sqrt{2}} + \frac{1}{2}(X + Y) + \frac{Z}{\sqrt{2}} \\
L_{LBU} &= \frac{W}{\sqrt{2}} + \frac{1}{2}(-X + Y) + \frac{Z}{\sqrt{2}} \\
L_{LBD} &= \frac{W}{\sqrt{2}} + \frac{1}{2}(X - Y) - \frac{Z}{\sqrt{2}} \\
L_{LFD} &= \frac{W}{\sqrt{2}} + \frac{1}{2}(X + Y) - \frac{Z}{\sqrt{2}} \\
L_{LBD} &= \frac{W}{\sqrt{2}} + \frac{1}{2}(-X + Y) - \frac{Z}{\sqrt{2}} \\
L_{RBD} &= \frac{W}{\sqrt{2}} - \frac{1}{2}(X + Y) - \frac{Z}{\sqrt{2}}
\end{aligned} \tag{9.7}$$

where the letters "U" and "D" refer respectively to the upward and downward loudspeakers.

Two localization criteria, introduced by Gerzon, are used to estimate the perceived direction of the virtual sound sources reproduced by a loudspeaker array: the *velocity vector* and the *energy vector* (Gerzon, 1992). These criteria are derived from the localization model of Makita

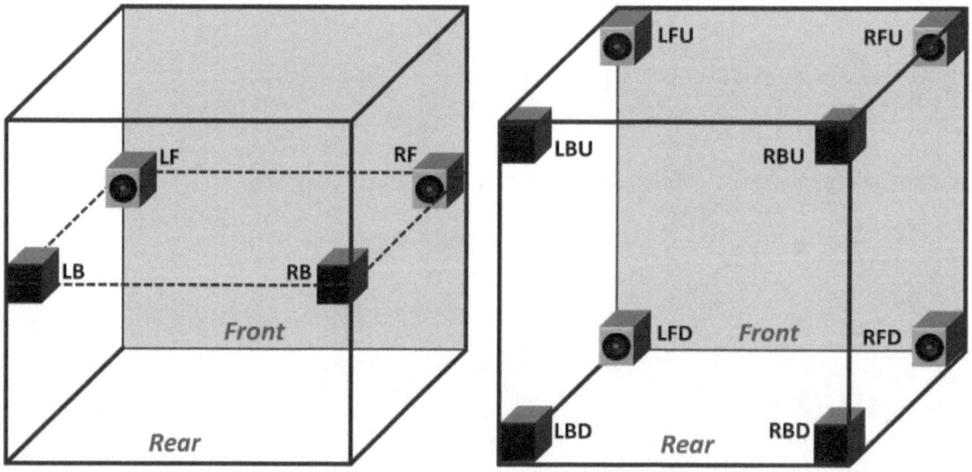

Figure 9.4 Loudspeaker configurations of the Square (left) and Cube (right), as described in Equations 9.6 and 9.7.

for a stereophonic system (Makita, 1962). If the unitary vector $\overrightarrow{x_l}$ refers to the direction of the lst loudspeaker and, $s(l,\omega)$, to its signal, the velocity ($\vec{V}$) and energy ($\vec{E}$) vectors are respectively defined by:

$$\vec{V} = \frac{\sum_{l=1}^{N_l} s(l,\omega)\overrightarrow{x_l}}{\sum_{l=1}^{N_l} s(l,\omega)}, \quad \vec{E} = \frac{\sum_{l=1}^{N_l} |s(l,\omega)|^2 \overrightarrow{x_l}}{\sum_{l=1}^{N_l} |s(l,\omega)|^2} = r_E \overrightarrow{x_E} \tag{9.8}$$

The direction of the pointed vector corresponds to the average direction of the energy arrival and can be interpreted as the spatial barycenter of the reproduced sound field. The norm of the vector reflects its spatial spread; if the norm is close to one, it means that the sound energy is focused on only a few loudspeakers. The velocity vector is a low-frequency criterium in the sense that the phase of the loudspeaker signals is taken into account, which is relevant only for low frequencies. The energy vector may be seen as its high-frequency version, in which the energy of the loudspeaker signals is considered instead, since the auditory system is not sensitive to phase for high frequencies.

The reproduction of a sound field raises some specific issues. The system has to create a complex acoustic wave over an extensive area characterized by various time, frequency and spatial properties. To do this, loudspeaker arrays are required. Each loudspeaker can be seen as a secondary source, which emits a wavelet, so that the sum of all the loudspeaker contributions leads to the reconstruction of the target sound field at any point in the reproduction area. The loudspeakers are considered secondary to the source in the sense that they create only a synthetic copy of the target sound field, as opposed to a real or virtual sound source, which would have created a sound field. The virtual or real sound source is called the primary source for this reason.

Both the amplitude and the phase of the signal feeding each secondary source are controlled to accurately create the proper features of the reproduced sound field. To compute the loudspeaker signals, sound field synthesis, which is defined by either Equation 9.34 or 9.35, can be used. One example is the Wave Field Synthesis (WFS) method (discussed in Chapter 10).

Another more general method is sound field control, where the loudspeaker signals are obtained using the constrained optimization solution. More precisely, the reproduction system is composed of L loudspeakers and M error sensors, which are distributed at control points over the reproduction array. The error is computed as the difference between the target sound field and the sound field synthesized by the loudspeaker array. An error vector is defined as the set of errors evaluated at the location of the M sensors. Ideally the error should be null. The objective is therefore to minimize the error vector with respect to the loudspeaker amplitude (Gauthier & Berry, 2006). The loudspeaker signals are then obtained as the solution that minimizes a given cost function, which is expressed as a function of the quadratic error and may include a regularization term to prevent it from causing any ill-conditioning problems.

Higher Order Ambisonics (HOA)

The tetrahedral microphone provides a tool allowing the capture of full 3D sound field. However, since it is only composed of four capsules, its spatial resolution is low, which means that the discrimination between the sound components is not accurate. Intuitively, it seems relevant to generalize its concept by using a spherical array of microphones with a higher number of sensors. To help this generalization, Bamford and Daniel pointed out the link between Ambisonics and Spherical Harmonics (Bamford, 1995; Daniel, 2001). Spherical Harmonics are spatial functions, which allow one to represent any sound wave as a linear sum of directional components (see Equations 9.25 and 9.34). The omnidirectional component (W) is the Spherical Harmonics of 0th order, whereas the bidirectional ones (i.e. figure-of-eight, namely X, Y, Z) are the three Spherical Harmonics of 1st order. The components of orders higher than 1 have more complex directivity (see Figure 9.5). The (W, X, Y, Z) representation of a sound field can thus be extended by including Spherical Harmonics of higher orders. As the number of microphones increases, it is less and less possible to arrange them in a coincident setup. Furthermore, directional microphones corresponding to the directivity of the Spherical Harmonics of the highest orders do not exist. In a similar way as for the tetrahedral microphone, a convenient solution for spatial sampling of the sound field is a spherical array of cardioid microphones. The spatial information is extracted from the microphone outputs, by a proper matrixing process, which is close to Equation 9.2, leading to the Higher Order Ambisonics (HOA) representation of the sound field, which is the extension of the 1st order Ambisonics (i.e. W, X, Y, Z components) to higher orders.

The number of microphones of the spherical array imposes the maximal order that can be extracted. One example is the Eigenmike® (see Figure 9.6), which is composed of 32 microphones and allows HOA encoding up to the 4th order. For sound reproduction, a second

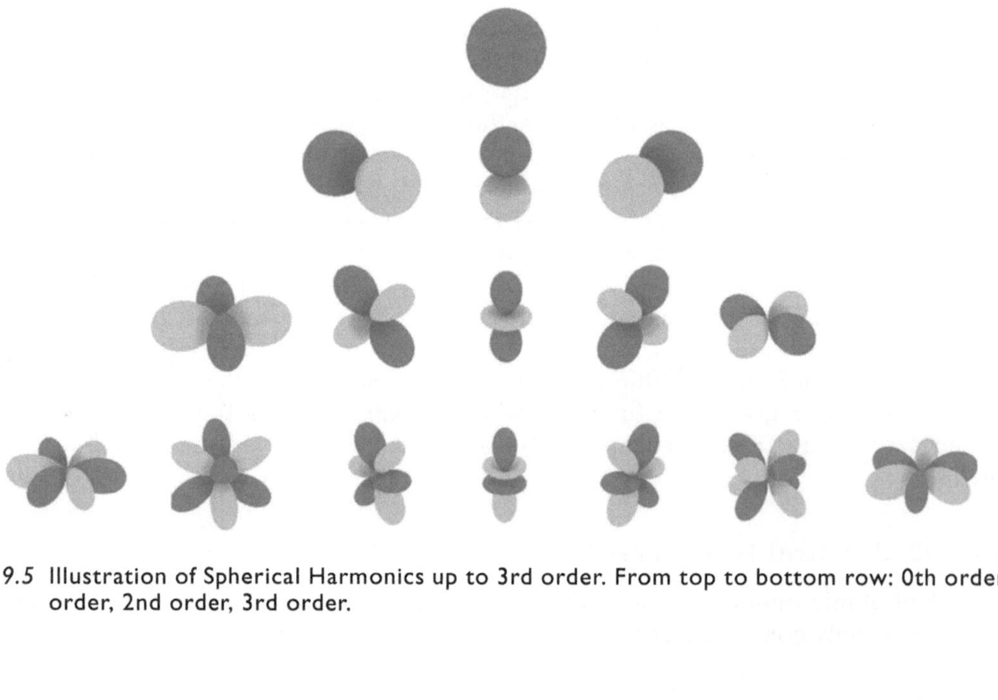

Figure 9.5 Illustration of Spherical Harmonics up to 3rd order. From top to bottom row: 0th order, 1st order, 2nd order, 3rd order.

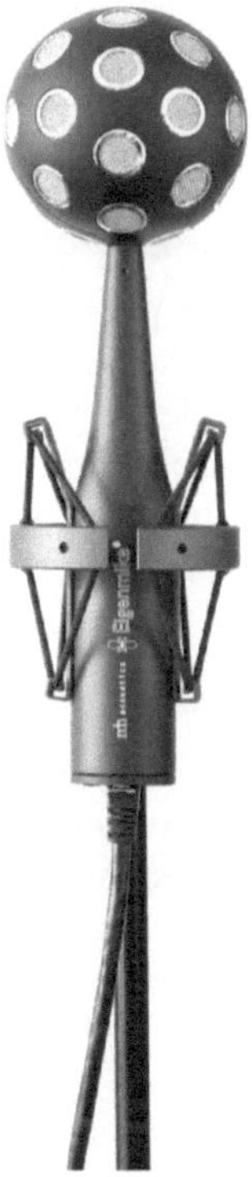

Figure 9.6 em32 Eigenmike® microphone.

step of appropriate decoding is needed to correctly map the spatial information contained in the HOA components to the loudspeaker array, in order to compute the loudspeaker input signals.

In its original definition, Ambisonics is based on the Spherical Harmonics expansion limited to 0th and 1st order components. HOA generalizes this concept by including components of order m greater than 1, as shown in (Bamford, 1995) and (Daniel, 2001). If the Spherical Harmonics expansion is truncated to the order m = M, the HOA representation of the acoustic pressure is composed of $(M + 1)^2$ components, which are the $(M+1)^2\, B^{\sigma}_{mn}(\omega)$ coefficients of the Spherical Harmonics expansion (see Equation 9.34). These HOA components convey spatial variation as a function of the azimuth and elevation angles (see Figure 9.5). Each order m is composed of $(2m + 1)$ components with various directivities. It should be noted that some components are characterized by a null response in the horizontal plane. The consequence is that they do not contribute to any spatial horizontal information. By contrast, the directivity of the remaining components is symmetrical to the horizontal plane. These latter components are referred to as the "2D Ambisonics components", in the sense that, if the sound field reproduction is restricted to the horizontal plane (i.e. the loudspeaker setup is limited to the horizontal plane), only these components must be considered. On the contrary, if a full 3D reproduction is expected, all the components are used and the reproduction setup requires both horizontal and elevated loudspeakers to render spatial height information.

The component of 0th order corresponds to the spatial equivalent of the DC component and is characterized by no spatial variation. In other words the 0th order component, W, is the monophonic recording of a sound field by a pressure microphone. The three 1st order components are characterized by a figure-of-eight variation (i.e. cosine or sine function). As the order increases, the spatial variation is faster and faster as a function of the angle, as illustrated in Figure 9.5. A first benefit of including components of higher order is therefore to enhance the spatial accuracy and the spatial definition (resolution) of the sound field representation. This is due to the increase of the high frequency cutoff of the associated spatial spectrum.[3] The resulting effect on the reproduced sound field is complex: both the size of the listening area and the bandwidth of the "time spectrum" are affected. Indeed, 1st order Ambisonics reproduction is penalized by a phenomenon of "sweet spot": the sound field is correctly reproduced only at the close vicinity of the center of the loudspeaker setup. In addition, for a given reproduction area, low frequencies, which are linked to large wavelengths and therefore to slow spatial variations, are better reconstructed than high frequencies. Adding Ambisonics components of order higher than M = 1 increases both the size of the listening area and the high-frequency cutoff of the time spectrum. Thus, small movements of the listener are then allowed. In Figure 9.7, the sound field reproduced by Ambisonics systems of various orders is illustrated in the case of a plane wave. It is observed that a low-frequency plane wave (f = 250 Hz) is well reconstructed over a wide area by a 4th order system. If the frequency increases up to 1 kHz, the area of accurate reproduction shrinks considerably. An upgrade to a 19th order system is needed to achieve a listening area the size of which is equivalent to that obtained by the 4th order system at f = 250 Hz. Thus, if the sound field reproduced is observed over a fixed area, the high-frequency cutoff decreases as a function of the maximal Ambisonics order M. In the same way, if the sound field is observed at a fixed frequency, the size of accurate reproduction decreases as a function of the maximal

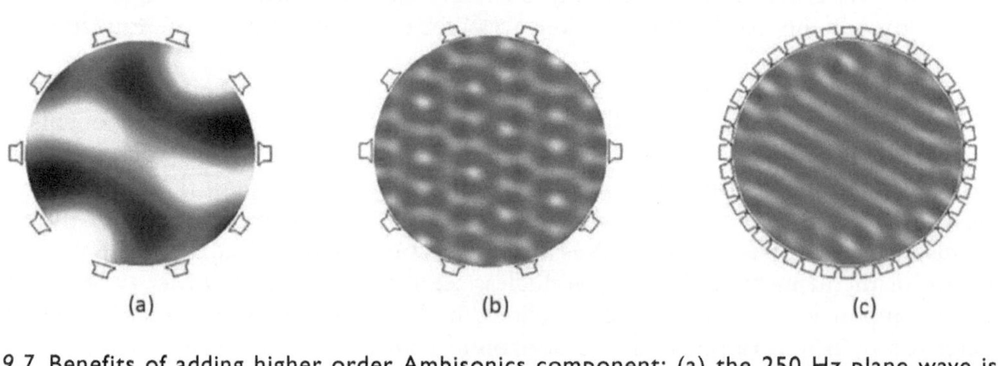

Figure 9.7 Benefits of adding higher order Ambisonics component: (a) the 250 Hz plane wave is well reconstructed over a 4th order system; (b) increasing the frequency to 1 kHz yields a poorer reconstruction using the same 4th order system; (c) to reconstruct the 1 kHz plane wave with the same accuracy as in (a), the system must be increased to a 19th order.

Ambisonics order M. In (Ward & Abhayapala, 2001), a rule of thumb was proposed to estimate the reproduction order as a function of the wave number k and the radius r of the reproduction sphere, to achieve a maximum threshold of the truncation error equal to 4%. The order M is obtained as:

$$M = \lceil kr \rceil$$

where $\lceil . \rceil$ denotes rounding up to the nearest integer. For instance, if we consider a radius of the reproduction sphere equal to 8.5 cm, which is close to the average radius of a human head, 1st order Ambisonics achieves valid reconstruction of the sound field (i.e. truncation error lower than 4%) only up to 637 Hz. To increase the frequency cutoff up to 16 kHz for the same area, it is needed to include HOA components up to the M = 25th order.

HOA Microphones

The concept of HOA is to represent the sound field by a series of signals $B^{\sigma}_{mn}(\omega)$, which are the coefficients of the Spherical Harmonics expansion of the sound field (see Equation 9.34). The tetrahedral microphone is the most convenient solution to record 0th order and 1st order components $B^{\sigma}_{mn}(\omega)$, but how can we record the components of order greater than 1 to upgrade to HOA? The solution is very close to the strategy adopted for the tetrahedral microphone. Intuitively, the $B^{\sigma}_{mn}(\omega)$ components could be recorded by a set of directional microphones, the directivity of which is defined by the directivity of the Spherical Harmonics (see Figure 9.5). This solution, that is not feasible for the 0th and 1st order components, is even less possible for the higher components, because the number of coincident microphones is higher and their directivity more complex. For HOA, it is suggested to use instead a spherical array of microphones (i.e. a tetrahedron array) and then to derive the Ambisonics components from the microphone outputs by appropriate matrixing (see Equation 9.2). This is the inspiration for the general concept of HOA microphones. The recording system is conceptually a spherical array of microphones, by which the acoustic pressure and the pressure gradient (or the acoustic velocity) over a sphere is captured. In addition, the microphone array is coupled to an encoding matrixing process to obtain the HOA components.

More precisely, the solution is based on the mathematical definition of HOA components. They are defined as the coefficients of the Spherical Harmonics expansion. Any sound field can be developed as a linear and weighted sum of Spherical Harmonics, since Spherical Harmonics are the eigenfunctions of the equation of the acoustic wave, in the same way as any time function can be expressed as a linear and weighted sum of sine and cosine functions, which is called a "Fourier series expansion". In other words, Spherical Harmonics are the equivalent of sines and cosines for space variations. Therefore, by definition of eigenfunctions, the coefficients of Spherical Harmonics expansion are computed from the projection of the sound field over the orthonormal basis of Spherical Harmonics (see Equation 9.27). If $U^{\sigma}_{mn}(\omega)$ is the result of the projection of the acoustic pressure $p(r, \varphi, \theta, \omega)$ on the Spherical Harmonic $Y^{\sigma}_{mn}(\phi, \theta)$ over the sphere (see Equation 9.29), it is shown that the component $B^{\sigma}_{mn}(\omega)$ is given by:

$$B^{\sigma}_{mn}(\omega) = E_q(m, kr)U^{\sigma}_{mn}(\omega) \tag{9.9}$$

where the term $E_q(m,kr) = \dfrac{1}{i^m \mathrm{j}_m(kr)}$ can be interpreted as an equalization. Thus, to obtain Ambisonics components, the first step is to measure the acoustic pressure over the sphere of radius r, from which the signals $U_{mn}^{\sigma}(\omega)$ are computed. Then Equation 9.9 gives the $B_{mn}^{\sigma}(\omega)$ signals.

Challenges of HOA

However, this process raises two main problems: one is the zeroes of the spherical Bessel function of first kind (i.e. $\mathrm{j}_m(kr)$), the other is the spatial sampling of the sound field. Indeed, whenever the function $\mathrm{j}_m(kr)$ is equal to zero, the equalization term $E_q(\mathrm{m}, kr)$ does no longer exist. What's more, as soon as the function $\mathrm{j}_m(kr)$ is close to zero, the equalization term $E_q(\mathrm{m}, kr)$ increases dramatically, which leads to a considerable amplification of the signals, and consequently to an alteration of the audio quality of the signals because of the resulting expansion of the microphone noise. One solution is to replace the acoustic pressure by another acoustic variable to describe the sound field, in order to modify the equalization term. For instance, instead of pressure microphones, cardioid microphones can be used. As already mentioned, these latter may be seen as a linear sum of a pressure microphone and a gradient pressure microphone. Thus the equalization term becomes (Moreau, Daniel & Bertet, 2006):

$$E_c(m,kr) = \frac{1}{i^m \left[\mathrm{j}_m(kr) + k\dfrac{\partial \mathrm{j}_m(kr)}{\partial r} \right]} \tag{9.10}$$

In that case, the denominator is never null. More generally, the directivity function of the microphone can be modified (e.g. by introducing acoustic diffraction through a solid structure) in order to design an equalization term $E_q(\mathrm{m}, kr)$ in accordance with expected properties (Epain & Daniel, 2008).

As for the second issue to solve, ideally, Ambisonics components should be derived from the knowledge of the continuous sound field over the sphere of radius r. However, in practice, the acoustic signals cannot be measured at any point, but only at a finite set of locations defined by a microphone array, which involves spatial sampling. The microphones are distributed over a sphere and form a spherical array. The main question is to choose properly their location in order to catch optimally the sound field information so that the signals $B_{mn}^{\sigma}(\omega)$ can be accurately estimated. The solution is the best compromise between the following constraints: minimizing the error of the estimation of the signals $B_{mn}^{\sigma}(\omega)$, minimal number of microphones, achieving a feasible geometry of the microphone array. By analogy with time sampling, spatial sampling is possible under the assumption that the sound field is spatially band-limited, i.e. that the components $B_{mn}^{\sigma}(\omega)$ are null for any order m greater than a maximal value m_{max}. If this condition is satisfied, the sound field can be correctly sampled and the signals $B_{mn}^{\sigma}(\omega)$ exactly estimated, provided that the azimuth and elevation angles are regularly and separately sampled (Driscoll & Healy, 1994). The drawback of this solution is a high number of microphones: for instance, to record the sound field up to the order M, at least $4(\mathrm{M}+1)^2$ microphones are required. To decrease the number N_c of microphones, an approximation is proposed. Assuming that cardioid microphones are used, the output of the qth cardioid microphone located at the location (r, φ_q, θ_q) is given by, in accordance with the theoretical definition of the cardioid directivity:

$$c(q,\omega) \equiv c(r,\phi_q,\theta_q,\omega) = p(r,\phi_q,\theta_q,\omega) - \frac{\vec{\nabla} p(r,\phi_q,\theta_q,\omega)\cdot \vec{n}}{ik}, q \in [1,\cdots,N_c] \tag{9.11}$$

Instead of computing the signals $B^{\sigma}_{mn}(\omega)$ from the projection over the orthonormal basis of Spherical Harmonics, the Ambisonics components are derived by replacing the acoustic pressure by its Spherical Harmonics expansion (see Equation 9.34) in Equation 9.11:

$$c(q,\omega) = \sum_{m=0}^{M} i^m \left[\mathrm{j}_m(kr) + k\frac{\partial \mathrm{j}_m(kr)}{\partial r} \right] \sum_{n=0}^{m} \sum_{\sigma=\pm 1} B^{\sigma}_{mn}(\omega) Y^{\sigma}_{mn}(\phi_q,\theta_q) \tag{9.12}$$

This equation defines a system of linear equations, which can be reformulated in a matrix form:

$$\boldsymbol{C} = \boldsymbol{Y_c W_c B} \tag{9.13}$$

where the terms C, B, Y_c and W_c are given by:

$$c = \begin{bmatrix} c(1,\omega) \\ c(2,\omega) \\ \vdots \\ c(N_c,\omega) \end{bmatrix}, B = \begin{matrix} B^1_{00}(\omega) \\ B^1_{10}(\omega) \\ \vdots \\ B^{-1}_{MM}(\omega) \end{matrix}, Y_c = \begin{bmatrix} Y^1_{00}(\phi_1,\theta_1) & Y^1_{10}(\phi_1,\theta_1) & \dots & Y^{-1}_{MM}(\phi_1,\theta_1) \\ Y^1_{00}(\phi_2,\theta_2) & Y^1_{10}(\phi_2,\theta_2) & \dots & Y^{-1}_{MM}(\phi_2,\theta_2) \\ \vdots & \vdots & \vdots & \vdots \\ Y^1_{00}(\phi_{N_c},\theta_{N_c}) & Y^1_{10}(\phi_{N_c},\theta_{N_c}) & \dots & Y^{-1}_{MM}(\phi_{N_c},\theta_{N_c}) \end{bmatrix},$$

$$W_c = \begin{bmatrix} \mathrm{j}_0(kr) + k\frac{\partial \mathrm{j}_0(kr)}{\partial r} & 0 & \dots & 0 \\ 0 & i\left[\mathrm{j}_1(kr) + k\frac{\partial \mathrm{j}_1(kr)}{\partial r}\right] & \dots & 0 \\ \vdots & \vdots & \vdots & \vdots \\ 0 & 0 & \dots & i^M\left[\mathrm{j}_M(kr) + k\frac{\partial \mathrm{j}_M(kr)}{\partial r}\right] \end{bmatrix}$$

Provided that the number of microphones N_c (i.e. the number of equations) is greater than or equal to the number of unknowns (i.e. the number of Ambisonics components: $(M + 1)^2$), Equation 9.13 can be solved by using the Moore-Penrose pseudoinverse of Y_c:

$$\hat{B} = E_c(Y_c^t Y_c)^{-1} Y_c^t C \tag{9.14}$$

where $E_c = \begin{bmatrix} E_c(0,kr) & 0 & \cdots & 0 \\ 0 & E_c(1,kr) & \cdots & 0 \\ \vdots & \vdots & \vdots & \vdots \\ 0 & 0 & \cdots & E_c(M,kr) \end{bmatrix}$ and Y_c^t refers to the transpose conjugate of Y_c.

It should be kept in mind that this solution is just an approximate estimate of the Ambisonics components $B_{mn}^{\sigma}(\omega)$. Errors are introduced for instance by the internal noise, or possible mispositioning of the microphones. The matrix E_c can cause instability, which is minimized by a regularization method (Moreau et al., 2006). Besides, in Equation 9.14, the term $\boldsymbol{Y}_c^t \boldsymbol{Y}_c$ is of particular interest: if the spatial sampling of the sound field (i.e. the geometry of the microphone array) keeps the orthonormality property of Spherical Harmonics (see Equation 9.27), this term can be simplified to $\boldsymbol{Y}_c^t \boldsymbol{Y}_c = \mathbf{1}$ (where $\mathbf{1}$ is the identity matrix). In that case, Equation 9.14 turns out to be the sampled version of Equation 9.9, in which cardioid microphones are considered instead of pressure microphones. If the geometry of the microphone array is arbitrarily chosen, the orthonormality property is generally not satisfied and the term $\boldsymbol{Y}_c^t \boldsymbol{Y}_c$ quantifies the amount of spatial aliasing (i.e. orthonormality error). However it is difficult to find a geometry for which $\boldsymbol{Y}_c^t \boldsymbol{Y}_c = 1$. Regular and semi-regular polyhedrons provide solutions which are valid only up to a maximal order max_{max} (Moreau et al., 2006). Thus the issue of the microphone array geometry must be carefully examined.

HOA in Practice

In practice, when designing an HOA microphone, the first step is to choose the maximal order M of the Spherical Harmonics expansion, which is expected. Then the value of M imposes the minimal number of microphones: $N_c = (M + 1)^2$. The third step is to find the array geometry (preferably a semi-regular polyhedron) composed of at least N_c elements, which minimizes the orthonormality error. The fourth step concerns the radius r of the microphone array, which affects the equalization term of Equation 9.10. The optimal radius is a difficult compromise to minimize spatial aliasing by decreasing r, while keeping a reliable estimation of Ambisonics components for low frequencies, which is better for larger r (Moreau et al., 2006). The tetrahedral system is one example of such a design for $M = 1$. It is composed of $N_c = (M + 1)^2 = 4$ microphones arranged in a tetrahedron, which is the simplest regular polyhedron. This setup implicitly provides a rough spatial sampling of the sound field and thus allows one to estimate the 0th and 1st order Ambisonics components.

An alternative solution was proposed in (Zotkin, Duraiswami & Gumerov, 2010). Broadly speaking, the idea is to find the set of plane waves that best explains the sound pressure collected by the set of microphones. The process is very close to Ambisonics encoding: a weighting matrix is applied to the microphone signals to derive the coefficients of the plane wave expansion.

HOA: A Promising Format for Sound Field Description

The HOA representation of the sound field (i.e. the signals B_{mn}^{σ}) is an attractive format to describe a sound scene. This format has three valued properties; it is generic, universal and scalable.

First, the matrixing process to compute the loudspeaker signals for a given layout may be interpreted as a transcoding of Ambisonics components into the domain of the loudspeakers (see Equations 9.15 and 9.19). Thus the HOA representation is really a generic format, where generic means that it is independent of both the recording format (i.e. microphone signals) and the reproduction format (i.e. loudspeaker signals). The advantage is that any editing or post-production of the sound scene (e.g. spatial transformations, such as rotation, angular distortion like forward dominance) is done preferably in the HOA domain (Daniel, 2009), and thus will not require to compute a new set of loudspeaker signals. The loudspeaker input signals are decoded only once—during sound field reproduction. HOA signals are therefore relevant for data storage.

Second, the HOA format is universal in that it is an exact representation of the acoustic wave, provided that all the terms of the Spherical Harmonics expansion are kept up to infinity.

Additionally, it is valid for any acoustic wave, whatever its spatial and propagation properties. Errors come only from the limitation of the Spherical Harmonics expansion to a finite order M, and from the estimation of the $B^{\sigma}_{mn}(\omega)$ signals by HOA microphones.

Third, scalability means that the HOA components of the lowest order convey a full description of the sound scene. In the extreme, the 0th order HOA component (i.e. the W component which is equivalent to a monophonic recording) is a full representation of the sound field (though with absolutely no spatial information), in the sense that a listener is able to listen to it and to interpret it as a coherent sound scene composed of various acoustic sources.

Adding the components of higher orders improve only the spatial definition of the sound field. Consequently, at any time, it is possible to discard the highest-order components, in order to adapt the maximal order M of HOA signals to the available bitrate of transmission/storage, or to the configuration of the listening setup (number of loudspeakers).

Ambisonics and HOA Reproduction

HOA reproduction aims at synthesizing the original sound field *p* by a loudspeaker array. For this, the Ambisonics components $B^{\sigma}_{mn}(\omega)$ have to be mixed together to build the proper input signal for each loudspeaker, so that the resulting synthetic sound field *p* matches the original one as close as possible. The input loudspeaker signals are derived from the $B^{\sigma}_{mn}(\omega)$ signals by using a decoding $N_l \times (M+1)^2$ matrix **D**, defined by:

$$S = DB \tag{9.15}$$

where *S* is the vector composed of the N_l loudspeaker signals $S = \begin{bmatrix} s(1,\omega) \\ \vdots \\ s(l,\omega) \\ \vdots \\ s(N_l,\omega) \end{bmatrix}$. The decoding matrix

D is computed by equating the synthetic sound field to the original one. The former is given by:

$$\hat{p}(\vec{r},\omega) = \sum_{l=1}^{N_l} s(l,\omega)\, p_l(\vec{r},\omega) \tag{9.16}$$

where p_l refers to the elementary acoustic wave emitted by the lst loudspeaker. This elementary wave can be developed over the Spherical Harmonics basis in accordance with Equation 9.34, which leads to:

$$\hat{p}(\vec{r},\omega) = \sum_{l=1}^{N_l} s(l,\omega) \sum_{m=0}^{+\infty} i^m j_m(kr) \sum_{n=0}^{m} \sum_{\sigma=\pm 1} L_{mn}^{\sigma}(l,\omega) Y_{mn}^{\sigma}(\phi,\theta) \tag{9.17}$$

Thus, the synthetic wave exactly matches the target sound field p, if and only if the coefficients of their respective Spherical Harmonics expansion are equal, that is to say:

$$B = LS \tag{9.18}$$

$$\text{where } L = \begin{bmatrix} L_{00}^{1}(\phi_1,\theta_1) & L_{00}^{1}(\phi_2,\theta_2) & \ldots & L_{00}^{1}(\phi_{N_l},\theta_{N_l}) \\ L_{10}^{1}(\phi_1,\theta_1) & L_{10}^{1}(\phi_2,\theta_2) & \ldots & L_{10}^{1}(\phi_{N_l},\theta_{N_l}) \\ \vdots & \vdots & \vdots & \vdots \\ L_{MM}^{-1}(\phi_1,\theta_1) & L_{MM}^{-1}(\phi_2,\theta_2) & \ldots & L_{MM}^{-1}(\phi_{N_l},\theta_{N_l}) \end{bmatrix}.$$

This matrix **L** represents the coefficients of the Spherical Harmonics expansion of the wave emitted by each loudspeaker. In other words, the matrix **L** contains the information about the spatial coordinates of the loudspeakers. This is needed for the remapping of the spatial information of the sound field over the loudspeaker array.

Equation 9.18 defines a system of $(M + 1)^2$ linear equations with N_l unknowns (the loudspeaker input signals). In practice, it is recommended to choose the number of loudspeakers as $N_l = (M + 1)^2$, which ensures an optimal reproduction of the sound field (Poletti, 2005). Consequently, if $N_l < (M + 1)^2$, it is preferable to discard the Ambisonics components of the highest orders $m_{max} < m < (M + 1)^2$ until $N_l = (m_{max} + 1)^2$. On the contrary, if $N_l > (M + 1)^2$, part of the loudspeakers should be muted to keep only $(M + 1)^2$ of them. Otherwise the reproduced sound field is likely to be unstable and some auditory artifacts like "phasiness" are observed (Daniel, 2009).

Decoding Matrix

Theoretically any arbitrary geometry of the loudspeaker array can be chosen, which is a remarkable advantage of HOA reproduction in contrast with channel-based formats, such as 5.1 or 22.2. The decoding matrix **D** is responsible for the adaptation of the HOA components, $B_{mn}^{\sigma}(\omega)$, to the loudspeaker signals (i.e. the "loudspeaker domain" versus the Spherical Harmonics domain), and is able to compensate for any loudspeaker layout. Indeed the decoding matrix takes into account both the location and the acoustic radiation of the loudspeaker through the matrix L (see Equation 9.18). For instance, if the waves emitted by the loudspeakers are assumed to be plane waves (far-field assumption), the matrix L is given by (see Equation 9.40):

$$L = Y_l = \begin{bmatrix} Y_{00}^{1}(\phi_1,\theta_1) & Y_{00}^{1}(\phi_2,\theta_2) & \ldots & Y_{00}^{1}(\phi_{N_l},\theta_{N_l}) \\ Y_{10}^{1}(\phi_1,\theta_1) & Y_{10}^{1}(\phi_2,\theta_2) & \ldots & Y_{10}^{1}(\phi_{N_l},\theta_{N_l}) \\ \vdots & \vdots & \vdots & \vdots \\ Y_{MM}^{-1}(\phi_1,\theta_1) & Y_{MM}^{-1}(\phi_2,\theta_2) & \ldots & Y_{MM}^{-1}(\phi_{N_l},\theta_{N_l}) \end{bmatrix} \tag{9.19}$$

Even if the geometry of the loudspeaker is free, a regular layout is preferably chosen, since a regular array leads to a decoding matrix that is mathematically more simple and stable. Therefore, the loudspeakers are generally distributed on the surface of a sphere of radius r_l. In that case, if spherical waves are considered instead of plane waves, the matrix $\boldsymbol{L}$ becomes $L = W_l Y_l$ (Morse & Feshback, 1953; Morse & Ingard, 1968; Daniel, 2001), with:

$$W_l = \begin{bmatrix} (-i) & 0 & \dots & 0 \\ 0 & -\frac{h_1^-(kr_l)}{k} & \dots & 0 \\ \vdots & \vdots & \vdots & \vdots \\ 0 & 0 & \dots & \frac{h_M^-(kr_l)}{k} i^{-(M+1)} \end{bmatrix} \tag{9.20}$$

The exact positioning of the loudspeakers on the sphere must be regular in the sense that the orthonormality property of Spherical Harmonics is preserved by the resulting spatial sampling (see Equation 9.27). In other words, the loudspeakers have to be arranged at the vertices of a regular or semi-regular polyhedron, so that ideally: $\boldsymbol{Y_l^t Y_l = 1}$. If the reproduction is restricted to the horizontal plane (i.e. 2D reproduction), this requirement is achieved by a circular array of equally spaced loudspeakers. In the case of a regular setup satisfying $\boldsymbol{Y_l^t Y_l = 1}$, the decoding matrix turns out to be: $\boldsymbol{D=L^t}$. under the assumption of plane waves.

Decoding Rules

The decoding matrix solving Equation 9.18 is one strategy of HOA reproduction, by which a perfect reconstruction of the sound field is intended (i.e. perfect match between the recorded sound field and the reproduced one), and which is termed *basic decoding*. There are many alternatives (Daniel, 2001). The velocity and energy vectors, which were presented in the first section (see Equation 9.8), are useful to optimize the sound field reproduction (Gerzon, 1992). Particularly when the frequency increases, perfect reconstruction is no longer achievable, at least over an extensive listening area. Instead, approximate solutions, which are computed by optimizing a given set of constraints, allow one to improve the sound field rendering. Thus, the velocity and energy criteria can be used by the optimization process as specific constraints. For example, during reconstruction, the *maximum* r_E constraint uses the loudspeakers which are the closest to the direction of the virtual sound source, by maximizing the norm of the energy vector. In the same way, the *in-phase* constraint imposes to mute the loudspeakers that are at the opposite of the direction of the virtual sound source, which improves the sound field rendering for off-centered listeners. These are examples of alternative decoding rules to compute the decoding matrix. In practice, different decoding rules can be applied as a function of frequency, e.g. basic decoding for low frequencies and *maximum* r_E decoding for high frequencies.

Sound Field Synthesis

In order to illustrate HOA reproduction of a sound field, Figure 9.8 depicts the acoustic pressure wave synthesized by a 2D circular loudspeaker array of radius $r_l = 3\ m$. The target sound field is a harmonic plane wave coming from azimuth $\phi = 60°$ at the frequency $f = 1$ kHz. The loudspeaker

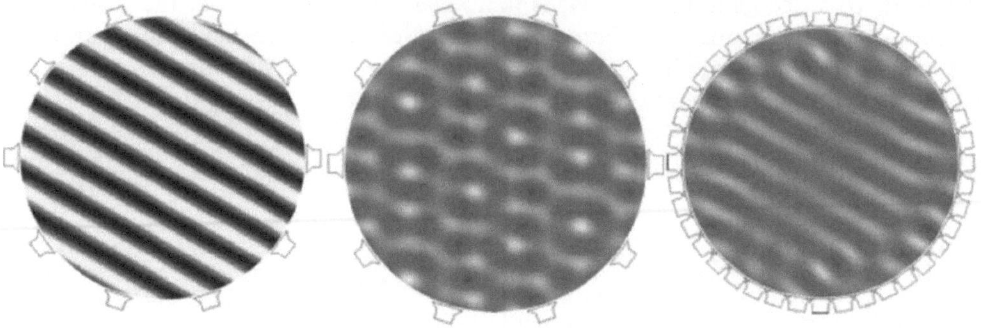

Figure 9.8 Plane wave (ϕ=60°, f = 1 kHz) synthesized by an HOA array as a function of the maximal order M of the Spherical Harmonics expansion (from left to right: target wave, HOA synthetic wave M = 4, HOA synthetic wave M = 19).

signals are computed from the theoretical HOA components (see Equation 9.40) through a basic decoding and assuming that loudspeakers are emitting plane waves. The number of the loudspeakers is fixed to the closest of the optimal value $N_l = (2M + 1)$. The wave synthesized by the 4th order HOA system is accurate only in the immediate vicinity of the center of the loudspeaker array. When upgrading the HOA synthesis up to order M = 19, the expected plane wave is correctly reproduced over almost all the area inside the loudspeaker array. The benefit of higher orders to enlarge the listening area is clearly demonstrated.

To assess the performances of the sound field reproduction in terms of the perceived localization of virtual sound sources, the ITD (Interaural Time Difference) and ILD (Interaural Level Difference) values are estimated for a set of locations within the listening area (see Figure 9.9). For this simulation, the wave propagation between each loudspeaker and the listener's ears includes HRTF (Head Related Transfer Function) to account for the acoustic diffraction of the acoustic wave by the listener's morphology (particularly the pinna). Ideally the ITD and ILD are respectively around –500 µs and –12 dB, assuming that the listener is facing the 0° direction. In Figure 9.9, it is observed that for most of the locations, the lateralization (i.e. perceived azimuth) of the virtual sound source is roughly correct, even though the exact localization is not accurate. However the ITD is affected by many spatial instabilities, and these artifacts are minimized when the HOA order M increases. In addition, a third criterion, the ISSD (Inter Subject Spectrum Difference), is estimated to quantify the spectrum distortion. The ISSD is a measure of the dissimilarity between two spectrum magnitudes (i.e. the target spectrum and the reproduced one at one given location) and is defined as the variance of the difference of the dB-magnitudes (Middlebrooks, 1999). Figure 9.9 shows that the spectrum distortion is high for almost all the locations in the listening area, which means timbre artifacts.

Figure 9.10 and Figure 9.11 illustrate the reproduction of a spherical wave, which would have been emitted by a sound source located outside the loudspeaker array. The 4th order HOA system does not succeed in correctly synthesizing the curvature of the wavefront. Higher orders up to around M = 19 are required for a better reconstruction of the spherical wave within the listening area. As for the localization cues, the ILD is better reproduced than the ITD, which exhibits

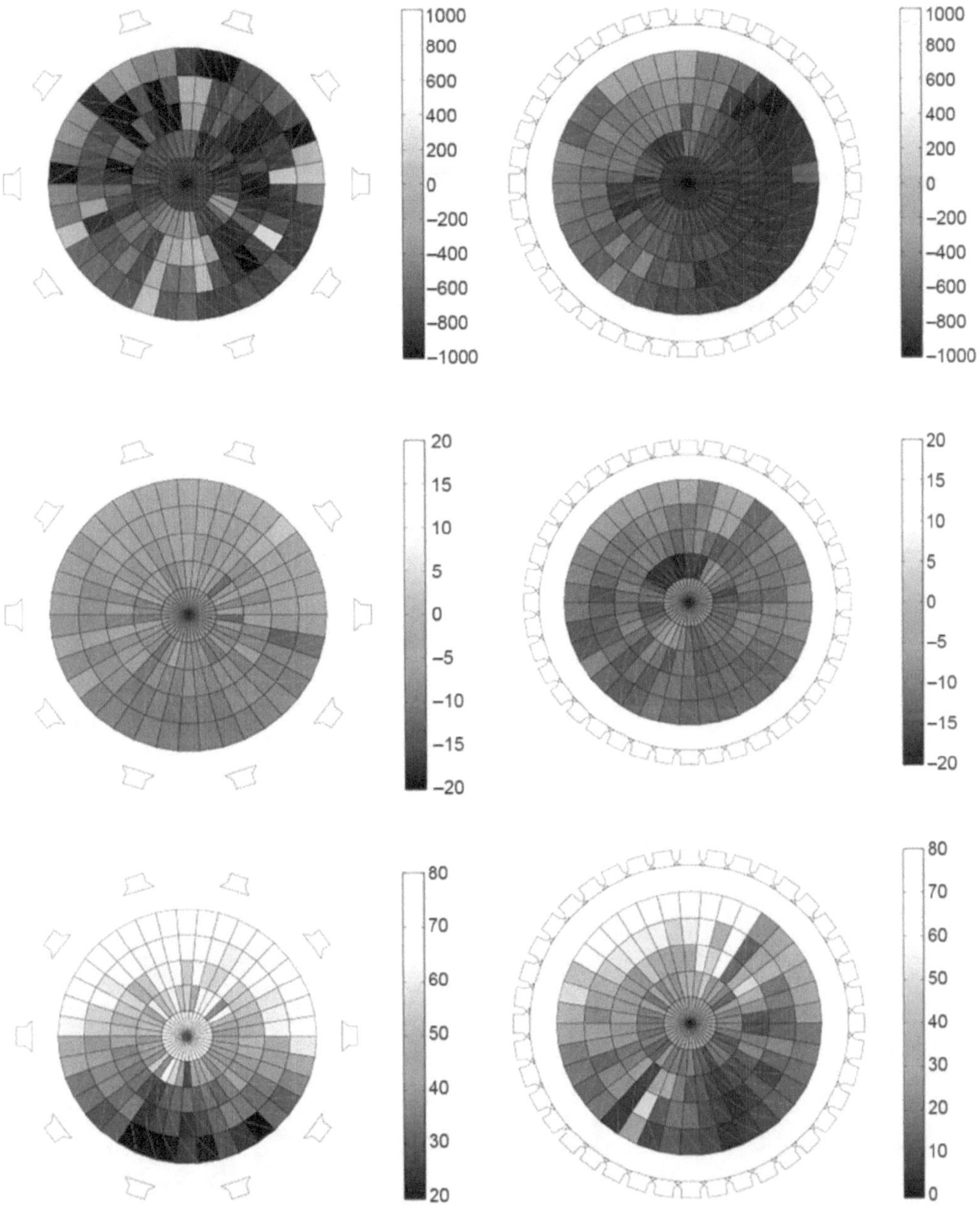

Figure 9.9 Estimate of the localization cues (ITD and ILD) and the spectrum distortion (ISSD) computed for a plane wave ($\phi=60°$, f = 1 kHz) synthesized by an HOA array (from top to bottom: ITD, ILD, ISSD, and from left to right: M = 4, M = 19).

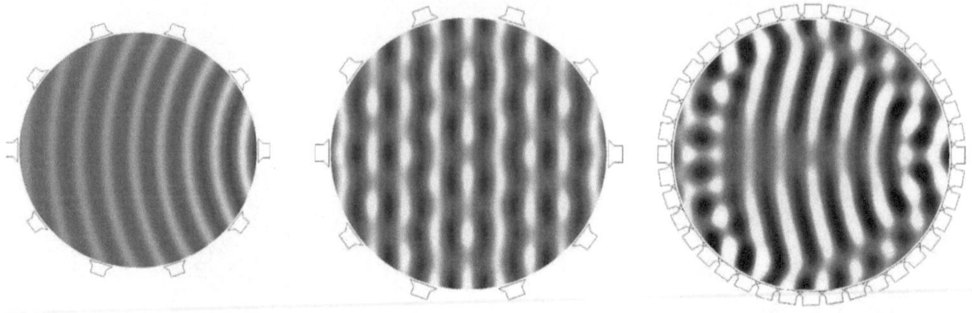

Figure 9.10 Spherical wave emitted by an external source (r = 3 m, ϕ=0°, f = 1 kHz) and synthesized by an HOA array as a function of the maximal order M of the Spherical Harmonics expansion (from left to right: target wave, HOA synthetic wave M = 4, HOA synthetic wave M = 19).

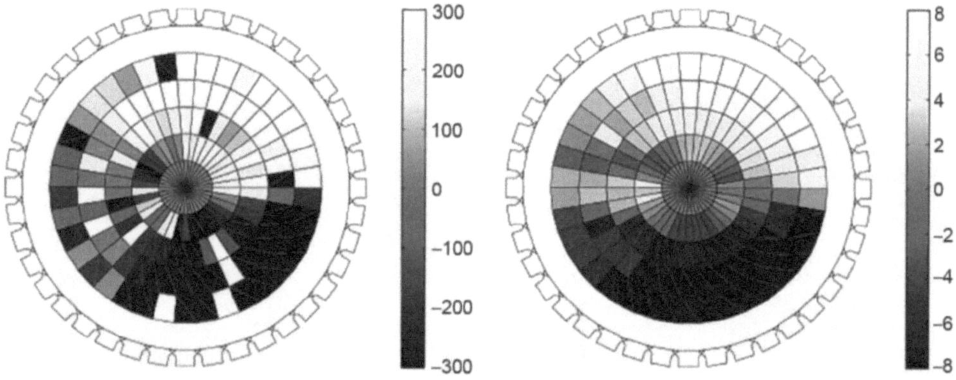

Figure 9.11 Estimate of the localization cues (ITD (left) and ILD (right)) computed for a spherical wave emitted by an external source (r = 3 m, ϕ=0°, f = 1 kHz) and synthesized by a 19th order HOA array.

Figure 9.12 Spherical wave emitted by an internal source (r = 1.25 m, ϕ=0°, f = 1 kHz) and synthesized by an HOA array as a function of the maximal order M of the Spherical Harmonics expansion (from left to right: target wave, HOA synthetic wave M = 4, HOA synthetic wave M = 19).

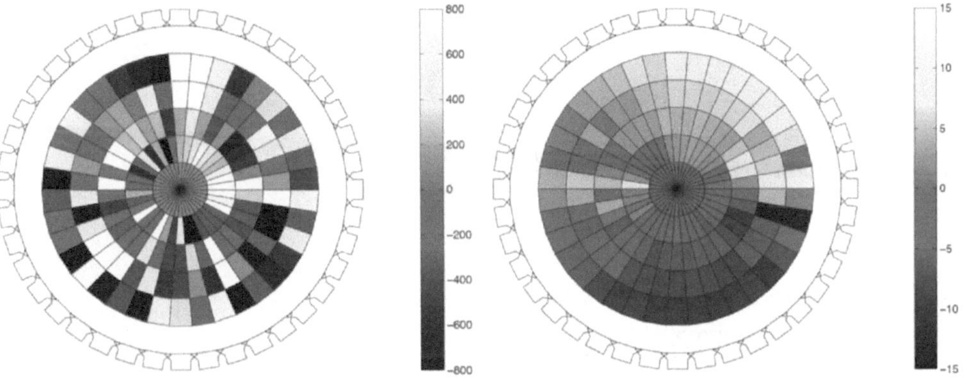

Figure 9.13 Estimate of the localization cues (ITD and ILD) computed for a spherical wave emitted by an internal source (r = 1.25 m, ϕ=0°, f = 1 kHz) and synthesized by a 19th order HOA array.

strong spatial inhomogeneities. The same behavior is observed in the case of the reproduction of a spherical wave emitted by a sound source located this time inside the loudspeaker array (see Figure 9.12 and Figure 9.13). Particularly, only the 19th order HOA system is able to accurately reconstruct the spherical wavefront.

Sound Field Formats

A-, B-, C- and D-Formats From Ambisonics Terminology

By format, it is meant any representation of the sound field by a set of signals. From recording to reproduction, there are potentially various formats. The output signals of the microphones can be seen as a *recording* format, which may be different from the *reproduction* format defined by the input signals of the loudspeakers. This is the case for Ambisonics and HOA technology. In the specific case of 1st order representation (i.e. Ambisonics), the output signals of the tetrahedral microphone (i.e. the LF, RF, LB, RB signals) are referred to as the *A-format*, which is a recording format and should be distinguished from the *B-format*, which consists of the components (W, X, Y, Z). This B-format is fully independent of the recording setup and the reproduction layout. The set of the input signals to the loudspeakers defines the D-format, also called G-format. Initially, the G-format corresponded to the specific loudspeaker signals computed for a 5.1 layout, but it found general use for any loudspeaker configuration.

The *C-format*, most often referred to as the UHJ format, is another Ambisonics format, which was introduced for broadcast and diffusion purpose (CD, DVD, television or radio), "C" meaning "Consumer" (Gerzon, 1985). The main goal of the UHJ format is to provide signals that are directly compatible with the conventional systems of reproduction, namely monophonic and stereophonic reproduction. It is composed at least of two signals: the left (L) and right (R) stereophonic signals. Two signals T and Q can be added to increase the spatial accuracy in the horizontal plane (T) and to convey height information (Q). More precisely, the encoding of B-format into UHJ format is based on a set of six signals, which are composed on the basis of the sum (S) and difference (D) signals:

$$\begin{aligned} S &= 0.9396926W + 0.18555740X \\ D &= j(-0.3420201W + 0.5098604X) + 0.6554516Y \end{aligned} \tag{9.21}$$

and, on the other hand, the (L, R, T, Q) signals:

$$\begin{aligned} L &= 0.5(S + D) \\ R &= 0.5(S - D) \\ T &= j(-0.1432W + 0.6512X) - 0.7071Y \\ Q &= 0.9772Z \end{aligned} \tag{9.22}$$

where j is + 90° phase shift.

To obtain the L and R stereophonic signals, the matrixing equations are inspired by the M/S technique (Equation 9.1). The monophonic reproduction only uses the signal S. The matrixing equations to go back to B-format from UHJ format are:

$$\begin{aligned} S &= 0.5(L + R) \\ D &= 0.5(L - R) \\ W &= 0.982S + 0.197j(0.828D + 0.768T) \\ X &= 0.419 - j(0.828D + 0.768T) \\ Y &= 0.187j\ S + (0.796D - 0.676T) \\ Z &= 1.023Z \end{aligned} \tag{9.23}$$

These concepts also hold for HOA. Recording is performed by a spherical array of microphones, which provides a spatial sampling of the acoustic pressure over a sphere, and which deliver the recording format. Then, from the microphone signals (see Equation 9.14) are derived the HOA components $B_{mn}^{\sigma}(\omega)$ (spatial encoding), which defines a kind of *kernel* format in the Ambisonics chain. This format is independent from both the recording and the reproduction setup, and can therefore be considered as a generic representation of the sound field. In the end, the loudspeaker inputs are computed by matrixing the signals $B_{mn}^{\sigma}(\omega)$ to recombine spatial information in an appropriate way (re-encoding or transcoding) for a proper synthesis of the sound field by the loudspeaker array. In the current terminology of audio formats, the signals $B_{mn}^{\sigma}(\omega)$ are typically referred to as a *sound field–based format*, as opposed to the *channel-based* format (e.g. 5.1, 10.2 or 22.2), which is composed of the loudspeaker signals, and the *object-based* format, in which the sound scene is described as a set of elementary components (i.e. sound sources) in combination with their spatial coordinates and trajectory (Geier, Ahrens & Spors, 2010; Bleidt et al., 2014).

Conclusion

This chapter was dedicated to sound field, which means that our main concern was the reproduction of an acoustic wave over an extended listening area. Three main questions were investigated:

how to record, how to represent and how to reproduce a sound field. After an overview of the general concepts, these investigations were illustrated in the specific case of HOA technology (microphones, format, encoding, decoding matrix, loudspeaker setup). Examples of HOA reproduction were finally given for the case of plane and spherical waves.

In practice, the quality of the actual source capture in an HOA recording will strongly affect the reproduced outcome. Microphone parameters like phase accuracy, matching the amplitude to frequency response and the signal-to-noise ratio are all important factors to consider. The quality of the HOA playback systems and the listening environments will have a great effect as well. Currently, the high demand for microphones and playback systems to support new immersive audio consumer formats is driving innovation and improvements to HOA technology as engineers, content providers and consumers look to sound field technology. Further assessment of HOA reproduction in terms of perceived quality is also needed. In most of the existing systems, the highest order is M = 4 or 5. One question, which is still unsolved, concerns the limit order, beyond which the improvement of the sound field reconstruction is not perceptible. But to investigate this issue, methods to assess the multi-dimensional perception of sound field should progress.

Notes

1 Ribbon microphones with figure-of-eight pickup patterns are sensitive to particle velocity. A figure-of-eight pattern can also be achieved with two coincident, opposing 180° in orientation, cardioid pressure gradient microphones, with one of them inverted in phase.

2 However, it should be noticed that the first practical implementation of the Soundfield Microphone employed sub-cardoid capsules rather than cardioids.

3 Like for time variations, spatial variations may be represented in two alternative domains: either in the domain of spatial coordinates or in the dual domain of spatial frequencies. The spatial frequency can be interpreted as the inverse of the wavelength in the case of a harmonic plane wave. Thus the spatial spectrum is the representation of the sound field in the domain of spatial frequencies. As an example, the $B_{mn}^{\sigma}(\omega)$ coefficients of HOA representation is the spatial spectrum obtained in the dual domain of Spherical Harmonics. The spatial spectrum is opposed to the time spectrum which is derived from the Fourier Transform of a time signal.

References

Bamford, J. S. (1995). *An Analysis of Ambisonics Systems of First and Second Order*, Ph.D. Thesis, University of Waterloo, Ontario, Canada.

Bleidt, R., Borsum, A., Fuchs, H., & Merrill Weiss, S. (2014). Object-based audio: Opportunities for improved listening experience and increased listener involvement. *SMPTE Conference Proceedings*, October 2014.

Blumlein, A. D. (1931). U.K. Patent 394325.

Craven, P. G., & Gerzon, M. A. (1977). U.S. Patent 4,042,779.

Daniel, J. (2001). *Représentation de champs acoustiques, application à la transmission et à la reproduction de scènes sonores complexes dans un contexte multimédia*, Ph.D. Thesis, University of Paris VI, France.

Daniel, J. (2009). Evolving views on HOA: From technological to pragmatic concerns. *Ambisonics Symposium 2009*, June 25–27, Graz.

Daniel, J., Nicol, R., & Moreau, S. (2003). Further investigations of higher order Ambisonics and wavefield synthesis for holophonic sound imaging. *114th AES Convention*, April 2003. Amsterdam.

Driscoll, J. R., & Healy, D. M. (1994). Computing Fourier transforms and convolutions on the 2sphere. *Advances in Applied Mathematics*, *15*, 202–250.

Epain, N., & Daniel, J. (2008). Improving spherical microphone arrays. *124th AES Convention*, May 2008. Amsterdam, Netherlands.

Farrar, K. (1979a). Soundfield microphone. *Wireless World*, October 1979, 48–50.

Farrar, K. (1979b). Soundfield microphone—2. *Wireless World*, November 1979, 99–103.

Gallo, E., Tsingos, N., & Lemaitre, G. (2007). 3D-Audio matting, postediting, and rerendering from field recordings. *EURASIP Journal on Advances in Signal Processing*, *2007*(1), 047970.

Gauthier, P. A., & Berry, A. (2006). Adaptive wave field synthesis with independent radiation mode control for active sound field reproduction: Theory. *Journal of the Acoustical Society of America*, *119*(5), May 2006, 2721–2737.

Geier, M., Ahrens, J., & Spors, S. (2010). Object-based audio reproduction and the audio scene description format. *Organised Sound*, *15*(3), 219–227.

Gerzon, M. A. (1973). Periphony: With-height sound reproduction. *Journal of Audio Engineering Society*, *21*(1), 2–10.

Gerzon, M. A. (1985). Ambisonics in Multichannel broadcasting and video. *Journal of Audio Engineering Society*, *33*(11), 859–871.

Gerzon, M. A. (1992). General metatheory of auditory localisation. *Proceedings of the A.E.S. 92nd Convention*, 1992.

Hibbing, M. (1989). XY and MS microphone techniques in comparison. Presented at 86th AES Convention, Hamburg. Preprint 2811 (A-5).

Makita, Y. (1962). On the directional localisation of sound in the stereophonic sound field. *E.B.U. Review*, June 1962, 102–108.

Middlebrooks, J. (1999). Virtual localization improved by scaling nonindividualized external-ear transfer functions in frequency. *Journal of Acoustical Society of America*, *106*(3), 1493–1510.

Moreau, S., Daniel, J., & Bertet, S. (2006). 3D Sound Field Recording with higher order ambisonics—objective measurements and validation of a 4th order spherical microphone. *120th AES Convention*, May 2006. Paris, France.

Morse, P. M., & Feshback, H. (1953). *Methods of Theoretical Physics*. New York: McGraw-Hill.

Morse, P. M., & Ingard, K. U. (1968). *Theoretical Acoustics*. New York: McGraw-Hill.

Nicol, R., & Emerit, M. (1999). 3D-sound reproduction over an extensive area: A hybrid method derived from Holophony and Ambisonic. *AES 16th International Conference on Spatial Sound Reproduction*, April 1999, Rovaniemi.

Poletti, A. (2005). Three-dimensional surround sound systems based on spherical harmonics. *Journal of Audio Engineering Society*, *53*(11), 1004–1024.

Ward, D. B., & Abhayapala, T. D. (2001). Reproduction of a plane wave sound field using an array of loudspeakers. *IEEE Transactions on Speech and Audio Processing*, *9*(6), September 2001, 697–707.

Zotkin, D. N., Duraiswami, R., & Gumerov, N. A. (2010). Plane-wave decomposition of acoustical scenes via spherical and cylindrical microphone arrays. *IEEE Transactions on Audio, Speech and Language Processing*, *18*(1), 2–16.

Appendix A

Mathematics and Physics of Sound Field

Equation of Acoustic Waves

Any acoustic wave is the solution of the general problem, defined as follows for a space-time domain $\Omega \times [t_1, t_2]$ by:

- The equation of acoustic waves, given by:

$$\left(\Delta - \frac{1}{c^2}\frac{\partial^2}{\partial t^2}\right)\psi(\vec{r},t) = - \ s(\vec{r},t)\forall\vec{r} \in \Omega, \forall t \in [t_1, t_2] \tag{9.24}$$

where $\psi(\vec{r},t)$ refers to the potential velocity (at location $\vec{r}$ and time t), which is linked to the acoustic pressure $p(\vec{r},t)$ and the particle velocity $\vec{v}(\vec{r},t)$ by the relations: $p(\vec{r},t) = \rho_0 \frac{\partial\psi(\vec{r},t)}{\partial t}$, and $\vec{v}(\vec{r},t) = -\vec{\nabla}\ \psi(\vec{r},t)$. In these equations, c and ρ_0 are respectively the speed of sound and the volumetric mass density of the propagation medium. The term $s(\vec{r},t)$ refers to the presence of acoustic sources.

- In combination with boundary conditions, which specify the values of either $\psi(\vec{r},t)$ or $\vec{\nabla}\ \psi(\vec{r},t)$ (or even both) on the boundary $\partial\Omega$ of the domain Ω, and can be used for instance to introduce the effect of walls in a room, with phenomena of acoustic reflection, scattering and diffusion, or diffraction.
- And in combination with initial conditions, which express the values of $\psi(\vec{r},t)$ and $\frac{\partial\psi(\vec{r},t)}{\partial t}$ at the starting time.

In this formulation, all the variables are expressed as a function of time. By taking the Fourier Transform of the equations, it is possible to derive the equivalent problem in the frequency domain. There are many ways to solve this problem.

Deriving the Solution of the Equation of the Acoustic Waves With Spherical Harmonics

One way to solve the equation of acoustic waves is to use eigenfunctions which constitute an orthonormal basis and which are able to represent any acoustic wave. If the coordinate system is spherical (i.e. any location $\vec{r}$ is described by a radius r, an azimuth angle ϕ and an elevation angle θ, see Figure 9.14), the eigenfunctions of the equation of acoustic waves are composed of Spherical Bessel functions (namely Spherical Bessel function of first kind $j_m(kr)$ and second kind $n_m(kr)$, and/or Spherical Hankel function of first kind $h^+_m(kr)$ and second kind $h^-_m(kr)$, respectively for decreasing r and increasing r propagating waves) and Spherical Harmonics $Y^{\sigma}_{mn}(\phi,\theta)$ (Morse & Feshback, 1953; Morse & Ingard, 1968). Spherical Bessel functions account for spatial variations as a function of radius, whereas Spherical Harmonics convey spatial variations as a function of azimuth and elevation angles (see Figure 9.5). The Spherical Harmonics are given by:

$$Y^{\sigma}_{mn}(\phi,\theta) = \sqrt{(2m+1)\epsilon_n \frac{(m-n)!}{(m+n)!}} P_{mn}(\sin\theta) \times \begin{cases} \cos(n\phi)\, si\ \sigma = +1 \\ \sin(n\phi)\, si\ \sigma = -1 \end{cases} \tag{9.25}$$

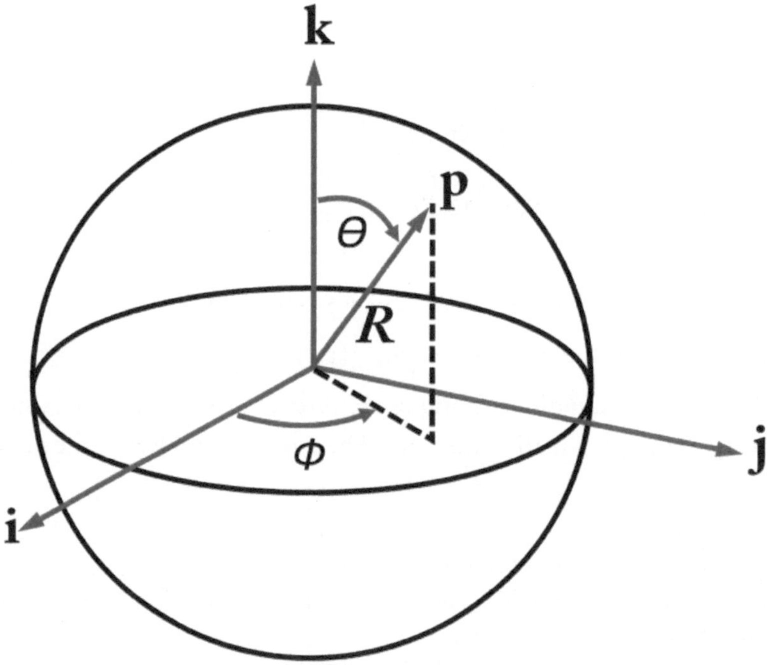

Figure 9.14 Vertical-polar spherical coordinates.

where the coefficient ϵ_n is equal to 1 if $n = 0$ and 2 if $n > 0$. The functions $P_{mn}(\sin\theta)$ are Legendre polynomials defined by:

$$P_{mn}(\sin\theta) = \frac{d^n P_m(\sin\theta)}{d(\sin\theta)^n} \tag{9.26}$$

where the function P_m is the Legendre polynomial of first kind of order n. Spherical Harmonics form a complete set of orthonormal functions for any square-integrable function on the unit sphere. Therefore they fulfill the orthonormality property in the sense of the product defined by:

$$\frac{1}{4\pi}\int_{\phi=0}^{2\pi}\int_{\theta=\frac{-\pi}{2}}^{\frac{\pi}{2}} Y_{mn}^{\sigma}(\phi,\theta)Y_{m,n}^{\sigma'},(\phi,\theta)\cos\theta d\theta d\phi = \delta_{mm'}\delta_{nn'}\delta_{\sigma\sigma'} \tag{9.27}$$

where $\delta_{mm'}$ denotes the Kronecker delta, equal to 1 if $m = m'$ and to 0 otherwise.

To express the acoustic pressure at a given location $\vec{r}$, a kind of "listening" area is created around this point. This area is delimited by two spheres of radius R_1 and R_2, such as $R_1 < r < R_2$, and is totally free of any acoustic source. Then the eigenfunctions previously introduced can be used to expand the acoustic wave. It is thus shown that the acoustic pressure p at location $\vec{r}$ resulting from any sound wave generated by acoustic sources located outside of the so-called listening area can be expressed as a linear sum of Spherical Bessel functions in combination of Spherical Harmonics:

$$p(\vec{r},\omega) = \sum_{m=0}^{+\infty} i^m h_m^-(kr)\sum_{n=0}^{m}\sum_{\sigma=\pm1} A_{mn}^{\sigma}(\omega)Y_{mn}^{\sigma}(\phi,\theta) + \sum_{m=0}^{+\infty} i^m j_m(kr)\sum_{n=0}^{m}\sum_{\sigma=\pm1} B_{mn}^{\sigma}(\omega)Y_{mn}^{\sigma}(\phi,\theta) \tag{9.28}$$

where ω denotes the pulsation (i.e. $\omega = \frac{2\pi}{f}$, where f is the frequency). The coefficients A_{mn}^{σ} and B_{mn}^{σ} are the weights of the eigenfunctions and thus define the representation of the acoustic wave in the associated basis. In other terms, these coefficients are the equivalent of the coefficients of a Fourier series, but in the present case spatial variations are considered instead of time variations. It should be noted that the coefficients A_{mn}^{σ} and B_{mn}^{σ} depend on frequency f, and by this way convey spectral/time information. For an exact representation of the sound field, an infinite series of terms is required in Equation 9.28. However this series may be truncated to a finite number: for instance if only the Spherical Harmonics components up to order $m = M$ are kept, the series is composed of $(M + 1)^2$ terms. This result forms the fundamental idea of Ambisonics representation of a sound field.

As the signals $A_{mn}^{\sigma}(\omega)$ and $B_{mn}^{\sigma}(\omega)$ are the coefficients of the Spherical Harmonics expansion (see Equation 9.28), they are obtained by using the orthonormality property of Spherical Harmonics (see Equation 9.27). The amplitude of each component $A_{mn}^{\sigma}(\omega)$ and $B_{mn}^{\sigma}(\omega)$ is derived by computing the result, $U_{mn}^{\sigma}(\omega)$, of the projection of the acoustic pressure $p(r, \varphi, \theta, \omega)$ on the associated Spherical Harmonic $Y_{mn}^{\sigma}(\phi,\theta)$ in accordance with the product defined by Equation 9.27:

$$U_{mn}^{\sigma}(\omega) = \frac{1}{4\pi r^2}\int_{\phi=0}^{2\pi}\int_{\theta=\frac{-\pi}{2}}^{\frac{\pi}{2}} p(r,\phi,\theta,\omega)Y_{mn}^{\sigma}(\phi,\theta)\cos\theta d\theta d\phi \tag{9.29}$$

To compute $U_{mn}^{\sigma}(\omega)$, it is only required to know the acoustic pressure $p(r, \varphi, \theta, \omega)$ at any point on the whole surface of the sphere of radius r and centered at the origin of the coordinate system. The objective is to derive the signals $A_{mn}^{\sigma}(\omega)$ and $B_{mn}^{\sigma}(\omega)$ from $U_{mn}^{\sigma}(\omega)$. For this, in Equation 9.29, the acoustic pressure is replaced by its Spherical Harmonic expansion defined by Equation 9.28, leading to, through the orthonormality property:

$$U_{mn}^{\sigma}(\omega) = i^{m}\mathrm{h}_{m}^{-}(kr)A_{mn}^{\sigma}(\omega) + i^{m}\mathrm{j}_{m}(kr)B_{mn}^{\sigma}(\omega) \tag{9.30}$$

Unfortunately this equation is not sufficient to compute the signals $A_{mn}^{\sigma}(\omega)$ and $B_{mn}^{\sigma}(\omega)$. A second equation is needed. In addition to the acoustic pressure, the knowledge of the radial acoustic velocity $v_r(r, \varphi, \theta, \omega)$ will be used to get the quantity $V_{mn}^{\sigma}(\omega)$ (Daniel et al., 2003):

$$V_{mn}^{\sigma}(\omega) = \frac{1}{4\pi r^2}\int_{\phi=0}^{2\pi}\int_{\theta=\frac{-\pi}{2}}^{\frac{\pi}{2}} v_r(r,\phi,\theta,\omega)\,Y_{mn}^{\sigma}(\phi,\theta)\cos\theta d\theta d\phi \tag{9.31}$$

In the same way as for the acoustic pressure, by using the Spherical Harmonic expansion (see Equation 9.28) and the Euler Equation, which allows one to obtain the acoustic velocity from the acoustic pressure, Equation 9.31 becomes:

$$V_{mn}^{\sigma}(\omega) = \frac{i^{m-1}}{cr}\frac{\partial h_m^-}{\partial r}(kr)A_{mn}^{\sigma}(\omega) + \frac{i^{m-1}}{cr}\frac{\partial j_m}{\partial r}(kr)B_{mn}^{\sigma}(\omega) \tag{9.32}$$

From Equations 9.30 and 9.32, it is now possible to extract both the signals $A_{mn}^{\sigma}(\omega)$ and $B_{mn}^{\sigma}(\omega)$:

$$\begin{aligned} A_{mn}^{\sigma}(\omega) &= i^{-m}\frac{\frac{\partial j_m}{\partial r}(kr)U_{mn}^{\sigma}(\omega) - icrj_m(kr)V_{mn}^{\sigma}(\omega)}{\frac{\partial j_m}{\partial r}(kr)h_m^-(kr) - j_m(kr)\frac{\partial h_m^-}{\partial r}(kr)} \\ B_{mn}^{\sigma}(\omega) &= i^{-m}\frac{\frac{\partial h_m^-}{\partial r}(kr)U_{mn}^{\sigma}(\omega) - icrh_m^-(kr)V_{mn}^{\sigma}(\omega)}{j_m(kr)\frac{\partial h_m^-}{\partial r}(kr) - \frac{\partial j_m}{\partial r}(kr)h_m^-(kr)} \end{aligned} \tag{9.33}$$

Equation 9.33 shows how to compute the representation of any acoustic wave in the Spherical Harmonics domain from the knowledge of solely the acoustic pressure and the acoustic velocity on the surface of the sphere of radius r. Let us go back to Equation 9.28, which can be reinterpreted as the expansion of the acoustic wave as a superposition of wavelets of type $\mathrm{h}_m^-(kr)Y_{mn}^{\sigma}(\phi,\theta)$ and $\mathrm{j}_m(kr)Y_{mn}^{\sigma}(\phi,\theta)$. The amplitude of the former is $A_{mn}^{\sigma}(\omega)$, and that of the latter is $B_{mn}^{\sigma}(\omega)$. The wavelets of the first type propagates in the direction of increasing r and are therefore due to acoustic sources which are located inside the sphere of radius R_1, whereas those of the second type result from acoustic sources which are located outside the sphere of radius R_2

(Daniel, 2003). Thus, Spherical Harmonics expansion allows one to separate inside and outside components in the sound field. In most of the cases, there is no source inside the sphere of radius R_1, which leads to consider that all the signals $A^{\sigma}_{mn}(\omega)$ are null. Equation 9.28 becomes then:

$$p(\vec{r}, \omega) = \sum_{m=0}^{+\infty} i^m j_m(kr) \sum_{n=0}^{m} \sum_{\sigma=\pm 1} B^{\sigma}_{mn}(\acute{E}) Y^{\sigma}_{mn}(\phi, \theta) \tag{9.34}$$

By default, Spherical Harmonics expansion refers to this expression. In consequence, the sound field is fully described only by the signals $B^{\sigma}_{mn}(\omega)$.

Deriving the Solution of the Equation of the Acoustic Waves With Green Functions

To solve the equation of acoustic waves (Equation 9.24), an alternative to Spherical Harmonics expansion is to use the Green function related to the problem (Morse & Feshback, 1953). In a similar way as previously, the considered domain Ω is divided into two subareas, Ω_1 and Ω_2. All the acoustic sources are exclusively in Ω_1, so that Ω_2 defines a "listening" area which is disturbed by no source. By applying the Green theorem, it is shown that the acoustic pressure p at any location $\vec{r} \in \Omega_2$ may be expressed as a 2D integral called the "Kirchhoff-Helmholtz" integral:

$$p(\vec{r}, \omega) = \iint_{\partial\Omega_0} \left[g(\vec{r} - \vec{r_0}, \omega) \vec{\nabla} p(\vec{r_0}, \omega) - p(\vec{r_0}, \omega) \vec{\nabla} g(\vec{r} - \vec{r_0}, \omega) \right] \cdot \vec{n} dS_0 \tag{9.35}$$

where g is the associated Green function, $\partial\Omega_0$ is the boundary separating the two subareas Ω_1 and Ω_2, and $\vec{n}$ is the unit vector perpendicular to the surface $\partial\Omega_0$. Similarly to Equation 9.28, in Equation 9.35 the acoustic wave is a sum of elementary components (another type of wavelets), which reminds us of the Huyghens' Principle. A second observation which is common with the Spherical Harmonics expansion is that both the acoustic pressure $p(\vec{r_0}, \omega)$ and the pressure gradient $\vec{\nabla} p(\vec{r_0}, \omega)$ (in other words the acoustic velocity through Euler's formula) are needed to fully describe the sound wave (Daniel, 2003). It should be kept in mind that both Equation 9.28 and 9.35 are equivalent and exact representations of the acoustic pressure. What's more it was shown in Nicol and Emerit (1999) that Equation 9.28 can be deduced from Equation 9.35 under some assumptions (namely the acoustic wave is a plane wave, and the boundary $\partial\Omega_0$ is a circle which extends into the infinite).

Appendix B

Mathematical Derivation of W, X, Y, Z

There is another interpretation of the components (W, X, Y, Z). The component W is recorded by an omnidirectional microphone, which, in practice, corresponds to a pressure microphone. In other words, the component W is assimilated to the acoustic pressure. In the same way, the components (X, Y, Z) are recorded by figure-of-eight microphones, which are obtained by combining pressure gradient microphones. Because of the Euler's relation (see Equation 9.37) between the pressure gradient and the particle velocity, the components (X, Y, Z) can therefore be assimilated to the x-, y- and z-components of the particle velocity (X, Y, Z). Now, from the point of view of sound field capture, what is the meaning of these signals (W, X, Y, Z)? Let us take the case of an acoustic plane wave. The acoustic pressure is given by:

$$p(\vec{r},\omega) = p_0 e^{j\vec{k}\cdot\vec{r}} \tag{9.36}$$

where p_0 is the wave amplitude and $\vec{k}$ the wave vector. Through Euler's formula, the particle velocity is derived from the pressure gradient:

$$\vec{v}(\vec{r},\omega) = j\vec{k}p_0 e^{j\vec{k}\cdot\vec{r}} \tag{9.37}$$

Let us consider the four signals (W, X, Y, Z) as the acoustic pressure and the particle velocity measured at the origin $\vec{r} = \vec{0}$:

$$\begin{aligned} W &= p(\vec{r},\omega) = p_0 \\ X &= v_x\left(\vec{0},\omega\right) = jp_0k_x \\ Y &= v_y\left(\vec{0},\omega\right) = jp_0k_y \\ Z &= v_z\left(\vec{0},\omega\right) = jp_0k_z \end{aligned} \tag{9.38}$$

The direct interpretation of this result is that the signal W is in fact the wave amplitude, whereas the signals (X, Y, Z) are the x-, y- and z-components of the wave vector, which means that these signals convey the information of propagation direction of the wave. In other words, they contain the spatial information. Thus, the signals (W, X, Y, Z) form a full representation of the acoustic wave. If the propagation direction of the wave is defined by the azimuth angle ϕ_0 and the elevation angle θ_0, the signals (W, X, Y, Z) become:

$$\begin{aligned} W &= p_0 \\ X &= jp_0 k \cos\theta_0 \cos\phi_0 \\ Y &= jp_0 k \cos\theta_0 \sin\phi_0 \\ Z &= jp_0 k \sin\theta_0 \end{aligned} \tag{9.39}$$

It is time to prove in the specific case of a plane wave that the signals (W, X, Y, Z) are the 0th and 1st order components of the Spherical Harmonics expansion. Like any acoustic wave, the plane wave defined by Equation 9.36 can be expressed as a linear sum of Spherical Harmonics $Y_{mn}^{\sigma}(\phi, \theta)$ (see Equation 9.34). In the case of a plane wave, the coefficients of this Spherical Harmonics expansion, i.e. the signals $B_{mn}^{\sigma}(\omega)$, are given by Morse and Feshback (1953) and Morse and Ingard (1968):

$$B_{mn}^{\sigma}(\omega) = p_0 Y_{mn}^{\sigma}(\phi_0, \theta_0) \tag{9.40}$$

If the Spherical Harmonics expansion is restricted up to order M = 1, only four terms are kept, corresponding to the 0th order and the three 1st order components:

$$\begin{aligned} B_{00}^{1}(\omega) &= p_0 Y_{00}^{1}(\phi_0, \theta_0) = p_0 \\ B_{11}^{1}(\omega) &= p_0 Y_{11}^{1}(\phi_0, \theta_0) = p_0 \cos\theta_0 \cos\phi_0 \\ B_{11}^{-1}(\omega) &= p_0 Y_{11}^{-1}(\phi_0, \theta_0) = p_0 \cos\theta_0 \sin\phi_0 \\ B_{10}^{1}(\omega) &= p_0 Y_{10}^{1}(\phi_0, \theta_0) = p_0 \sin\theta_0 \end{aligned} \tag{9.41}$$

This result is achieved by defining the expression of the Spherical Harmonics $Y_{mn}^{\sigma}(\phi_0, \theta_0)$ in accordance with Equations 9.25 and 9.26. A comparison of Equations 9.39 and 9.41 shows that the signals (W, X, Y, Z) are indeed the 0th and 1st order $B_{mn}^{\sigma}(\omega)$ coefficients of the Spherical Harmonics expansion of the plane wave, which gives a further insight into the physical meaning of the signals (W, X, Y, Z), and confirms their relevance as a full (though rough because of the truncation up to order M = 1) representation of the sound field.

Appendix C

The Optimal Number of Loudspeakers

Equation 9.18 defines a system of $(M + 1)^2$ linear equations with N_l unknowns (the loudspeaker input signals). Three cases must be distinguished (Poletti, 2005). First, if $N_l < (M + 1)^2$, the problem is overdetermined and is solved by quadratic minimization. Second, if $N_l = (M + 1)^2$, the matrix **L** is square. Provided that its inverse L^{-1} exists, the decoding matrix is given by $D = L^{-1}$. Third, if $N_l > (M + 1)^2$, the problem is underdetermined and has an infinity of solutions. The solution that minimizes the signal energy is achieved through the pseudoinverse of **L** and the decoding matrix is then $D = L^t(LL^t)^{-1}$. In practice, it is recommended to choose the number of loudspeakers as $N_l = (M + 1)^2$, which ensures an optimal reproduction of the sound field (Poletti, 2005).

Chapter 10

Wave Field Synthesis

Thomas Sporer, Karlheinz Brandenburg, Sandra Brix, and Christoph Sladeczek

Motivation and History

Perhaps the earliest example of an immersive sound system is the acoustic curtain developed by Steinberg and Snow at AT&T Bell Laboratories and published in 1934 (Steinberg & Snow, 1934). Several microphones placed in a row in the recording room were wired 1:1 to loudspeakers in the reproduction room. This is depicted in Figure 10.1.

A few years later, researchers at Bell Labs suggested reducing the number of loudspeakers of the acoustic curtain down to three channels. Later, in 1953/1955 Snow reported that "the number of channels will depend upon the size of the stage and listening rooms, and the precision in localization required" (Snow, 1955, p. 48). The effect of adding channels is not great enough to justify additional technical and economical efforts (Snow, 1955). Snow's suggestion indeed came to pass, as 2-channel stereo systems did eventually become the standard in many homes.

As discussed in Chapter 6, stereo systems can lack precise virtual sound source positioning especially when working with phantom images. Further, the existence of a stereo *sweet-spot* limits the area where accurate spatial impressions and immersion can be experienced within a listening room. There have been a number of proposals for systems that use more than two loudspeakers to overcome these limitations, and Wave Field Synthesis (WFS) is one such system.

As proposed by Gus Berkhout in 1988 (Berkhout, 1988), the idea for WFS originally comes from seismic research and oil exploration. Sound waves created by explosions traveling through different layers of soil are reflected and bend at boundaries. These wavefronts are recorded with an array of microphones to analyze certain properties of the layers. Based on this theory, Berkhout had the idea to replace the microphones with loudspeakers making it possible to regenerate a sound field.

WFS can be considered a *holophonic* method of sound reproduction, capable of generating sound fields that maintain the temporal and spatial properties that represent virtual sound sources within an area bounded by loudspeakers. Using WFS, virtual sources can be placed not only along the speaker array, but behind and in front of the array as well. This is a remarkable feature that sets WFS apart from conventional stereo and surround systems. Due to its ability to enable the free positioning of virtual sound sources, WFS can be classified as an object-based audio method and therefore has several advantages when compared to discrete channel-based audio reproduction methods. With the recent progress made in microelectronics and the decreasing

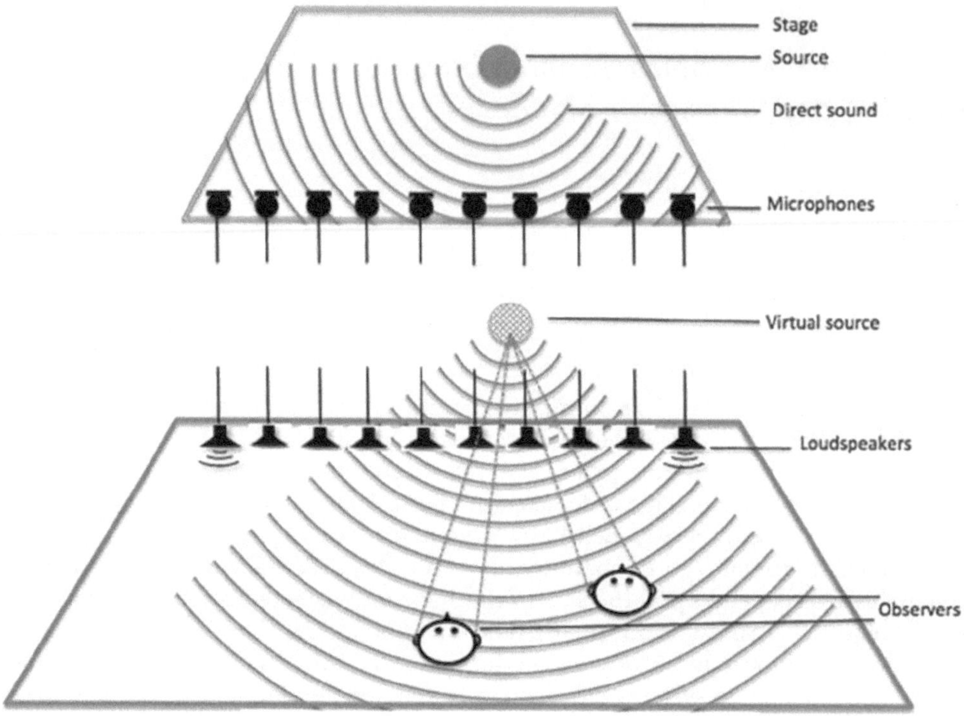

Figure 10.1 The acoustic curtain based on the original concept from 1934 (Snow, 1955). See Figure 6.1 for original drawing.

costs of computing power, loudspeakers, and power amplifiers, WFS systems have become more accessible and have appeared in the commercial marketplace.

Mathematical Background—From the Wave Equation to Wave Field Synthesis

Since an array of loudspeakers is used to synthesize a sound field representing virtual sources in WFS, a driving function is needed to calculate the individual speaker signals. In this section, an introduction to the mathematical concept behind WFS is given. The reader who would like to get a deeper mathematical background in WFS is referred to the referenced literature.

The concept of wave field synthesis is based on Huygen's Principle, published in 1690, which is shown in Figure 10.2.

Huygens' Principle states that a propagating wave front of a *primary source* Ψ can be synthesized by an infinite number of so-called *secondary sources*, placed on the primary source's wave front. To reconstruct the wave front, all secondary sources are fed by the signal emitted from the primary source. The superposition of all secondary source signals results in an accurate copy of the primary source wave front. Looking at Figure 10.3, we can assume that the primary

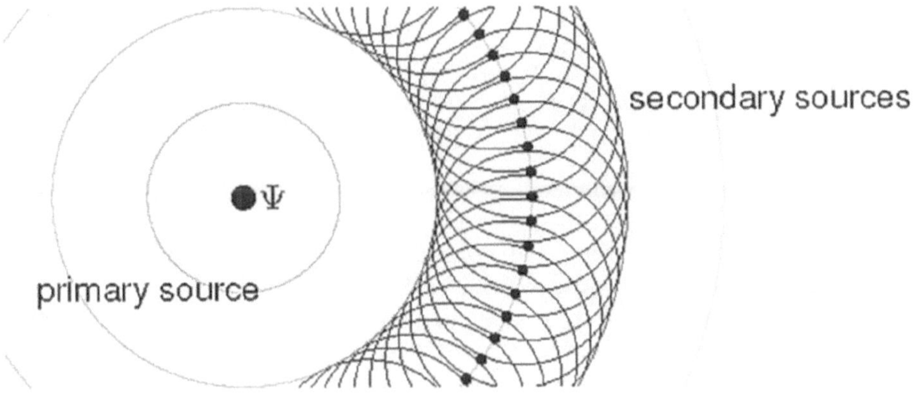

Figure 10.2 Huygens' Principle. The wave front of a primary source Ψ can be synthesized by an infinite number of secondary sources.

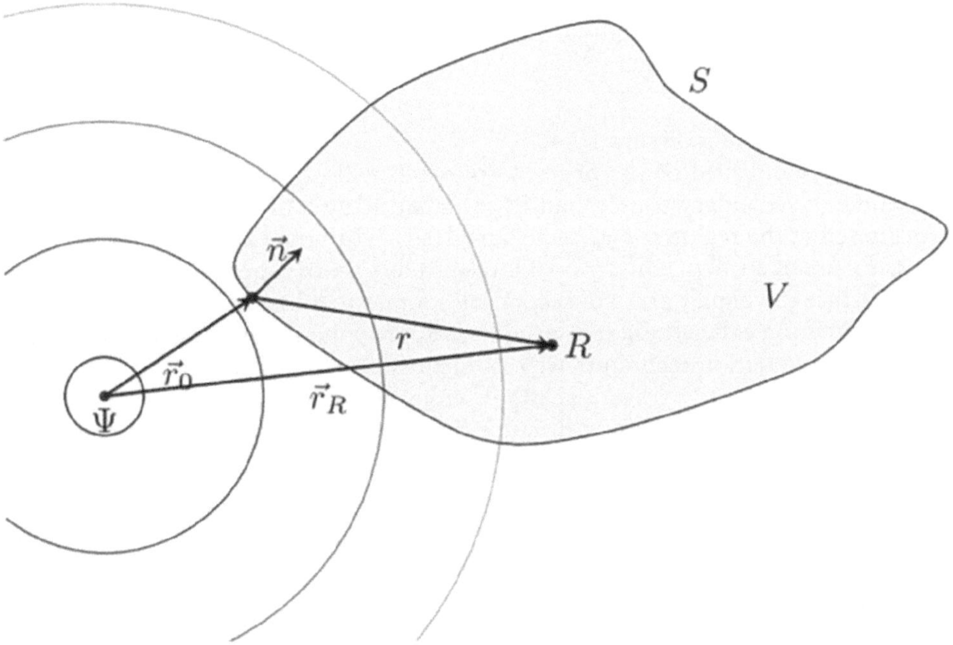

Figure 10.3 Geometry of the Kirchhoff-Helmholtz integral.

sound source Ψ causes a sound pressure $P(\vec{r}_R, \omega)$ at the listening position R inside a volume V. If the sound pressure and velocity, caused by the primary source, is known on the surface S, then the sound pressure field at R can be reconstructed by using an infinite number of monopole and dipole sources on the three-dimensional surface S. The general mathematical description of this principle in the frequency domain was given by Kirchhoff in 1883, which is known as the Kirchhoff-Helmholtz integral.

$$P(\vec{r}_R,\omega)=\frac{1}{4\pi}\iint_{-\infty}^{\infty}\left[\left(\underbrace{-\nabla P(\vec{r}_0,\omega)\frac{-jk|\vec{r}|}{|\vec{r}|}}_{\text{monopole sources}}+\underbrace{P(\vec{r}_R,\omega)\nabla\frac{-jk|\vec{r}|}{|\vec{r}|}}_{\text{dipole sources}}\right)\right]dS_0 \tag{10.1}$$

To calculate the sound pressure at the wave front ω is the angular frequency, $k=\omega/c$ is the wave number, c is the speed of sound, $\vec{n}$ is the normal vector to S and $\vec{r}=\vec{r}_R-\vec{r}_0$.

The realization of this integral is hard to achieve because it means covering the volume around the listening space with an infinite number of monopole and dipole loudspeakers. By eliminating one secondary source type and choosing a special geometry of V the Kirchhoff-Helmholtz integral can be reduced to Rayleigh's integrals incorporating some further limitations (Start, 1997). Assuming only secondary sources with monopole characteristics, the Rayleigh integral is given as:

$$P(\vec{r}_R,\omega)=\iint_{-\infty}^{\infty}\underbrace{\frac{1}{2\pi}\left(-\vec{n}.\nabla P(\vec{r}_R,\omega)\right)}_{Q(\vec{r},\omega)}\frac{e^{-jk\Delta r}}{\Delta r}dxdz. \tag{10.2}$$

with the geometry depicted in Figure 10.4.

In this case the sound field of the primary source Ψ will no longer be synthesized by a surrounding volume of secondary sources but by a planar array S of monopoles. Furthermore, the synthesis is limited to the region $y < y_s$ (see Figure 10.4). With the exception of some very special applications, the use of a full planar array of loudspeakers is not practical (Reussner et al., 2013).

In order to reduce the planar array of secondary sources to a line array of transducers, a mathematical approximation called the stationary-phase approximation is applied to Equation (10.2). Based on this approximation technique we can identify the loudspeakers along y that have the greatest impact on the sound pressure at the listening position R. As the energy radiated by a sound source decreases with increasing distance from the source, we find that speakers that are far enough away can be neglected. In order to derive a simple wave field synthesis driving function, the primary source is assumed to be omnidirectional. Mathematically this is described by the following formula:

$$P_\Psi(\vec{r},\omega)=S(\omega)\frac{e^{-jkr}}{r}. \tag{10.3}$$

Inserting Equation (10.3) into Equation (10.2) and applying the simplification, the secondary source driving function for linear loudspeaker arrays yields this equation (Verheijen, 1998):

$$Q(\vec{r},\omega)=S(\omega)\sqrt{\frac{jk}{2\pi}}\sqrt{\frac{\Delta r}{r+\Delta r}}\cos\varphi\frac{e^{-jkr}}{\sqrt{r}}, \tag{10.4}$$

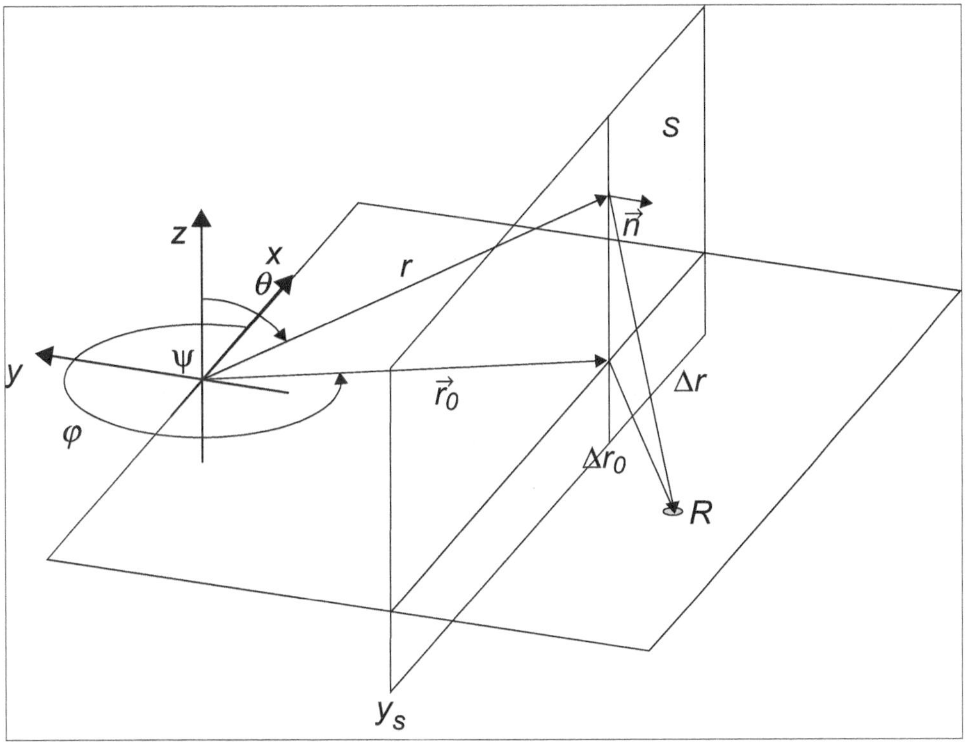

Figure 10.4 Geometry used for Rayleigh's integral and derivation of the 2.5D secondary source driving function.

The synthesis integral is as follows:

$$P(\vec{r}_R, \omega) = \int_{-\infty}^{\infty} Q(\vec{r}, \omega) \frac{e^{-jk\Delta r}}{\Delta r} dx. \tag{10.5}$$

Equation (10.4) is called the 2.5D synthesis operator whereas $S(\omega)$ is the input signal of the virtual source in the frequency domain. The prefix 2.5D, as opposed to 3D, indicates that the sound field of the virtual source is only synthesized on the horizontal plane when $z = 0$. The underlying geometry is depicted in Figure 10.4. As a result of the approximation, the distance-dependent pressure loss of the virtual source does not match the real source. This can be seen by comparing the last term of the synthesis operator and Equation (10.3). While the distance-dependent attenuation of the virtual source $1/\sqrt{r}$ is equal to a line source, a behavior of $1/r$ was desired. To compensate for this the individual pressure behavior, described by the term $\sqrt{\frac{\Delta r}{r + \Delta r}}$, matches both on a reference line. Another result of the mathematical approximation is the term $\sqrt{(jk/2\pi)}$ that represents a static high-pass filter since it does not depend on the virtual source position. The cosine term describes another gain, depending on the direction of the primary source according to the secondary source. The last part, $e^{(-jkr)}$, defines a frequency-independent delay, related to the distance of the primary source to the secondary source in the time domain. Using this synthesis technique, a virtual source from the perspective of the listener located behind the loudspeaker array can be created. Depending on the position of these *distant sources* called *objects*, the virtual wave front is constructed. Figure 10.5 shows a simulation of two individual positions. Image (a) corresponds to a virtual source positioned 1 m behind the loudspeaker array whereas image (b) shows the virtual source is placed 1,000 m behind the loudspeaker array.

Focused Sound Sources

The principle of time reversal is a common technique in signal processing (Fink, 1992). It is based on the assumption that the path between source and receiver, in this case the virtual sound source and listener, is reversible. Verheijen (1998) applied this technique to the 2.5D WFS operator resulting in the focusing operator:

$$Q_f(\vec{r},\omega) = S(\omega)\sqrt{\frac{k}{2\pi j}}\sqrt{\frac{\Delta r}{\Delta r - r}}\cos\varphi\frac{e^{jkr}}{\sqrt{r}}, \tag{10.6}$$

with the synthesis integral modified to:

$$P(\vec{r}_R,\omega) = \int_{-\infty}^{\infty} Q_f(\vec{r},\omega)\frac{e^{-jk\Delta r}}{\Delta r}dx \tag{10.7}$$

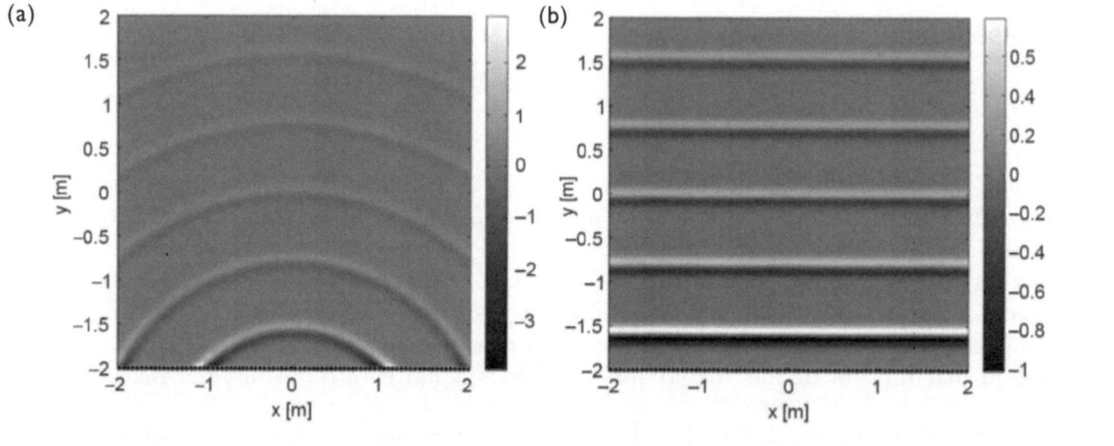

Figure 10.5 Synthesis of virtual distance sources using driving function $Q(\vec{r},\omega)$. In (a) the virtual source is placed 1 m behind the loudspeaker array; in (b) the source is placed 1,000 m behind the secondary source array.

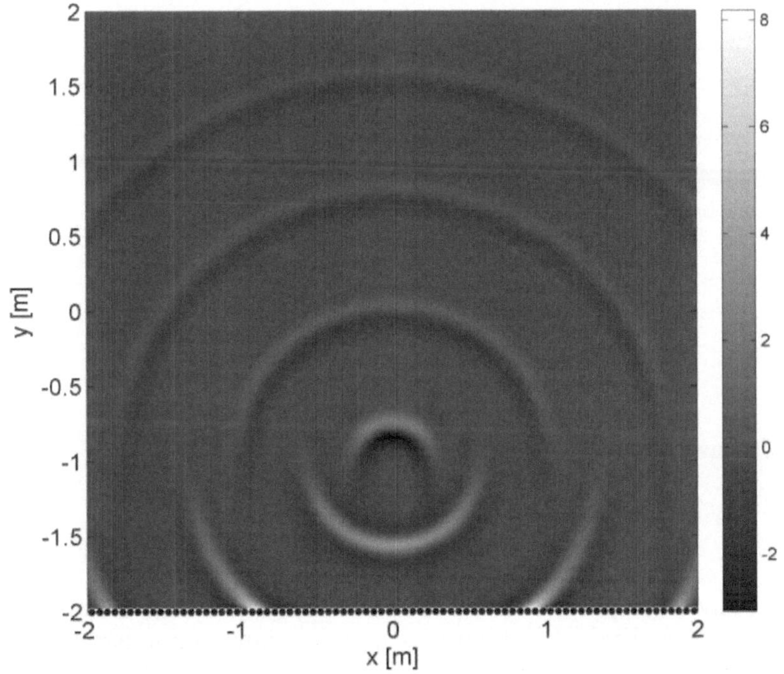

Figure 10.6 Simulated sound field of a focused virtual source using the 2.5D-focusing operator.

Figure 10.6 shows a simulation of a *focused sound* source that is created in front of the loudspeaker array.

Arbitrarily Shaped Loudspeaker Distributions

The 2.5D synthesis operator derived is only valid if a linear loudspeaker array is used. This also holds true for the focusing operator. In the derivation of the 2.5D synthesis operator the stationary-phase approximation was used. To reduce the number of loudspeakers for a distribution on a plane with a linear distribution, each column of speakers was analyzed, using Rayleigh's integral, according to their contribution to the overall pressure at the listener position R. To obtain a 2.5D synthesis operator, the loudspeaker with the shortest distance is the most relevant. If this concept is applied to the linear loudspeaker array, it can be shown that, for a single listener on the reference line, one loudspeaker will have the main influence on the synthesized sound pressure level (Start, 1996; Verheijen, 1998). This is visualized in Figure 10.7. The loudspeaker that contributes the most is always the one that is directly between the virtual source and the receiver position. The secondary source line, and the reference line, can be shaped arbitrarily using gain to

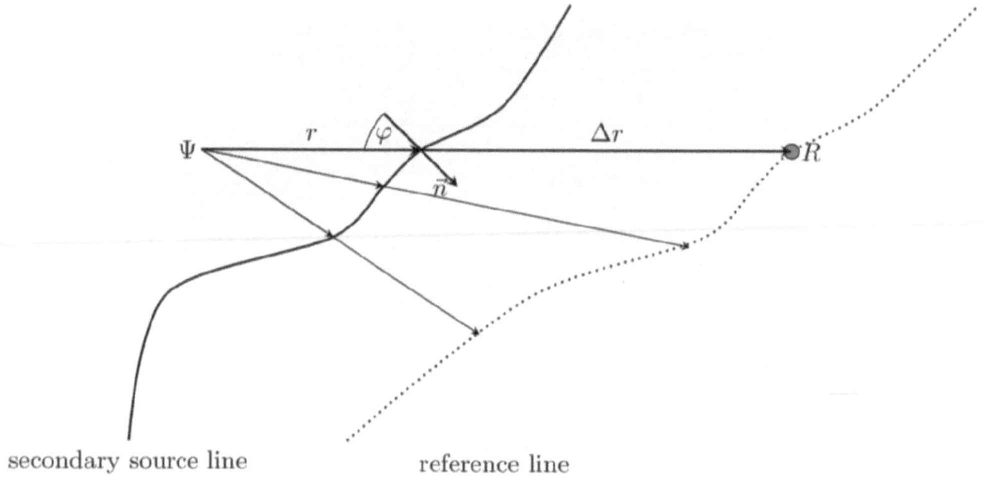

Figure 10.7 Geometry that can be used with 2.5D synthesis operator.

compensate for the incorrect pressure level of the virtual source. This can be determined individually for each loudspeaker. The only limitation is that one intersection between the line Δr and a secondary source and the reference line exists. The same is true for the path between secondary source and virtual source position. Because of this, loudspeaker array configurations can be in a line, or arranged with a soft bend.

Separation of Sound Objects and Room

Using a WFS system, it is possible to reproduce both point sources and plane waves. The sound field of a natural environment usually contains foreground sound objects that carry information about speech, objects close by, and music, as well as some information about the room. In general, foreground sound objects require a well-defined position in the reproduction space. During the recording process, they are often recorded with dedicated microphones to best represent point sources. The room information typically consists of early reflections and some diffuse reverberation. The pattern for the early reflections is influenced by the position of the sound source with respect to the listener within the room. In most applications, especially in small reproduction rooms, the listeners are not placed close to the walls of the virtual room. Therefore, the precise listener position is not that critical when creating virtual early reflections.[1] Similarly, creating diffuse reverberation is a stochastic process. It has been shown in perceptual experiments at TU Delft that for small rooms, eight plane waves are sufficient to encode the whole reflection pattern information (de Bruijn, Piccolo & Boone, 1998). Using eight plane waves to represent the virtual room acoustics, it is possible to separate the virtual foreground objects and room information.

Using the method below, it is possible to modify the position or level of the foreground objects to some degree in post-production.

- Foreground objects are recorded with close-up spot-microphones and their audio information together with the position of each sound object is stored and used in reproduction.
- In the production process the reflection pattern of all sound sources is recorded simultaneously or simulated separately and merged to form eight equally distributed plane waves. Any ambient (background) sound would be added to these eight plane waves.

However, larger modifications made to the foreground objects do not cause a proper modification of the room reflection patterns. A way to overcome this is to change the storage and reproduction strategy as follows (Brix, Sporer & Plogsties, 2001):

- Foreground objects are recorded with close-up spot-microphones and their audio information together with the position of each sound object is stored and used in reproduction.
- Background (ambient) sound is stored in eight plane waves.
- The room information is stored as sets of eight impulse responses for a set of expected positions of objects in the reproduction room. These impulse responses have been recorded and processed as eight plane waves.
- For the reproduction, the sound information of each sound object is convolved with the eight plane waves closest to the intended reproduction position of this object.

Using this approach it is possible to modify the position and level of sound sources. The room impression can be changed completely as well thus making it possible to apply the room acoustics of a target space to sound objects recorded at another location. To be as precise as possible, it would be necessary to record the room acoustics at many locations. This procedure may prove to be too expensive or time-consuming. Room reproduction in WFS can also be based on room simulations or hybrid schemes. Based on the position of an object in a room, the reflection pattern can be generated by a room simulation module. For example, if the image model (mirror source) simulation technique is used, it is useful to include some low-order mirror sources directly as point sources in the rendering. Abiding to the mirror source simulation method, the virtual sound source is mirrored at each room plane. Image sources are mirrored again, which leads to second, third, etc., order image sources. As a result, the original room can be represented by an infinite pattern. Higher-order mirror sources, or measured diffuse reverberation tails, are included as plane waves (Melchior, 2011).

Separation of Capturing and Reproduction

The acoustic curtain of Steinberg and Snow used a 1:1 relationship between microphones and loudspeakers. Using this technique in post-production becomes impractical. The mathematical background of WFS described herein only considers the reproduction of audio objects, sound capture is not considered. Several practical methods may be used to generate content which can be reproduced by WFS.

Spot-Microphones

A spot-microphone is located close to each audio source to be recorded. If the audio source is at a fixed location, the position of the microphone is measured once and used as metadata for this object. If the position of the object changes over time, either automatic or manual tracking is necessary to create the metadata associated with this object. In both cases, the position and acoustic properties of each object can be modified individually. Very often this method is used in combination with either some main microphones to capture the room reflections caused by all audio objects, or with some room simulation enabling to reproduce individual reflections for each audio object. The latter has the benefit that the positions of early reflections are changed correctly if the position of an audio object is changed in post-production. When recording complex scenes, it is often not possible to avoid crosstalk between microphones capturing different objects. This crosstalk can be minimized through microphone type selection or in post-production. For sound sources with non-uniform directivity or which have some spatial extension, like a choir, virtual panning spots are used (see below).

Acoustic Scene Analysis

Using the acoustic scene analysis technique, a number of microphones are placed around the acoustic space. The first step in post-production is to separate the signals into sound objects and residual sound. The residual sound usually contains ambient noise, room reflections, and diffuse reverberation. The number and position of microphones generally limits the separation of objects. Also, each microphone creates some noise. By increasing the number of microphones, the total noise will increase as well. It is therefore recommended to capture each object with the least number of microphones as possible.

Virtual Panning Spots

Somewhere between the spot-microphones and acoustic scene analysis lays the concept of virtual panning spots. Large sound sources, such as choirs, can be recorded with a small number of microphones resulting in a 2-channel stereo recording. Instead of reproducing these stereo signals via real loudspeakers, these signals are used as point sources in a WFS reproduction. In that way, virtual loudspeakers create virtual phantom sources (Theile, Wittek & Reisinger, 2002). By adjusting the position of the virtual loudspeakers, the width of large sound objects can be changed.

WFS Reproduction: Challenges and Solutions

A perfect reproduction can be obtained if all assumptions of the WFS driving functions are met; however, in practice, there are a number of restrictions that limit the performance of the WFS system. Some of the limitations are inherent to WFS, while other limitations are due to more practical issues. This section explains what challenges exist using WFS and how special algorithms can improve the performance of the system.

Distance of Loudspeakers and Alias Frequency

The Kirchhoff-Helmholtz integral and the Rayleigh integral are both based on a continuous driving function, assuming that an infinitely large number of infinitely small loudspeakers exist. Obviously, in practice, loudspeakers are not infinitely small and the distance between loudspeakers is limited by the size of the loudspeakers and enclosures. More often, the distance between loudspeakers has to be increased in order to reduce parts and installation costs.

The use of discrete loudspeaker positions is analogous to sampling in analog-to-digital conversion (ADC). In ADC, a continuous audio signal is discretized. The alias terms are in the frequency domain. In WFS theory, the mapping of audio objects to an infinite number of loudspeakers is continuous. In WFS systems, sampling is the process that leads to a discrete number of loudspeakers. This reduction introduces errors due to the spatial sampling of the wavelengths that causes spatial aliasing. It should be noted that in ADC, the aliasing occurs when capturing the content whereas in WFS, the alias occurs in the reproduction. In contrast to analog-digital conversion where the alias frequency is only dependent on the sampling rate, in spatial sampling, the alias frequency is also dependent on the position in the reproduction room and the direction of the wave front. The alias frequency $f_A = \frac{\omega_A}{2\pi}$ depends on the distance of the loudspeakers *seen* by the wave front. For plane waves this is shown in Figure 10.8. The following equation holds to determine the aliasing frequency:

$$f_A = \frac{c}{2\Delta x \sin\alpha} \tag{10.8}$$

where Δx is the distance between the loudspeakers and α is the angle between loudspeaker array and the wave front. It can be seen that the lowest alias frequency occurs, albeit the worst case, if the wave

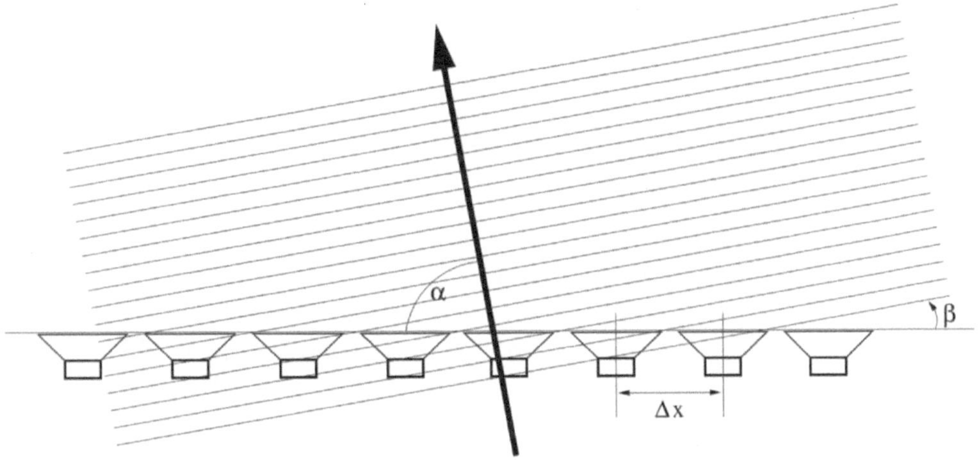

Figure 10.8 WFS reproduction of a plane wave. The aliasing frequency depends on the angle of incidence. In the example α is 80°.

fronts are parallel to the loudspeaker array. To generate wave fronts with a direction rectangular to the array, the alias frequency would have to be infinite.

Figure 10.9 shows a snapshot of the amplitude in space for a virtual sound source with a frequency higher than the alias frequency. Distortions are largest close to the loudspeaker array. The sound field becomes smoother with increased distance between the listener and the loudspeaker array (Corteel, 2006).

The distance between the centers of the adjacent membranes are used for the calculation of the distance between loudspeakers. A typical spacing of 17 cm results in a worst-case alias frequency of about 1 kHz. Calculations based on numerical simulations suggest that above the alias frequency, the sound field is severely distorted. However, listening tests show that distortion due to aliasing is often inaudible. There are several reasons for this effect:

- Spatial aliasing causes dips rather than peaks in the sound pressure. Dips are much less annoying than peaks, but simulations usually only consider the absolute difference between the natural sound field and the generated sound field. Therefore, in general, the perceptual effect of the error is overestimated.

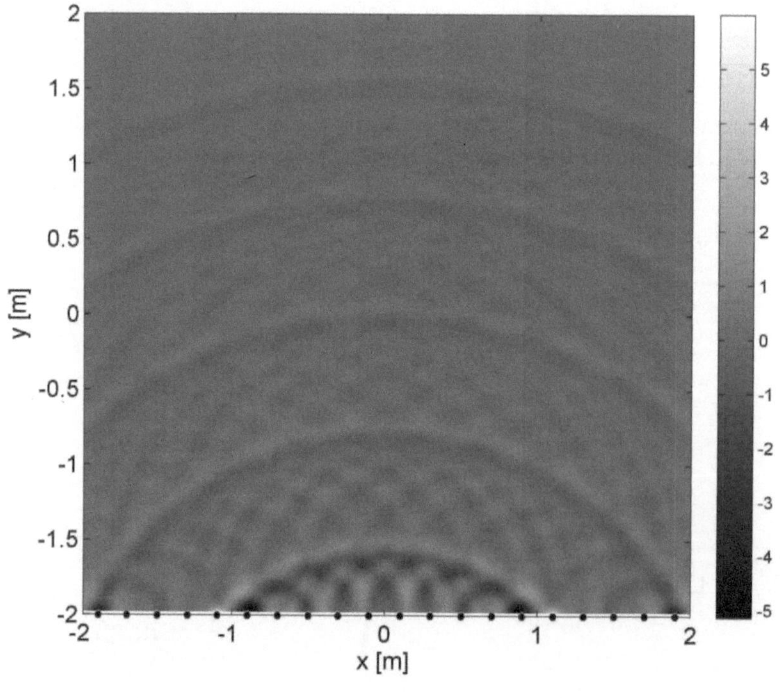

Figure 10.9 WFS reproduction: Sound field for a point source with a frequency above the alias frequency located 2 m behind the loudspeaker array. The black dotted line on the bottom represents the loudspeaker array.

- Just above the alias frequency the errors are still small. The largest errors occur at higher frequencies. At higher frequencies the position-dependent dips in the spectrum are very narrow band, but at these frequencies, the resolution of the human auditory system is poor. Narrow band dips are therefore only detectable if the input signal is extremely narrow band, and this rarely happens in practice.
- Just above the alias frequency, the dips are rather broad and not very deep. At higher frequencies, the dips are more deep and narrow. While simulations are based on ideal microphones with infinitely small sensors, the human auditory system averages the sound around the outer ear. Therefore, small-spaced frequency dips are inaudible.
- In natural indoor environments, there is an abundance of reflections causing narrow band spectral coloration of the sound. Humans usually adapt to environments and therefore do not perceive such constant sound colorations.

There are two situations when colorations due to spatial aliasing become audible; one is if the colorations are distinctly different in the two ears, and the other is if the listener and/or sound sources are moving quickly through the room.

In the past, several attempts have been made to reduce the problem of spatial aliasing. The OPSI[2] concept, invented by Helmut Wittek (2007), combines WFS at low frequencies with traditional 2-channel stereo at higher frequencies. Below the alias frequency, the loudspeaker array is controlled via WFS. Above the alias frequency, only the two loudspeakers closest to the sound source position are used with amplitude panning. Wittek proved that the sound coloration based on this approach is less audible than for pure WFS. Although, there are some disadvantages using the OPSI method. First, the objects with dominant high frequencies always sound like they are between the loudspeakers, and second, the distance and size of objects is not reproduced.

Other attempts are based on the assumption that listeners are usually not located throughout the entire listening room, but are restricted to a smaller listening area. Based on this assumption it is possible to control the sound field in a way that reduces coloration for that area (Franck et al., 2007; Spors, 2006; Spors, 2007; Ahrens & Spors, 2008; Melchior et al., 2008).

Limited Length of the Loudspeaker Array—Level

The mathematical derivation for WFS assumes an infinitely long array of loudspeakers. In practice, the size of the reproduction room and the length of the array are limited. If a virtual sound source is in proximity to the loudspeaker array, the loudspeakers closest to the virtual sound source provide the major part of the sound energy. The *missing* loudspeakers outside the room would only contribute with an insignificant energy to the sound at the listener's position. If a sound source is far behind the loudspeaker array, all loudspeakers have to emit almost the same energy. The amount of sound energy not reproduced, due to the missing loudspeakers, is significant. In audio-visual installations, where virtual sound sources are moved around, this may cause a mismatch between the visual and the auditory position because the virtual sound source is losing loudness at a rate faster than it would in a natural acoustic environment.

An easy way to overcome this problem is to use an analysis-by-synthesis approach as follows: The renderer first simulates and compares the sound pressure level of the existing loudspeakers to the sound pressure level of an ideal reproduction system at a reference position in the reproduction room. Then, the driving signal of all existing loudspeakers is amplified to achieve the same reproduction level in both simulations. Due to the fact that the correction factor is only dependent on the geometry of the loudspeaker array and the position of the virtual sound source, the correction factor can be pre-calculated at once and stored. This analysis-by-synthesis approach also solves the problem of gaps in loudspeaker arrays. In many installations, there are places where loudspeakers cannot be installed such as at the sides of the screen in movie theaters where curtains are located. If a virtual sound source is moved in the proximity of such a gap, it would be reproduced too softly. The analysis-by-synthesis method described here automatically allocates the energy to the loudspeakers closest to the gap.

Limited Length of the Loudspeaker Array—Truncation

In addition to the level problem described above, the limited length of the loudspeaker array can also cause artifacts. A simulation plot of this effect can be seen in Figure 10.10. Shadow waves are emitted from the edges of the array. The effect is similar to a discrete Fourier transform (DFT) with a rectangular window. In a DFT, the solution is to use windowing functions that reduce the level of the samples at the beginning and at the end of each transform block. Examples of such windowing functions are the Hann window or the cosine taper. A similar solution can be applied to WFS. Here, the driving function for the loudspeakers close to the end of the array is modified by a window function. In most practical implementations, WFS is not used as a straight array alone, rather, systems have several arrays around the listener. Here, the loudspeakers on the side arrays can compensate for the missing contributions of the loudspeakers in the front. To avoid problems with moving sound sources around the *corner*, rectangular arrangements of arrays should be avoided. Very often, a small number of loudspeakers at 45° are inserted at the corners of the room (see Figure 10.11). Together, the analysis-by-synthesis approach mentioned earlier and these corner arrays avoid the spatial distortions caused by truncation.[3]

Position-Dependent Filtering

The driving function for pure WFS systems contain the term $\sqrt{(jk/2\pi)}$. This term represents a −3 dB per octave shelving filter designed to compensate for the excess bass amplification caused by sampling. The derivation of the driving function assumes that virtual sound sources are in the far field, not close to the loudspeaker array. If a virtual sound source is close to the array, the bass amplification becomes less. If the sound source is placed exactly at the position of a loudspeaker, no compensation filter is necessary. The necessary filter is dependent on the source position, the direction of the wave front, and finally the loudspeaker geometry. An analysis-by-synthesis approach is the most practical solution here to find the correct filter. Most implementations do not use different filter settings for each loudspeaker; instead, a common filter for each sound object is used. Due to the fact that the calculation of the filter coefficients is computationally complex, the values are often pre-calculated and stored. In general, a set of filter coefficients for each spatial region is sufficient.

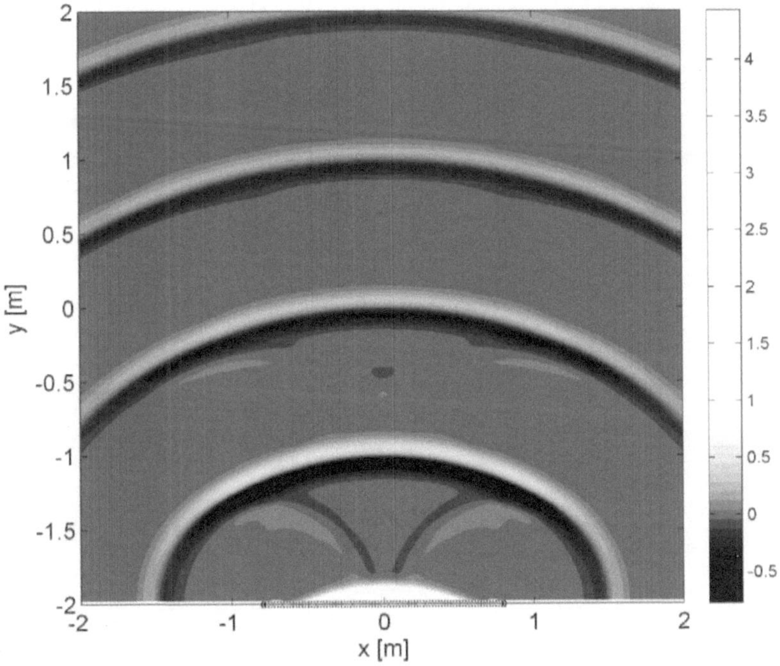

Figure 10.10 Distortions due to truncation (limited length of the loudspeaker array). The loudspeaker array is from –0.8 to 0.8.

Directivity of Loudspeakers

As described in the WFS theory section, the Kirchhoff-Helmholtz integral implies that an infinite number of monopoles and dipoles encircling the reproduction space are necessary to achieve *perfect* results. This assumes that the reproduced sound field outside the listening space does not extend behind the speakers. Practical implementations of WFS are usually based on monopole loudspeakers. The directivity of real loudspeakers is neither an ideal monopole, nor an ideal dipole, but rather a cardioid. Depending on the size of the membrane, the loudspeaker behaves like a monopole below some frequency, but becomes more directional at higher frequencies. Therefore, for stereo reproduction, 2-way and 3-way systems are used with separate drivers for the low, mid, and high frequencies, with a crossover system to split the input signal accordingly. In general, problems with phase coherence can occur at the crossover area. WFS is based on the phase coherent super-positioning of loudspeakers. Therefore, unclear phase behavior of the individual loudspeakers can be harmful. However, the perceptual trouble caused by the directivity of

Figure 10.11 Typical arrangement of loudspeakers arrays to reduce truncation artifacts.

sound sources is much more disturbing than problems due to unclear phase coherence, especially with moving sound sources (Klehs & Sporer, 2003).

In the past, attempts have been made to compensate for symptomatic directivity issues associated with loudspeakers (de Vries, 1996; de Vries, 2009; Ahrens & Spors, 2008). However, these approaches either lack flexibility to be used in real systems, are computationally very complex, or both.

Influence of the Reproduction Room

Using only monopoles has another unwanted side effect; the loudspeakers send as much energy to the back as towards the listener. Behind the loudspeakers there is often a reflective wall that sends part of the energy back to the listener. These unwanted reflections cause a super-positioning of reflected wave fronts with the intended signal from the front side of the loudspeakers. The reflection pattern depends on the position of the loudspeaker respective to the walls of the reproduction room. The perceptual process of the listener is influenced by this reflection pattern when localizing the position of the loudspeakers. If this happens, the super-positioned wave front from the loudspeakers' signal is distorted and the *perceived* position of the virtual sound source is closer to the loudspeaker array than intended. This effect is especially strong for focused sound sources.

The simplest way to overcome this problem is to acoustically treat the reproduction room. If there are no hard reflections from the walls, ceiling, and floor, localization of the individual loudspeakers will not occur. Acoustically treating the reproduction room also solves a more general problem. Often, the recording room (e.g. a concert hall) is larger than the reproduction room (e.g. a living room). If the reflection pattern of such a reproduction room contains hard reflections, they would arrive before the reflections intended in the recording. Strong early reflections lead to the perception of a small room first, and therefore instead of the intended large room a mixture of both rooms is perceived.

On the other hand, it has been found that some diffuse reverberation caused by the reproduction room can improve the quality of the overall experience. The diffuse components enrich the envelopment and may mask some otherwise audible spatial aliasing artifacts.

Time Domain Effects in Large Loudspeaker Arrays

In large auditoria, like Bregenz lakeside open-air stage in Austria with its 7,000 seats, pure WFS causes audible distortions. This effect in general is not audible for steady state sounds but is audible for transients. To understand this effect it is necessary to look at the impulse response and at perceptual factors. A plane wave that is parallel to the loudspeaker array is emitted by firing all of the loudspeakers at the same time. For long arrays, the signal from loudspeakers near the center is received earlier than the signal from loudspeakers near the end of the array. If this time difference is above the detection threshold for echoes,[4] the signals from the loudspeakers are no longer merged to form a single sound event. The detection threshold for this effect is greater than the detection threshold for a pure reflection because there are many loudspeaker signals filling gaps between the closest and the farthest loudspeakers, and also because the farthest loudspeakers provide less energy due to their greater distance. A solution to this problem is to subdivide the long array into subsections thus avoiding plane waves in long arrays.

WFS With Elevation

In traditional channel-based reproduction without height channels, a small number of loudspeakers are located in a single plane. The sweet-spot (place or area of optimal listening) is limited, therefore a critical listener needs to be close to or in the sweet-spot. Information about the elevation of sound sources and/or reflections from the ceiling are mixed in with the normal loudspeaker channels. Due to the fact that the position of the listener is well defined, binaural cues of the original recording are preserved and the listener has the impression that there is also sound from above. If the listener moves away from the sweet-spot the spatial information becomes distorted and the wrong elevation information is perceived. Because of the large deviations in the horizontal plane, the errors concerning reflections are not noticed. In WFS, the effective listening area encompasses almost the whole reproduction room. When listeners experience WFS, they often notice that information from above is missing. The improved spatial localization along the azimuth makes the missing information in the elevation direction more apparent. Several attempts have been made to solve this problem.

The most natural solution would be to extend the WFS system to a 3D WFS system. The additional loudspeakers would be located at several different elevations separated by a small distance. However this approach is very expensive and not adequate for most applications. There are three factors which help reduce the number of loudspeakers. First, the ears of humans are on the side of the head at about the same height; therefore, the precision of sound localization for elevation is much less than it is in the azimuthal direction. Second, in most situations, minor sound components, like ceiling reflections, are presented at the same time as sounds from the horizontal plane. Third, if a dominant sound is coming from above, humans turn their head to look in the direction of the sound. If the sound source is not visible, there is no indication of a wrong position; therefore, the loose perception that a sound source is somewhere above is sufficient. The number of height loudspeakers needed depends on the size of the reproduction space.

Newer WFS installations tend to use a small number of additional loudspeakers. To make content scalable, the position of audio objects is stored as 3D Cartesian coordinates (x, y, z) or as polar coordinates (azimuth, elevation, distance). The mapping of elevated sound sources to the actual loudspeaker setup is done in the rendering. While sound objects close to the horizontal plane are reproduced via WFS only (i.e. as point and focused sources) elevated sound objects can be reproduced as a combination of WFS and Vector Base Amplitude Panning (VBAP) for the elevated loudspeakers. For small reproduction sites like home cinemas, where there is no space for ceiling loudspeakers, binaural cues are used to simulate height loudspeakers. The signal that would be coming from the ceiling speaker, that does not exist, is filtered according to the Blauert's directional bands and reproduced via the existing loudspeakers in the horizontal plane. A detailed description of the algorithm is part of the MPEG-H 3D audio standard (ISO/IEC 23008–3, 2015).

Audio Metadata and WFS

There are three types of objects in WFS; point sources, focused sources, and plane waves. The main difference between a point source and a focused source is the algorithm used to calculate the signals that drive the loudspeakers. Depending on the size of the reproduction system, an object might be inside the encircling loudspeaker array or outside. Therefore, the metadata does not distinguish between these two sets. For both a point source and a focused source, the metadata contains the position of the object. For plane waves, just the direction is stored. In the past, Cartesian coordinates have been used (Brix et al., 2001), but today most systems are based on polar coordinates where the azimuth angle starts in the front and goes counterclock-wise with a range from 0° to 360° or from –180° to 180°. The elevation angle starts at the horizontal plane with a range from –90° to 90°. The distance is stored as well. A reasonable number of bits for the resolution of azimuth and elevation angles is 8 and 6, respectively. Some metadata sets use the same set for point sources and plane waves. The maximum distance is reserved as a flag indicating that the source is a plane wave. This format is possible because listeners cannot detect the curvature of nearly flat wave fronts with an acoustic horizon about 15 m away. Therefore, it is unnecessary to encode larger distances precisely. In many applications, a few additional values/flags are useful (Ruiz, Sladeczek & Sporer, 2015):

- Distance-Dependent Delay: When rendering moving objects, WFS generates a natural Doppler shift. Such an effect is not welcomed by most mix engineers. Therefore, a flag is set to change the behavior of the renderer to avoid the Doppler effect, and the arrival time of an object traveling to the center of the reproduction space is changed. By switching the distance-dependent delay off, the time of arrival becomes independent of the distance. It should be noted that when distance-dependent delays are switched off, the sound source close to the listener has a greater curvature than a sound source far away.

- Distance-Dependent Level: When the distance between an object and the listener position is increased, WFS creates a natural decrease in level. A flag tells the renderer to compensate for this decrease in level.
- Sound Object Visible: In cinema applications, sound objects might be also visible. For such applications, it is advisable to adjust the position of the audio object to the screen size. This flag indicates whether an object should be adjusted or not.

An additional element related to metadata is the calibration of reproduction systems. WFS rendering creates proper driving signals when the position of each loudspeaker is known. However, loudspeakers need to be calibrated as close as possible. As a result of the standardization of MPEG-H 3D audio, a calibration procedure beneficial for WFS was defined as follows:

- In the center of the reproduction system is an omnidirectional microphone.
- Each loudspeaker is equalized so that the frequency response at the microphone is flat.
- The delay of each loudspeaker is adjusted in a way that an impulse emitted by any loudspeaker arrives at the same time at the microphone.
- The level of each loudspeaker is adjusted so that the level of the signal emitted by any loudspeaker arrives with the same sound pressure level at the microphone.

In general, equalization, time, and level calibration are implemented as FIR filters for each loudspeaker.

Applications Based on WFS and Hybrid Schemes

Early developers of WFS thought of many applications and related systems. The European Union co-funded project CARROUSO (Brix et al., 2001) that demonstrated a complete system chain from the distributed microphones to the encoding of the object-based audio using MPEG-4 and WFS-based rendering. A similar system was demonstrated in 2003 at an Ilmenau movie theater. The 5.1 channel audio that was then available on any film was mapped to virtual loudspeakers and rendered in real time via a 192-speaker WFS system in the cinema. In addition native object-based demo content was reproduced, too.

Applications of WFS include cinema, concerts, planetariums, exhibitions, theme parks, automotive, medical rehabilitation, and VR caves. While the application of WFS to home cinema has been a goal from early on, there have only been a small number of prototype implementations of WFS in home cinema. On one hand, this is due to missing content for a wider audience; on the other hand, home acoustics are far from ideal for immersive audio rendering. One big obstacle for the implementation of a large number of WFS venues is the number of required loudspeakers. As discussed above, a good distance between speakers to reduce the amount of spatial aliasing below audibility is 17 cm or less. This translates to dozens of loudspeakers for living rooms and hundreds of loudspeakers for large auditoria. While in the latter case this has been done, it is prohibitive in terms of cost and, not to be neglected, the visual effects on the room. To overcome this limitation, a number of hybrid systems using both WFS theory and ideas derived from Vector Base Amplitude Panning (VBAP) (Pulkki, 1997) have been developed. Nearly all of the immersive sound applications in the past few years have fallen into this category.

WFS and Object-Based Sound Production

Multichannel audio systems have the potential to reproduce spatial sound. The production task includes creating impressions of sound coming from certain directions. One way to achieve this is by panning the audio signal. Mixing desks and digital audio workstations provide dedicated tools for this task. However, these tools assume that the audio signal is being played back over a standardized loudspeaker setup like stereo, 5.1, 7.1, and so forth. When varying existing loudspeaker setups or introducing new reproduction formats, this assumption will not be met. As a consequence of this, every new reproduction format will need a new panning tool.

In the field of audio post-production for motion pictures, the audio production process is currently highly parallel and segregated. The introduction of new spatial audio systems will require even more production steps, including spatial authoring, to accomplish a rich auditory experience for the audience.

Currently, most mixing and sound design processes are based on the channel paradigm, where the coding format defines the reproduction setup. For these systems, any changes will require doing the complete mix again. As described below, if the WFS processor used for reproduction knows the target setup, it can render the loudspeaker signals in a way that fits the targeted setup perfectly.

The mixing process for wave field synthesis is based on a sound object paradigm. This method overcomes the limitations of the channel-based approach. The position of the sound object in an audio scene is needed and determined during the mixing process. Tracks and channels, which are indirectly considered in the process, form a sound object, which can be moved in an audio scene using a WFS authoring tool. Besides the audio information, the sound objects can have directivity information and interact with the virtual spatial environment. Based on the virtual object position, information for the direct sound, early reflections, and diffuse reverberations can be calculated, processed, and rendered to any loudspeaker configuration. Room acoustic information can be based on real or artificial spatial data. Therefore, the final WFS mix does not contain loudspeaker-related material. All audio signals that represent sound objects in the final mix are sent to the WFS rendering processor, which calculates the signals for all loudspeakers for the reproduction.

Notes

1 As known from psychoacoustics listeners merge the first room reflections with the direct sound to localize sound objects. These early reflections are dependent on source and listener position. In small reproduction rooms the listener position is of minor influence.
2 Optimised Phantom Source Imaging of the high-frequency content of virtual sources in Wave Field Synthesis.
3 In research also circular arrays of loudspeakers have been used. While circular arrays do not have any problems with truncation, such geometry is not adequate for most practical applications.
4 The echo threshold for transients is between 50 ms (for speech) and 100 ms (for music) (Blauert, 1996).

References

Ahrens, J., & Spors, S. (2008). Notes on rendering focused directional virtual sources in wave field synthesis. *34: Jahrestagung für Akustik (DAGA)*. Dresden, Germany, Deutsche Gesellschaft für Akustik (DEGA).

Berkhout, A. J. (1988). A holographic approach to acoustic control. *Journal of the Audio Engineering Society (JAES)*, *36*(12), 977–995.

Blauert, J. (1996). *Spatial Hearing*. Cambridge: The MIT Press.

Brix, S., Sporer, T., & Plogsties, J. (2001). Carrouso—an European approach to 3d-audio. *Proceedings of the 110th Audio Engineering Society (AES) Convention*. Amsterdam.

Bruijn, W. de, Piccolo, T., & Boone, M. M. (1998). Sound recording techniques for wavefield synthesis and other multichannel sound systems. *Proceedings of the 104th Audio Engineering Society Convention*. Amsterdam.

Corteel, E. (2006). On the use of irregularly spaced loudspeaker arrays for Wave Field Synthesis, potential impact on spatial aliasing frequency. *Proceedings of the International Conference on Digital Audio Effects (DAFx-06)*. Montreal, Quebec, Canada.

de Bruijn, W., Piccolo, T., Boone, M.M. (1998). "Sound recording techniques for wavefield synthesis and other multichannel sound systems," *In: Proceedings of the 104th Audio Engineering Society Convention*, Amsterdam.

de Vries, D. (1996). "Sound reinforcement by wave_eld synthesis: Adaptation of the synthesis operator to the loudspeaker directivity characteristics," *Journal of the Acoustical Society of America (JASA)*, *44*(12): 1120–1131.

de Vries, D. (2009). *Wave Field Synthesis, Audio Engineering Society (AES), 2009*, AES Monograph.

Fink, M. (1992). Time reversal of ultrasonic fields. i. basic principles. *Ultrasonics, Ferroelectrics, and Frequency Control, IEEE Transactions on*, *39*(5), 555–566.

Franck, A., Gräfe, A., Korn, T., & Strauß, M. (2007). Reproduction of moving sound sources by wave field synthesis: an analysis of artifacts. *Proceedings of the 32nd International Conference of the Audio Engineering Society (AES)*. Hillerrod, Denmark.

Huygens, C. (1690). *Traité de la Lumière*. Leiden: Pieter van der Aa.

ISO/IEC 23008–3. (2015). Information technology: High efficiency coding and media delivery in heterogeneous environments-Part 3: 3D audio. *Standard, International Organization for Standardization*. Geneva, CH, October 2015.

Kirchhoff, G. (1883). Zur Theorie der Lichtstrahlen. *Annals of Physics*, *254*, 663–695.

Klehs, B., & Sporer, T. (2003). Wavefield synthesis in the real world: Part 1-in the living room. *Proceedings of the 114th Audio Engineering Society Convention*. Amsterdam.

Melchior, F. (2011). *Investigations on Spatial Sound Design Based on Measured Room Impulse Responses*, Ph.D. thesis, Technische Universiteit Delft, June 2011.

Melchior, F., Brix, S., Sporer, T., Roder, T., & Klehs, B. (2003). Wave field syntheses in combination with 2d video projection. *Proceedings of the 24th International Audio Engineering Society Conference: Multichannel Audio, the New Reality*, Banff, Canada.

Melchior, F., Sladeczek, C., de Vries, D., & Fröhlich, B. (2008). User-dependent optimization of wave field synthesis reproduction for directive sound fields. *Proceedings of the 124th Convention of the Audio Engineering Society (AES)*. Amsterdam.

Pulkki, V. (1997). Virtual sound source positioning using vector base amplitude panning. *Journal of Audio Engineering Society*, *45*(6), 456–466.

Reussner, T., Sladeczek, C., Rath, M., Brix, S., Preidl, K., & Scheck, H. (2013). Audio network based massive multichannel loudspeaker system for flexible use in spatial audio research. *Journal of the Audio Engineering Society (JAES)*, *61*(4), 235–245.

Ruiz, A., Sladeczek, C., & Sporer, T. (2015). A description of an object-based audio workflow for media productions. *Proceedings of the 57th Audio Engineering Society Conference: The Future of Audio Entertainment Technology—Cinema, Television and the Internet.*

Snow, W. B. (1955). Basic Principles of Stereophonic Sound. *Audio, IRE Transactions on Audio*, *3*(2), 42–53.

Spors, S. (2006). Spatial Aliasing Artifacts produced by Linear Loudspeaker Arrays Used for Wave Field Synthesis. *Proceedings of the 2nd International Symposium on Communications, Control and Signal Processing (ISCCSP)*. IEEE Signal Processing Society.

Spors, S. (2007). Extension of an analytic secondary source selection criterion for wave field synthesis. *Proceedings of the 123rd Audio Engineering Society (AES) Convention*. New York, USA.

Start, E. (1996). Application of curved arrays in wave field synthesis. *Proceedings of the 100th Audio Engineering Society (AES) Convention*. Copenhagen, Denmark.

Start, E. (1997). *Direct Sound Enhancement by Wave Field Synthesis*, Ph.D. thesis, Technische Universiteit Delft, June 1997.

Steinberg, J. C., & Snow, W. B. (1934). Physical factors. *Bell System Technical Journal*, *13*(2), 245–258.

Theile, G., Wittek, H., & Reisinger, M. (2002). Wellenfeldsynthese-Verfahren: Ein Weg für neue Möglichkeiten der räumlichen Tongestaltung. *22: Tonmeistertagung, Hannover, Germany, November 2002.* Verband Deutscher Tonmeister e.V.

Verheijen, E. (1998). *Sound Reproduction by Wave Field Synthesis*, Ph.D. thesis, Technische Universiteit Delft, January 1998.

Weinzierl, S. (2008). Handbuch der Audiotechnik. *Springer Science & Business Media, in German.*

Wittek, H. (2007). *Perceptual Differences Between Wave Field Synthesis and Stereophony*, Ph.D. thesis, University of Surrey, 2007.

Chapter 11

Applications of Extended Multichannel Techniques

Brett Leonard

Current immersive audio technologies offer today's content producers more creative and recreative possibilities than ever before. These systems can provide an enhanced sense of realism to accompany video, a multi-sensory immersive experience with video games, or rich, luscious musical soundscapes. The ability to place sound, not just in front, but 360° around the listeners, or even above and below them, drastically expands the acoustic space within which a recording engineer or mixer can work and create.

Even so, the fundamentals for creating high-quality audio content are the same regardless of format. The need for appropriate frequency spectrum content, proper dynamic range, and appropriate ambiance all carry over from as far back as the days of monophonic production. On the other hand, state-of-the-art immersive audio systems present the engineer with unparalleled creative possibilities. But, as exciting as these developments in immersive audio are, new systems necessitate new approaches to audio throughout the production process. For some engineers this may involve working with new and unique tool sets, but it is far more common to find engineers using existing tools in alternate manners, often unintended by the tools' designers, to achieve the desired result. The constantly evolving technological landscape coupled with an increasing demand for content often outpaces the development of tools needed for content creation.

Therefore, it then falls to the engineers to make use of their knowledge of the immersive audio possibilities and the system in question, its operating principles, and the tools at hand to fit the ever-changing immersive audio landscape.

Based on the experiences of the author, fellow engineers, and researchers, this chapter serves to present both concrete ideas and broader concepts that engineers working with immersive sound will find both creatively and pragmatically useful. These topics include issues surrounding sound source spreading, the use of preexisting content generated for stereo or lower-order surround systems, creative use of panning, the use of time-based effects and reverberation, and mixing basics for the newest generation of immersive audio systems. Although some of the topics discussed in this chapter may have been addressed previously in this book, we will now look at them from the practical standpoint of a mixing or recording engineer. By leveraging knowledge of human auditory perception, mixing and panning technologies, acoustics, and standard production practices, each engineer can create their own toolset for generating truly immersive audio.

Source Panning and Spreading

One of the first things engineers may attempt to do when presented with the opportunity to mix for an immersive system is to "spread the sound around". However, the ways to accomplish this deserve considerable discussion because the techniques used for positioning a sound source or spreading a sound source over an area of the soundscape will have a significant effect on the resulting mix.

Panning and Source Imaging Techniques for Immersive Mixing

There are almost as many systems for source panning and placement as there are engineers. Despite advances in source panners, many engineers still rely on time-proven techniques from conventional surround production, in part because they translate between many new immersive audio formats. An understanding of these techniques can help provide creative options regardless of format or system.

Use of Phantom Imaging

The use of the phantom image is by no means a novel concept, even in multichannel surround systems. The idea that a virtual or phantom sound source can be placed between two speakers (or between more with careful manipulation, such as vector-based amplitude panning) has been a key part of stereo and various surround sound formats since their inception (Blumlein, 1933). There is, however, some debate as to the role and general usefulness of phantom imaging within today's state-of-the-art, dimensionally expanded multichannel immersive audio systems.

The primary use of phantom imagery is to explore the possibility of different localization cues across the listening area. Immersive audio allows the use of the entire spatial envelope engulfing the listener, for placing sources anywhere in azimuth, elevation, and distance. Although, if a listener is in close proximity to a loudspeaker in a multichannel system, the phantom image source will be overtaken by the closest loudspeaker. The result is a breakdown of the phantom image, thereby relocating the source. This phenomenon is of particular concern in home and commercial theater environments where some audience members will no doubt be closer to one side and/or a rear loudspeaker than others. Cinema mixers and sound designers take particular care to ensure that sources are stable in their perceived location within the room, and that surround sources are not overwhelming to the audience members closest to those loudspeakers. By placing sources exclusively in a single loudspeaker, the listener's *perspective* may change depending on their proximity to a given loudspeaker, but the source will remain anchored to the fixed point of the loudspeaker in the room. Likewise, distributing a sound to a great number of speakers (without some signal alteration or decorrelation) can increase the likelihood of noticeable phase cancelations, particularly when a listener moves within the speaker setup.

On the other hand, some engineers specifically try to *avoid* placing a source in only one loudspeaker. Historically, some practical concerns were raised with single-speaker sources, namely the ability of the consumer to easily isolate a source at home. This was a particularly prevalent concern for vocals in early surround music mixes where some artists are purported to have refused to allow themselves to be exclusively present in the center channel/loudspeaker. More practical to today's engineer is the concern over loudspeaker placement and radiation pattern varying from playback system to playback system. While these factors are by no means eliminated by choosing to create phantom image sources, they theoretically maintain source positions more reliably in any *properly* aligned playback system. While timbre differences will undoubtedly still exist between loudspeakers, the phantom image should still appear in the same relative position between loudspeakers. This concept is drawn to its logical conclusion in the world of object-based audio systems, where source reproduction is essentially playback system agnostic (see Chapter 8 for a complete discussion of object-based immersive systems). Rendering of object-audio sources is carried out through loudspeaker channels according to a defined layout of loudspeakers in a given listening room.

To the creative engineer, however, either of the above hard-line approaches to source panning are somewhat limiting. A world of creative opportunity lies in the combination of both speaker-specific and phantom image-based source positioning. The effects that engineers may try to avoid by using or disusing phantom imaging can actually act as tools to differentiate between two sources that occupy relatively the same position. The audible difference primarily manifests itself as a difference in perceived depth of the source. Typically, the source placed in a single loudspeaker appears to be more present and concrete in its position than a phantom-imaged source in the same location. This leads to the appearance of a more diffused phantom source located *behind* the source emanating from a single loudspeaker.

A typical application of the phantom image versus specific speaker panning paradigm is in the panning of kick drums, snare drums, bass guitars, and vocals in a multichannel/immersive pop or rock mix. It is a common practice to pan all of these elements to the center of a stereo mix. Placing all of these items in the same location can often lead to competition for level, frequency bandwidth, and ultimately for the listener's attention between these important sources in a conventional mix. Instead, an engineer may choose to phantom-image the kick and snare drums and bass guitar to the center of the mix, while placing only the vocals in a discrete center-channel loudspeaker (Figure 11.1). This helps the vocals to stand out slightly, despite being in the same physical position as the other three sources. In addition, there is an advantage realized in lowering the intermodulation distortion and increasing the headroom in the given channel. This approach is also quite typical in surround film and television mixes, where the center channel can be used almost exclusively for dialog while the adjacent loudspeaker channels on either side can be used to place sound effects, underlying ambiance, or even the soundtrack in a similar position.

Delay Panning

The use of short temporal delays between loudspeakers can also be useful in panning sources in immersive audio systems. Placing an identical signal in two loudspeakers and delaying one of these by anywhere between a few microseconds and 20 msec exploits the precedence effect, shifting the source's perceived location towards the loudspeaker reproducing the sound earlier. In essence, this technique emulates the operation of a purely time-of-arrival-based stereo microphone system (e.g. a pair of spaced omnidirectional microphones). The technique becomes even more powerful as a mixing tool when also imposing slight spectral alteration on one or more channels of the sound. Often a very mild subtractive high-shelf filter or low-pass filter can reinforce the realism of a delay-panned source, as they mimic the loss of high-frequency energy at the contralateral ear when a source's energy diffracts around the head.

Delay panning can also be extraordinarily useful in large-scale surround and immersive mixes, such as presented in the NHK 22.2 format. Delay panning is most useful when applied to

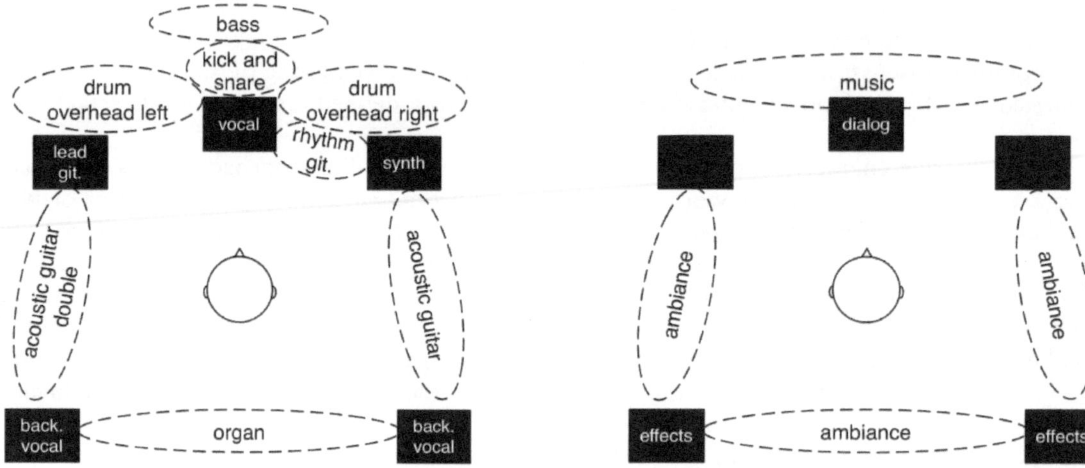

Figure 11.1 Examples of overlapping phantom image and discrete source positioning. Phantom sources are shown in dashed enclosures, while discrete sources are labeled within their assigned channels. Left: A 5-channel surround mix of popular music with the kick drum, snare drum, and bass guitar co-located with vocals. By routing the vocals directly to the center channel loudspeaker, a sense of depth is created, with solid localization yielded to the musically important vocal track. Right: Position of stereo music across the frontal image, but using only the left and right channels. This allows dialog to be centrally located, without possible interference from center-panned music sources.

important sources that must remain at the forefront of the mix, despite not being located front and center (Figure 11.2). The delay panning approach to these sources allows for their overall higher output level and perceived loudness even when a listener turns their head away from the primary source location. In essence, this forms a sparse wave field synthesis system, where the source is presented by a great number of transducers with precisely calculated delays. While most immersive systems do not have the precision of control to fulfill the Huygen-Fresnel principle (see Berkhout, 1988 and Spors et al., 2008), a course approximation can be made, creating a sense of stable source locations from anywhere within the loudspeaker array.

Delay panning also offers the advantage of being achievable in almost any modern digital audio workstation or on nearly any digital mixing console. Using either minuscule delays or plugins intended for delay compensation and source alignment, simple busing and short time adjustments can create a profound effect.

Panning for Height

One of the greatest challenges in today's three-dimensional immersive audio systems involves dealing with the dimension of height. Manufacturing a source position above a listener becomes difficult, even in many state-of-the-art systems. Our relatively coarse perception of elevation

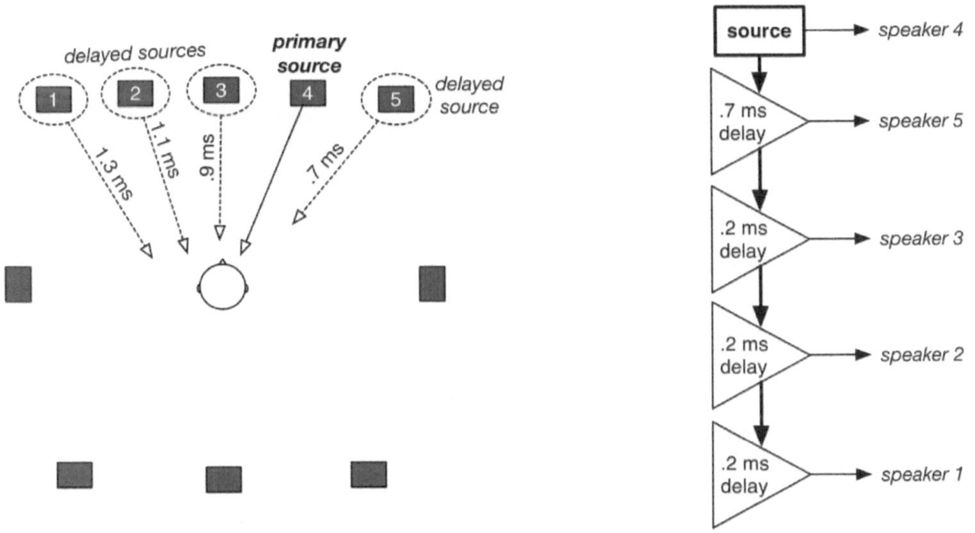

Figure 11.2 Example of delay panning in the center layer of an NHK 22.2 setup. Left: In this case, the source is located in the center right speaker, but is reinforced by the four adjacent front speakers (labeled "delayed sources"). This will help further cement the source's location, regardless of listener position. Right: Block diagram for creating the panning scheme illustrated. Example delay times are indicated, but would be different depending on the source's frequency content and the depth of source desired.

necessitates creative approaches to panning and source placement by the engineer to highlight and enhance sources that are to appear above the listener.

Creating an image above the plane of the ears is not a shortcoming of the reproduction system, but instead a challenge to our localization ability. Whenever an engineer chooses to position a source above the listener, they should keep in mind the following: inter-aural level difference cues (ILD) and inter-aural time of arrival difference cues (ITD) do not readily affect height perception (Williams, 2012). That is to say, the cues we are used to exploiting to create and alter a source's panning cannot be relied upon for height perception. Besides the obvious placement of a source in a speaker or object physically located above the listener, the engineer must reinforce the sense of height in some manner.

One of the main ways to reinforce a sense of height in a panned source is to consider and potentially alter its frequency content. Anecdotal evidence from leading immersive audio mixers suggests that higher-frequency sources and those that are generally brighter feel more convincing when panned above the head. Alternately, strong low-frequency sources seem less capable of existing concretely above the plane of the listener. It seems that many listeners have a hard time envisioning a bass guitar amplifier or kick drum flying over their head! This is quite likely due to the slight reflected energy off of the shoulder and torso from sounds emanating from an elevation above the ears. Lower frequencies, however, diffract around the shoulders and torso, bypassing this localization cue altogether. Early research found that substantial high-frequency energy, above 7,000 Hz according to Roffler (Roffler & Butler, 1968), is needed from a source to be easily locatable in the vertical plane (as discussed in Chapters 1 and 7). Keeping this in mind may lead to a more convincing source placement when dealing with elevation.

A large-scale version of delay panning may also be useful when trying to create an elevated source position. As stated earlier with regards to source spreading, some movement in source position can help enhance or draw attention to a source's elevated position. One such technique is to employ a series of delays moving downward from the source's intended elevated position. The sound will start at some position above the listener and then move downwards, drawing further attention to the initial position of the source. This is generally most effective with short delays in the range of 10–80 ms, and typically at a significantly reduced level. This can be combined with low-pass or subtractive high-shelf filter on the input to the delay to further enhance the effect of the sound "washing over" the listener from above. This is even further enhanced by using reverb (which will be covered more thoroughly later in this chapter) instead of or in tandem with delay to fill out the movement from top to bottom.

While the greatest focus on elevation in immersive audio is always on raising sources above the listener, that does not mean that *lowering* sources can't be equally useful to the mixing engineer. A number of current immersive systems neglect the area below the listener, perhaps for reasons of lower return on investment per speaker/channel/object from the audience perspective; however, this area can be used to great effect. While the techniques used to move sources to an area below the listener are similar to those used to elevate a source, it is worth pointing out the benefit of having some control over this lower elevation. Sound effects like footsteps, ground movement, in-flight motion or wind, and anchoring rhythmic musical sources (like kick drum and bass) can greatly benefit from having their own spatial region, leaving the main area in front of the listener for more critical sources such as dialog or lead instruments.

Panning in Object-Based Systems

Object-based immersive audio systems, however, can break from the paradigms discussed above. Rather than the engineer creating source localization through traditional panning methods, object-based systems offer simple assignment of X, Y, and Z coordinates to a sound with all panning or matrixing provided by the system. Chapter 8 deals in-depth with the theory and technology of object-based audio that has been established, but mixing paradigms for these systems are still emerging.

Much of the discussion and evolution of object-based mixing techniques surrounds the issue of *bed tracks*. Bed tracks are groupings of standard multichannel tracks without the X, Y, Z positioning system unique to object-based audio. Typically, a bed track is either a 5.1 or 7.1 track that are addressed in a traditional manner, and exist throughout a project, unlike audio objects, which appear as needed and are absent when unused. These bed tracks offer some significant advantages for mixers, namely the ability to start from a traditional release format and build upwards without having to build an entirely new mix when an object-based format is requested. While this can offer a significant time saving for the content creator, it can also be a source of confusion. How should a mixer determine what should be an object, and which content is better suited to the bed track?

A safe approach to object and bed tracks is to divide sound sources into the following two categories: critical sources and special effects. In this approach, all story-critical material is constrained to the bed track. Dialog, music, and audio effects essential to the telling of a story are mixed in a traditional channel-based method and committed to the bed track. The *ear candy* effects, such as arrows flying over the audience's head, or antiphonal attention-grabbing noises, are then assigned as objects. This mixing philosophy carries the distinct advantage of not relying upon objects for basic story telling. If the mixed material is shown in a theater without proper object rendering, the audience will likely still receive the production's content. A secondary advantage is in the time spent generating an object-based mix. Most productions still require a traditional surround (5.1 or 7.1) mix, so this approach adds minimal work to generate the object-based mix.

However, many mixers prefer to be more adventurous with their use of objects. Even so, these productions will likely start like others with music and dialog committed to the bed track, but it is not uncommon to branch out from the bed either. Some mixers choose to make additional space for music by using objects to help elevate some or all of the music above the screen with objects located directly above and/or slightly outside of the main left and right speakers. For these mixers, moving the music out of the main speakers behind the cinema screen can allow for greater flexibility in dialog placement with minimized risk of masking. Likewise, objects tend to be more conducive to panned dialog than traditional phantom imaging, since object renderings are more predicable than channel-based surround panning techniques. With this approach, many sound effects can also be converted to objects, taking advantage of sharp or attention-grabbing effects that can now be placed in more isolated areas (e.g. above the listener) through the use of objects. Also, keep in mind that creating moving effects is somewhat easier with objects. By treating these dynamically positioned effects as objects, they can be lifted out of the horizontal soundscape, helping them stand out from the bed, while also highlighting their movement. Likewise, the ability to spread environmental sounds to the sides, rear, and above the listener can free up space in the horizontal plane for the bed tracks and the more central elements of the mix without sacrificing clarity and density in the sonic environment. Some sound designers, however, take a hybrid approach to object usage. With composite sounds (sound effects created by mixing a number of preexisting sounds), a portion of the sound can be placed into the bed tracks to maintain clarity and power, while another component of the sound effect can be placed in an object, allowing the mixer to create a more dynamic presentation, possibly including subtle panning or movement effects.

The treatment of ambiance in object-based environments can be quite interesting as well. Opinions differ widely on where ambiance and reverb should be located. Some mixers constrain the majority of their reverberant energy to the bed tracks in a manner more consistent with non-object formats. Other mixers have experimented with dedicated objects for reverberation and ambiance. Given the availability of truly immersive reverberators, there are great benefits to positioning some reverberant energy around and above the listener. For example, some mixers have used more than 10 objects as statically positioned reverb returns which remain active throughout a mix.

Spreading and Expanding of Source Image

In some situations, panning monophonic sources to a specific location in the surround sound scape may be sufficient, but for many engineers, an immersive mix may include increasing the perceived size of reproduced sound sources to spread over a larger area of the soundscape than a single speaker can represent. There are virtues and drawbacks to each approach that are dealt with more thoroughly later in this section, but it is advisable for all audio engineers to understand the methods commonly employed for increasing the perceptual size of a monophonic source.

Often the first instinct of an engineer who is new to surround or immersive audio mixing is to spread a source by simply placing a monophonic sound into multiple adjacent loudspeakers. Routing the output of the source into two or more adjacent speakers does physically increase the area from which it radiates, as well as the perceived loudness of the source, but has relatively little effect on perceived source size. Instead, the effect of this multi-speaker approach is the creation of a new phantom source location formed between the adjacent loudspeakers. This can be an effective panning technique as discussed later in this chapter, but does little to increase the perceived size of a source. To truly increase the source's size, it is necessary to decorrelate the sound radiated from these adjacent loudspeakers. That is to say, the sound emanating from these loudspeakers must be related, but slightly different, as not to form a single phantom image between the loudspeakers but to diffuse it slightly. The introduction of slight differences between signals reproduced in multiple loudspeakers can help to broaden or enlarge a sound source (Kendall, 1995). Time-based alterations (e.g. delay or reverberation) are often used to decorrelate signals, but will be dealt with separately in the second section of this chapter. There are, however, a number of frequency-based methods for signal decorrelation.

Frequency-Based Decorrelation

One of the simplest ways to decorrelate two or more loudspeaker outputs from a monophonic source is to alter the frequency content of each output. By differing the frequency content fed to two adjacent loudspeakers, the source now seems to take up a larger space, with different physical areas corresponding to different frequency ranges. While the signal remains highly correlated in the time domain, this frequency shift is enough to spread the source between the two loudspeakers, avoiding a single phantom image between them. These techniques are extremely effective for expanding sources with a wide fundamental frequency range, but which were recorded as a monophonic source. The classic example is a monophonic electric piano, which may sound diminutive or unimpressive when represented through a single loudspeaker, especially in contrast to a similar instrument (e.g. acoustic piano) represented through multiple loudspeakers. The beauty of the techniques described below is found in not only their effectiveness, but also in their simplicity; all can be realized using only standard digital audio workstation busing and the basic filters and equalizers found in all modern audio production software.

The most common and easy to implement of these frequency alterations is to employ a crossover-like filter to each output channel. This is often referred to as "piano" stereo, as the effect closely resembles the impression obtained when listening to a piano from a close distance with the notes seeming to move from one side to the other as they increase in pitch. The filter itself is quite reminiscent of a multi-driver loudspeaker crossover, with overlapping high- and low-pass filters used to assign certain frequency ranges to certain drivers, or in this case loudspeaker channels (Figure 11.3). The resulting effect is a slight amount of movement with changing frequency, giving a sense of an enlarged source.

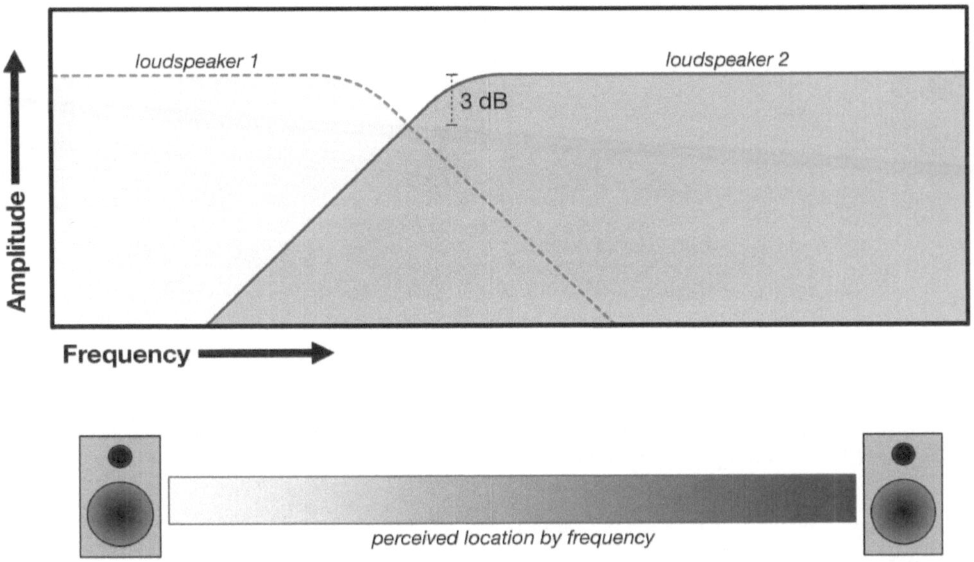

Figure 11.3 Example filter curves for source spreading based on the "piano" or crossover method. In the given example, the crossover point is at −3 dB, ensuring no loss in energy across the input signal, but only spreading of the source by frequency. The crossover point in this set of filters potentially weights the energy unevenly, with the larger portion of the frequency spectrum directed towards loudspeaker 2. The lower grayscale bar shows the transition of energy from loudspeaker 1 to loudspeaker 2 for the given filter set.

Similar methods can be applied using more than two adjacent loudspeakers. The addition of a band-pass filter between the low-pass and high-pass filters allows for further distribution of a monophonic source to multiple loudspeaker channels. This can be particularly useful, as the width of the center pass-band can determine how anchored a sound source is, despite adding width. Given a source with a broad frequency range, a wide pass-band assigned to a loudspeaker would keep the source focused around that loudspeaker, while a narrow pass-band could allow the source to spread more, as seen in Figure 11.4.

Some sources, however, will be degraded by a wide spatial spread across their frequency range. For these sources, a single filter overlapping full-range reproduction may be advisable. Figure 11.5 shows such a configuration with a high-pass filter, but the technique could just as easily employ a low-pass or band-pass filter. The effective result of such a configuration is that the source is only spread when certain frequencies are activated. This can be highly effective for sources such as a monophonic drum overhead microphone or room microphone. In the high-pass configuration, the kick and snare would be heavily centered, while cymbal hits would be perceptually spread and widened. Distorted electric guitar can be similarly effective, with the "fuzz" and noise being spread over a wider space, but with the body of the sound remaining anchored in a single loudspeaker.

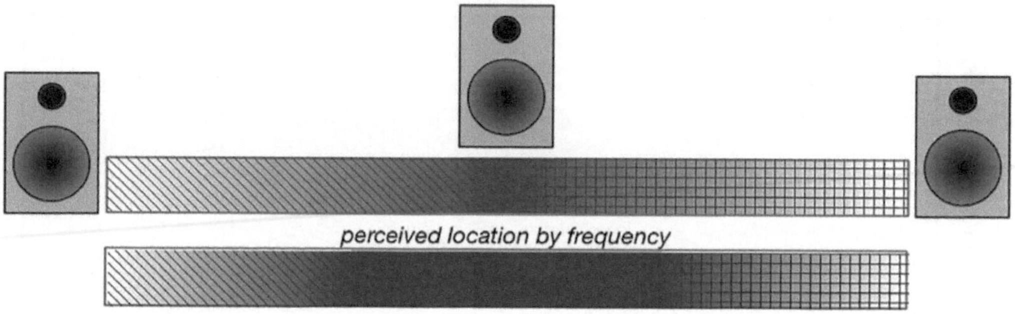

Figure 11.4 Grayscale/texture bar showing two options for a three-way source spread technique employing a band-pass filter to feed the center loudspeaker. The top bar has roughly equal width on the high-pass (diagonally striped pattern), band-pass (solid gray), and low-pass filters (horizontally striped pattern), while the bottom illustrates a markedly wider band-pass filter, keeping the majority of a full-range source's energy in the center loudspeaker.

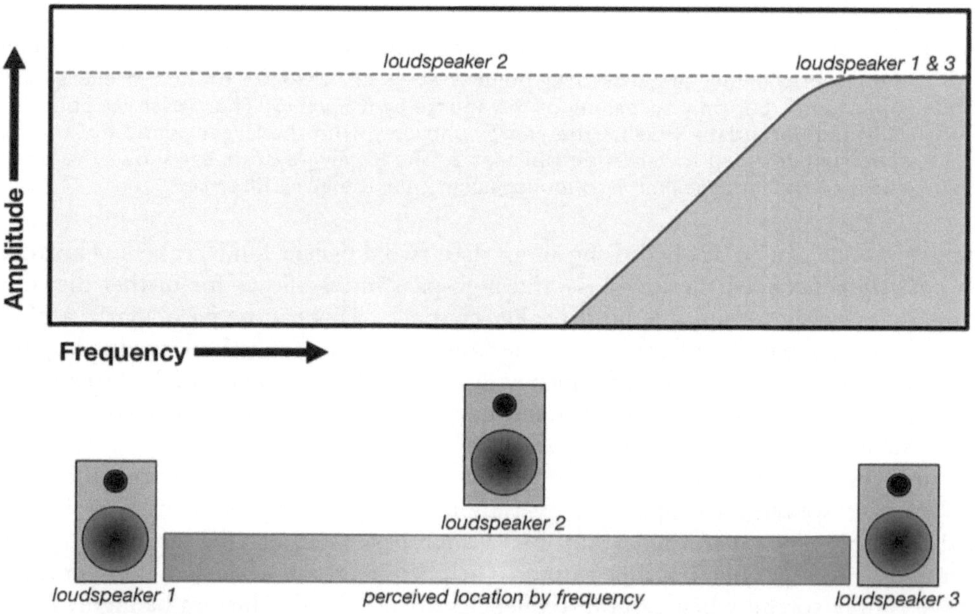

Figure 11.5 Overlapping a full-range loudspeaker channel with two band-limited speakers creates a source with a strong center anchor, but with the addition of some spread at a given frequency range. The example shown illustrates the use of a high-pass filter feeding two outside loudspeakers, with the center loudspeaker being fed the full-range signal at all times, creating a temporary widening when higher frequencies are activated by the source.

Spreading Stereo Sources

Stereo sources present a slightly different challenge when an engineer desires greater source size. The above techniques can certainly be applied to individual channels of a stereo (or larger) source, but may lead to an incoherent, disjointed sound. Instead, other methods of spreading may be more appropriate for preexisting stereo material. The combination of sum and difference processing with the frequency-based spreading methods described above leads to a number of interesting source expansion methods.

Sum and difference processing (also commonly known as mid/side or simply M/S) allows the center information to be extracted from a stereo signal. The information that is common to both the original left and right channels will appear in the center of the stereo image, and can therefore be referred to as the sum of the two channels or the middle information. The information that is unique in each of the two signals (i.e. the difference information) constitutes the side information. An in-depth explanation of mid/side processing and stereo capture can be found in Chapter 6. Even this basic decoding can help to distribute a 2-channel source (left/right) into a 3-channel output (left, center, and right). However, it is the further processing of the output of the extracted mid/side data that can be more interesting in the spreading of stereo sources.

The easiest way to combine mid/side decoding and frequency-based source spreading is to apply a filter or filter set (high-pass/low-pass filter, or high pass and band pass on one side and band pass and low pass on the other) to the individual left and right side channels and distribute them to adjacent loudspeakers. The net result is an increase in signal distribution from two loudspeaker channels to five loudspeaker channels, with the mid signal applied to a single loudspeaker and the left and right side channels spread to two loudspeakers on either side of the mid information's location (Figure 11.6). It is of note that the image will lose some precision in this setup, making it less than ideal for a detailed main capture or similar stereo source, but can lead to an interesting increase in source size and subtle source movement for less detailed stereo input signals. This technique may also play havoc with some more sensitive signals, such as voice, in which the phase distortion introduced may be too audible for effective use. This extended sum and difference spreading is equally viable without use of filters, but may suffer from a noticeable increase in phase problems.

Creative Applications of Source Spreading

Up to this point all discussion of source spreading has been limited to adjacent loudspeakers. The potential to spread sources to non-adjacent speakers opens up entirely new creative possibilities. By employing the above spreading methods using diagonally located, non-adjacent loudspeakers, sound can be drawn into the environment, rather than existing only along the periphery of the soundscape. As long as the two loudspeakers in use are close enough to create a coherent source, the resulting image can spread in any direction within the soundscape. This can be particularly effective in the height domain where our localization ability is somewhat more coarse (less precise), as discussed in Chapters 1 and 7. Source spreading techniques, which create subtle frequency-dependent motion in the elevation plane, can be even more perceptible than the exact source position itself.

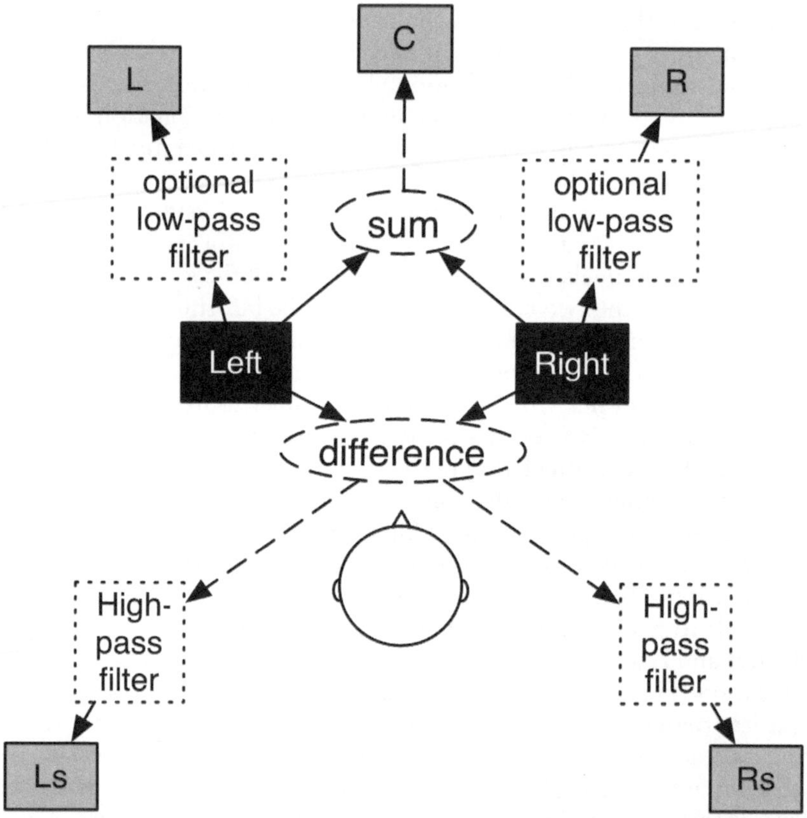

Figure 11.6 Example mapping of a stereo source to 5-channel surround employing both sum and difference techniques and complementary filtering. The sum of the stereo input feeds the center channel, while the unmatrixed left and right are fed to a filter to complete the frontal image. The rear loudspeakers are fed from the difference information, which is in turn filtered to complement the front left and right channels, as shown in Figure 11.3. As an additional means of separating, a polarity inversion can be added to the information in the right surround channel, creating a mirror of the sum-difference technique applied to the front.

An Immersive Overhaul for Preexisting Content

One of the largest practical problems facing recording and mixing engineers in immersive content production roles is the disparity in format between raw audio generated through sound design or recording and the immersive formats of the deliverables required for a project. This is particularly the case when dealing with remixing or remastering *existing* content in non-immersive formats for new, immersive distributions. The volume of material recorded and mixed in mono and stereo far outstrips even that produced in lower-order immersive formats, let alone the newest object-based formats. This proves to be particularly problematic for producers of audiovisual content moving into the world of immersive audio, who are used to relying on the ubiquitous rerun of content to fill the airwaves. For television studios and broadcasters, the amount of work involved in reproducing or remixing all of the commonly aired preexisting content in a new immersive format is cost-prohibitive. On the other hand, consumers often feel slighted by receiving any content, or even legacy material, in less than the most up-to-date distribution format. As the old adage goes: nobody buys extra speakers to *not* hear them!

Upmixing

The immediate solution to the problem of preparing existing, non-immersive material for new immersive formats comes in the form of the *upmix*. Simply defined, the term *upmix* refers to the process of converting a finished mix with a low number of channels or speakers to a higher number (Rumsey, 1998). For example, a stereo to 5.1 upmix takes a completed two-track stereo mix and converts it to a 5.1 surround format, duplicating, separating, or generating information to push content into the speaker channels not present in the original product. In today's parlance, the term *upmix* generally implies that this process is heavily or completely automated, rather than relying exclusively on an engineer's efforts to generate the new, (it is hoped) more immersive, content.

There is no shortage of ways to deal with the most common upmixing tasks, especially those that have targeted 5.0 and 5.1 surround as the delivery format. Many of these systems or formulae for upmixing have been driven by the consumer demand to experience multichannel surround sound from older content while maintaining real-time rendering, which requires low memory and computational overhead. In other terms, the consumer desires an immersive experience, but not at the cost of convenience. While the specifics of various available commercial systems fall outside of the scope of this book, the basic operational principles are worth discussing, as they inform many of the techniques used for both upmixing and source spreading in new immersive audio systems.

One of the most common, and coincidentally easiest to reproduce, methods for upmixing is the *nearest speaker* method. Aside from monophonic source materials, most existing audio content has some dissimilar material in each channel, or even completely unique material in a given channel. Likewise, most 2-channel or greater content employs phantom imaging to some extent. The easiest way to "fake" immersive content is to simply bleed content from preexisting channels into the adjacent channels within the target release format. This does little to create a new experience for the listener familiar with the previous release format, but it does meet the criteria of using all channels and surrounding the listener.

A more interesting but still easily achievable option is to employ a scheme of sum and difference (i.e. mid/side) processing between all adjacent channels of the source material. The sum content is placed into the nearest loudspeaker to its previous phantom location, while the difference material remains in the loudspeaker closest to its source channel in the original format, as shown in Figure 11.7. Some experimentation has been done to further extract sum and difference information from non-adjacent channels for use in generating height content, but has been thus far unconvincing to content producers and casual listeners.

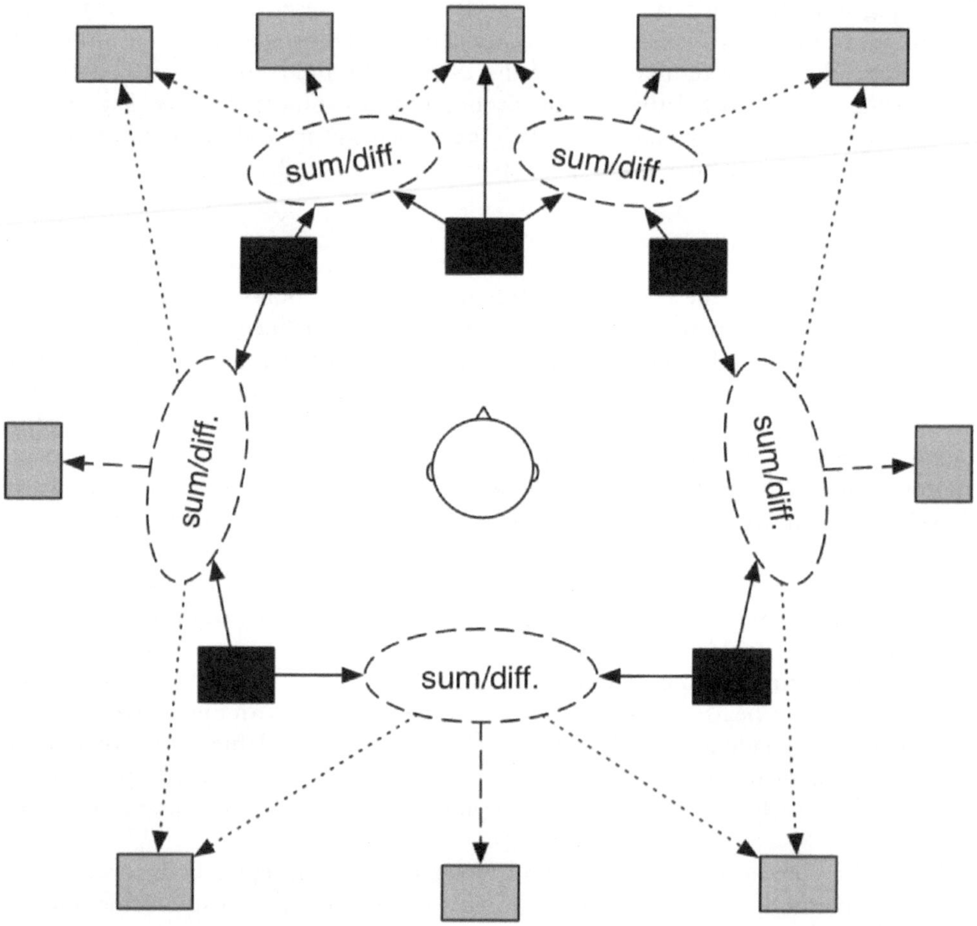

Figure 11.7 Diagram showing a standard 5-channel surround mix upmixed using the sum and difference technique. In this example, the sum and difference information found in the original five channels shown in the interior in black are mapped to the 10 channels found in the middle layer of the NHK 22.2 format, shown in light grey. Solid black lines between loudspeakers illustrate direct mapping of channel information, while dashed ovals represent sum and difference processing from adjacent channels. The dashed lines originating from these ovals represent the resulting sum information, and the dotted lines represent the mapping of difference information. Direct mapping of the center channel is often used, as to avoid unwanted alteration of critical sources such as dialog of vocals.

Considerations in Mixing for Film and Games

Immersive audio systems can yield an unparalleled range of possibilities to the content creator, but it is important to not lose sight of the goal of the content. It can be easy to allow creativity to interfere with the focus and progress of the story being told. Regardless of format, there are some guidelines to remember when mixing for mixed media content.

First and foremost, dialog is king. In most films, video games, or even in music, the human voice is the central story-telling element. Without intelligibility and clarity in the voice, the listener/viewer can become confused, fatigued, or lose interest due to the effort required to follow dialog. It should be noted that even absolute sound quality is sometimes sacrificed to intelligibility; isolated dialog tracks may sound odd or filtered in isolation, but are processed to maximize intelligibility in the context of the entire mix. Deliberate obfuscation of dialog can be used sparingly as an artistic technique, but should be avoided in most cases. Regardless of format, dialog is generally placed at or near the center of the frontal sound image, with surprisingly little deviation. The center channel speaker is the typical destination for dialog in systems with a dedicated center loudspeaker. This is due in part to clarity afforded sounds arriving from 0° azimuth, and due to considerations of loudspeaker coloration when off-axis and when phantom images are employed. While those are certainly important considerations, it is also important to acknowledge the cognitive dissonance that can be caused by having visible action on the screen while the dialog is placed to the side or rear. As a character moves across even a large screen, the visual sense can overcome a static voice position to a great extent, far outweighing the potential hazards of panning the dialog between speakers. While this is not an absolute rule, it is a time-tested guideline that will often help avoid unwanted voice coloration, phase cancellation, and mistaken localization over larger audience areas. Character's movement can be convincingly reinforced by change in auditory perspective, through tonal and spatial alteration, as was successfully accomplished for years in films having just monophonic sound.

A secondary advantage of constraining dialog primarily to the center channel is the ability to avoid masking. With a dedicated dialog channel, other centrally located sound sources can be reproduced with phantom center images with minimal interference to the ever-important dialog. This is particularly true for music. Generally film and game scores are positioned frontally, perhaps with some rear reverberant energy to enhance the sense of immersion, but with most concrete sources positioned in the front ± 30°. Unlike in many music-only applications, the center material in music for film may be somewhat recessed or subdued. Even with a dedicated dialog channel, keeping the center of the phantom image free from distracting or masking material (particularly in the frequency range of the voice) will help ensure dialog intelligibility even in the presence of music (as shown in Figure 11.1). If desired, some music can be spilled gently into the center channel to keep it from migrating completely to one side of the screen for viewers seated near the side walls.

Sound effects are often presented in a similar manner as music, but with a greater likelihood of non-central source placement. Much of the engagement with surround in the context of film, television, or video games comes from effects panned to the side and rear of the listener. This helps draw the listener in, and potentially place them at the center of the action, rather than as a passive observer. Sound effects also provide a way to stretch the apparent size of the scene, by exaggerated frontal left/right placement, exceeding that which is visually present on screen.

Ambiance and environmental sound can also help to create the sense of immersion in an audiovisual presentation. A great majority of environmental sounds will be found to the sides and rear of the listener to help engage them in the soundscape, but without interfering with the frontal material. It is however important to note that a complete lack of ambiance or environmental sound in the frontal image can destroy the illusion of a particular environment. Typical mixes will place some ambiance and environmental sound in the frontal image to create a continuous soundscape, but will either thin the texture or lower the level of the effect in the front speakers to reduce competition for the dialog, music, and scene-specific effects.

Envelopment

The goal of most immersive audio content is to create a sense of envelopment. The term *envelopment* was originally used to describe the feeling of *surroundedness* or *engulfment* in sound experienced in a well-designed concert hall (Beranek, 1996 and Barron, 1988). This sense of envelopment in the concert hall is created by reflected energy arriving from the sides following the arrival of the direct sound (Morimoto & Maekawa, 1989). In the present era, envelopment can just as easily be created by immersing a listener in sound originating from all angles of azimuth and including elevation.

One way to create envelopment is to position sources in 360° around the listener. This form of envelopment is particularly effective for music-only content. By moving instruments and effects from the frontal presentation to surround, the listener can get the impression that they are sitting on stage with the band, or perhaps that they are even a member of the ensemble! This can create a very engaging soundscape, but can still occasionally lack a true sense of envelopment due to a potential for disjointedness. To help smooth the presentation or fill the gaps between sources, we must look to the same effect that provides acoustic envelopment: reverberation.

Reverberation

Up to this point in the chapter there has been a conspicuous lack of discussion of reverberation. Reverberation plays directly into the topics of source spreading and upmixing, but is unique and important enough to warrant a discussion unto itself. Reverberation is certainly an important effect in conventional stereo and multichannel production, but it becomes an even more powerful tool when full envelopment is called for and in systems where additional dimensions (e.g. height) are added.

Source Spreading with Reflected Energy

Source spreading through the use of acoustic reflection has, intentionally or otherwise, been a staple of recording since its inception. A recording engineer will typically take great care in choosing the location of a source within the recording studio (or even choosing the studio itself) to generate and capture a pleasing set of reflections. This careful use of acoustic reflections, combined with prudent microphone technique, can actually create a larger-than-life source right from the recording phase. The application of this technique is commonly exemplified in drumset recording, where positioning the drum kit on different floor surfaces, against walls, or even in a corner of the studio can all have a significant effect on the size and shape of the drums' pickup, particularly for the snare drum, in both the overhead microphones and any room microphones. Likewise, the strong lateral early reflections in concert halls increase apparent source width, broadening the sound's source and increasing its loudness. It is reasonable that this same effect could be exploited in processing to aid in artificial source spreading.

The judicious application of early reflections from either an algorithmic or convolution reverb to a source can have the same effect as capturing the acoustic affect of a source in a room described above. Reflections naturally spread a source by providing delayed copies of the original signal, each with altered amplitude/frequency balance, redirected to arrive from *slightly different locations*. The last portion of that statement is the critical component of using reflected energy to enlarge sound images. With few exceptions, the acoustic reflections in a typical room provide a location-coherent set of delays and angles of incidence that reinforce the original source's location. The precedence effect states that the direct sound will offer the strongest cues to the source's location, leaving the closely spaced reflections to add body to the perceived width of the source. If the reflections are spread in space but remain coherent with the original sound source's location, the source can be enlarged, yet retain a common perceived origin.

Basic source enlargement can be achieved using standard stereo or larger reverberators of any design. Preference certainly goes to multichannel reverberators which model the early reflections based on the input source's location, but some spreading can be obtained with even basic, input-agnostic (single-channel input) stereo reverberators. In the case of a stereo reverberator, the output of each channel can be placed in adjacent loudspeakers to the source's initial position, or phantom imaged on either side of the source. Use of the "pre-delay" or input delay can be very helpful in creating reinforcement of the source without altering its location. Slight adjustments of only a few milliseconds can prove critical in maintaining the source's initial position and leaving the timbre of the source relatively unaltered. This essentially imitates the commensurately longer travel time needed for sound to reach boundaries located farther from the source, despite the reverberator's fixed reflection pattern. Typically, as the reflections and source are being spread farther apart, a slightly longer delay time between source and reflections will be needed to ensure the sound remains coherent and natural. Continuing the example of a stereo reverberator, if the reverberation's channels are spread too widely (perhaps to loudspeakers not directly adjacent to the source), the illusion can break down into a single source with two clearly discernible albeit diffused secondary sources. In this instance, it may be prudent to duplicate the reverberator's outputs with varying pre-delay times assigned to each loudspeaker. Given a set of five speakers with the original sound source located in the middle, the adjacent loudspeakers would receive a very slight pre-delay, while the outer, non-adjacent loudspeakers would receive a longer input delay. This creates the illusion of the sound spreading out, rather than a single wall of sound impacting the listener.

Reverberation and the Height Dimension

Just as reflected energy can help spread a source in the horizontal plane, early reflections and reverberation can aid in creating a sense of height. Traditional reverberators used in immersive audio did not include a true height dimension, but instead repurposed existing horizontal plane channels to fill in overhead speakers, such as chaining multiple quadraphonic Yamaha SREV1s (sampling reverberation devices). The ideal situation, however, involves capturing true three-dimensional room reflection data to be used in immersive audio mixes. While there is no shortage of "surround" reverberators, which capture, model, or emulate reflections in the horizontal plane, there are few that venture into the height dimension. This is sure to change as immersive systems with height information proliferate, because the height dimension contains interesting and unique acoustic information. Consider a concert hall, where the ceiling might be the farthest reflecting surface from a listener in the proverbial *best seat* in the house, but the ceiling above the orchestra reflects and directs sound outward to strengthen the audience's impression of the performer. Behind the audience is a complex series of reflections off of balconies and mezzanines. From the hard reflectors above the stage to the deep and diffuse sound field directly above the audience, the reverberant information above the listener is truly unlike the surrounding horizontal reflections.

Convolution Reverberation With Height

The currently available reverberators including height data tend to be algorithmic. There are, however, a number of researchers and companies exploring the idea of convolution reverberators that include capture of the height dimension, including new packages for TC Electronic's System 6000 reverberator and reverberators built into the Auro 3D and Atmos production tools. The wide variety of ceilings, domes, organ lofts, and mezzanines alone included in architectural enclosures ensure that this information imparts unique and interesting new characteristics to each and every room measured. By simply adding two or more elevated, upward-angled channels to an impulse response capturing setup, an accurate, enveloping, and natural-sounding set of overhead information can be collected. Ideally, these elevation channels focus on the information uniquely related to the ceiling, elevated sidewalls, and overhead reflectors, excluding much of the correlated data from the horizontal plane. By excluding data similar to that found at ear level, the decorrelated elevation data is further separated perceptually, rather than being dragged downwards by the typically stronger horizontal reflections. In order to ensure proper decorrelation and to highlight the upper frequency range needed for elevation localization, unidirectional or bidirectional microphones are recommended for height impulse response capture, rather than the more standard omnidirectional mics used in capturing the horizontal plane.

Synthesizing Height Reverberation

In suboptimal situations, an engineer could simply employ the same source spreading techniques described above rotated 90° to expand a source in the height dimension. Given this option, the use of additional pre-delay, equalization, and some low-level reverberator controls can still yield convincing results. First and foremost of these is the employment of proper time delay values. It is important to implement a travel time from source to reflection and back to the listener representative of the ceiling height in the space being emulated. As discussed above, the typical concert hall has longer delay times from overhead reflections than anywhere in the horizontal plane. When trying to emulate a smaller room, however, a pre-delay shorter than that of the horizontal plan may be more appropriate.

The application of diffusion can also be quite useful when emulating height reflections and reverberation. In large, ornate churches, concert halls, castles, and ballrooms, the intricate carvings, stone work, and so forth found along the upper perimeter of the room and often on the ceiling itself lend a unique diffusion to elevated reflections. Most algorithmic reverberators employ some type of "diffusion" or "spread" control which may be able to mimic a diffusive ceiling's reflections. On the other hand, reducing the diffusion may lead to a more convincing emulation of a small residential or commercial space with a hard, flat ceiling.

High-pass filtering may also aide in synthesizing height reverberation. As stated previously, high-frequency information is of utmost importance in height localization, while lower-frequency information can cause muddiness and clutter above the listener. While many reverberators attempt to create a *warm* or *rich* reverberant sound field, having such information elevated above the listener can actually detract from the height sensation with an impression of certain heaviness keeping the sound at ground level.

Upmixing with Reflected Sound

Reflected sound also offers the possibility of alleviating some of the problems seen in the simplistic upmix paradigms presented earlier, namely in the height dimension. For the previous upmixing paradigms, filling in space between existing horizontal channels is rather trivial compared to synthesizing the entire height dimension. As we have seen, though, reflected energy can easily generate new information to fill the vertical dimension using elevated loudspeakers. By applying a set of early reflections, the energy from single- or multichannel content can be spread into new channels in both the horizontal and vertical dimensions.

Initial work on reflection-based upmixing employed full room impulse responses to expand existing content and fill the greater space afforded by new immersive audio systems. It was quickly confirmed that the additional reverberation tail imposed on previously mixed content led to muddiness and loss of intelligibility. The early reflection energy, however, does offer some interesting possibilities. A set of early reflections in the form of an impulse response captured with full horizontal and vertical data can be used to spread preexisting surround content into larger immersive systems with little work on the part of the content producer (Woszczyk et al., 2010).

This is similarly true, albeit with less impressive results, for preexisting stereo content. In this upmixing scenario, the choice of early reflections is key. Some reverberators allow control of only a very short span of first reflections (typically algorithmic reverberators), while many convolution reverberators allow for little or no control of early reflections. For upmixing to be achieved without significant content degradation, a short, sonically neutral set of reflections is required. In the case of convolution reverberators, an industrious engineer can simply edit the impulse response to include only the desired length of early reflections, removing the reverberant tail completely. Impulse responses of rooms and temporal regions with less dense reflections typically work well in the upmixing role. While highly dense reflections may add to a source's perceived width, they are also more likely to reduce clarity and intelligibility in preexisting content. Strongly colored reflections are likewise problematic when transparency and neutrality in the upmix are desired. Despite being somewhat counterintuitive, the early reflections of larger spaces often end up lending size and spread to the sound without radically altering the timbre or clarity of the source material.

Other Time-Based Effects

Reverberation is not, by any means, the only useful time-based effect in immersive audio. Any engineer who has worked on a modern rock or pop mix knows the power of delay effects. In today's large-scale immersive audio formats, spatialized delay effects can be leveraged to create a dynamic, interesting mix. Ping pong delays (a delay where each echo appears in the channel opposite the previous echo) take on new life when more than two channels are available for use.

One of the most dynamic uses of delay in immersive audio is cross-panning. Rather than a traditional left-right ping pong delay, panning a short delay with minimal repeats across the listener can help to draw the source out from the speakers and attract attention to an otherwise unremarkable source (Figure 11.8, left). This cross-delay also has the effect of adding energy to a source in a non-adjacent loudspeaker, which perceptually pulls the source out from the plane of the loudspeaker towards the listener. This effect can be quite useful on lead elements, solos, or even panned vocals. This technique can also be extended to multiple loudspeakers. Using two or more loudspeakers to present a delay can fill an entire spatial region with a single source (Figure 11.8, right). The effect can be expanded by using non-adjacent loudspeakers, e.g. on diagonal axes, or even using a greater number of loudspeakers, for reproduction of the delayed signal. A somewhat counterintuitive benefit for the use of multiple loudspeakers is the possibility of *increased* subtlety. Lower level is required for audibility when the delay is reproduced in multiple loudspeakers, ensuring the delay is more of an effect than concrete source unto itself. When a delay is positioned in a great number of loudspeakers, a pulsing sensation can be created, even when the delay is presented at very low levels. This technique can be both enveloping to the listener and enhancing the rhythmic character or "groove" of a mix.

Source Movement

Another distinct advantage of immersive audio systems is the ability to move a source. While movement of sound is not traditionally thought of as a source of envelopment, it can help yield a more engaging and immersive experience for the listener. Even when sound is not consistently

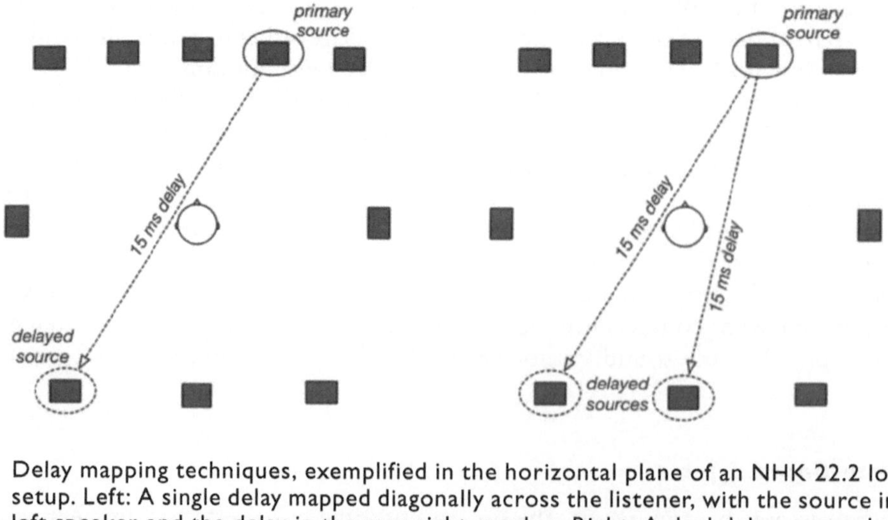

Figure 11.8 Delay mapping techniques, exemplified in the horizontal plane of an NHK 22.2 loudspeaker setup. Left: A single delay mapped diagonally across the listener, with the source in the front left speaker and the delay in the rear right speaker. Right: A dual delay, mapped to multiple speakers on the opposite side of the listener as the direct sound.

present around the front of, behind, or above the listener, having a source move through these dimensions can draw the listener into the mix. Once the listener experiences movement in a certain dimension, they become keenly aware of their position within the soundscape. Immersive audio provides a much larger range of possible motion for sound sources. With the inclusion of overhead loudspeakers in particular, the ability for sound to perform "flyovers" creates an enticing new effect, particularly for sound designers. There are some techniques which content creators may find useful when trying to construct dynamically moving sounds, especially when augmenting action visible on a screen.

More loudspeakers aren't always better when trying to move a source. The temptation of having a large reproduction array is to constantly involve all channels. When moving a source, however, it can be more effective to pan or travel between two or three loudspeakers rather than a larger number of channels. In the latter situation, it is often the case that sound can be localized in each individual channel as it passes through and is handed off to the adjacent loudspeaker (Figure 11.9, left). It is somewhat counterintuitive, but a movement through *fewer* loudspeakers may be more fluid throughout the sound's travel path (Figure 11.9, right).

A judicious use of spectral coloration can also aid in creating smooth, convincing sound movement. As with source position discussed earlier, subtle alteration of frequency content can augment the actual movement of sound amongst loudspeakers. In particular, decreasing the amount of low-frequency content in sources moving over the listeners can help to reinforce their upward position. Similarly, a decrease in high frequency when a source moves to the rear of a listener can enhance the realism of a final source position behind the listener by exaggerating

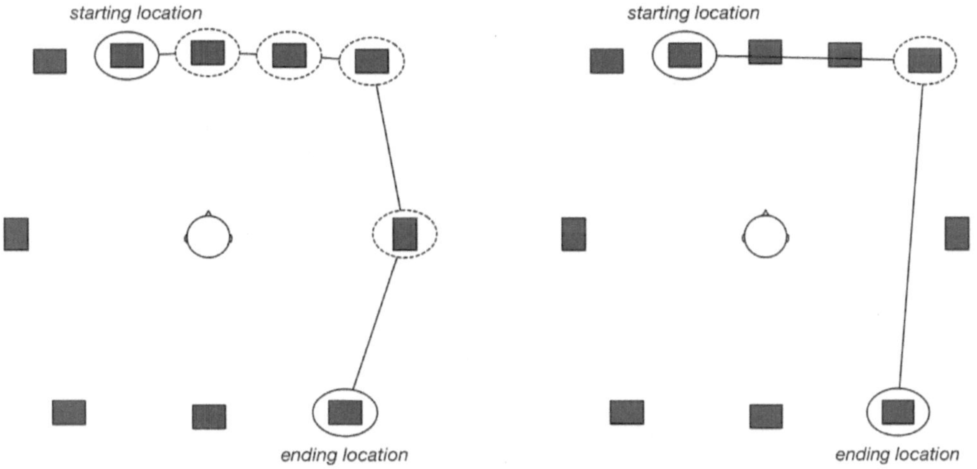

Figure 11.9 Two possible paths for a clockwise movement of sound from the front center-left to the rear-right within an NHK 22.2 loudspeaker array. Left: The sound is moved from channel to channel around the perimeter of the loudspeaker array, raising the potential for localization in each discrete channel along its trajectory. Right: The sound is moved between the fewest possible channels in its journey around the listener. This minimizes the stability of the sound source along its path, especially midway.

the high-frequency attenuation caused by head and pinna shadowing of sources behind the ears. Lastly, the Doppler effect can be simulated by applying a pitch shift to a moving source to mimic a real sound source passing a listener (Ahrens & Spors, 2008). While these effects are insufficient to create a sense of movement alone, they can help to create a more convincing sense of movement.

Musings on Immersive Mixing

The world of immersive audio presents engineers and content producers with an incredible array of creative options for crafting mixes and soundscapes. We are continuously at the forefront of the next generation of immersive audio technology, which means we get to make and break the rules! Much of the work done in the immersive realm is still reliant upon the same skill set that makes an engineer successful at producing mono, stereo, or traditional surround content. Attention to detail, care for the delicate timbre of sources, and maintenance of proper dynamic range all exist as primary concerns regardless of format. Beyond those basic concerns lies a world of creative options.

It is clear that the role of reverberation is becoming more prominent with the advent of larger immersive audio systems, particularly those with height. Any realistic presentation of audio in three dimensions will require similarly realistic ambiance to reinforce and expand the dry sound. Due to limited current options, engineers may need to generate their own height information, or better yet, capture it themselves. The creativity involved with capturing height information, whether during tracking or through impulse response capture, opens new avenues for the engineer to explore. It is also quite likely that the proliferation of large, height-inclusive immersive systems will bring with it new three-dimensional reverberation tools for use in the studio.

The challenging side to the present state of immersive audio is the lack of production tools. Advanced reverberation, delay, panning, and even multichannel equalization and dynamics processing tools have been slow to develop and reach content producers. The economics of this means slower workflow for engineers, fulfilling the dire predictions of project funders who claimed immersive audio productions are too expensive or time-consuming. The inclusion of immersive sound source panners in standard production environments (digital audio workstations) will allow for an incredible increase in production speed and will allow a far greater number of engineers to move into immersive content production. Until then, this makes it all the more critical that the successful immersive audio content producer is resourceful and creative using the existing tools. Exploiting the techniques presented above and in other chapters of this book can bring high-quality results with minimal additional processors and specialized production tools.

References

Ahrens, J., & Spors, S. (2008, May 1). Reproduction of moving virtual sound sources with special attention to the Doppler Effect. *Proceedings of the 124th Convention of the Audio Engineering Society*. Amsterdam, Netherlands: Audio Engineering Society.

Barron, M. (1988). Subjective study of British symphony concert halls. *Acoustica*, *66*(1), 114.

Beranek, L. (1996). *Concert and Opera Halls—How They Sound*. Woodbury, NY: Acoustical Society of America/American Institute of Physics.

Berkhout, A. J. (1988). A holographic approach to acoustic control. *The Journal of the Audio Engineering Society*, *36*(12), 977–995.

Blumlein, A. (1933, June 14). *Improvements in and Relating to Sound-transmission, Sound Recording and Sound-reproducing Systems*. British Patent Office.

Kendall, G. (1995). The decorrelation of audio signals and its impact on spatial imagery. *The Computer Music Journal*, *19*(4), 71–87.

Morimoto, M., & Maekawa, Z. (1989). Auditory spaciousness and envelopment. *Proceedings of the 13th International Congress on Acoustics*, 2, 215–218. Belgrade, Serbia: Dragan Srnic Press, Sabac.

Roffler, S., & Butler, R. (1968). Factors that influence the localization of sound in the vertical plane. *The Journal of the Acoustical Society of America*, *43*(6), 1255–1259.

Rumsey, F. (1998). Synthesised multichannel signal levels versus the MS rations of 2-Channel programme items. *Proceedings of the AES 104th Convention of the Audio Engineering Society*. Amsterdam, Netherlands: Audio Engineering Society.

Spors, S., Rabenstein, R., & Ahrens, J. (2008, May 1). The theory of wave field synthesis revisited. *Proceedings of the 124th Convention of the Audio Engineering Society*. Amsterdam, Netherlands: Audio Engineering Society.

Williams, M. (2012, April 26). Microphone array design for localization with elevation cues. *Proceedings of the 132nd Convention of the Audio Engineering Society*. Budapest, Hungary: Audio Engineering Society.

Woszczyk, W., Leonard, B., & Ko, D. (2010, October 8). Space builder: An impulse response based tool for immersive 22.2 channel ambiance design. *Proceedings of the 40th International Conference of the Audio Engineering Society on Spatial Audio*. Tokyo, Japan: Audio Engineering Society.

Index

Note: Page numbers in italics indicate a figure on the corresponding page.